Mobile Computing Deployment and Management

Real World Skills for CompTIA Mobility+™ Certification and Beyond

Robert J. Bartz

Acquisitions Editor: Kenyon Brown
Development Editor: Kelly Talbot
Technical Editor: Denny Hughes
Production Editor: Christine O'Connor
Copy Editor: Judy Flynn
Editorial Manager: Pete Gaughan
Production Manager: Kathleen Wisor
Associate Publisher: Jim Minatel
Media Supervising Producer: Richard Graves
Book Designers: Judy Fung and Bill Gibson
Proofreader: Josh Chase, Word One New York
Indexer: Ted Laux
Project Coordinator: Patrick Redmond
Cover Designer: Wiley

Copyright © 2015 by John Wiley & Sons, Inc., Indianapolis, Indiana

Published simultaneously in Canada

ISBN: 978-1-118-82461-0
ISBN: 978-1-118-82464-1 (ebk.)
ISBN: 978-1-118-82466-5 (ebk.)

No part of this publication may be reproduced, stored in a retrieval system or transmitted in any form or by any means, electronic, mechanical, photocopying, recording, scanning or otherwise, except as permitted under Sections 107 or 108 of the 1976 United States Copyright Act, without either the prior written permission of the Publisher, or authorization through payment of the appropriate per-copy fee to the Copyright Clearance Center, 222 Rosewood Drive, Danvers, MA 01923, (978) 750-8400, fax (978) 646-8600. Requests to the Publisher for permission should be addressed to the Permissions Department, John Wiley & Sons, Inc., 111 River Street, Hoboken, NJ 07030, (201) 748-6011, fax (201) 748-6008, or online at http://www.wiley.com/go/permissions.

Limit of Liability/Disclaimer of Warranty: The publisher and the author make no representations or warranties with respect to the accuracy or completeness of the contents of this work and specifically disclaim all warranties, including without limitation warranties of fitness for a particular purpose. No warranty may be created or extended by sales or promotional materials. The advice and strategies contained herein may not be suitable for every situation. This work is sold with the understanding that the publisher is not engaged in rendering legal, accounting, or other professional services. If professional assistance is required, the services of a competent professional person should be sought. Neither the publisher nor the author shall be liable for damages arising herefrom. The fact that an organization or Web site is referred to in this work as a citation and/or a potential source of further information does not mean that the author or the publisher endorses the information the organization or Web site may provide or recommendations it may make. Further, readers should be aware that Internet Web sites listed in this work may have changed or disappeared between when this work was written and when it is read.

For general information on our other products and services or to obtain technical support, please contact our Customer Care Department within the U.S. at (877) 762-2974, outside the U.S. at (317) 572-3993 or fax (317) 572-4002.

Wiley publishes in a variety of print and electronic formats and by print-on-demand. Some material included with standard print versions of this book may not be included in e-books or in print-on-demand. If this book refers to media such as a CD or DVD that is not included in the version you purchased, you may download this material at http://booksupport.wiley.com. For more information about Wiley products, visit www.wiley.com.

Library of Congress Control Number: 2014937183

TRADEMARKS: Wiley, the Wiley logo, and the Sybex logo are trademarks or registered trademarks of John Wiley & Sons, Inc. and/or its affiliates, in the United States and other countries, and may not be used without written permission. CompTIA and Mobility+ are trademarks or registered trademarks of CompTIA Properties LLC. All other trademarks are the property of their respective owners. John Wiley & Sons, Inc. is not associated with any product or vendor mentioned in this book.

10 9 8 7 6 5 4 3 2 1

Acknowledgments

I would like to thank my wife for her support and patience during the many, many hours that were dedicated to writing this book. Even though our two adult children are grown and out of the house, I know they were thinking about me writing and cheering me on during the entire process, even with their busy schedules.

I would also like to thank everyone at Sybex who helped with the creation of this book, including acquisitions editors Jeff Kellum and Kenyon Brown, production editor Christine O'Connor, copy editor Judy Flynn, editorial assistant Connor O'Brien, and editorial manager Pete Gaughan. I owe all these individuals a lot of gratitude for their patience with me and the several delays encountered while working with me on this book. The developmental editor for this book is Kelly Talbot. Many thanks go to Kelly for his time and his work in helping with the flow, organization, and suggestions that allowed me to make this book an easy read. His editorial skills and attention to detail were an enormous help to me.

The technical editor for this book is Denny Hughes. I want to thank Denny for his timely reviews, comments, and suggestions that helped make this book a nice read and a valuable reference source. His many years of experience as a technical trainer, engineer, and content developer were a great contribution in creating a book I am sure you will enjoy reading.

A special thank you goes to contributing author Sebastian Coe for his help with Chapter 17. Sebastian's expertise in information systems security was instrumental in his contribution.

I would also like to thank the thousands of students who have taken the time to attend the computer networking classes that I have had the opportunity to teach over the past 20 years. Educating, mentoring, and entertaining so many of these individuals gave me the inspiration to author this and other books about wireless networking.

Finally, I would like to thank the manufacturers, vendors, organizations, and individuals that provided the subject matter, allowing me access to the technology and tools needed to write this book:

AirMagnet/Fluke Networks (www.flukenetworks.com)

Cloudpath Networks (cloudpath.net)

CradlePoint (www.cradlepoint.com)

Ekahau (www.ekahau.com)

IEEE (www.ieee.org)

L-com Global Connectivity (www.l-com.com)

MetaGeek (www.metageek.net)

Ruckus Wireless (www.ruckuswireless.com)

SolarWinds (www.solarwinds.com)

TamoSoft (www.tamos.com)

Wi-Fi Alliance (www.wi-fi.org)

About the Author

Robert J. Bartz is an engineer, technical instructor, and computer networking consultant. He is a graduate of California State University, Long Beach, College of Engineering, with a Bachelor of Science degree in Industrial Technology. Prior to becoming a computer networking engineer and technical instructor, Robert was employed as an aerospace test engineer working with aircraft radar systems and satellite communications. He has attained many technical certifications over the years, including Master Certified Novell Engineer (MCNE), Master Certified Novell Instructor (MCNI), Microsoft Certified Systems Engineer (MCSE), Microsoft Certified Trainer (MCT), Certified Wireless Network Trainer (CWNT), Certified Wireless Network Expert (CWNE), and CompTIA Mobility+, to name a few. He has over 25 years' experience with computers and computer networking technology.

Robert has taught computer and wireless networking technology to thousands of people from various industries and markets across the United States and abroad. He is the founder of Eight-O-Two Technology Solutions, LLC, a computer networking technical training and consulting services company that provides technical education and computer networking services. He spends his spare time learning new technology, having fun outside, and enjoying the beauty of his surroundings at his home in Colorado.

Contents at a Glance

Introduction   *xxvii*

Chapter 1	Computer Network Types, Topologies, and the OSI Model	1
Chapter 2	Common Network Protocols and Ports	29
Chapter 3	Radio Frequency and Antenna Technology Fundamentals	75
Chapter 4	Standards and Certifications for Wireless Technology	145
Chapter 5	IEEE 802.11 Terminology and Technology	183
Chapter 6	Computer Network Infrastructure Devices	253
Chapter 7	Cellular Communication Technology	299
Chapter 8	Site Survey, Capacity Planning, and Wireless Design	319
Chapter 9	Understanding Network Traffic Flow and Control	409
Chapter 10	Introduction to Mobile Device Management	441
Chapter 11	Mobile Device Policy, Profiles, and Configuration	461
Chapter 12	Implementation of Mobile Device Technology	479
Chapter 13	Mobile Device Operation and Management Concepts	509
Chapter 14	Mobile Device Technology Advancements, Requirements, and Application Configuration	531
Chapter 15	Mobile Device Security Threats and Risks	559
Chapter 16	Device Authentication and Data Encryption	577
Chapter 17	Security Requirements, Monitoring, and Reporting	601
Chapter 18	Data Backup, Restore, and Disaster Recovery	625
Chapter 19	Mobile Device Problem Analysis and Troubleshooting	647
Appendix	The CompTIA Mobility+ Certification Exam	669

Index   *689*

Contents

Introduction   *xxvii*

Chapter 1   **Computer Network Types, Topologies, and the OSI Model**   **1**

    Network Types   2
        The Local Area Network   3
        The Wide Area Network   3
        The Metropolitan Area Network   5
        The Campus Area Network   5
        The Personal Area Network   6
    Network Topologies   7
        The Bus Topology   8
        The Ring Topology   9
        The Star Topology   10
        The Mesh Topology   10
        Ad Hoc Connections   11
        Point-to-Point Connections   13
        Point-to-Multipoint Connections   13
    The OSI Model   14
        Layer 1 – The Physical Layer   14
        Layer 2 – The Data Link Layer   15
        Layer 3 – The Network Layer   16
        Layer 4 – The Transport Layer   17
        Layer 5 – The Session Layer   17
        Layer 6 – The Presentation Layer   18
        Layer 7 – The Application Layer   18
        How the Layers Work Together   19
        Peer Layer Communication   20
        Data Encapsulation   20
    Device Addressing   20
        Physical Addressing   21
        Logical Addressing   23
    Summary   26
    Chapter Essentials   27

Chapter 2   **Common Network Protocols and Ports**   **29**

    The Internet Protocol Suite   30
    Request for Comments (RFC)   32
    Common Network Protocols and Services   32

Transmission Control Protocol (TCP)	33
User Datagram Protocol (UDP)	34
Dynamic Host Configuration Protocol (DHCP)	35
Simple Network Management Protocol (SNMP)	40
Network Address Translation (NAT)	42
Port Address Translation (PAT)	44
Domain Name System (DNS)	44
Internet Control Message Protocol (ICMP)	45
Understanding Ports, Protocols, and Sockets	46
Secure Sockets Layer (SSL)	48
Common Protocols for File Transfer	48
File Transfer Protocol (FTP) – 20/21	49
File Transfer Protocol Secure (FTPS) – 990	57
Secure File Transfer Protocol (SFTP) – 22	57
Secure Copy Protocol (SCP) – 22	58
Common Protocols Used with Messaging	58
Simple Mail Transfer Protocol (SMTP) – 25	58
Message Retrieval Protocols	60
Apple Push Notification Service (APNS) – 2195 / 2196	63
Extensible Messaging and Presence Protocol (XMPP) – 5223	65
Google Cloud Messaging (GCM) – 5228	65
Other Common Protocols	66
Hypertext Transfer Protocol (HTTP) – 80	66
Microsoft Desktop AirSync Protocol (AirSync) – 2175	68
Remote Desktop Protocol (RDP) – 3389	68
Server Routing Protocol (SRP) – 4101	69
Summary	71
Chapter Essentials	72

Chapter 3   Radio Frequency and Antenna Technology Fundamentals   75

Radio Frequency Fundamentals	76
RF Regulatory Governing Bodies and Local Regulatory Authorities	79
United States: Federal Communications Commission	80
Europe: European Telecommunications Standards Institute	81
Radio Frequency Characteristics	81
Wavelength	82
Frequency	83
Amplitude	84
Phase	84
Bandwidth	85
Modulation	86

	Radio Frequency Range and Speed	87
	Attenuation	88
	Environmental Factors, Including Building Materials	89
	Interference from Wi-Fi and Non-Wi-Fi Sources	94
	Measurement Units of Radio Frequency	95
	Absolute Measurements of Radio Frequency Power	95
	Relative Measurements of Radio Frequency Power	96
	Radio Frequency Antenna Concepts	99
	Radio Frequency Lobes	100
	Antenna Beamwidth	101
	Antenna Gain	103
	Antenna Polarization	105
	Antenna Types	107
	Omnidirectional Antennas	107
	Omnidirectional Antenna Specifications	108
	Semidirectional Antennas	111
	Highly Directional Antennas	121
	Highly Directional Antenna Specifications	122
	Minimizing the Effects of Multipath Using Antenna Diversity	124
	Distributed Antenna System (DAS)	125
	Leaky Coax DAS	127
	DAS and IEEE 802.11 Wireless Networking	127
	DAS Antennas	128
	DAS Antenna Specifications	129
	Radio Frequency Cables and Connectors	131
	Impedance and Voltage Standing Wave Ratio	131
	RF Cable Types	131
	RF Cable Length	132
	RF Cable Cost	132
	RF Cable Connectors	133
	Antenna Installation Considerations	134
	Maintaining Clear Communications	134
	Wind and Lightning	138
	Installation Safety	140
	Antenna Mounting	140
	Summary	143
	Chapter Essentials	144
Chapter 4	**Standards and Certifications for Wireless Technology 145**	
	The IEEE	146
	IEEE Standards for Wireless Communication	147
	The IEEE 802.11 Standard	147
	The IEEE 802.11a Amendment	148
	The IEEE 802.11b Amendment	150

The IEEE 802.11g Amendment	150
The IEEE 802.11n Amendment	153
The IEEE 802.11ac Amendment	155
Additional IEEE 802.11 Amendments	157
Wireless Networking and Communications	159
Wireless Personal Area Networks and the IEEE 802.15 Standard	160
Wireless Local Area Networks	160
Wireless Broadband, Metropolitan Area Networks, and the IEEE 802.16 Standard	161
Wireless Wide Area Networks	161
Interoperability Certifications	162
The Wi-Fi Alliance	162
Wi-Fi Protected Access Certification	162
Wi-Fi Protected Access 2 Certification	163
Wi-Fi Protected Setup Certification	164
Wi-Fi Multimedia Certification	165
Wi-Fi Multimedia Power Save Certification	165
The Wi-Fi Voice-Enterprise Certification	166
Common IEEE 802.11 Deployment Scenarios	166
Small Office, Home Office	167
Enterprise Deployments: Corporate Data Access and End-User Mobility	168
Extension of Existing Networks into Remote Locations	169
Public Wireless Hotspots	169
Carpeted Office Deployments	170
Educational Institution Deployments	173
Industrial Deployments	173
Healthcare Deployments	175
Last-Mile Data Delivery: Wireless ISP	177
High-Density Deployments	177
Municipal, Law Enforcement, and Transportation Networks	178
Building-to-Building Connectivity Using Wireless LAN Technology	178
Summary	180
Chapter Essentials	180
Chapter 5 IEEE 802.11 Terminology and Technology	**183**
Wireless LAN Operating Frequencies and RF Channels	184
US (FCC) Unlicensed Frequency Bands	185
Radio Frequency Channels	186
Radio Frequency Range	189
Wireless LAN Modes of Operation	190
The Independent Basic Service Set (IBSS)	191
The Basic Service Set (BSS)	196

The Extended Service Set (ESS)	198
The Mesh Basic Service Set (MBSS)	200
IEEE 802.11 Physical Layer Technology Types	201
Spread Spectrum Technology	201
IEEE 802.11g Extended Rate Physical	205
IEEE 802.11n High Throughput	206
IEEE 802.11a, 802.11g, 802.11n, and 802.11ac Orthogonal Frequency Division Multiplexing (OFDM)	207
IEEE 802.11a, 802.11g, 802.11n, and 802.11ac OFDM Channels	207
Wireless LAN Coverage and Capacity	210
Wireless LAN Coverage	210
Wireless LAN Capacity	212
Radio Frequency Channel Reuse and Device Colocation	215
Radio Frequency Signal Measurements	217
Receive Sensitivity	217
Radio Frequency Noise	217
Received Signal Strength Indicator (RSSI)	218
Signal-to-Noise Ratio (SNR)	219
Connecting to an IEEE 802.11 Wireless Network	220
IEEE 802.11 Frame Types	221
Wireless Network Discovery	222
IEEE 802.11 Authentication	225
IEEE 802.11 Association	230
IEEE 802.11 Deauthentication and Disassociation	231
The Distribution System	232
Data Rates	233
Throughput	234
Dynamic Rate Switching	242
Wireless LAN Transition or Roaming	242
IEEE 802.11 Power Save Operations	244
Active Mode	245
Power Save Mode	245
Automatic Power Save Delivery	246
IEEE 802.11 Protection Modes and Mechanisms	247
IEEE 802.11g Extended Rate Physical Protection Mechanisms	247
IEEE 802.11n High-Throughput Protection Mechanisms	248
Summary	250
Chapter Essentials	251
Chapter 6 Computer Network Infrastructure Devices	**253**
The Wireless Access Point	254
Autonomous Access Points	256
Controller-Based Access Points	266

Cloud-Based Access Points	268
Wireless Branch Router/Remote Access Point	269
Wireless Mesh	270
Wireless Bridges	271
Wireless Repeaters	272
Wireless LAN Controllers and Cloud-Managed Architectures	274
Centralized Administration	274
Virtual Local Area Network	275
Power over Ethernet (PoE) Capability	275
Improved Mobile Device Transition	275
Wireless LAN Profiles and Virtual WLANs	275
Advanced Security Features	276
Captive Web Portals	276
Built-in RADIUS Services	276
Predictive Modeling Site Survey Tools	277
Radio Frequency Spectrum Management	277
Firewalls	277
Quality of Service (QoS)	277
Infrastructure Device Redundancy	277
Wireless Intrusion Prevention System (WIPS)	278
Direct and Distributed AP Connectivity	278
Layer 2 and Layer 3 AP Connectivity	278
Distributed and Centralized Data Forwarding	279
Power over Ethernet	280
Power Sourcing Equipment	281
Powered Devices and Classification Signatures	282
Benefits of PoE	283
Virtual Private Network (VPN) Concentrators	284
Point-to-Point Tunneling Protocol	286
Layer 2 Tunneling Protocol	286
Components of a VPN Solution	287
Network Gateways	292
Proxy Servers	293
Summary	296
Chapter Essentials	297
Chapter 7 Cellular Communication Technology	**299**
The Evolution of Cellular Communications	300
First Generation (1G)	302
Second Generation (2G)	303
Third Generation (3G)	303
Fourth Generation (4G)	304
Comparing the Features of the Generations	304

	Channel Access Methods	305
	Frequency Division Multiple Access	306
	Time Division Multiple Access	307
	Code Division Multiple Access	308
	Circuit Switching vs. Packet Switching Network Technology	309
	Circuit Switched Data	309
	Packet Switching	310
	Mobile Device Standards and Protocols	311
	Global Standard for Mobile Communications	311
	General Packet Radio Service	312
	Enhanced Data Rates for GSM Evolution	312
	Universal Mobile Telecommunications System	312
	Evolution Data Optimized	312
	High Speed Downlink Packet Access	313
	High Speed Uplink Packet Access	313
	High Speed Packet Access	313
	Evolved High Speed Packet Access	313
	Long Term Evolution	314
	WiMAX	314
	Roaming between Different Network Types	315
	Summary	316
	Chapter Essentials	317
Chapter 8	**Site Survey, Capacity Planning, and Wireless Design**	**319**
	Wireless Site Surveys	320
	Size of the Physical Location	322
	Intended Use of the Wireless Network	322
	Number of Mobile Devices	322
	Wireless Client Device Capabilities	323
	The Environment	324
	Understanding the Performance Expectations	324
	Bring Your Own Device Acceptance	324
	Building Age and Construction Materials	325
	Network Infrastructure Devices	326
	Gathering Business Requirements	327
	General Office/Enterprise	328
	Interviewing Stakeholders	328
	Manufacturer Guidelines and Deployment Guides	333
	Gathering Site-Specific Documentation	333
	Floor Plans and Blueprints	333
	Furnishings	334
	Electrical Specifications	335

Documenting Existing Network Characteristics	335
Identifying Infrastructure Connectivity and Power Requirements	337
Understanding Application Requirements	339
Understanding RF Coverage and Capacity Requirements	340
Client Connectivity Requirements	341
Antenna Use Considerations	342
The Physical Radio Frequency Site Survey Process	345
Radio Frequency Spectrum Analysis	346
Spectrum Analysis for IEEE 802.11 Wireless Networks	347
Wi-Fi and Non-Wi-Fi Interference Sources	349
Wi-Fi Interference	355
Spectrum Analysis for Cellular Communications	356
Received Signal Strength	362
Performing a Manual Radio Frequency Wireless Site Survey	364
Obtaining a Floor Plan or Blueprint	366
Identifying Existing Wireless Networks	366
Testing Access Point Placement	368
Analyzing the Results	369
Advantages and Disadvantages of Manual Site Surveys	369
Software-Assisted Manual Site Survey	370
Manual Site Survey Toolkit	378
Performing a Predictive Modeling Site Survey	380
Internet-Based Predictive Modeling Site Survey	387
Performing a Post-Site Survey	388
Protocol Analysis	388
RF Coverage Planning	396
Infrastructure Hardware Selection and Placement	396
Testing Different Antennas	397
Testing Multiple Antenna Types	398
Choosing the Correct Antennas	399
Wireless Channel Architectures	400
Multiple-Channel Architecture	400
Single-Channel Architecture	401
Wireless Device Installation Limitations	402
Site Survey Report	403
Summary	404
Chapter Essentials	405

Chapter 9   Understanding Network Traffic Flow and Control   409

Local Area Network and Wide Area Network Traffic Flow	410
Local Area Network Traffic Flow	411
Wide Area Network Traffic Flow	414

	Network Subnets	416
	IP Address Classes	416
	Special Use IP Addresses	418
	IP Subnetting	418
	Subnet Mask	419
	Creating Subnets	425
	Routing Network Traffic	433
	Network Traffic Shaping	434
	Backhauling Network Traffic	435
	Bandwidth and User Restrictions	436
	Quality of Service	438
	Summary	439
	Chapter Essentials	440
Chapter 10	**Introduction to Mobile Device Management**	**441**
	Mobile Device Management Solutions	442
	The Software as a Service (SaaS) Solution	444
	The On-Premise Solution	444
	Common Mobile Device Operating System Platforms	446
	The Mobile Application Store	447
	Pushing Content to Mobile Devices	451
	MDM Administrative Permissions	451
	Understanding MDM High Availability and Redundancy	452
	MDM Device Groups	453
	Location-Based Services	453
	Geo-fencing	454
	Geo-location	455
	Mobile Device Telecommunications Expense Management	456
	Captive and Self-Service Portals	456
	The Captive Portal	456
	The Self-Service Portal	457
	Summary	458
	Chapter Essentials	459
Chapter 11	**Mobile Device Policy, Profiles, and Configuration**	**461**
	General Technology Network and Security Policy	462
	Industry Regulatory Compliance	463
	Security Policy Framework	464
	Information Technology and Security Policy Implementation and Adherence	465
	Acceptable Use Policy	466
	Balancing Security and Usability	467

		Backup, Restore, and Recovery Policies	467
		Operating System Modifications and Customization	468
		Operating System Vendors	468
		Original Equipment Manufacturer (OEM)	469
		Vendor Default Device Applications	469
		Technology Profiles	470
		Mobile Device Profiles	470
		Directory Services Integration	471
		Issuing Digital Certificates	471
		End-User License Agreement (EULA)	472
		Understanding Group Profiles	473
		Corporate-Owned Mobile Device Profiles	474
		Employee-Owned Mobile Device Profiles	474
		Outside Consultants' and Visitors' Mobile Device Profiles	475
		Policy and Profile Pilot Testing	475
		Summary	476
		Chapter Essentials	476
Chapter	**12**	**Implementation of Mobile Device Technology**	**479**
		System Development Life Cycle (SDLC)	480
		Pilot Program Initiation, Testing, Evaluation, and Approval	482
		Training and Technology Launch	483
		Training for Network Administrators	483
		Training for End Users	483
		Technology Launch	483
		Documentation Creation and Updates	484
		Mobile Device Configuration and Activation	484
		Wireless LAN, Cellular Data, and Secure Digital Adapters	486
		USB Wireless LAN Adapters	487
		Cellular Data Adapters	493
		Secure Digital (SD) Adapters	494
		Mobile Device Management Onboarding and Provisioning	495
		Manual Method	496
		Self-Service Methods	496
		Certificate Enrollment Methods	496
		Role-Based Enrollment	497
		Other Enrollment and Identification Methods	497
		Voucher Methods	503
		Mobile Device Management Offboarding and Deprovisioning	504
		Employee Terminations and Mobile Technology	505
		Mobile Device Deactivation	505
		Mobile Device Migrations	505
		Removing Applications and Corporate Data	505

		Asset Disposal and Recycling	506
		Summary	506
		Chapter Essentials	507
Chapter	**13**	**Mobile Device Operation and Management Concepts**	**509**
		Centralized Content, Application Distribution, and Content Management Systems	510
		Server-Based Distribution Models	511
		Cloud-Based Distribution Models	511
		Content Permissions, Data Encryption, and Version Control	512
		Content Permissions and Data Encryption	512
		Version Control	513
		Mobile Device Remote Management Capabilities	513
		Mobile Device Location Services	514
		Remote Lock and Unlock	515
		Device Remote Wipe	516
		Remote Control	519
		Reporting Features	519
		Life Cycle Operations	519
		Certificate Expiration and Renewal	520
		Software Updates	521
		Software Patches	521
		Device Hardware Upgrades	522
		Device Firmware Updates	522
		Information Technology Change Management	525
		The Technology End-of-Life Model	525
		Device Operating Systems	526
		Device Hardware	526
		Device Applications	526
		Deployment Best Practices	526
		Summary	527
		Chapter Essentials	528
Chapter	**14**	**Mobile Device Technology Advancements, Requirements, and Application Configuration**	**531**
		Maintaining Awareness of Mobile Technology Advancements	532
		Mobile Device Hardware Advancements	533
		Mobile Operating Systems	533
		Third-Party Mobile Application Vendors	533
		New Security Risks and Threat Awareness	535
		Configuration of Mobile Applications and Associated Technologies	535

	In-House Application Requirements	535
	Application (App) Publishing	536
	Mobile Device Platforms	537
	Device Digital Certificates	537
	Mobile Application Types	537
	Native Apps	538
	Web Apps	538
	Hybrid Apps	538
	Configuring Messaging Protocols and Applications	539
	Simple Mail Transfer Protocol (SMTP)	540
	Configuring Internet Message Access Protocol (IMAP)	541
	Post Office Protocol 3 (POP3)	546
	Messaging Application Programming Interface (MAPI)	546
	Configuring Network Gateway and Proxy Server Settings	547
	Configuring Network Gateway Settings	547
	Configuring Proxy Server Settings	549
	Push Notification Technology	551
	Configuring Apple Push Notification Service (APNS)	551
	Google Cloud Messaging (GCM)	554
	Information Traffic Topology	555
	On-Premise Network Operations Center	555
	Third-Party Network Operations Center	556
	Hosted Network Operations Center	556
	Summary	556
	Chapter Essentials	557
Chapter 15	**Mobile Device Security Threats and Risks**	**559**
	Understanding Risks with Wireless Technologies	560
	Public Wi-Fi Hotspots	561
	Rogue Access Points	561
	Denial of Service (DoS) Attacks	562
	Radio Frequency (RF) Jamming	564
	Cell Tower Spoofing	565
	Wireless Hijack Attacks	565
	Wireless Man-in-the-Middle Attacks	566
	Weak Security Keys	567
	Warpathing	567
	Understanding Risks Associated with Software	568
	Malware	568
	Mobile App Store Usage	570
	Mobile Device Jailbreaking	570
	Android Rooting	570

		Understanding Risks Associated with Mobile Device Hardware	571
		Mobile Device Loss or Theft	571
		Mobile Device Cloning	572
		Understanding Risks Affecting Organizations	572
		Bring Your Own Device (BYOD) Ramifications	572
		Risks with Removable Media	573
		Wiping Personal Data	573
		Unknown Devices on Network or Server	575
		Summary	575
		Chapter Essentials	576
Chapter	**16**	**Device Authentication and Data Encryption**	**577**
		Access Control	578
		Username and Password	578
		Personal Identification Numbers (PINs)	579
		Security Tokens	579
		Certificate Authentication	579
		Biometrics	579
		Multifactor Authentication	580
		Authentication Processes	580
		Wireless Passphrase Security	580
		Wireless Networking WPA and WPA2 Enterprise Security	582
		IEEE 802.1X/EAP	582
		Remote Authentication Dial-In User Service (RADIUS)	585
		Encryption Methods for Wireless Networking	587
		Wired Equivalent Privacy (WEP)	587
		Temporal Key Integrity Protocol (TKIP)	588
		Counter Mode with Cipher Block Chaining Message Authentication Code Protocol (CCMP)	589
		Cipher Methods	589
		Stream Ciphers	589
		Block Ciphers	590
		Secure Tunneling	591
		Virtual Private Networking (VPN)	592
		Secure Shell (SSH)	594
		Web-Based Security	594
		Public Key Infrastructure (PKI) Concepts	595
		Physical Media Encryption	596
		Full Disk Encryption	596
		Block-Level Encryption	596
		Folder and File-Level Encryption	597
		Removable Media Encryption	598
		Summary	598
		Chapter Essentials	598

Chapter 17 Security Requirements, Monitoring, and Reporting — 601

- Device Compliance and Report Audit Information — 602
- Third-Party Device Monitoring Applications (SIEM) — 604
- Understanding Mobile Device Log Files — 605
- Mitigation Strategies — 606
  - Antivirus — 606
  - Software Firewalls — 607
  - Access Levels — 608
  - Permissions — 608
  - Host-Based and Network-Based IDS/IPS — 609
  - Anti-Malware — 611
  - Data Loss Prevention (DLP) — 612
  - Device Hardening — 615
  - Physical Port Disabling — 616
  - Firewall Settings — 617
  - Port Configuration — 618
  - Application Sandboxing — 618
  - Trusted Platform Modules — 619
  - Data Containers — 619
  - Content Filtering — 620
  - Demilitarized Zone (DMZ) — 621
- Summary — 623
- Chapter Essentials — 623

Chapter 18 Data Backup, Restore, and Disaster Recovery — 625

- Network Server and Mobile Device Backup — 626
- Backing Up Server Data — 627
- Backing Up Client Device Data — 631
  - Mobile Device Backup for Corporate and Personal Data — 636
  - Backing Up Data to a Device Locally — 636
  - Frequency of Backups — 637
  - Testing Your Backup — 637
- Disaster Recovery Principles — 637
  - Disaster Recovery Plan — 638
  - Disaster Recovery Locations — 641
- Maintaining High Availability — 642
- Restoring Data — 643
  - Restoring Corporate Data — 644
  - Restoring Personal Data — 644
- Summary — 644
- Chapter Essentials — 645

| Chapter | 19 | **Mobile Device Problem Analysis and Troubleshooting** | **647** |

The Troubleshooting Process 648
Understanding Radio Frequency Transmitters and Receivers 649
    Steps in the Troubleshooting Process 651
    Gathering Information 651
    Understanding the Symptoms 652
    Changes to the Environment 652
    Documenting Findings, Actions, and Outcomes 653
Troubleshooting Specific Problems 653
    Problem: Shortened Battery Life 653
    Problem: Synchronization Issues 654
    Problem: Power Adapter/Supply 654
    Problem: Password-Related Issues 655
    Problem: Device Crash 656
    Problem: Power Outage 656
    Problem: Missing Applications 656
    Problem: Email Issues 657
    Problem: Profile Authentication and
       Authorization Issues 657
Common Over-the-Air Connectivity Problems 658
    Wireless Latency 659
    No Cellular Signal 659
    No Wireless Network Connectivity 660
    Roaming and Transition Issues 662
    Cellular Device Activation Issues 663
    Access Point Name Issues 663
    Wireless Network Saturation 664
Troubleshooting Common Security Problems 664
    Authentication Failures 664
    Non-expiring Passwords 665
    Expired Passwords 665
    Expired Certificates 665
    Misconfigured Firewalls 666
    False Negatives 666
    False Positives 666
    Content Filtering Misconfigurations 666
Summary 667
Chapter Essentials 667

Appendix	**The CompTIA Mobility+ Certification Exam**	**669**
	Preparing for the CompTIA Mobility+ Exam	670
	Taking the Exam	671
	Reviewing Exam Objectives	672
Index		*689*

Table of Exercises

Exercise	1.1	Viewing Device Address Information on a Computer	26
Exercise	2.1	Viewing DHCP Information	40
Exercise	2.2	FTP Server Software Installation	50
Exercise	2.3	FTP Client Software Installation	54
Exercise	3.1	Demonstrating Passive Gain	105
Exercise	3.2	Antenna Polarization Example	106
Exercise	3.3	Demonstrating Fresnel Zone and Blockage	137
Exercise	3.4	Installing a Pole/Mast Mount	141
Exercise	5.1	Measuring Throughput of a Wireless Network	236
Exercise	6.1	Setting Up a VPN	288
Exercise	8.1	Using Spectrum Analysis Tools for Wi-Fi Networks	350
Exercise	8.2	Using Spectrum Analysis Tools for Cellular Communications	358
Exercise	8.3	Installing Ekahau HeatMapper	372
Exercise	8.4	Installing RF3D WiFiPlanner2	382
Exercise	8.5	Installing a Protocol Analyzer	390
Exercise	9.1	Using an IP Subnet Calculator	428
Exercise	10.1	Installing Android Emulator Software in Microsoft Windows	448
Exercise	11.1	Resetting the Default Application—Android	470
Exercise	12.1	Installing a USB 2.0 Wireless LAN Adapter	488
Exercise	12.2	Installing Mini PCI and Mini PCIe Cards	492
Exercise	13.1	Enable Android Device Manager	516
Exercise	13.2	Update an Android Device to the Latest Firmware Version	523
Exercise	14.1	Configure an Android Device to Access Email Using IMAP	542
Exercise	14.2	Enabling APNS on an Apple iPhone	551
Exercise	15.1	Steps to Wipe Various Mobile Devices	574
Exercise	16.1	Encrypting a folder in Microsoft Windows	597
Exercise	18.1	Viewing the Archive Attribute	630
Exercise	18.2	Installing a Backup Program	632

Introduction

The pace at which information technology is progressing seems to be getting faster all the time. When the concept of personal computing was introduced about three decades ago, one would never have imagined that we would be where we are today. From using a computer and a monitor that sat on a desk and together weighed about 30 pounds to using a mobile device that has more computing power and fits in the palm of your hand, we have come a long way. Computer networking and mobile technology fascinates people from all walks of life. The advancements in mobile technology allow people of all professions and ages to access information in ways they would have never imagined possible.

The purpose of this book is to provide an introduction to the exciting and emerging world of wireless and mobile computing, mobile device management (MDM), and mobile technology. Reading this book will teach you the fundamentals of computer networking and protocols, radio frequency communication principles, and IEEE standards based wireless technology and give you an overview of hardware and software components, cellular communications, wireless site surveys, mobile device management, troubleshooting, and security principles for both wireless networking and mobility. In addition, this book will help you to prepare for the Mobility+ certification exam available from CompTIA. The Mobility+ certification is geared toward candidates that have CompTIA Network+ certification or equivalent experience and working knowledge of and at least 18 months of work experience in the administration of mobile devices.

Who Should Read This Book

This book is a good fit for anyone who wants to learn about or increase their knowledge level of wireless computer networking and wireless mobility. Help desk personnel, network administrators, network infrastructure design engineers, and most people who work in the information technology sector will benefit from the information contained in this book. It provides an understanding of how computer networking technology and radio frequency technology work together to create wireless networks. In addition, by closely following the exam objectives and using Appendix A to see how the exam objectives are covered in each chapter, this book will assist in preparation for the Mobility+ certification exam from CompTIA.

What You Will Learn

The opening chapter in the book is about network types, topologies and includes an introduction to the OSI model. This is a great topic for those who may be new to computer networking, and it's a nice review for those who already have experience. You will then read about common networking protocols and ports with a focus on those used with mobile device technology. Many individuals who work in the information technology and computer networking field have minimal experience with wireless technology and radio frequency.

For the reader who does not have this RF experience, Chapter 3 will explore radio frequency principles and antenna technology. This is a great introduction for those who want to gain knowledge of the radio frequency concepts that are needed to design and manage a mobile wireless network infrastructure. The reader will learn about wireless networking standards and the various devices that are used to create a wireless network infrastructure. Next the reader will get an overview of wireless cellular technology, including the different generations of cellular communications and how the technology is implemented. This is an important component within the topic of mobility because many mobile devices contain multifunction capabilities such as cellular and wireless LAN connectivity.

One very important part of a successful wireless deployment includes proper design. The design process also includes understanding wireless site surveys. This book explores wireless site survey and design for both wireless LANs and wireless cellular technology. Chapter 10 will introduce the reader to mobile device management (MDM) and the next several chapters will provide an in-depth look at mobile device policies, profiles, configuration, implementation, operations, management concepts, and mobile device technology advancements.

With information so readily available from anyplace in the world with an Internet connection, mobile device security cannot be underestimated or overlooked. The reader will learn about common security threats and risks that may have an impact on the mobile device user. Security concepts such as device and user authentication, data encryption, security monitoring, and reporting are covered next. The reader will then learn about the importance of data backup, restore, and disaster recovery as it pertains to computer networking, and mobile device technology. Finally, the book will explore the concept of troubleshooting from a networking, radio frequency, and mobile device technology perspective.

What You Need

This book contains various exercises that help to reinforce the topics that you read and learn about. To complete the exercises, you'll need a computer running the Microsoft Windows operating system and a mobile device such as an Android device or an iPhone or iPad. Some exercises require evaluation software that may be included on the companion website for this book at www.sybex.com/go/mobilityplus. If you do not have an Android device or choose not to use your Android device for the related exercises, you can download an Android emulator program that can be installed on a computer running the Microsoft Windows operating system. You can download the Android emulator program from YouWave at youwave.com. Evaluation software such as packet analyzers and site survey programs can be downloaded directly from the manufacturer's website as specified in the introduction of some exercises.

What Is Covered in This Book

Mobile Computing Deployment and Management: Real World Skills for CompTIA Mobility+ Certification and Beyond will help you learn about common networking protocols, standards-based wireless networking, cellular technology, mobility, mobile device management, security, device backup, and troubleshooting. This book is based on the exam objectives for the CompTIA Mobility+ certification exam (MB0-001). Reading the book, performing the exercises, and using the Sybex test engine to run the flashcards and practice exam will help you to prepare for the CompTIA Mobility+ certification exam (MB0-001). Here is a brief explanation of what is included in each chapter:

Chapter 1, "Computer Network Types, Topologies, and the OSI Model" If you are new to networking or just need a refresher, this chapter provides an overview of basic computer networking concepts, including foundational computer networking topics such as computer network types, computer topologies, the OSI model, and network device addressing.

Chapter 2, "Common Network Protocols and Ports" This chapter will take a deeper look at some of the common protocols that are contained in the layered suite of networking protocols. You will also learn about some of the common services that are used today in most computer networks and for Internet connectivity.

Chapter 3, "Radio Frequency and Antenna Technology Fundamentals" Understanding the basics of radio frequency (RF) technology is an important component of wireless networking for both wireless LAN and cellular technologies. This chapter will explore some of the basic RF concepts and provide you with a better understanding of the technology. Antennas are an essential part of a successful wireless deployment. You will learn about antenna technology and see various antenna types that are used with different wireless technologies.

Chapter 4, "Standards and Certifications for Wireless Technology" This chapter takes an in-depth look at the IEEE 802.11 standard and its amendments, including those associated with the communications and functional aspects of wireless networking. You will also learn about the IEEE 802.15 and IEEE 802.16 standards. In addition, we will explore interoperability certifications for IEEE 802.11 wireless networking technology.

Chapter 5, "IEEE 802.11 Terminology and Technology" Here you will learn about the terminology used in IEEE 802.11 wireless networking and about ad hoc and infrastructure models, RF channels, and the frequencies of the unlicensed RF bands. This chapter also covers RF signal measurements, including received signal strength indicator (RSSI) and signal-to-noise ratio (SNR), as well as other topics related to wireless LAN technology.

Chapter 6, "Computer Network Infrastructure Devices" This chapter explores a variety of infrastructure devices, including wireless access points, wireless mesh devices, wireless bridges, wireless repeaters, hardware wireless LAN controllers, and cloud-managed wireless systems. We will explore the concepts of Power over Ethernet (PoE) and of other network infrastructure devices, including virtual private network (VPN) concentrators, network gateways, and network proxy devices.

Chapter 7, "Cellular Communication Technology" It is important to understand that IEEE 802.11 wireless networking and cellular technology are both key components of mobile computing deployment and management. In this chapter, we will explore the common communications methods used with wireless mobile devices and cellular technology, including how cellular technology has evolved and common access methods that are used with it.

Chapter 8, "Site Survey, Capacity Planning, and Wireless Design"   This chapter explores wireless site surveys for both IEEE 802.11 wireless networks and indoor cellular connectivity. You will learn about the components of wireless network site survey and design, including the types (manual and predictive modeling) of site surveys. You will also learn how to determine areas of RF coverage and interference by using a spectrum analyzer.

Chapter 9, "Understanding Network Traffic Flow and Control"   No book on computer networking would be complete without a discussion of the basics of traffic flow for local area networks and wide area networks, including Network layer (Layer 3) logical addressing, IP addresses, subnetting, subnet masks, and how to subnet a network. In this chapter we will also explore traffic shaping techniques, bandwidth restrictions, and quality of service.

Chapter 10, "Introduction to Mobile Device Management"   With the advancements in mobile technology and the increased acceptance of the bring your own device (BYOD) philosophy, managing mobile devices is becoming a big concern in an enterprise. In this chapter, we will explore the basics of mobile device management (MDM) options, including both on-premise and cloud-based Software as a Service (SaaS) solutions and many of the related features of MDM solutions.

Chapter 11, "Mobile Device Policy, Profiles, and Configuration"   This chapter will provide a basic outline of some of the more common policy components that fit in as a framework for most organizations, and we will explore some of the basic components of a network security policy.

Chapter 12, "Implementation of Mobile Device Technology"   Knowledge of proper implementation techniques will help to provide a successful technology deployment of any type. The System Development Life Cycle (SDLC) and pilot programs are only part of the entire process. In this chapter we will explore some of these techniques and how they relate to a mobile device management deployment.

Chapter 13, "Mobile Device Operation and Management Concepts"   In this chapter we explore solutions that are available for mobile device content management and distribution, which includes enterprise-server-based and cloud-based solutions. You will also learn about mobile device remote management capabilities such as remote control, remote lock, and remote wipe. We will also explore change management and the end-of-life process.

Chapter 14, "Mobile Device Technology Advancements, Requirements, and Application Configuration"   This chapter explores topics that pertain to the awareness of mobile technology advancements, which includes understanding the importance of changes to the actual hardware devices (such as computers, smartphones, and tablets) and the mobile operating systems that are used on the devices. You will also learn about the requirements for application (app) types that may be used within an organization's deployment: in-house, custom, and purpose-built apps.

Chapter 15, "Mobile Device Security Threats and Risks"   Security threats are present with all types of technology, and mobile devices are no exception. In this chapter we will explore the security risks and threats that may have an impact on mobility, including the risks associated with wireless (radio frequency) technology, software, and hardware and the risks within an organization itself.

Chapter 16, "Device Authentication and Data Encryption" Technology is available to lessen the possibility of intrusion or hacking of computer networks and devices. In this chapter you will learn about various methods of access control, the authentication process, and encryption types used with mobile devices and wireless computer networking.

Chapter 17, "Security Requirements, Monitoring, and Reporting" The use of mobile device technology has added an additional, highly complex variable to the mix because regulators do not see any difference between data breaches on corporate-owned machines or those on personal devices that employees also use for work. In this chapter we will explore the available options that provide security controls for mobile devices, which in turn will help secure the corporate network as a whole.

Chapter 18, "Data Backup, Restore, and Disaster Recovery" Performing data backup and planning for various disasters is not a new concept. Data and configuration information backup is an essential part of all aspects of information technology, and mobile devices are no exception. In this chapter we will explore the concept of data backup and recovery solutions for both network servers and client or mobile devices and common disaster recovery procedures, high availability, backup, and restore for both the server side and the client device side.

Chapter 19, "Mobile Device Problem Analysis and Troubleshooting" Troubleshooting in any sense can be considered an acquired skill. Those who are tasked with troubleshooting wireless and mobile device technology will encounter many of the same problems that occur with wired networking plus others that are a result of the fact that wireless technologies use radio frequency. In this chapter, you will learn about common problems associated with wireless and mobile technology and how to identify problems based on the symptoms. We will explore many common problems that are associated with mobility.

If you think you've found a technical error in this book, please visit http:/sybex.custhelp.com. Customer feedback is critical to our efforts at Sybex.

Interactive Online Learning Environment and Test Bank

The interactive online learning environment that accompanies *Mobile Computing Deployment and Management: Real World Skills for CompTIA Mobility+ Certification and Beyond: Exam MB0-001* provides a test bank with study tools to help you prepare for the certification exam—and increase your chances of passing it the first time! The test bank includes the following:

Practice Exam Use the questions to test your knowledge of the material. The online test bank runs on multiple devices.

Flashcards   Questions are provided in digital flashcard format (a question followed by a single correct answer). You can use the flashcards to reinforce your learning and provide last-minute test prep before the exam.

Other Study Tools   Several bonus study tools are included:

Glossary   The key terms from this book and their definitions are available as a fully searchable PDF.

Videos   The videos enable you to see in action some of the products that are used in the exercises in the book.

Software   You can download trial versions of software such as CommView for Wifi and TamoGraph–wireless and mobile computing tools you'll find useful

Whitepapers   The whitepapers provide authoritative perspectives on topics to help you understand issues, solve problems, and make informed decisions.

Go to http://sybextestbanks.wiley.com to register and gain access to this interactive online learning environment and test bank with study tools.

How to Use This Book

This book uses certain typographic styles and other elements to help you quickly identify important information and to avoid confusion. In particular, look for the following style:

- *Italicized text* indicates key terms that are described at length for the first time in a chapter. (Italic text is also used for emphasis.)

In addition to this text convention, a few conventions highlight segments of text:

Tips will be formatted like this. A tip is a special bit of information that can make your work easier.

Notes are formatted like this. When you see a note, it usually indicates some special circumstance to make note of. Notes often include out-of-the-ordinary information about the subject at hand.

Warnings are found within the text to call particular attention to a potentially dangerous situation.

Sidebars

This special formatting indicates a sidebar. *Sidebars* are entire paragraphs of information that, although related to the topic being discussed, fit better into a stand-alone discussion. They are just what their name suggests: a sidebar discussion.

 Real World Scenario

Real World Examples

These special sidebars are used to give real-life examples of situations that actually occur in the real world. This may be a situation I or somebody I know has encountered, or it may be advice on how to work around problems that are common in real, working environments.

EXERCISES

An exercise is a procedure you should try out on your own to learn about the material in the chapter. Don't limit yourself to the procedures described in the exercises though! Work through as many procedures as you can to become more familiar with mobile computing deployment and management.

How to Contact the Author

If you have any questions regarding this book, the content, or any related subjects, you can contact Robert directly by email at robert@eightotwo.com.

Chapter 1

Computer Network Types, Topologies, and the OSI Model

TOPICS COVERED IN THIS CHAPTER:

✓ Computer Network Types

✓ Computer Network Topologies

✓ The OSI Model

✓ Peer Layer Communication

✓ Data Encapsulation

✓ Device Addressing

It is important to have an understanding of basic personal computer networking concepts before you begin exploring the world of over-the-air (wireless) networking technology, wireless terminology, and mobility. This chapter looks at various topics surrounding foundational computer networking, including computer network types, computer topologies, the OSI model, and network device addressing. It is intended to provide an overview of basic computer networking concepts as an introduction for those who need to gain a basic understanding or for those already familiar with this technology and want a review of these concepts.

You will look at the various types of wireless networks—including wireless personal area networks (WPANs), wireless local area networks (WLANs), wireless metropolitan area networks (WMANs), and wireless wide area networks (WWANs)—in Chapter 4, "Standards and Certifications for Wireless Technologies."

Network Types

Personal computer networking technology has evolved at a tremendous pace over the past couple of decades, and many people across the world now have some type of exposure to the technology. Initially, personal computers were connected, or networked, to share files and printers and to provide central access to the users' data. This type of network was usually confined to a few rooms or within a single building and required some type of cabled physical infrastructure. As the need for this technology continued to grow, so did the types of networks. Computer networking started with the local area network (LAN) and grew on to bigger and better types, including wide area networks (WANs), metropolitan area networks (MANs), and others. The following are some of the common networking types in use today:

- Local area networks (LANs)
- Wide area networks (WANs)
- Metropolitan area networks (MANs)
- Campus area networks (CANs)
- Personal area networks (PANs)

You may also come across the term *storage area network (SAN)*. The SAN is basically a separate subnet for offloading of large amounts of data used within an enterprise network. High-speed connections are used, so the data is easily accessible because it appears to be part of the network. The connections are commonly Fibre Channel or iSCSI utilizing the TCP/IP protocol.

Most computer networks now contain some type of wireless connectivity or may consist of mostly wireless connectivity. The need for wireless networking and mobility continues to be in great demand and is growing at a rapid pace.

The Local Area Network

A *local area network (LAN)* can be defined as a group of devices connected in a specific arrangement called a topology. The topology used depends on where the network is installed. Some common legacy topologies such as the bus and ring and more modern topologies such as the star and mesh are discussed later in this chapter. Local area networks are contained in the same physical area and usually are bounded by the perimeter of a room or building. However, in some cases a LAN may span a group of buildings in close proximity that share a common physical connection.

Early LANs were mostly used for file and print services. This allowed users to store data securely and provided a centralized location of data for accessibility even when the user was physically away from the LAN. This central storage of data also gave a network administrator the ability to back up and archive all the saved data for disaster recovery purposes. As for print services, it was not cost effective to have a physical printer at every desk or for every user, so LANs allowed the use of shared printers for any user connected to the local area network. Figure 1.1 illustrates a local area network that includes both wired and wireless networking devices.

The Wide Area Network

As computer networking continued to evolve, many businesses and organizations that used this type of technology needed to expand the LAN beyond the physical limits of a single room or building. These networks covered a larger geographical area and became known as *wide area networks (WANs)*. As illustrated in Figure 1.2, WAN connectivity mostly consists of point-to-point or point-to-multipoint connections between two or more LANs. The LANs may span a relatively large geographical area. (Point-to-point and point-to-multipoint connections are discussed later in this chapter.) The WAN has allowed users and organizations to share data files and other resources with a much larger audience than a single LAN would.

FIGURE 1.1 Example of a local area network (LAN)

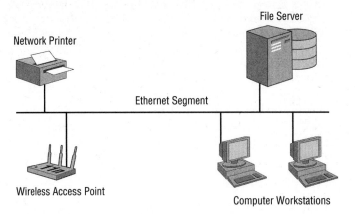

WANs can use leased lines from telecommunication providers (commonly known as *telcos*), fiber connections, and even wireless connections. The use of wireless for bridging local area networks is growing at a fast pace because it can often be a cost-effective solution for connecting LANs.

FIGURE 1.2 Wide area network (WAN) connecting two LANs

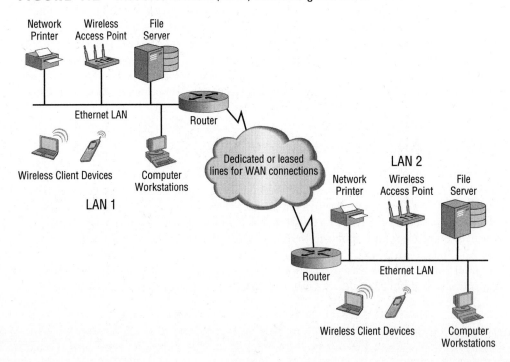

The Metropolitan Area Network

The *metropolitan area network (MAN)* interconnects devices for access to computer resources in a region or area larger than that covered by local area networks (LANs) but yet smaller than the areas covered by wide area networks (WANs). A MAN consists of networks that are geographically separated and can span from several blocks of buildings to entire cities (see Figure 1.3). MANs include fast connectivity between local networks and may include fiber optics or other wired connectivity that is capable of longer distances and higher capacity than those in a LAN.

MANs allow for connections to outside larger networks such as the Internet. They may include cable television, streaming video, and telephone services. Devices and connectivity used with metropolitan area networks may be owned by a town, county, or other locality and may also include the property of individual companies. Wireless MANs are also becoming a common way to connect the same type of areas but without the physical cabling limitations.

The MAN is growing in popularity as the need for access in this type of environment also increases.

FIGURE 1.3 Example of a metropolitan area network connecting a small town

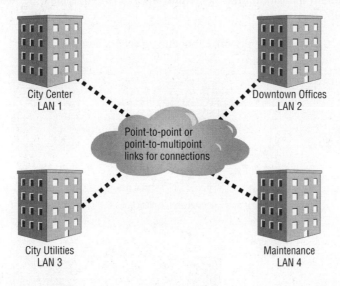

The Campus Area Network

A *campus area network (CAN)* includes a set of interconnected LANs that basically form a smaller version of a wide area network (WAN) within a limited geographical area, usually an office or school campus. Each building within the campus generally has a separate LAN. The LANs are often connected using fiber-optic cable, which provides a greater distance than copper wiring using IEEE 802.3 Ethernet technology. However, using wireless connections between the buildings in a CAN is an increasingly common way to connect the individual LANs. These wireless connections or wireless bridges provide a quick, cost-effective way to connect buildings in a university campus, as shown in Figure 1.4.

In a university campus environment, a CAN may link many buildings, including all of the various schools—School of Business, School of Law, School of Engineering, and so on—as well as the university library, administration buildings, and even residence halls. Wireless LAN deployments are becoming commonplace in university residence halls. With the rapidly increasing number of wireless mobile devices on university campuses, the number of wireless access points and the capacity of each need to be considered.

As in the university campus environment, a corporate office CAN may connect all the various building LANs that are part of the organization. This type of network will have the characteristics of a WAN but be confined to the internal resources of the corporation or organization. Many organizations are deploying wireless networks within the corporate CAN as a way to connect various parts of the business together. As with the university CAN, in the corporate world wireless can be a quick, cost-effective way to provide connectivity between buildings and departments.

All of the physical connection mediums and devices are the property of the office or school campus, and responsibility for the maintenance of the equipment lies with the office or campus as well.

FIGURE 1.4 Campus area network connecting a school campus

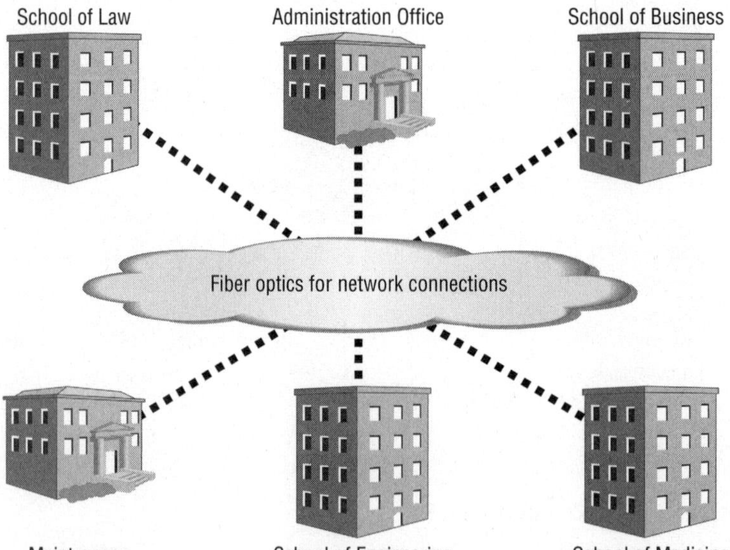

The Personal Area Network

Personal area networks (PANs) are networks that connect devices within the immediate area of individual people. PANs may consist of wired connections, wireless connections, or both. On the wired side, this includes universal serial bus (USB) devices such as printers, keyboards, and computer mice that may be connected with a USB hub. With wireless technology, PANs are short-range computer networks and in many cases use Bluetooth wireless technology. Wireless Bluetooth technology is specified by the IEEE 802.15 standard and is not IEEE 802.11 wireless local area technology. Bluetooth will be discussed in more detail in Chapter 4. Like wired PANs, wireless PANs are commonly used in connecting an individual's wireless personal communication accessories such as phones, headsets, computer mice, keyboards, tablets, and printers and are centered on the individual personal workspace without the need for physical cabling. Figure 1.5 illustrates a typical wireless PAN configuration.

FIGURE 1.5 Wireless Bluetooth network connecting several personal wireless devices

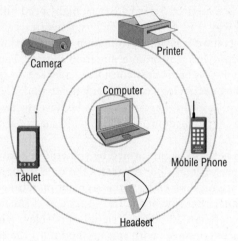

Network Topologies

A computer physical network topology is the actual layout or physical design and interconnection of a computer network. A topology includes the cabling and devices that are part of the network. In the following sections you will learn about several different types of network topologies:

- Bus
- Ring
- Star
- Mesh
- Ad-hoc
- Point-to-point
- Point-to-multipoint

The bus, ring, star, mesh, and ad-hoc topologies are typically what make up the local area network (LAN) you learned about previously. Point-to-point and point-to-multipoint topologies can be commonly used for connecting LANs and are mostly used for wide area network (WAN) connections. The size of your network will determine which topologies will apply. If your network is a single building and not part of a larger corporate network, the LAN topologies may be the extent of the technologies used. However, once that LAN connects to a different LAN, you are moving up and scaling to a wide area network.

The Bus Topology

A *bus* topology consists of multiple devices connected along a single shared medium with two defined endpoints. It is sometimes referred to as a high-speed linear bus and is a single collision domain in which all devices on the bus network receive all messages. Both endpoints of a bus topology have a 50 ohm termination device, usually a Bayonet Neill-Concelman (BNC) connector with a 50 ohm termination resistor. The bus topology was commonly used with early LANs but is now considered a legacy design.

One disadvantage to the bus topology is that if any point along the cable is damaged or broken, the entire LAN will cease to function. This is because the two endpoints communicate only across the single shared medium. There is no alternative route for them to use in the event of a problem.

Troubleshooting a bus network is performed by something known as the half-split method. A network engineer "breaks" or separates the link at about the halfway point and measures the resistance on both ends. If the segment measures 50 ohms of resistance, there is a good chance that side of the LAN segment is functioning correctly. If the resistance measurement is not 50 ohms, it signals a problem with that part of the LAN segment. The engineer continues with this method until the exact location of the problem is identified.

Figure 1.6 illustrates an example of the bus topology.

FIGURE 1.6 Example of the bus topology

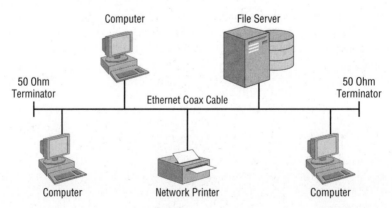

> ### Real World Scenario
>
> **Troubleshooting the Bus Topology**
>
> Many years ago I was called to troubleshoot a problem on a small local area network using a bus topology. The network consisted of a network file server, about 20 client stations, and a few network printers. The users complained of intermittent connection problems with the network. After spending some time looking over the network, I decided to test the bus using the half-split method and checked to verify that the cable was reporting the correct resistance using a volt-ohm-milliamp (VoM) meter. Sure enough, one side of the network cable reported the correct resistance reading, but the other side was giving intermittent results.
>
> After spending some time repeating the troubleshooting method, I was able to determine the problem. It turns out that someone had run the coax (bus) cable underneath a heavy plastic office chair mat and one of the little pegs used to protect the flooring was causing the intermittent connection as it struck the cable when the user moved their chair around the mat. I quickly replaced and rerouted the section of cable in question. It is a good thing I was there during the normal business operating hours when the person was moving around in the chair or I might have never found the problem. Ah, the joys of troubleshooting a bus topology.

The Ring Topology

The *ring* topology is rarely used with LANs today, but it is still widely used by Internet service providers (ISPs) for high-speed, resilient backhaul connections over fiber-optic links. In the ring topology, each device connects to two other devices, forming a logical ring pattern.

Ring topologies in LANs may use a token-passing access method, in which data travels around the ring in one direction. Only one device at a time will have the opportunity to transmit data. Because this access method travels in one direction, it does not need to use collision detection and often outperforms the bus topology, achieving higher data transfer rates than are possible using a collision detection access method. Each computer on the ring topology can act as a repeater, a capacity that allows for a much stronger signal.

The IEEE standard for LANs is IEEE 802.5, specifying Token Ring technology. IEEE 802.5 Token Ring technology used in LANs was a very efficient method used to connect devices, but it was usually more expensive than the bus or star topologies. Because of the token-passing method used, early 4 Mbps Token Ring networks could sometimes outperform a 10 Mbps IEEE 802.3 collision-based Ethernet network. Token Ring technology speeds increased to 16 Mbps but decreased in popularity as Ethernet speeds increased. Even though this is a ring topology, devices are connected through a central device and appear to be similar to devices on an Ethernet hub or switch. Figure 1.7 shows an example of the ring topology.

FIGURE 1.7 An example of the ring topology

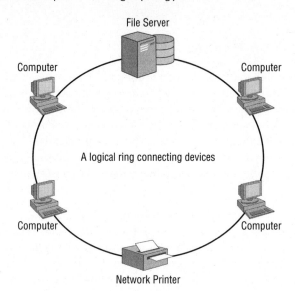

The Star Topology

The *star* topology, as shown in Figure 1.8, is the most commonly used method of connecting devices on a LAN today. It consists of multiple devices connected by a central connection device. Hubs, switches, and wireless access points are all common central connection devices, although hubs are rarely used today. The hub provides a single collision domain similar to a bus topology. However, the Ethernet switch and wireless access point both have more intelligence—the ability to decide which port specific network traffic can be sent to. Note that in Figure 1.8, the wireless star topology includes an Ethernet switch, which could also have extended devices connected to it with wires. In that sense, it is possible to have a wired/wireless hybrid topology.

A big advantage to the star over the bus and some ring topologies is that if a connection is broken or damaged, the entire network does not cease to function; only a single device in the star topology is affected. However, the central connection device such as a switch or wireless access point can be considered a potential central point of failure.

The Mesh Topology

A device in a mesh network will process its own data as well as serving as a communication point for other mesh devices. Each device in a *mesh* topology (see Figure 1.9) has one or more connections to other devices that are part of the mesh. This approach provides both network resilience in case of link or device failure and a cost savings compared to full redundancy. Mesh technology can operate with both wired and wireless infrastructure network devices. Wireless mesh networks are growing in popularity because of the potential uses in outdoor deployments and the cost savings they provide.

FIGURE 1.8 A common star topology using either wired or wireless devices

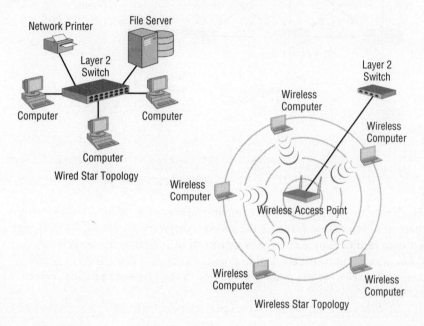

From an IEEE 802.11 wireless perspective, wireless mesh technology has now been standardized, although most manufacturers continue to use their proprietary methods. The amendment to the IEEE 802.11 standard for mesh networking is 802.11s. This amendment was ratified in 2011 and is now part of the latest wireless LAN standard, IEEE 802.11-2012. In addition to IEEE 802.11 networks, mesh is also standardized in IEEE 802.15 personal area networks for use with Zigbee and IEEE 802.16 Wireless MAN networks. Wireless standards will be discussed in more detail in Chapter 4.

As mentioned earlier, IEEE 802.11 wireless device manufacturers currently continue to use proprietary Layer 2 routing protocols, forming a self-healing wireless infrastructure (mesh) in which edge devices can communicate. Manufacturers of enterprise wireless networking infrastructure devices provide support for mesh access points (APs) such that the mesh APs connect back to APs that are directly wired into the network backbone infrastructure. The APs, wireless LAN controllers or software-based cloud solutions in this case, are used to configure both the wired and mesh APs.

Ad Hoc Connections

In the terms of computer networking, the *ad hoc* network is a collection of devices connected without a design or a plan for the purpose of sharing information or resources. Another term for an ad hoc network is *peer-to-peer network*.

FIGURE 1.9 Mesh networks can include either wired or wireless devices.

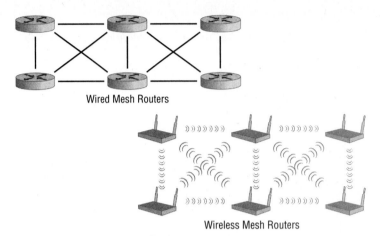

In a wired peer-to-peer network, all computing devices are of equal status. In other words, there is no server that manages the access to network resources. All peers can either share their own resources or access the resources of their devices on the network.

An ad hoc wireless network is one that does not contain a distribution system, which means no wireless access point is contained in the system to provide centralized communications.

Figure 1.10 shows an example of a wired peer-to-peer network and a wireless ad hoc network.

FIGURE 1.10 Wired peer-to-peer and wireless ad hoc networks

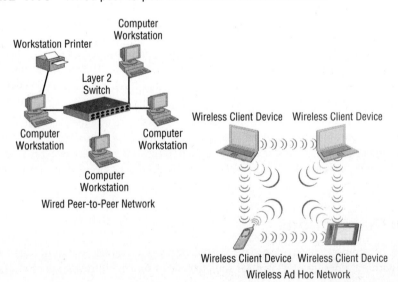

Point-to-Point Connections

When at least two LANs are connected, it is known as a *point-to-point connection* or link (see Figure 1.11). The connection can be made using either wired or wireless network infrastructure devices and can include bridges, wireless access points, and routers. Wireless point-to-point links can sometimes extend very long distances depending on terrain and other local conditions. Point-to-point links provide a connection between LANs, allowing users from one LAN to access resources on the other connected local area network.

FIGURE 1.11 Point-to-point connections using either wired or wireless

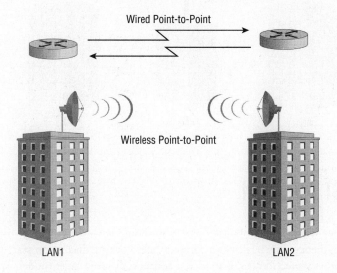

Wired point-to-point links consist of fiber-optic connections or leased lines from local telecommunication providers. Wireless point-to-point links typically call for semi-directional or highly directional antennas. Wireless point-to-point links include directional antennas and encryption to protect the wireless data as it propagates through the air from one network to the other. With some regulatory domains such as the Federal Communications Commission (FCC), when an omnidirectional antenna is used in this configuration it is considered a special case, called a point-to-multipoint link.

Point-to-Multipoint Connections

A network infrastructure connecting more than two LANs is known as a *point-to-multipoint connection* or link (see Figure 1.12). When used with wireless, this configuration usually consists of one omnidirectional antenna and multiple semidirectional or highly directional antennas. Point-to-multipoint links are often used in campus-style deployments, where connections to multiple buildings or locations may be required. Like point-to-point connections; wired point-to-multipoint connections can use either direct wired connections such as fiber-optic cables or leased line connectivity available from telecommunication providers.

FIGURE 1.12 Point-to-multipoint connections using either wired or wireless connections

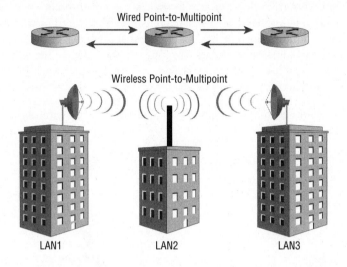

The OSI Model

Before we continue with other mobility topics, you should have some background on computer networking theory. The basics of a computer networking discussion start with the *Open Systems Interconnection (OSI)* model, a conceptual seven-layer model. The OSI model has been around for decades. It came about in 1984 and was developed by the International Organization for Standardization (ISO). The ISO is a worldwide organization that creates standards on an international scale. The OSI model describes the basic concept of communications in the computer network environment. Be careful not to confuse the two.

There are seven layers to the OSI model. Each layer is made up of many protocols and serves a specific function. You will take a quick look at all seven layers of the OSI model. Some wireless-specific functionality of the OSI model will be discussed later in Chapter 5, "IEEE 802.11 Terminology and Technology." Figure 1.13 illustrates the seven layers of the conceptual OSI model.

The following sections describe how each layer is used.

Layer 1 – The Physical Layer

The *Physical layer* (sometimes referred as the PHY) is the lowest layer in the OSI model. The PHY consists of bit-level data streams and computer network hardware connecting the devices together. This hardware that connects devices includes network interface cards, cables, Ethernet switches, wireless access points, and bridges. Keep in mind some of these hardware devices, such as Ethernet switches and bridges, actually have Data Link layer (Layer 2) functionally and operate at that layer but also make up the actual physical connections. In the case of wireless networking, radio frequency (RF) uses air as the medium for wireless communications. With respect to wireless networking, the Physical layer consists of two sublayers:

- Physical Layer Convergence Protocol (PLCP)
- Physical Medium Dependent (PMD)

The PLCP, the higher of the two layers, is the interface between the PMD and Media Access Control (MAC) sublayer of the Data Link layer. This is where the Physical layer header is added to the data. The PMD is the lower sublayer at the bottom of the protocol stack and is responsible for transmitting the data onto the wireless medium. Figure 1.14 shows the two sublayers that make up the Physical layer.

FIGURE 1.13   Representation of the OSI Model

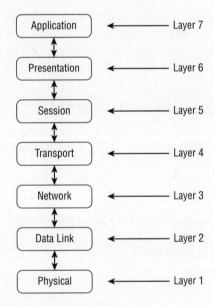

Layer 2 – The Data Link Layer

The *Data Link layer* is responsible for organizing the bit-level data for communication between devices on a network and detecting and correcting Physical layer errors. This layer consists of two sublayers:

- Logical Link Control (LLC)
- Media Access Control (MAC)

FIGURE 1.14 Physical layer sublayers, PMD and PLCP

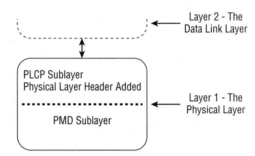

The bit-level communication is accomplished through Media Access Control (MAC) addressing. A *MAC address* is a unique identifier of each device on the computer network and is known as the physical or sometimes referred to as the hardware address. (MAC addresses are discussed later in this chapter.) Figure 1.15 illustrates the two sublayers of the Data Link layer, Layer 2.

FIGURE 1.15 Data Link layer sublayers, LLC and MAC

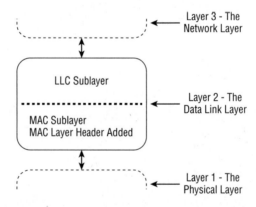

Layer 3 – The Network Layer

The *Network layer* is where the Internet Protocol (IP) resides. The Network layer is responsible for addressing and routing data by determining the best route to take based on what it has learned or been assigned. An IP address is defined as a numerical identifier or logical address assigned to a network device. The IP address can be static, manually assigned by a user, or it can be dynamically assigned from a server using Dynamic Host Configuration Protocol (DHCP). (IP addresses are discussed later in this chapter.) Figure 1.16 illustrates the Layer 2 MAC address translation to a Layer 3 IP address.

FIGURE 1.16 Data Link layer (Layer 2) to Network layer (Layer 3) address translation

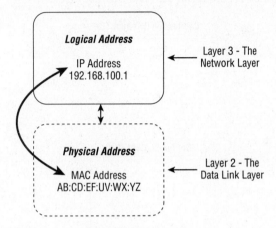

Layer 4 – The Transport Layer

The *Transport layer* consists of both connection-oriented and connectionless protocols providing communications between devices on a computer network. Although there are several protocols that operate at this layer, you should be familiar with two commonly used Layer 4 protocols:

- Transmission Control Protocol (TCP)
- User Datagram Protocol (UDP)

TCP is a connection-oriented protocol and is used for communications that require reliability, analogous to a circuit-switched telephone call.

UDP is a connectionless protocol and is used for simple communications requiring efficiency, analogous to sending a postcard through a mail service. You would not know if the postcard was received or not. UDP and TCP port numbers are assigned to applications for flow control and error recovery. Figure 1.17 represents the relationship between the Transport layer protocols TCP and UDP.

Layer 5 – The Session Layer

The *Session layer* opens, closes, and manages communications sessions between end-user application processes located on different network devices. The following protocols are examples of Session layer protocols:

- Network File System (NFS)
- Apple Filing Protocol (AFP)
- Remote Procedure Call Protocol (RPC)

FIGURE 1.17 Comparison between TCP and UDP protocols

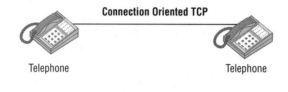

Layer 6 – The Presentation Layer

The *Presentation layer* provides delivery and formatting of information for processing and display. This allows for information that is sent from one device on a network (the source) to be understood by another device (the destination) on the network.

Layer 7 – The Application Layer

The *Application layer* can be considered the interface to the user. Application is another term for a program that runs on a computer or other networking device and that is not what we are looking at here. Protocols at this layer are for network operations such as, for example, transferring files, browsing web pages, and sending email. The following list includes some of the more common examples of Application layer protocols we use daily:

- File Transfer Protocol (FTP) for transfering data
- Hypertext Transfer Protocol (HTTP) for web browsing
- Post Office Protocol v3 (POP3) for email

Common Application layer protocols will be discussed further in Chapter 2, "Common Network Protocols and Ports."

How the Layers Work Together

In order for computers and other network devices to communicate with one another using the OSI model, a communication infrastructure of some type is necessary. In a wired network, such an infrastructure consists of cables, repeaters, bridges, and Layer 2 switches. In a wireless network, the infrastructure consists of access points, bridges, repeaters, radio frequency, and the open air. Some of these devices will be discussed in more detail in Chapter 6, "Computer Network Infrastructure Devices."

Wireless networking functions at the two lowest layers of the OSI model, Layer 1 (Physical) and Layer 2 (Data Link). However, to some degree Layer 3 (Network) plays a role as well, generally for the TCP/IP protocol capabilities.

OSI Model Memorization Tip

One common method you can use to remember the seven layers of the OSI model from top to bottom is to memorize the following sentence: All people seem to need data processing. Take the first letter from each word and that will give you an easy way to remember the first letter that pertains to each layer of the OSI model.

- **A**ll (**A**pplication)
- **P**eople (**P**resentation)
- **S**eem (**S**ession)
- **T**o (**T**ransport)
- **N**eed (**N**etwork)
- **D**ata (**D**ata Link)
- **P**rocessing (**P**hysical)

Here's another one, this time from the bottom to the top:

- **P**lease (**P**hysical)
- **D**o (**D**ata Link)
- **N**ot (**N**etwork)
- **T**hrow (**T**ransport)
- **S**ausage (**S**ession)
- **P**izza (**P**resentation)
- **A**way (**A**pplication)

Peer Layer Communication

Peer layers communicate with other layers in the OSI model and the layers underneath are their support systems. Peer layer communication is the "horizontal" link between devices on the network. Figure 1.18 shows three examples of *peer layer communication*. Keep in mind, however, that this principle applies to all seven layers of the OSI model. This allows for the layers to communicate with the layer to which a device is sending or receiving information.

FIGURE 1.18   Peer communication between three of the seven layers

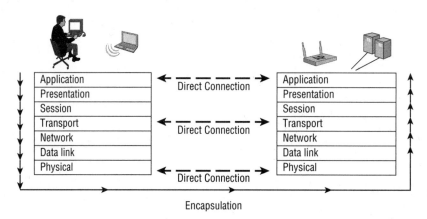

Data Encapsulation

The purpose of *encapsulation* is to allow Application layer data communication between two stations on a network using the lower layers as a support system. As data moves down the OSI model from the source to the destination, it is encapsulated. As data moves back up the OSI model from the source to the destination, it is de-encapsulated. Some layers will add a header and/or trailer when information is being transmitted and remove it when information is being received. Encapsulation is the method in which lower layers support upper layers. Figure 1.19 illustrates this process.

Device Addressing

Every device on a network requires unique identification. This can be accomplished in a couple of ways:

- Physical addresses
- Logical addresses

FIGURE 1.19 Information is added at each layer of the OSI model as data moves between devices

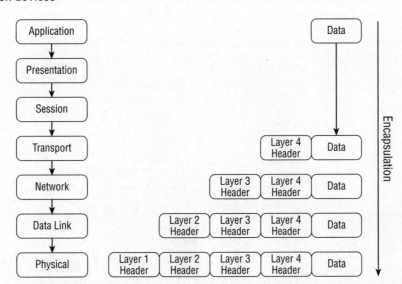

The *physical address* of a network adapter is also known as the Media Access Control (MAC) address. As shown in Figure 1.20, every device on a network (like every street address in a city) must have a unique address. The physical address is required in order for a device to send or receive information (data). An analogy to this is sending a package to be delivered via a courier service. Before you hand over the package to the courier, you would write the name and physical street address of the recipient on the package. This would ensure that the package is delivered correctly to the recipient.

The *logical address* is also known as the Internet Protocol (IP) address. Each device on a Layer 3 network or subnet must have a unique IP address (like every city's zip code). The IP address can be mapped to the physical address by using the Address Resolution Protocol (ARP).

The streets shown in Figure 1.20—1st, Main, and 2nd—represent local area network subnets. The street addresses—10, 20, and so on—represent the unique address of each structure on a street as a MAC address would a device on a LAN.

Physical Addressing

The physical address of a network device is called a MAC address because the *MAC sublayer* of the Data Link layer handles media access control. The MAC address is a 6-byte (12-character) hexadecimal address in the format AB:CD:EF:12:34:56. The first 3 bytes (or octets) of a MAC address are called the organizationally unique identifier (OUI). Some manufacturers produce many network devices and therefore require several OUIs. A table of all OUIs is freely available from the IEEE Standards Association website at

http://standards.ieee.org/develop/regauth/oui/oui.txt

MAC addresses are globally unique; an example is shown in Figure 1.21. The first 3 bytes or octets (6 characters) are issued to manufacturers by the IEEE. The last 3 bytes or octets (6 characters) are incrementally assigned to devices by the manufacturer.

FIGURE 1.20   The MAC address is analogous to the address of buildings on a street.

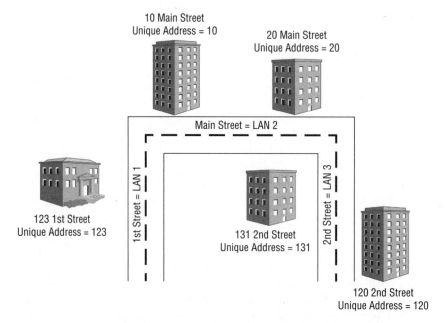

The streets shown—1st, Main, and 2nd—represent local area networks. The numbers 10, 20, 123, 131, and 120 represent the unique address of each structure on the streets just as MAC addresses would represent devices on a LAN.

FIGURE 1.21   Example of a Layer 2 MAC address shows the OUI and unique physical address

The MAC address of a device is usually stamped or printed somewhere on the device. This allows the device to be physically identified by the MAC address. By typing the simple

command **ipconfig /all** in the command-line interface of some operating systems, you can view the physical address of the network adapter. Figure 1.22 shows an example of the information displayed by using this command-line utility in the Microsoft Windows operating system.

FIGURE 1.22 The **ipconfig** command-line utility displaying a physical/MAC address in Microsoft Windows

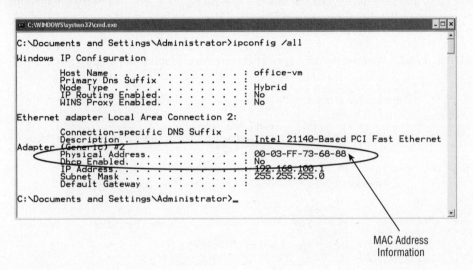

MAC Address Information

Logical Addressing

Network devices can also be identified by a logical address, known as the Internet Protocol (IP) address. The Layer 3 IP protocol works with a Layer 4 transport protocol, either User Datagram Protocol (UDP) or Transport layer Protocol (TCP). You learned earlier in this chapter that UDP is a connectionless protocol, and using it is analogous to sending a postcard through the mail. The sender has no way of knowing if the card was received by the intended recipient. TCP is a connection-oriented protocol, used for communications analogous to a telephone call, and provides guaranteed delivery of data through acknowledgements. During a telephone conversation, communication between two people will be confirmed to be intact, with the users acknowledging the conversation. Routable logical addresses such as TCP/IP addresses became more popular with the evolution of the Internet and the Hypertext Transfer Protocol (HTTP) that is used with the World Wide Web (WWW) service. IP moves data through an internetwork such as the Internet one router (or hop) at a time. Each router decides where to send the data based on the logical IP address. Figure 1.23 shows a basic network utilizing both Layer 2 and Layer 3 data traffic.

Logical addresses (IP addresses) are 32-bit dotted-decimal addresses usually written in the form www.xxx.yyy.zzz. Figure 1.24 illustrates an example of a logical Class C, 32-bit IP address. Each of the four parts is a byte, or 8 digital bits. There are two main IP address types: private addresses and public addresses. Private addresses are unique to an internal network, and public addresses are unique to the Internet. These addresses consist of two main parts: the network (subnet) and the host (device). Logical addresses also require a subnet mask and may have a gateway address depending on whether the network is routed. IPv4 addresses fall under three classes: Class A addresses, Class B addresses, and Class C addresses.

FIGURE 1.23 A network with Layer 3 network device logical addressing

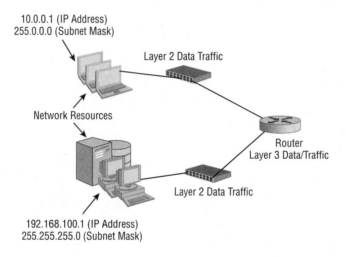

FIGURE 1.24 Example of a Class C logical IP address

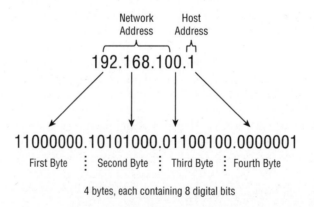

The logical IP addresses you just learned about are known as IPv4 addresses. Newer addresses called IPv6 addresses also exist and are discussed in Chapter 2.

Unlike a MAC address, an *IP address* is logical and can be either specified as a static address assigned to the device manually by the user or dynamically assigned by a server. However, the same command-line utility used to identify the physical address of a device can be used to identify the logical address of a device. Typing **ipconfig** at a command prompt displays the logical address, including the IP address, subnet mask, and default gateway (router) of the device. The ipconfig /all command illustrated earlier in the chapter will yield additional information, including the physical or MAC address of the device's network adapter. This command is for a computer using the Microsoft Windows operating system. For some Apple and Linux devices, the ifconfig command will yield similar information. Figure 1.25 shows the ipconfig utility displaying the logical address information, including the IP address and subnet mask.

FIGURE 1.25 The Microsoft Windows **ipconfig** command-line utility showing logical address information

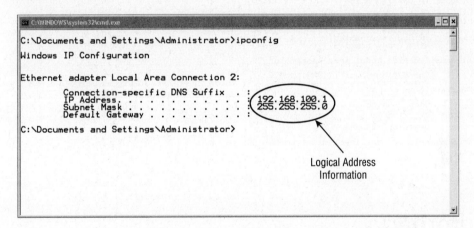

In Exercise 1.1, you will use the ipconfig utility from a command prompt on a computer using the Microsoft Windows operating system. This will allow you to see the address information for any available network adapters within the device.

 Exercise 1.1 was written using a computer with the Microsoft Windows 7 operating system. If you're using another version of the operating system, the steps may vary slightly. Keep in mind that there are many different shortcuts and ways to get to a command prompt in the Microsoft operating systems. The steps in this exercise use one common method.

EXERCISE 1.1

Viewing Device Address Information on a Computer

1. Click the Start button.
2. Mouse over the All Programs arrow. The All Programs window appears in the left pane.
3. Navigate to and click on the Accessories folder. The accessories programs appear.
4. Click the Command Prompt icon. The command window will appear.
5. In the command window, type **ipconfig /all**.
6. View the results in the command window. Notice the physical address of the network adapter as well as other information. The results should look similar to that shown here for Microsoft Windows 7 but may vary slightly based on the OS version in use.

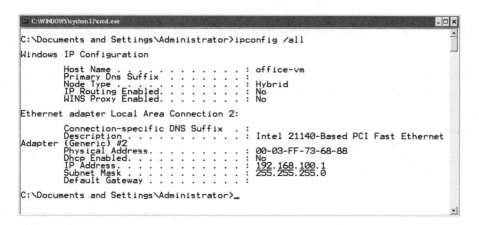

Summary

This chapter provided a survey of networking topics to help you understand the basics of computer networking as an introduction or a simple review. It began with an outline of the common network technology types:

- Local area networks (LANs)
- Wide area networks (WANs)
- Metropolitan area networks (MANs)

- Campus area networks (CANs)
- Personal area networks (PANs)

The next fundamental networking concept discussed was computer network topologies. You learned about network topologies ranging from the legacy high-speed linear bus and ring to the current star topology, the most common topology used today with both wired and wireless networks. You looked at the following various topologies:

- Bus
- Ring
- Star
- Mesh
- Ad hoc
- Point-to-point
- Point-to-multipoint

You then reviewed the basics and different layers of the OSI model, including a brief overview of each layer illustrating the different protocols and sublayers where applicable. Then I discussed the basics of peer communications and data encapsulation.

The chapter's final topic was device addressing. You explored the concepts of physical (MAC Sublayer) and logical (Network layer) addressing, including the IP address and subnet mask. A simple exercise using a computer with the Microsoft Windows operating system showed how to view device addressing information.

Chapter Essentials

Understand the components of a local area network (LAN).   A local area network is a group of computers connected by a physical medium in a specific arrangement called a topology.

Know the different types of networks.   The basic networks types are LAN, WAN, CAN, MAN, and PAN.

Become familiar with various networking topologies.   Bus, star, ring, mesh, and ad hoc are some of the topologies used in computer networking. Bus is considered legacy, and the star topology is one of the most common in use today.

Understand point-to-point and point-to-multipoint connections.   These can consist of both wired and wireless connections and will connect two or more LANs.

Understand the OSI model basics.   Each of the seven layers of the OSI model serves a specific function. It's beneficial to have an overall understanding of all seven layers.

Remember the details of the lower two layers of the OSI model. The Physical layer and Data Link layer are the two lowest layers in the OSI model. Wireless networking technology operates at these layers. The Data Link layer consists of two sublayers: the Logical Link Control (LLC) sublayer and the Media Access Control (MAC) sublayer.

Understand device addressing. Devices are assigned a unique physical address by the manufacturer. This address is known as the MAC address. MAC addresses consist of two parts, the organizationally unique identifier (OUI) and the unique physical address. A logical address may also be assigned at the Network layer to identify devices on different internetworks using the Internet Protocol (IP).

Chapter 2

Common Network Protocols and Ports

TOPICS COVERED IN THIS CHAPTER:

- ✓ The Internet Protocol Suite
- ✓ Request for Comments (RFC)
- ✓ Common Network Protocols and Services
- ✓ Common Network Protocols and Ports
- ✓ Secure Sockets Layer (SSL)
- ✓ Common Protocols for File Transfer
- ✓ Common Protocols Used with Messaging
- ✓ Other Common Protocols

In Chapter 1, "Computer Network Types, Topologies and the OSI Model," you were introduced to the conceptual OSI model and the basics of its different layers. This chapter will take a look at the Internet protocol suite and common networking protocols and ports. Like the OSI model, the Internet protocol suite consists of several layers. You will learn about some of the common protocols that are contained in this layered suite of protocols. We will also look at some of the common networking services that are used in most computer networks and for Internet connectivity today.

The Internet Protocol Suite

In this section you will learn about some of the common computer network protocols that are used for computer and mobile device communications, applications, and services. In Chapter 1 you learned about the conceptual seven-layer OSI model and how it works with computer communications. Here you will see how the Internet protocol suite plays a role with communications and its relationship to the OSI model. Like the OSI model, the *Internet protocol suite* consists of layers of protocols, each used for a specific purpose. It consists of four layers:

- The Application layer
- The Transport layer
- The Internet layer
- The Network Interface (Link) layer

 You might see documentation that refers to the Internet protocol suite as being a five-layer model. In such instances, the Network Interface (Link) layer is considered to be two separate layers, the Data Link and the Physical layer.

As with the OSI model, each layer of the Internet protocol suite specifies certain protocols that are used for a specific function in computer communications. However, the Internet protocol suite was placed in service years before the OSI model came into existence. Figure 2.1 shows the Internet protocol suite and how it logically maps to the OSI model.

FIGURE 2.1 OSI to Internet protocol suite mapping

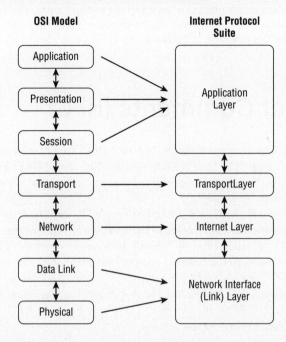

The four layers of this model represent the different protocols used for device communications on a computer network and over the Internet. I will now define each layer starting with the lowest:

The Network Interface (Link) Layer   This layer is the lowest layer and is responsible for the communication of information (data), including how data bits are organized and sent and the physical medium that is used for the transmission.

The Transport Layer   Session management between devices and familiar protocols, including Transmission Control Protocol (TCP) and User Datagram Protocol (UDP), reside here. You will learn more about these protocols later in this chapter. This layer provides the connection between two hosts on a network.

The Internet Layer   Operating at this layer are protocols responsible for addressing, resolving addresses, and routing, including Internet Protocol (IP), Address Resolution Protocol (ARP), and Reverse Address Resolution Protocol (RARP).

The Application Layer   The protocols that reside at the top layer are those related to applications and processes such as File Transfer Protocol (FTP), Simple Network Management Protocol (SNMP), Hypertext Transfer Protocol (HTTP), and many others. This is the layer that is closest to the user device level and the interface to the user.

You may sometimes see the Internet protocol suite referred to as the TCP/IP model or even the DoD model because it was funded by the United States Department of Defense.

Request for Comments (RFC)

It is important to have a basic understanding of how networking protocols originated and are defined and documented. The Internet Engineering Task Force (IETF) and the Internet Society (ISOC) are the organizations responsible for creating Internet standards and promoting Internet technology and usage.

A *Request for Comments (RFC)* is a document created by engineers and scientists and designed to define innovation and technology that works with the Internet. If an RFC is approved by the IETF, it will eventually become an Internet standard. The IETF also works with standards related to the Internet protocol suite and with the World Wide Web Consortium (W3C).

All of the protocols and services described in this chapter are based on RFCs, and the documents are easily accessible on the Internet. There may be many RFCs related to some of the protocols described in this chapter.

 It is important to understand that due to advancements in technology and innovation, new RFCs may be written that affect protocols already defined in an existing RFC. The result is that the existing RFCs may become obsolete and the new RFC is assigned a new RFC number. There is a website available that will allow you to search for RFCs based on the RFC number or titles and keywords. You can visit www.rfc-editor.org/search/rfc_search.php to search for RFCs and for additional information.

Common Network Protocols and Services

The definition of *protocol* in any context of the word is "a set of rules." When it comes to computer networking, the same definition applies but the rules pertain to the exchange of messages between connected systems or computing devices, commonly identified as "host-to-host" communications. In the following sections, we will survey some of the common protocols and services used with everyday computer networking and mobile device technology:

- Transmission Control Protocol (TCP)
- User Datagram Protocol (UDP)

- Dynamic Host Configuration Protocol (DHCP)
- Simple Network Management Protocol (SNMP)
- Network Address Translation (NAT)
- Domain Name System (DNS)
- Internet Control Message Protocol (ICMP)

Transmission Control Protocol (TCP)

If you remember from Chapter 1, *Transmission Control Protocol (TCP)* operates at the Transport layer of the OSI model. TCP is a connection-oriented protocol that provides reliable guaranteed message delivery between two devices on a computer network. The devices may reside on a local area network (LAN), on a wireless local area network (WLAN), on a wide area network (WAN), or even across an Internet connection. This guaranteed packet delivery does come at the cost of some overhead because it is a reliable, guaranteed delivery.

TCP is commonly used in conjunction with the Internet Protocol (IP), which is a Network Layer protocol. IP is responsible for the addressing and routing of packets sent across the network. When a device sends information or data to another device on a network, all of the data might not be sent at one time; it might be broken down into smaller packets, depending on the amount of data. When packets are used, network traffic flows from host to host efficiently. When the data is received by the destination device, it is reassembled into the complete message that was originally sent by the transmitting device. Figure 2.2 shows a trace from a packet analyzer displaying the TCP information from an FTP file transfer.

FIGURE 2.2 Wireshark packet analyzer showing a FTP file transfer and TCP protocol information

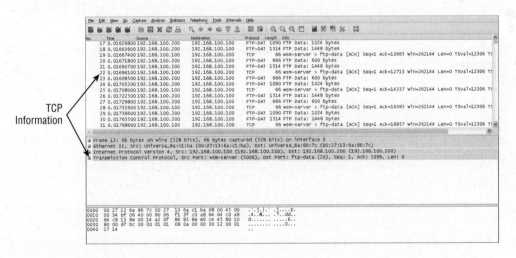

TCP's guarantee of delivery to the destination device is accomplished through an acknowledgment process. Once all the data is received at the destination, the session is ended. You can see in Figure 2.2 that several TCP packets are sent prior to an

acknowledgment. File Transfer Protocol (FTP), discussed later in this chapter, is one example that uses the TCP protocol.

> Wireshark was used to create the information for the screen capture in Figure 2.2 and others in this chapter. Wireshark is a free, open-source packet analyzer program you can download and use if you would like see what happens on a network and all of the frames that traverse the medium. You can download Wireshark at www.wireshark.org/download.html.

User Datagram Protocol (UDP)

User Datagram Protocol (UDP) is also a Transport layer protocol. There are some similarities between TCP and UDP. Both function at Layer 4 of the OSI model. Like TCP, UDP is also used in conjunction with the Internet Protocol (IP).

However, unlike TCP, UDP is a connectionless protocol, which means the data sent does not need to be acknowledged by the destination device. Therefore, UDP can be much faster than TCP because there is less overhead. In Chapter 1 I said that using UDP is like sending a postcard through a mail service. When a postcard is sent, you have no way of knowing if it arrives at the destination, so there is no guaranteed packet delivery.

You will learn that some of the common protocols and services discussed in this chapter may use either TCP or UDP for communications. One good example for the use of UDP is streaming video. Video and voice data use real-time communications (RTC). If a packet of this type is dropped, you don't want it arriving later. RTC typically uses UDP for the speed. TCP is used for non-real-time communications like file transfer, where reliability is more important than speed. With this type of data, you would want speed, and UDP will provide that over TCP. TFTP is an example that uses UDP for communications. Figure 2.3 shows a trace from a packet analyzer displaying the UDP information from a Trivial File Transfer Protocol (TFTP) file transfer.

FIGURE 2.3 Wireshark packet analyzer showing a TFTP file transfer and UDP information

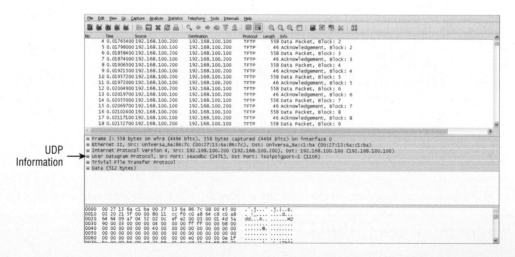

Dynamic Host Configuration Protocol (DHCP)

Dynamic Host Configuration Protocol (DHCP) is a networking protocol/service that provides the dynamic mapping and assignments of logical Layer 3 IP addresses of a network device to the physical Layer 2 MAC addresses of a network device. This simplifies the process of IP address assignments, allowing for network communications that use the Internet protocol suite. If DHCP is not used, static IP addresses and other information have to be manually assigned to the devices in order for them to communicate using TCP/IP.

Figure 2.4 illustrates the `ipconfig` command on a Microsoft-based computer showing the IP address of the server hosting DHCP services and other assigned IP information.

FIGURE 2.4 A Microsoft Windows computer shows the DHCP IP address information assigned from a DHCP server.

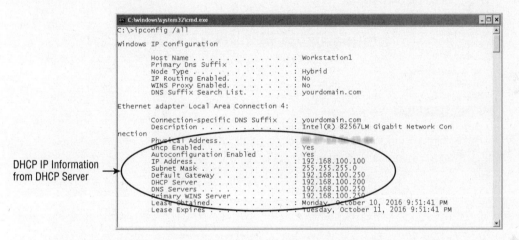

A computer or other network device that is running the DHCP service is known as the DHCP server. This can be a dedicated server, an appliance, or even a software program installed on a computer. Other infrastructure devices such as wireless LAN access points and home wireless routers can also operate as a DHCP server if they have that functionality. There are some very basic DHCP server programs available for free, and with a simple search of the Internet you can find them. These free programs are great for small networks, home labs, and experimentation with the DHCP protocol.

A DHCP server requires configuration, which will include the range of IP addresses that will be issued to the requesting client devices. This range of addresses is sometimes called the address range, the address pool, or the address scope, or it may be referred to by another term that identifies the range of addresses (from the starting address to the end address) that will be issued to requesting client devices.

In addition to the IP address, a DHCP server can also be configured to deliver other IP information if required or needed. This additional information may include the following:

- Subnet mask address
- Default gateway or router address

- DNS server address
- Domain name information
- Additional options

Most DHCP servers have many other available options. The installation and use of the server will determine which options will need to be configured and issued. Figure 2.5 shows a screen capture of a basic DHCP server configuration screen.

FIGURE 2.5 A Microsoft Windows Server 2008 DHCP server configuration screen displays some of the possible DHCP settings issued from a server to a client device.

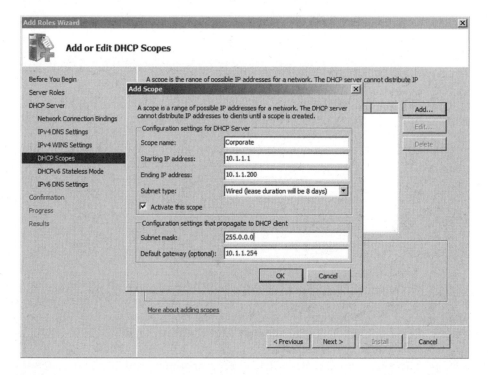

 If you experiment with DHCP servers on your network, it is important that you consider the potential consequences. When a DHCP client device requests DCHP information, it will send out a broadcast message on the network. All DHCP servers that hear the message will reply with an offer. The client device is not very picky and will choose the first offer it receives. Therefore, if the DHCP server does not contain all the correct parameters or is not configured correctly, the DHCP client device may receive a duplicate IP address or not be able to connect to hosts outside the local subnet. These connectivity issues may be time consuming and difficult to troubleshoot. If you do set up DHCP for practice, I recommend that you install the server on an isolated network to prevent potential issues and unnecessary troubleshooting. The entire DHCP process is discussed in the upcoming material.

When a network device is configured to get an IP address, it will automatically be assigned one address from a DHCP server. In addition to the IP address, the device will receive any other properties that were assigned to it. Figure 2.6 shows a screen capture of how a Microsoft Windows–based device is configured to receive an IP address automatically from a DHCP server.

FIGURE 2.6 The TCP/IP Properties window on a Microsoft Windows 8 computer shows where to configure a device for DHCP.

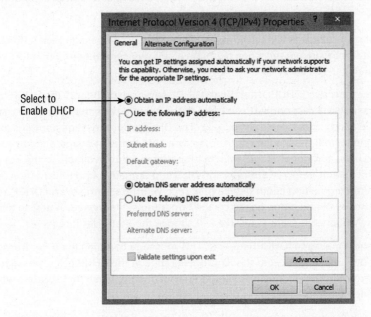

There are several steps involved in the process of assigning and maintaining IP addresses on network devices. These are the four basic frames used with the DHCP process to obtain an IP address:

- DHCP discover message
- DHCP offer message
- DHCP request message
- DHCP acknowledgment message

Real World Scenario

Obtaining an IP Address Using Dynamic Host Configuration Protocol (DHCP)

The following steps show the process, through a series of exchanged messages between a client device and a DHCP server, that the client device will use to obtain an IP address using the DHCP protocol:

1. DHCP discover message (DHCPDiscover). This is the first message sent in the DHCP process. It is a broadcast message from the client device requesting IP information from available DHCP servers.

2. DHCP offer message (DHCPOffer). Every DHCP server that "hears" the DHCP discover message will respond with an offer to the client device that made the request if it has the ability to service that request. The DHCP offer message contains an IP address and subnet mask address from an available range or scope of addresses. It will also have additional TCP/IP information that is made available to the device, such as the default gateway address and DNS and domain name information. Keep in mind that the requesting client device may receive offers from several DHCP servers and will choose the best one based on the operating system used. A device using the Microsoft Windows operating system will choose the first offer it receives.

3. DHCP request message (DHCPRequest). This message contains the IP address that the client device selected from the IP address range in the DHCP offer message that was sent from the server. It is also used for renewing the lease that it chose when it first obtained the IP address.

4. DHCP acknowledgment message (DHCPAck). This is a message sent from the DHCP server to the client device that will acknowledge the previous DHCP request message. Any additional configured TCP/IP options are also sent with this message. Once this process is complete, the client device will be able to send and receive data using TCP/IP.

There are several other DHCP messages that may be used for various reasons. These are DHCPNAK, DHCPDecline, DHCPRelease and DHCPInform.

The DHCP server will be configured to provide IP addresses for a specific period of time known as a *lease*. The time of the lease will depend on the specific application or environment where the DHCP server is used. A device configured to automatically receive an IP address will attempt to renew its lease after 50 percent of the lease time has expired. This is the default renew time. However, the DHCP server may have an option to allow a different renew time. If for some reason the client device cannot renew the lease after 50 percent of the time has elapsed, the renewal period will shorten for subsequent renewal attempts. This can be a disadvantage for devices that have short lease periods because it may cause a lot of extra network traffic. Short lease periods are often used where there may be a limited number of IP addresses available or in areas where devices may be connected to the network for shorter periods of time. Figure 2.7 shows the ipconfig /all command displaying the lease time of an assigned DHCP IP address.

FIGURE 2.7 Using the ipconfig /all command-line utility on a Microsoft Windows computer to display the DHCP server IP information lease time

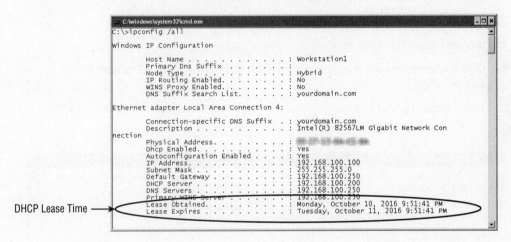

As you can see in Figure 2.7, the lease time in this case is set for 24 hours. If the pool of addresses you have is larger than the number of client devices, a longer lease period should not be a problem.

The computer operating system in use will determine what action the client device takes if an IP address cannot be obtained from a DHCP server. Microsoft Windows uses a service called Automatic Private IP Addressing (APIPA). This service became available with Windows 98 and is now available in all versions of the Microsoft Windows operating system. APIPA is designed to provide an IP address automatically to any computer network device requesting one that is connected to a common LAN. APIPA does not require the use of a DHCP server. It also eliminates the need for users to manually set up IP addressing on all the devices. An APIPA address will be in the 169.254.yyy.zzz range. For example, a device such as a notebook computer may have an APIPA address such as 169.254.100.20. If DHCP services are running on the LAN and a device cannot obtain an address from the server, or if DHCP services are not available, an APIPA address will be issued.

In exercise 2.1 you will use the Microsoft Windows ipconfig.exe utility to view DHCP information.

EXERCISE 2.1

Viewing DHCP Information

In this exercise you will view the DHCP information on a Microsoft Windows computer. This exercise assumes you are using a Windows-based computer connected to a network running a DHCP server service. The following steps were created using the Microsoft Windows 7 operating system. If you are using a different version of the operating system, these steps may vary slightly.

1. Click the Start button on the lower-left side of the Desktop.
2. Click the All Programs arrow.
3. Navigate to and click the Accessories folder icon.
4. Click the command prompt icon. A command prompt window will appear on the screen.
5. Type **ipconfig /all** at the command prompt window and press the Enter key. You will see a result similar to the following screen capture.

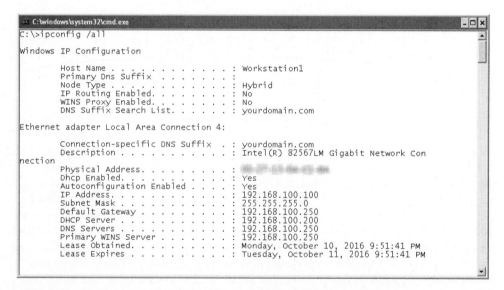

6. Scroll through the command prompt and observe the DHCP-related information.
7. Type **exit** and press the Enter key to close the command prompt window.

Simple Network Management Protocol (SNMP)

Managing devices on a TCP/IP network can be accomplished by using *Simple Network Management Protocol (SNMP)*, an Application layer protocol. The following infrastructure devices can be managed:

- Network servers
- Network routers
- Ethernet switches
- Wireless LAN controllers
- Wireless access points
- Printers

SNMP uses several components to aid in the management of these devices:

The SNMP Manager This is the device that will perform the setup and management of devices using the SNMP protocol and provides communication to the other network devices to be managed. This may be a server running SNMP management software or an appliance. The SNMP manager will send commands and also receive traps.

The Devices to Be Managed These are the devices mentioned earlier and include network servers, routers, Ethernet switches, wireless LAN controllers, and others. They will be controlled or managed by the SNMP manager.

The SNMP Agent Program The SNMP agent can be local or remote. The agent is on the SNMP device to be managed and will respond to commands sent from the SNMP manager.

The Management Information Base (MIB) The MIB is on the managed device and contains information about that device and how it can be managed. There are standard MIBs, but there are also proprietary MIBs that are unique to the manufacturer of the device.

Figure 2.8 shows SNMP managed devices on a local area network.

FIGURE 2.8 Example of an SNMP managed network

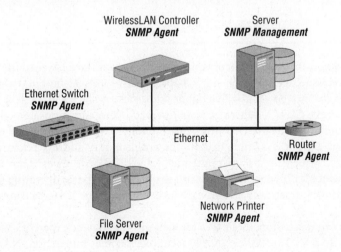

There have been several different versions of SNMP, some more secure than others. Depending on the version used, anyone with the proper knowledge and tools can monitor a network and possibly compromise its security.

SNMPv1   This was the first version of SNMP and uses UDP for communications. SNMP v1 uses community-based security and therefore is not considered secure because the community string value that equates to a password is transmitted in cleartext.

SNMPv2c   This version of SNMP provided some enhancements. However, like SNMPv1, this version uses community-based security.

SNMPv3   This is the newest version of the protocol and provides secure communications. SNMP v3 requires user-based authentication and is therefore the most secure. This version should be used when possible.

Network Address Translation (NAT)

In the mid-1990s, the use of the Internet along with the TCP/IP protocols was fueled by the release of the Hypertext Transfer Protocol (HTTP) and the World Wide Web. The heavily used Internet Protocol version 4 (IPv4) of assigning IP addresses was in jeopardy because of the increased use of TCP/IP networks. Because the IPv4 mechanism has a finite number of addresses, there was a possibility of running out of IP addresses. Table 2.1 shows the number of IPV4 addresses available based on the IP address class.

TABLE 2.1   IPv4 IP addresses

Class Type	1st Octet – Decimal	Max Number of Hosts
A	1–126	16,777,214
B	128–191	65,534
C	192–223	254

Newer technology, Internet Protocol version 6 (IPv6), was the answer to the IPv4 address limitation issue. As you learned in Chapter 1, IPv4 uses 32-bit binary addresses; there is a limit to the number of possible public addresses, as shown in Table 2.1. Keep in mind that all of the Class A addresses were reserved by organizations and companies early on and therefore not available to the general public. The IPv6 method of addressing new IP technology is what was going to solve the problem of IP address exhaustion. IPv6 uses 128 bits (2^{128}), which equates to a potential 3.4×10^{38} IP addresses. That's a lot of IP addresses. The switch to IPv6 would not be an easy task. Computer operating systems, servers, and network infrastructure hardware lacked support for IPv6, causing delays in the implementation.

So why all of this discussion about IP addresses or potential lack thereof? Well, this is where *Network Address Translation (NAT)* technology comes into action. NAT allows the use of a "private" IP address network for internal use and mapping it to a single "public" IP address connected to the Internet. This is accomplished by modifying the header in the IPv4 packet, which is where the term *address translation* comes in. NAT allows for a home network, a very small organization, or even a very large organization to be represented

Common Network Protocols and Services

by a single public IP address on the Internet. It is addressed in RFC 2663 and allows for the continued use of IPv4 without having to worry about potentially running out of IPv4 addresses. NAT consists of different implementations. The following are two common implementations:

- One-to-one NAT
- One-to-many NAT

One-to-one NAT will map an IP address on the outside to an IP address on the inside. This can be accomplished through either a static assignment or a dynamic assignment. The static NAT entry is a permanent entry and needs to change manually. The dynamic process is useful when there are a limited number of addresses and the one-to-one assignment occurs when needed.

One-to-many NAT allows for the translation of one single public IP address on the outside of a network to service many devices with IP addresses on the inside network. This is accomplished by modifying the information in the IP packet. Figure 2.9 shows how NAT on a DSL router or cable modem will provide IP address translation. Notice that the private IP addresses on the private side of the router in this example are in the 192.168.0.X range. These private IP addresses will be translated to the single public IP address 172.10.5.20 on the public-facing side.

FIGURE 2.9 Network Address Translation (NAT) and Port Address Translation (PAT). The IP addresses shown are examples.

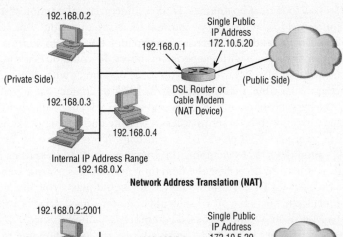

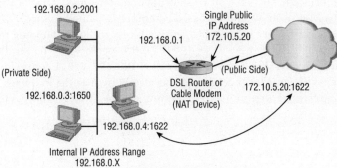

Port Address Translation (PAT)

Port Address Translation (PAT) is a technology that works in conjunction with NAT. The objective of PAT is to provide a unique TCP port to the IP address of each device that is sending information across the network translating device, such as a home broadband router, for example. This will ensure that the information that is received will be directed to the correct device on the private network. The TCP port becomes part of the device's TCP/IP address to uniquely identify that device on the private side of the NAT device so it can send and receive data using the single public IP address. You can see in Figure 2.9 that each device on the private network is assigned its own unique TCP port.

Domain Name System (DNS)

In Chapter 1, you learned that an IP address is a logical address that will identify a device on a network and utilizes the TCP/IP protocols. This could be limited to a single LAN, a private routed WAN, or even across a wide area public network like the Internet. The logical address is mapped to the device physical network MAC address. In some operating systems, a program exists that will allow a user to manipulate the ARP cache data. A simple command-line utility such as arp.exe will allow you to see this logical-to-physical-address mapping. In the Microsoft Windows operating system, for example, you can enter **arp.exe -a** in a command prompt window to see it.

So where does *Domain Name System (DNS)* fit in? DNS will allow for the mapping of Layer 3 IP addresses to logical hostnames. DNS is a hierarchical naming system for all TCP/IP devices connected to the Internet or internal to an organization. One commonly used analogy to help understand this concept is a telephone book or an address book that maps people and company names with telephone numbers. It would difficult for you to remember the telephone numbers of everyone you came in contact with or someone you wanted to contact at a later time, making an address book or contact list a great resource. You will look up the person's name in the index and see the telephone number to call. Could you imagine using a web browser to access a website and having to know the IP address of the target web server? It would be very difficult if not impossible for most people to remember. DNS simplifies this by mapping the web server's IP address to its associated domain name. All you need to do is type the domain name address in your web browser and DNS will do the rest. Figure 2.10 shows how DNS would play a role in a network.

In its simplest form, the hosts file on your local computer's hard drive is a form of DNS. The hosts file on a Microsoft Windows–based computer is located in the C:\Windows\System32\Drivers\Etc folder. This file has an IP-address-to-name mapping for the local computer. One common entry you will see is "127.0.0.1 localhost." The localhost identifies your computer directly. If you were to enter the PING 127.0.0.1 command, you would be communicating with your local device. This is also known as the loopback address. If you knew the IP address of a remote web server and mapped that IP address to the web server's domain name by typing it in your local hosts file, you would not need any DNS server on the network. So theoretically DNS is a really big hosts file. Please note that I am not trying to downplay or lessen the importance of DNS because it is a critical part of any network, but the objective here is to provide an overview and basic understanding of the service and process.

FIGURE 2.10 Example of Domain Name System (DNS) in use

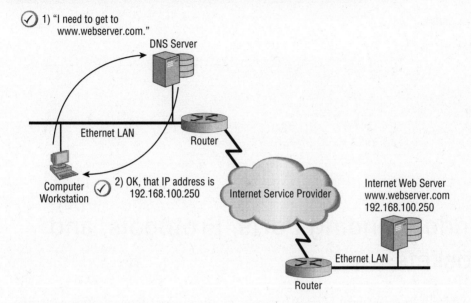

Internet Control Message Protocol (ICMP)

If you have ever used the PING command utility, you have used *Internet Control Message Protocol (ICMP)*. ICMP is used to send error messages, such as when a device cannot be reached using the PING command and for other diagnostic and troubleshooting purposes. ICMP is located at the same layer as the IP protocol and does not rely on IP as TCP does.

Although this protocol was designed with the good intentions of error identification, determining if a remote host is available, and helping to determine network congestion, a computer hacker could use it for malicious purposes. For example, a hacker could perform an ICMP denial of service (DoS) attack and cause dropped connections. These types of attacks are commonly prevented by through the use of firewall configurations and other security measures. For example, Microsoft Windows's built-in firewall software will not allow ICMP traffic unless it is enabled to do so.

Figure 2.11 shows a packet trace of the PING command and ICMP.

FIGURE 2.11 Wireshark packet analyzer showing an issued `PING` command and ICMP information

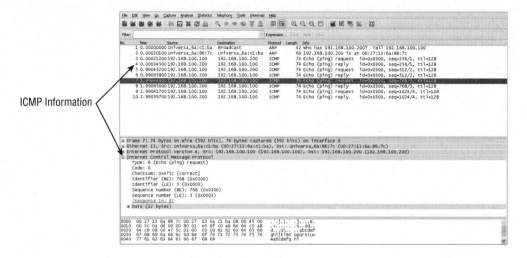

Understanding Ports, Protocols, and Sockets

Let's start by learning about ports. When network devices are communicating with the TCP/IP suite of protocols and using a protocol such as TCP or UDP, a port must also be defined. A port is a number that is assigned to a protocol. There are a total of 65,535 possible ports in the Internet protocol suite. The first 1,024 are reserved and assigned to some of the common protocols we use every day, such as SMTP, POP3, and IMAP. To understand network ports, think of office suite numbers in a multitenant office building. Suppose an office building has a street address of 100 Main Street and in the building there are 25 office suites, each housing a different business. For you to a send a package to the business in Suite 200, you would address it to 100 Main Street, Suite 200. 100 Main Street is equivalent to the IP address and 200 is equivalent to the port number. Figure 2.12 shows how TCP ports are used with specific protocols and IP addresses.

It is possible to change the default ports that are assigned to a protocol. However, if the default port number is changed, it may cause some communication issues. If you change the assigned port number, the users need to know the port number to use the protocol or service. As a general practice, you should avoid changing the default port numbers assigned to common protocols.

FIGURE 2.12 Graphical representation of a multitenant building with businesses showing how TCP ports are used with protocols

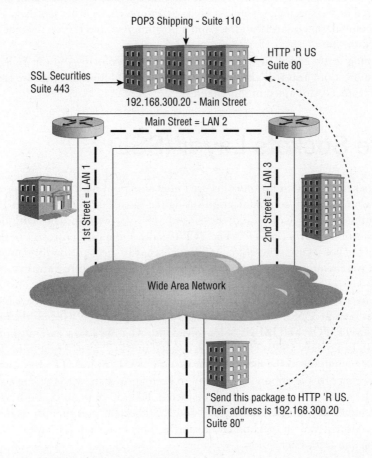

 If you are using a nonstandard port, you may need to identify the port number after the IP address, separated by a colon symbol. For example, a network infrastructure device such as a router has an IP address of 192.168.100.250 and is using port 8080 for the administration interface. To access the management page, the proper syntax for the browser would be 192.168.100.250:8080.

When dealing with TCP/UDP ports, you may run across the term *socket*. A socket is created from a combination of an IP address and a port number. This is analogous to a telephone extension number when you call a business. If you were to call the main telephone

number of a company, the answering service may require a specific extension number in order for you to reach a person directly. Upon entering the telephone extension number, you are directed to the specific person you are trying to reach. In other words, a TCP/UDP socket is the combination of an IP address and a port number and therefore is the endpoint of a specific connection.

The upcoming sections look at some common network protocols and the TCP ports they use for communication. You will also learn what the port numbers are that are assigned to the common protocols discussed in this chapter.

Secure Sockets Layer (SSL)

The Secure Sockets Layer (SSL) protocol is used in conjunction with other protocols to allow for secure network communications using the Internet Protocol (IP). In 1994, Netscape Communications Corporation developed SSL to provide security with the Hypertext Transfer Protocol (HTTP). The TCP/IP suite and native protocols of the Internet alone are not secure. Therefore, additional security measures to prevent attacks and intrusion were required to provide a secure method of transferring information across the Internet.

Secure Sockets Layer (SSL) provided some relief to these concerns. The development of SSL, now at version 3 (SSLv3), paved the way to secure protocols used on the Internet such as FTP, SMTP, POP, and IMAP. SSL evolved into what we know as Transport Layer Security (TLS). In January 1999, TLS 1.0 was defined in RFC 2246. At the time, it was intended to be an upgrade to the most current version of SSL, SSLv3. TLS has since been revised several times to add enhancements and to keep up-to-date with the latest security threats. TLS 1.2 is the most recent version, defined in RFC 5246 in 2008. Both SSL and TLS rely on a public key infrastructure (PKI) and X.509 certificates that use asymmetric cryptography. More detail on PKI and certificates will be discussed in Chapter 16, "Device Authentication and Data Encryption."

 The admin subnet website is a great online reference that provides several helpful tools, including IPv4 Subnet Calculator, Password Generator, MAC Address Finder, and TCP/UDP Port Finder. To use these tools and additional information, visit the admin subnet website at www.adminsub.net.

Common Protocols for File Transfer

Transferring files is a common use of a computer network. In many cases, a user may not even realize they are performing this function and using these protocols while they are connected to a network or to the Internet. In the following sections, you will learn about several file transfer protocols and the ports that they are assigned to operate with.

File Transfer Protocol (FTP) – 20/21

The *File Transfer Protocol (FTP)* is a commonly used protocol and has been for decades. As the name implies, FTP is used to transfer files from one device (or host) to another. One thing to keep in mind is that there are some potential security risks with FTP. These are addressed in the upcoming exercises and in the following sections, "File Transfer Protocol Secure (FTPS) – 990" and "Secure File Transfer Protocol (SFTP) – 22."

Unlike many of the other commonly used protocols, FTP requires two TCP ports to operate. One port is for the commands, or control, and the other port is for the data that is transferred. The default FTP server ports are 20 and 21.

Port 20 Used for the actual transfer of data. All FTP commands are processed over port 21.

Port 21 Used for control type messages and to establish an FTP session. When the FTP service is started, this is the port that listens for FTP requests. Once an FTP request is initiated by a remote host, messages with information such as the user's authentication credentials will be exchanged. Once the username and password is validated, a file transfer can be issued over port 20. Other FTP commands will also be sent over port 21. Figure 2.13 illustrates a common FTP setup.

FIGURE 2.13 LAN with FTP server providing services to an FTP client device

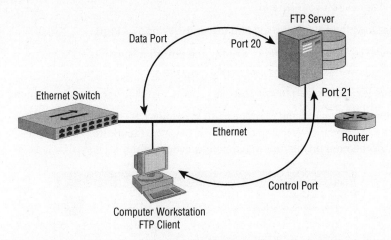

FTP can operate in one of two modes, active mode or passive mode. Selecting which mode to use is performed in the client-side FTP settings or by use of a command sent to the server.

FTP Active Mode In this operation mode, the client is in charge of which port the FTP server will use for control and transfer of information. In other words, the client device that initiated the FTP session will connect from a random port greater than 1023 to connect to the FTP server's port 21. This will identify to the server which client-side port to use for the data transfer.

FTP Passive Mode In this mode, the FTP server will decide which port the client will use for listening and tell the client device which port to use for the data transfer. Passive mode works much better with a client's firewall configuration because of potential ports blocked by the firewall.

To use FTP on a computer, you need to install the proper software. Fortunately, installing the software is pretty straightforward, regardless of whether you are setting up a client or a server. In exercise 2.2 you will install an open-source version of FTP server software. This will allow you to become familiar with how the FTP protocol operates from a server perspective. Exercise 2.3 teaches you how to install FTP client software.

EXERCISE 2.2

FTP Server Software Installation

In this exercise, you will download, install, and configure the FileZilla FTP server software. FileZilla is free, open-source, cross-platform FTP software, consisting of the FileZilla Client and FileZilla Server. The FileZilla FTP server software can be downloaded from `https://filezilla-project.org`.

1. Using your web browser, point to the `https://filezilla-project.org/` web page and click the Download FileZilla Server Windows Only icon. The Server Download page will appear on your screen.

2. Click the Download Now icon located under the Windows logo.

3. Click Save File. At the time of this writing, the program is named `FileZilla_Server-0_9_41.exe`.

4. Browse to your download folder and execute the install program you downloaded.

5. Depending on the version of the Microsoft Windows OS you are using, you may see an Open File – Security Warning dialog box. Click the Run button to continue. The License Agreement dialog box will appear on your screen.

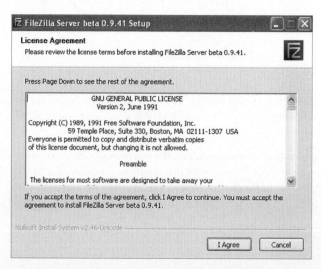

Common Protocols for File Transfer   51

6. Read the license agreement and click I Agree to continue the installation. The Choose Components dialog box will appear on your screen.

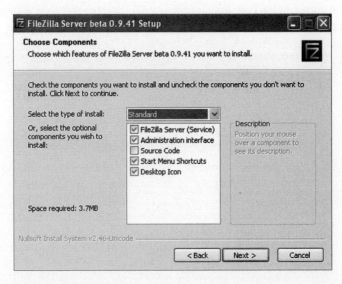

7. Review the options and click on the Next button to start the installation process. The Choose Install Location dialog box will appear on your screen.

8. Observe the Destination Folder location and click the Next button to continue with the installation. The Startup Settings dialog box will appear on your screen.

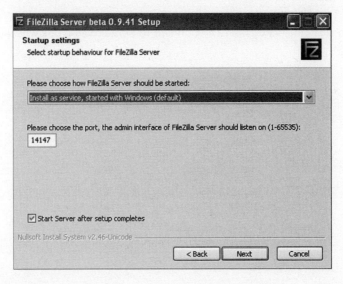

9. Click on the drop-down menu arrow and observe the available server startup options and the assigned port on which the admin interface of FileZilla Server should listen.

EXERCISE 2.2 *(continued)*

10. Click Next to continue the installation process. The Startup Settings dialog box will again appear on your screen.

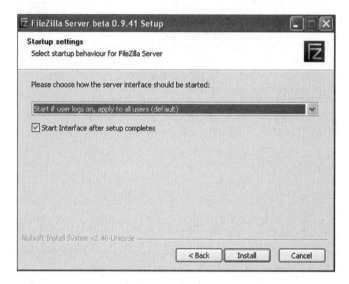

11. Click on the drop-down menu arrow and observe the possible options.
12. Click Next to complete the installation. The Installation Complete dialog box will appear on your screen.
13. Click OK to close the dialog box. The Connect To Server dialog box will appear on your screen.

Common Protocols for File Transfer   53

14. Enter the secure password of your choice in the Administration Password field and click OK. The installation is now complete, and the FileZilla Server administration console window will appear on your screen.

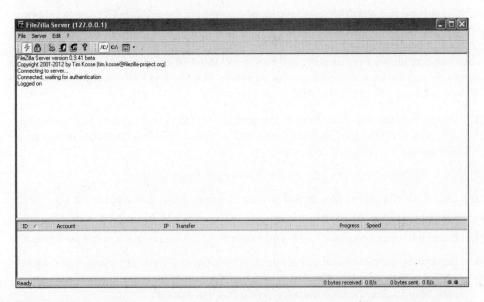

15. Browse through the menu options to familiarize yourself with the FileZilla FTP Server program.

 Exercise 2.2 is intended for you to download, install, and configure the FileZilla FTP Server software for educational purposes and to become familiar with an FTP server software program and the FTP protocol. If you plan to keep this software installed on your computer, it is important that you take proper security measures to ensure the integrity of your computer and only start the service as needed. You will install the FileZilla FTP client software in Exercise 2.3. If you do not plan to use this FTP client software, I recommend you uninstall the program after you finish these exercises. If you have a second computer, you can install the FTP client program in the next exercise and perform some simple file transfers between the two computers.

In Exercise 2.3 you will install an open-source version FTP client software that you can use to access an FTP server. This will enable you to become familiar with FTP from the client side perspective.

> **EXERCISE 2.3**
>
> **FTP Client Software Installation**
>
> In this exercise you will download, install, and configure the FileZilla FTP client software. FileZilla is free, open-source, cross-platform FTP software, consisting of the FileZilla Client and FileZilla Server. You installed the FTP server software in Exercise 2.2. If you have two computers to work with, you can install this FTP client software on the second computer and perform some simple file transfers. The FileZilla FTP client software can be downloaded from https://filezilla-project.org.
>
> 1. Using your web browser, point to the https://filezilla-project.org/ web page and click the Download FileZilla Client All Platforms icon. The Client Download page will appear.
>
> 2. Click the Download Now icon located under the Windows logo.
>
> 3. Click Save File. At the time of this writing, the program is named SFInstaller_SFFZ_filezilla_8992693_.exe.
>
> 4. Browse to your download folder and execute the install program you downloaded.
>
> 5. Depending on the version of the Microsoft Windows OS you are using, you may see an Open File – Security Warning dialog box. Click the Run button to continue. The Welcome To SourceForge Installer dialog box will appear.
>
>
>
> 6. Click Next to continue the download process. The Choose Installation Options dialog box will appear.

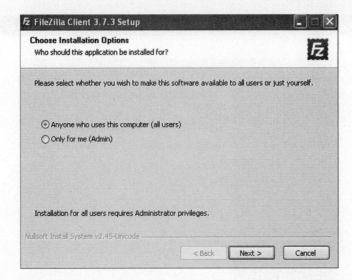

7. Click Next to use the default option. The Choose Components dialog box will appear. Review the options on this screen.

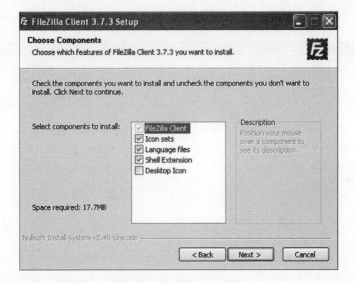

8. Click the Next button to start this installation process.

9. The Choose Install Location dialog box will appear. Observe the Destination Folder location.

10. Click the Next button to continue with the installation. The Choose Start Menu Folder dialog box will appear.

11. Click Install to continue. The Completing The FileZilla Client Setup dialog box will appear.

EXERCISE 2.3 (continued)

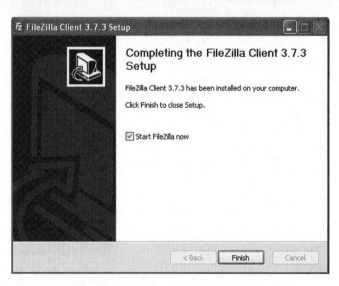

12. Click Finish to complete the installation. The FileZilla client screen will appear.

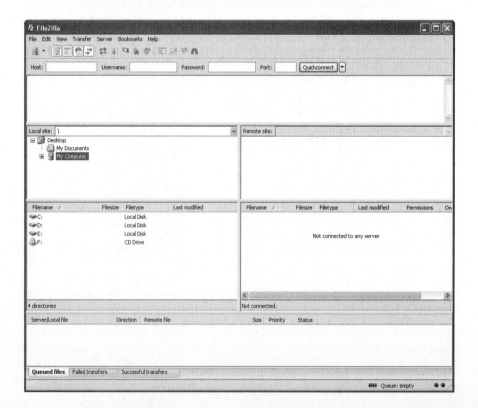

13. Browse through the menu options to familiarize yourself with the FileZilla FTP client program. If you installed the FTP server software on a different computer, you can perform simple transfers between the FTP client program you installed and the FTP server program you installed on the other computer. Otherwise, you can connect to a trusted known FTP server on the Internet and perform some FTP file transfers.

If you do not plan to use the FTP client software, you can uninstall the program.

File Transfer Protocol Secure (FTPS) – 990

FTP uses cleartext for both client authentication and data transfer. This means that anyone listening to the medium with the appropriate knowledge and with software tools such as a packet analyzer will be able to capture the client credentials and see the information that is transferred during the FTP session. This can be more of an issue if the device that is connected is using a wireless network because wireless is not bounded by any physical medium such as an Ethernet connection. All unsecured information traversing the air in cleartext can be seen if the traffic is captured.

This is where *File Transfer Protocol Secure (FTPS)* comes into action. (Do not to confuse this with SFTP, which is discussed in the next section.) FTPS is a secure method for transferring data using FTP. FTPS uses Secure Sockets Layer (SSL) and Transport Layer Security (TLS) to secure the communications and operates on port 990. You may also see this identified as FTP-SSL. You learned about SSL earlier in this chapter. FTPS will allow for a secure file transfer using FTP. This secure SSL connection will encrypt your login credentials (username and password) and the data that is being transferred. You may be familiar with SSL from using your browser. Any secure website you connect to, such as online banking and e-commerce websites and some web-based email services, use this technology.

The requirements for using FTPS are as follows:

- An FTP server configured for SSL connections
- An FTP client software program that will support SSL connections

Secure File Transfer Protocol (SFTP) – 22

Another secure method for performing file transfers using FTP is *Secure File Transfer Protocol (SFTP)*. As you learned in the previous section, FTPS uses port 990 and is based on Secure Sockets Layer (SSL) and Transport Layer Security (TLS). In contrast, SFTP uses Secure Shell (SSH) via port 22 to transfer files. SSH is a redirection over a specific port to ensure that the information (both control and data) that is sent and received is secure. TCP port 22 is the standard port assigned for communication with SSH servers.

SFTP is not the only protocol that uses SSH. Command-line utilities such as Telnet and other programs or protocols can use SSH to protect data. In addition to its use for secure file transfers, SSH port 22 is commonly used by network administrators to remotely manage network devices via a command-line interface.

Secure Copy Protocol (SCP) – 22

There is a protocol available that is designed to allow secure transfer of files between two hosts that can be a combination of a local and a remote host or two remote hosts. The Secure Copy Protocol (SCP) uses SSH port 22 and will provide a network administrator or a user with the ability to securely transfer files such as firmware for infrastructure devices over a secure SSH tunnel. However, it can be used to transfer any type of files. Copying files via the SCP protocol is done either by using a set of command options in an operating system such as Unix or by using a program. So the acronym *SCP* could mean either Secure Copy Protocol or Secure Copy Program. One example of an open-source Microsoft Windows–based SCP program is WinSCP.

Common Protocols Used with Messaging

In the following sections, you will learn about the different protocols used with messaging. The objective here is to provide an overview of some of the protocols used with email, notification, and instant messaging.

Simple Mail Transfer Protocol (SMTP) – 25

Not long ago, electronic mail (email) was new technology, and many people didn't use it and claimed they didn't need it. Now it is difficult to live without it; it's a big part of daily life, both business and personal. *Simple Mail Transfer Protocol (SMTP)* is the Internet standard that electronic mail services use to send and receive email messages. SMTP uses TCP port 25 for communications and is used to send email messages from client to server and to send and receive email messages between servers. The steps for a client device using SMTP to send an email message are as follows:

1. The user creates an email message using an email client software program.
2. The user will then send the email message to the email server using SMTP via port 25.
3. The SMTP email server queues the message it received from the sending client device.
4. The SMTP email server sends the message to the recipient's email server via SMTP.
5. The email message is now ready to be retrieved from the recipient's email server using the desired protocol: POP3, IMAP, or another protocol of choice.

Figure 2.14 illustrates the basic steps involved in the SMTP process.

FIGURE 2.14   Representation of the SMTP process

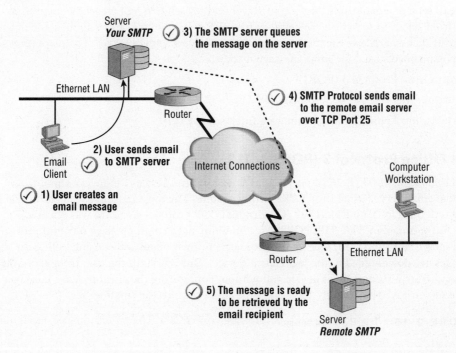

Many networks, broadband wireless providers, and Internet service providers (ISPs) block SMTP port 25. Blocking this default assigned port for SMTP helps prevent malicious attacks such as email spamming. There is an alternate port for SMTP that can be used by the email client device in some cases. This is TCP port 587. To enable this alternate SMTP port, simply enter port 587 instead of port 25 in your email client software configuration screen.

 Earlier in this chapter you learned about Secure Sockets Layer (SSL) and how it secures communications if used with certain protocols. It is possible to secure SMTP by using SMTP over SSL. This is accomplished by using TCP port number 465, which will secure the SMTP information using SSL. With SSL, this is known as SMTPS or SSMTP. Keep in mind this is a two-part process. In other words, both the client and the server must be enabled with SSL to use this secure method of communication. If you enable your email client software and the server is not configured for SMTPS, you will not be able to send email messages over a secure connection. Currently, not many ISPs support SMTPS port 465.

Message Retrieval Protocols

You just learned that Simple Mail Transfer Protocol (SMTP), port 25, is used for a client device to send an email to its local email server and for sending and receiving email messages from server to server on the network backend. Now you will learn about some of the protocols that can receive the messages queued at the user's email server. There are three common protocols used for email message retrieval:

- Post Office Protocol 3 (POP3)
- Internet Message Access Protocol (IMAP)
- Messaging Application Programming Interface (MAPI)

Post Office Protocol 3 (POP3) – 110

Post Office Protocol 3 (POP3), port 110, is used by client devices to retrieve email from a remote email server using the TCP/IP protocol suite. The most current specification for POP3 is defined in IETF RFC 1939. The original POP protocol standard was released in 1994 and specified in RFC 918. POP3 is a commonly used email message retrieval protocol.

As shown in Figure 2.15, when a user connects to their email server using POP3, the messages are downloaded to the local client device. This will limit the user from accessing the email from a second client device and without configuring the client to leave messages on the server it is valid for only for a one-time connection to the email server.

FIGURE 2.15 The POP3 process

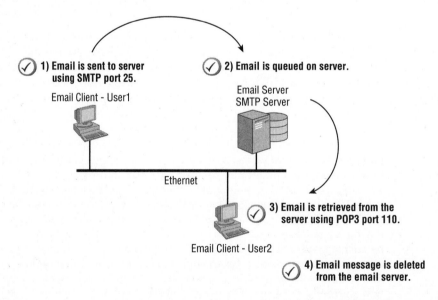

Some email software, such as Microsoft's Outlook email program, have advanced settings to allow messages to remain on the server permanently or for some extended period of time after they have been retrieved from a POP3 client device. Configuring the client software in this manner will allow a user to retrieve their email from a second device even though POP3 was not designed to work that way.

One advantage to using this protocol is that the email messages are stored locally on the user's device and can be viewed offline. Therefore, a connection to the Internet is not required for working with the email messages stored locally on the client device.

The steps for a client device to use POP3 are as follows:

1. The email message is sent to the server using SMTP port 25.
2. The email is queued on the server.
3. The email is retrieved from the server using POP3 port 110.
4. The email message is deleted from the email server.

Without any additional security measures in place, POP3 sends information in cleartext. This includes username, passwords, and all of the email data. This can pose a security risk, especially for devices that are connected to an open wireless network. If POP3 is used, I recommend that you verify that proper security measures such as a virtual private network (VPN) or SSL connection are in place. The TCP port for POP3S (POP3 over SSL) is port 995. Unfortunately, not many ISPs currently support POP3S.

Internet Message Access Protocol (IMAP) – 143

Like POP3, the *Internet Message Access Protocol (IMAP)* is another Application layer Internet protocol that allows a client device to access email on a remote email server. IMAP uses port 143 for communications, and IMAP4 is the latest version. IMAP4 is defined in IETF RFC 1730.

As shown in Figure 2.16, the main difference between IMAP and POP3 is that the messages are not downloaded locally to the client device's email software program. Instead, the email remains on the email server for retrieval from any device with an Internet connection and the proper credentials to authenticate to the email server. Depending on the email client software used and how it is configured, the email may be cached on the local client device while the user is using the email program. Any changes are then synchronized between the email client device and the email server. This will allow for current information to be available whenever the user connects to their email server from any client device.

FIGURE 2.16 The IMAP protocol process. Notice how it's different than POP3.

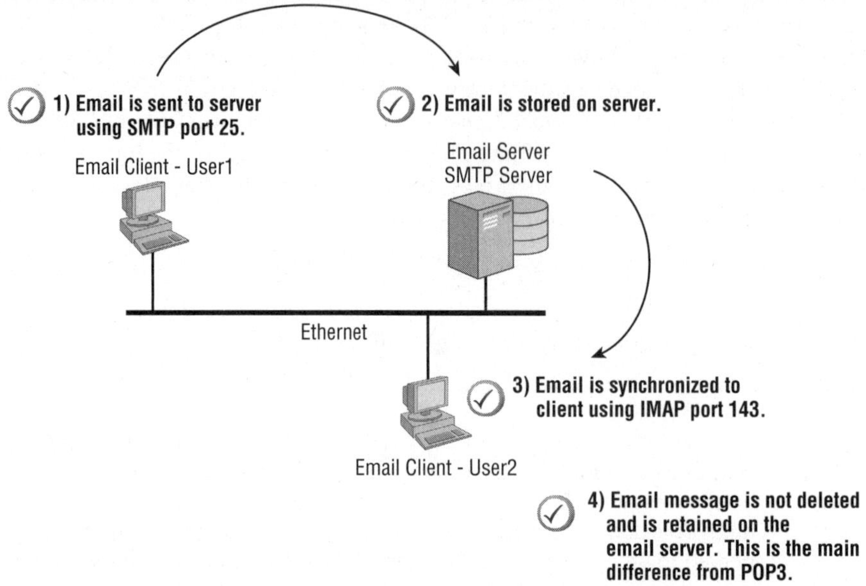

1) Email is sent to server using SMTP port 25.
Email Client - User1

2) Email is stored on server.
Email Server
SMTP Server

Ethernet

3) Email is synchronized to client using IMAP port 143.
Email Client - User2

4) Email message is not deleted and is retained on the email server. This is the main difference from POP3.

One disadvantage to IMAP is that the email messages may not be available on the local device for viewing without an active Internet connection. Therefore, if an Internet connection is not available, you may not be able to see or reply to messages and queue them for handling at a later time.

The steps for a client device to use POP3 are as follows:

1. The email message is sent to the server using SMTP port 25.
2. The email message is stored on the server.
3. The email is synchronized to the email client using IMAP port 143.
4. The email message is not deleted and is retained on the email server. This is the main difference from POP3.

As with POP3, without any additional security measures in place, IMAP sends information in cleartext. This includes all of the information transmitted—the username, passwords, and all of the email data. This can pose a security risk, especially for devices that are connected to an open wireless network. If IMAP is used, I recommend that you verify that proper security measures such as a virtual private network (VPN) or SSL connection are in place. The TCP port for IMAPS (IMAP over SSL) is port 993.

Messaging Application Programming Interface (MAPI) – 135

In the previous sections you learned that POP3 and IMAP are standards-based, commonly used email message retrieval protocols. Another available email retrieval protocol is *Messaging Application Programming Interface (MAPI)*. This is a proprietary protocol created by Microsoft. MAPI is a messaging architecture and an application programming interface (API) based on the Component Object Model (COM) for the Microsoft Windows operating system. It uses port 135 for communications.

If your network is configured to use a Microsoft Exchange email server and you are using Microsoft's Outlook email client, you will be able to use the MAPI protocol. However, other third-party email client software may also include MAPI protocol functionality. The following computer application programs can be used and integrated via email messaging with the MAPI protocol:

- Word processing applications
- Spreadsheet applications
- Graphics applications

The MAPI protocol is best used for internal or intranet email services with Microsoft Exchange Server and typically does not work well from outside an organization.

Figure 2.17 shows a screen capture for configuring an email account with MAPI and other email retrieval protocols. If you were to select Microsoft Exchange Server so that you can set up MAPI, you would be instructed to provide additional information to complete the setup process.

FIGURE 2.17 Microsoft Outlook and MAPI setup screen

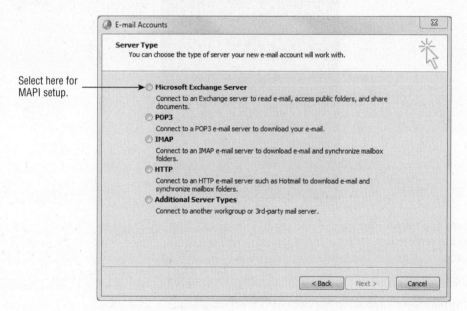

Apple Push Notification Service (APNS) – 2195 / 2196

Developed by Apple Inc., the *Apple Push Notification Service (APNS)* uses a native push notification technology. This technology allows a mobile device such as the iPad, iPhone, and others to receive information even when the destination application (app) is not open or running. APNS is a required protocol for Apple iOS mobile device management enrollments and therefore is dependent on the Apple iOS platform. Apple's APNS became available with iOS 3.0 in 2009. iOS apps can provide three types of push notifications:

- **Sounds:** An audible alert plays.
- **Graphical alerts/banners:** An alert or banner appears on the screen.
- **Badges:** An image or number appears on the application icon.

Figure 2.18 shows an example of an APNS banner alert.

FIGURE 2.18 APNS banner showing a received text message

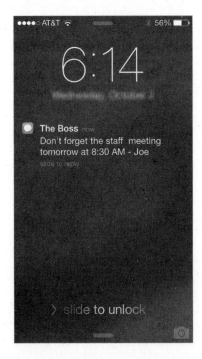

Each app on your device will need to be configured to use APNS. The following components are required:

- Application server
- X.509 certificate
- Apple Push Notification Service (APNS) server
- Mobile device
- App configured for APNS

A valid encrypted IP session is established between the mobile device and APNS. The device will receive all notifications over this SSL/TLS IP session. The device must obtain a token that contains information that will allow the service to locate the device where the app is installed.

Extensible Messaging and Presence Protocol (XMPP) – 5223

With the evolution of mobile technology, instant messaging (IM) has become a popular way to communicate. One common protocol used for instant messaging is *Extensible Messaging and Presence Protocol (XMPP)*. This protocol was originally known as Jabber and became XMPP after the IETF created the open standard. It is used with a variety of commonly available applications and services including these:

- Jabber
- Google Talk (now Google+ Hangouts)
- Jabber Extensible Communications Platform (XCP)

 Google Talk, now called Google+ Hangouts, has moved away from XMPP and now uses a proprietary communication technology.

Google Cloud Messaging (GCM) – 5228

Google Cloud Messaging (GCM) operates on port 5228. GCM is a free service regardless of your messaging needs. It allows for data to be sent from your server to your Android-based devices. It also allows you to receive messages from devices on the same connection.

Like Apple Push Notification Services, GCM is a push notification service. It handles all aspects of queuing of messages and delivery to the target Android application running on the target client device. Basically, GCM is used to send messages to Android-enabled devices.

Here are some specific details regarding GCM:

- It provides a free push notification service.
- It can handle a maximum of 4 KB of data.
- Data can be queued for up to four weeks.
- The application is not required to be active to receive messages.
- It does not poll servers for updated data.
- The device must have a Google account enabled.

The GCM server is the "middleman" between the Android device and the server hosting the application. Figure 2.19 shows the steps for the GCM process.

FIGURE 2.19 The Google Cloud Messaging (GCM) process

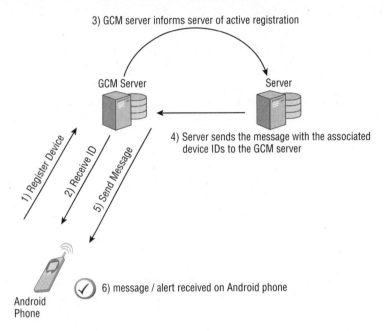

Other Common Protocols

In the following sections, you will learn about other common protocols in the TCP/IP suite. These protocols are used for accessing the Internet using a graphical user interface, synchronization of data between devices, remote computer access, and message routing.

Hypertext Transfer Protocol (HTTP) – 80

Prior to the mid-1990s, the Internet was text based and was not widely used by the general population. It consisted of a text-based interface and mainframe computer devices and was used mostly by educational organizations and the government. With the introduction of the World Wide Web (WWW) in the mid-1990s, the Internet exploded in popularity. This is largely due to the *Hypertext Transfer Protocol (HTTP)*, which operates using TCP port 80.

You are no stranger to this protocol. HTTP is an Application layer protocol providing the graphical user interface of the Internet that we have become accustomed to over the past 15+ years. HTTP uses a request/response model for communication. In other words, the client device will send an HTTP request to the server (such as a request to query to a database or submit an online form), and the server will provide a response to this request.

Every time a user clicks an object such as an OK or Submit button on the graphical interface, a request is sent to the server. At that point the server will respond in some fashion based on the context of the request.

The following components make up the HTTP process:

- A client device with a web browser
- A TCP/IP connection
- A server hosting a website

The server must host a web server software program of some type in order to respond to the HTTP requests from the client.

Figure 2.20 shows a basic HTTP network configuration with a local web server and remote web server request/response scenario.

FIGURE 2.20 The Hypertext Transfer Protocol (HTTP) process

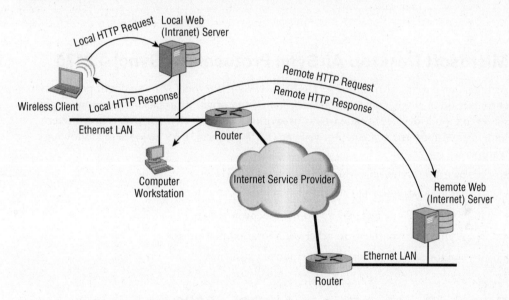

As with some of the other protocols you learned about in this chapter, such as POP3 and IMAP, information is transmitted in cleartext. Therefore, your username, password, and other information can potentially be seen by anyone with the knowledge and correct software tools to capture this information. Proper security mechanisms should be put in place to protect sensitive information and prevent malicious attacks.

One common way to secure the communication session when using HTTP is to use Secure Sockets Layer (SSL) or HTTPS. The SSL port for HTTPS is port 443. Many content providers such as online banking and email services now provide HTTPS as the default protocol when you connect to their web servers. The secure connection will remain intact for the entire session to provide adequate security for all information, not only the username and password during logon. Google by default will now send you to an HTTPS session for any device that uses the Google search engine, even if you aren't specifically logged in to Google's services.

By default, Hypertext Transfer Protocol (HTTP) information sent over port 80 is cleartext data. I highly recommend that you use HTTPS port 443 for accessing any web-based sensitive information over the Internet. In some cases, the HTTPS port 443 protocol is available only for a login session for the user to authenticate to a web server. After the credentials are verified, the SSL session is dropped and the remaining data is sent via HTTP port 80 in cleartext. With open 802.11 wireless hotspots, the threat is even greater because the information is sent over the air via radio frequency, which is an unbounded medium. Always read, understand, and comply with any warning messages you may see when they pertain to web server and SSL security.

Microsoft Desktop AirSync Protocol (AirSync) – 2175

As the name implies, the *Microsoft Desktop AirSync Protocol (AirSync)* will provide synchronization of email, contacts, calendar items, and other information from a messaging server to mobile devices using wireless networking connectivity. AirSync is now better known as Microsoft Exchange Active Sync (EAS) protocol. EAS includes the following features:

- It's based on Extensible Markup Language (XML).
- It communicates over HTTPS.
- It's considered a de facto standard for synchronization.
- It's licensed by Microsoft to other collaboration platforms.
- It's supported on a variety of smartphone platforms.

Remote Desktop Protocol (RDP) – 3389

With a need for remote access, Microsoft developed a proprietary protocol called *Remote Desktop Protocol (RDP)* to allow remote connections to other computers on a network. This protocol will enable a device running the Microsoft Windows operating system to allow other client devices to remotely connect and control the computer. To use RDP, two components are required:

RDP Server Software   RDP server software is built into the Microsoft Windows operating system starting with Windows XP. Formerly known as Microsoft Terminal Services, this is now called Remote Desktop Services.

RDP Client Software   RDP client devices are not required to be using the Windows Microsoft operating system to connect to a computer running the RDP server software. The following list shows the various devices that can be an RDP client:

- Windows Mobile
- Android
- iOS
- Mac OS X
- Linux
- Unix

Figure 2.21 shows a screen capture of how to enable RDP on a computer using the Microsoft Windows 8 operating system.

FIGURE 2.21   With the Microsoft Windows 8 OS, you can select to allow or not allow RDP connections.

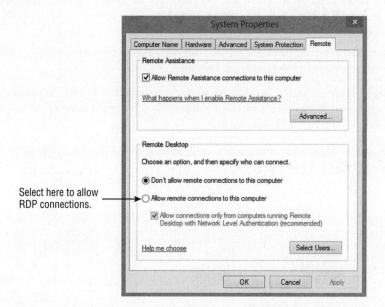

Server Routing Protocol (SRP) – 4101

TCP port 4101 is identified for official use by the Service Name and Transport Protocol Port Number Registry, www.iana.org, as a Braille protocol. However, an unofficial use is for the BlackBerry Enterprise Server (BES), which allows a BlackBerry mobile device user to communicate with their computer remotely to provide synchronization of information.

This port is typically blocked in firewalls and may need to be opened in order for this feature to work correctly, depending on the configuration.

The BES is designed for enterprise organizations that have their own onsite email services such as Microsoft Exchange or Novell Groupwise and require a high level of information technology (IT) control. This includes email redirection and synchronization of other information such as calendars and contacts.

Following is an overview of the basic steps involved in using the BlackBerry Enterprise Server. A fictitious company named XYZ using Microsoft Exchange email services is used. You are using the Microsoft Outlook email client program for email services:

1. An email message addressed to you arrives at your corporate email account from the Internet.
2. XYZ Company's Microsoft Exchange email server receives the email message.
3. You receive the email message via your Microsoft Outlook email client program on your corporate computer.
4. The BlackBerry Enterprise Server (BES) compresses the email message, encrypts it, and sends it to your BlackBerry device through the Internet and over your carrier's cellular wireless network.
5. Your BlackBerry mobile device receives, decrypts, and decompresses the email message and then alerts you via a notification message.

As shown in Figure 2.22, the BlackBerry router connects to the wireless carrier's mobile network. It sends and receives data to and from the BlackBerry infrastructure on behalf of the organization's BlackBerry Enterprise Server.

FIGURE 2.22 Message flow and synchronization with the BlackBerry Enterprise Server (BES) and BlackBerry router using port 4101

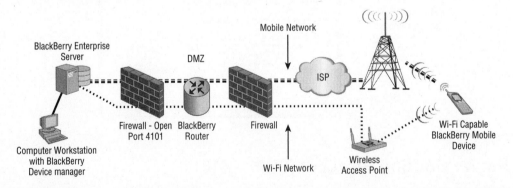

For security reasons, the BlackBerry router may be in the organization's demilitarized zone (DMZ). The mobile device can create a direct connection between the device and the BlackBerry router over a corporate Wi-Fi network using port 4101. This direct connection between the mobile device and the BlackBerry router is referred to as the "least-cost"

route because it eliminates the cost of using the BlackBerry network infrastructure. If the BlackBerry router is placed in the DMZ as shown in Figure 2.22, you must open port 4101 on the corporate firewall side to allow communication between the BlackBerry Device Manager and BlackBerry router. The end result is that a BlackBerry mobile device will be able to receive alerts from the BlackBerry Enterprise Server notifying the device's owner of email and other messages. This synchronization allows for a convenient way to receive messaging on the mobile device.

Summary

There are many protocols that work with computer and Internet technology. The list continues to grow as mobility and mobile device innovation advances. In this chapter you learned about the four-layer Internet protocol suite, how it maps to the conceptual OSI model, and some common protocols that are used in computer and mobile device technology. The following core protocols are among those that form the foundation of communications with mobile device technology:

- Transmission Control Protocol (TCP)
- User Datagram Protocol (UDP)
- Internet Protocol (IP)

These main protocols are responsible for the host-to-host connections that allow for the exchange of information (or data) between devices and are used with other Application layer protocols to provide the technology we use daily. You learned about some of the services that are used with these protocols for the distribution and management of TCP/IP information, computer network device management, IP address to hostname resolution, and more:

- Dynamic Host Configuration Protocol (DHCP)
- Simple Network Management Protocol (SNMP)
- Network Address Translation (NAT)
- Domain Name System (DNS)
- Internet Control Message Protocol (ICMP)

This chapter also provided a survey of various common protocols and ports that are used with computers and mobile device communication. Remember, there are a total of 65,535 possible ports in the Internet protocol suite. The first 1,024 are reserved and assigned to some of the common protocols we use every day. This chapter noted the specific ports used with the common protocols for the following technology and services:

- Transfer of files or data
- Messaging, both email and instant

- Secure protocols
- Web services
- Message synchronization
- Remote access

This chapter included several exercises to give you the opportunity to install and test some of the services and protocols you learned about.

Chapter Essentials

Understand the Internet protocol suite. The Internet Protocol suite is the backbone of Internet communications. This four-layer model consists of the protocols that allow for network and Internet communications worldwide using the TCP/IP suite of protocols. This model provides the Application layer protocols used with all forms of network communication.

Know the common core protocols and services. TCP, UDP, and IP are the main protocols used for device communications and are the foundational protocols on the Internet. You should know that these protocols are used for either reliable or unreliable communications between two network or Internet hosts and for the addressing and routing of network data. Be familiar with common network services, including DHCP, DNS, SNMP, and NAT. These services are responsible for issuing network address information, managing network devices, and providing easy name resolution to the address of the network hosts.

Know how to install FTP server and client software. A quick search of the Internet will allow you to find both FTP server and FTP client software programs. Several of these programs are available at no charge. Others are for a limited time or are functionality evaluation versions. Installing and configuring these programs will provide a better understanding of FTP and other file transfer protocols.

Understand the common protocols used for file transfer. FTP, TFTP, FTPS, and SFTP are file transfer protocols. FTP requires authentication and TFTP does not. These are commonly used protocols that are used in many areas of Internet communications. Understand that FTP and TFTP are protocols that send all information, including usernames and passwords, in cleartext and are a potential security threat. FTPS and SFTP will allow for the secure transfer of data.

Understand the common protocols used with messaging. Messaging across networks and worldwide communications is a major part of modern technology. Understand that SMTP, POP, IMAP, and MAPI are common standards based on proprietary protocols that are used with email services and technology. Knowing the pros and cons of these protocols will help you determine the best one to use based on your needs. APNS, XMPP, and GCM are some of the common notification and instant messaging protocols for Apple, Android, and other mobile devices. As with the email protocols, a basic understanding of how these operate is important.

Be familiar with additional common protocols. The use of the Internet grew at a tremendous pace with the development of HTTP and the World Wide Web in the mid-1990s. Almost all computers and mobile devices include technology that allows for web services. Because the Internet is accessible from virtually anywhere, security is a big concern. The development of Secure Sockets Layer technology helped alleviate some of these security concerns. If used correctly, SSL can help provide secure communications for HTTP and many other protocols. Know that RDP, AirSync, and SRP are protocols used for remote computer access and to provide synchronization of email messages, contacts, and other information from a messaging server.

Chapter 3

Radio Frequency and Antenna Technology Fundamentals

TOPICS COVERED IN THIS CHAPTER:

- ✓ Regulatory Governing Bodies and Authorities
- ✓ Radio Frequency Characteristics
- ✓ Bandwidth
- ✓ Modulation
- ✓ Radio Frequency Range and Speed
- ✓ Environmental Factors, Including Building Materials
- ✓ Faraday Cage
- ✓ Measurements of Radio Frequency Power
- ✓ Radio Frequency Antenna Concepts and Types
- ✓ Minimizing the Effects of Multipath Using Antenna Diversity
- ✓ Distributed Antenna System
- ✓ Radio Frequency Cables and Connectors
- ✓ Antenna Installation Considerations

Radio frequency (RF) plays an essential role in wireless computer networking and wireless cellular technology. Therefore, it is important to have an understanding of some of the basic RF concepts to help you better understand these wireless technologies. Radio waves are passed through the air (which is the medium) and are used to transfer information from one wireless device to another. Unlike physically wired computers and other network devices, which use a physical cable to communicate, wireless computer networks, cellular networks, and mobile devices use modulated radio waves and the air to communicate.

With respect to wireless data communications, RF consists of high-frequency alternating current (AC) signals passing over a copper cable connected to an antenna. The antenna then transforms the signal into radio waves that propagate through the open air from a radio transmitter to a radio receiver.

In this chapter you will learn about the characteristics of radio frequency and explore concepts such as the range and speed of RF transmissions. Range (how far radio waves will travel) and speed can both be affected by several environmental conditions or behaviors, such as reflection, refraction, and scattering.

Understanding the units in which RF is measured, such as watts (W), milliwatts (mW), and decibels (dB), is important to RF work. Antennas are an essential part of a successful wireless deployment, enabling modulated AC signals to send and receive across an unbounded medium (the air).

This chapter will examine many factors that are involved in determining the proper antenna to be used in an application or deployment of a wireless infrastructure such as the type of installation (indoor or outdoor), frequency, electrical characteristics and intended use.

Radio Frequency Fundamentals

Radio frequency (RF) waves are used in a wide range of communications, including AM and FM radio, television, cordless phones, cell phones, wireless broadband, wireless computer networks, and satellite communications. RF is around everyone and everything and comes in many forms. RF energy is emitted from numerous devices that use it for various types of communications. For the most part, it is invisible to humans. There is so much of it around that if you could actually see the RF it would probably frighten you. Figure 3.1 shows some of the many ways RF is used with technology and devices we use every day.

RF consists of high-frequency alternating current (AC) signals passing over a copper cable connected to an antenna. The antenna will then transform the signal into radio waves that propagate through the air. The most basic AC signal is a sine wave. The sine wave is the result of an electrical current that varies uniformly in voltage over a period of time. This sine wave cycle will repeat a specific number of times (cycles) over a period of 1 second. The number of cycles per second will result in different frequencies. Frequency is discussed in more detail throughout this chapter. Figure 3.2 shows an example of a basic sine wave.

FIGURE 3.1 Radio frequency is used in many different devices to provide wireless communications.

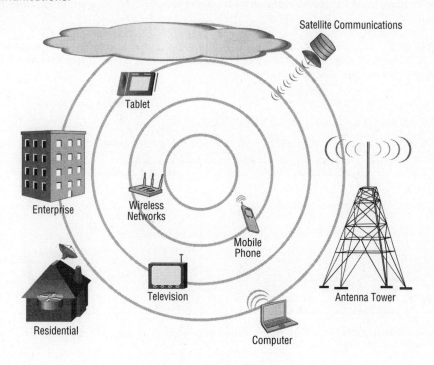

FIGURE 3.2 A basic sine wave, one complete cycle with the voltage varying over a point in time

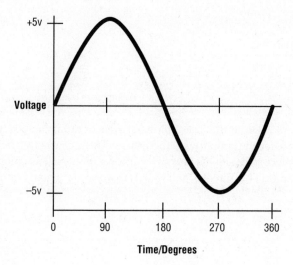

Successful radio transmissions consist of a minimum of two components: a radio *transmitter* and radio *receiver* (see Figure 3.3). With either wireless cellular technology or wireless computer networking, a wireless device or station that can transmit and receive RF signals is known as a *transceiver*. These two components work together: for every radio transmitter there must be one or more radio receivers in order for the communications to be successful.

FIGURE 3.3 RF transmitter and receiver. In wireless communications, the transmitter and receiver could be a wireless LAN access point and wireless client device or a mobile phone and a cellular site.

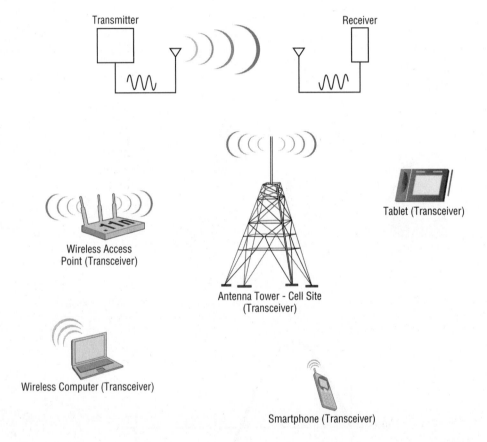

It is important to understand the basic characteristics of radio frequency transmissions such as wavelength, frequency, amplitude, and others. These characteristics work together to form the alternating current signals that carry information such as computer data from the device to the antenna. The antenna will then transform the signals into radio waves that travel through the air carrying the information from the radio transmitter to the radio receiver. This is accomplished in different ways, depending on the wireless technology in use. This theory will be discussed in this chapter.

RF Regulatory Governing Bodies and Local Regulatory Authorities

Wireless devices of all types use radio frequency (RF) to communicate. The RF spectrum we use needs to be regulated to ensure correct use of the available allocated frequency bands. At the global level, the International Telecommunication Union – Radiocommunication Sector (ITU-R) is responsible for global management of the terrestrial RF spectrum in addition to satellite orbits. This organization currently has over 700 sector members from 193 countries, which includes private-sector entities and academic institutions. It manages three different regions around the globe. Figure 3.4 shows the three administrative regions and the geographic area they encompass.

FIGURE 3.4 ITU-R region map

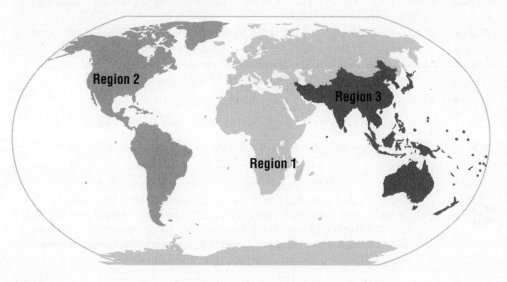

 For additional information, visit www.itu.int/ITU-R.

Table 3.1 describes the three administrative regions and the geographic areas they cover.

TABLE 3.1 ITU-R administrative regions and geographic locations

Region	Location
Region 1	Europe; Africa; the Middle East west of the Persian Gulf, including Iraq; the former Soviet Union and Mongolia
Region 2	Americas, Greenland, and some of the eastern Pacific Islands
Region 3	Most of non-former-Soviet-Union Asia, east of and including Iran, and most of Oceania

United States: Federal Communications Commission

The *local regulatory authority* that manages RF spectrum for the United States is the Federal Communications Commission (FCC). The FCC, founded in 1934, is (along with other local regulatory authorities) responsible for regulating both the licensed and unlicensed radio frequency spectrum.

Wireless cellular networks use licensed radio frequency, and IEEE 802.11 wireless networks may use unlicensed or licensed frequencies for communication between devices. A benefit of using the unlicensed radio spectrum is that there is no cost to the end user. IEEE wireless networks commonly use two unlicensed RF bands allowed by the FCC:

- The 2.4 GHz industrial, scientific, and medical (ISM) band
- The 5 GHz Unlicensed National Information Infrastructure (U-NII) Band

I will further discuss this and other technical details of standards-based wireless communications in Chapter 5, "IEEE 802.11 Terminology and Technology."

The FCC also allows the use of the unlicensed 900 MHz ISM band, but it is not used with standards-based IEEE 802.11 wireless networking. The 902-928 MHz ISM unlicensed band is not available worldwide and is only available in Region 2 (the Americas), with a few exceptions. Therefore, it was not specified for use with IEEE 802.11 wireless networking.

There are two licensed bands that can be used with IEEE 802.11 networking:

- 3.650–3.700 GHz band
- 4.940–4.990 GHz public safety band

In 2008, the IEEE ratified the IEEE 802.11y amendment to the standard. This amendment allows for the use of high-powered wireless LAN equipment to operate in the 3.650–3.700 GHz band. Within the United States, this is a licensed band that requires the user to pay some type of licensing fees.

The IEEE 802.11-2012 standard also specifies the use of the 4.940–4.990 GHz public safety band for use within the United States, consisting of 5 MHz, 10 MHz, and 20 MHz wide channels with both high and lower power limits.

Cellular networks in the United States use various frequency ranges depending on the wireless carrier and the geographic location. The following frequencies are among those used for cellular voice communications:

- 800 MHz
- 850 MHz
- 1700 MHz
- 1900 MHz
- 2100 MHz

I will discuss additional details of the radio frequency bands used for cellular communications in Chapter 7, "Cellular Communication Technology."

For additional information regarding radio frequency use in the United States, visit www.fcc.gov.

Europe: European Telecommunications Standards Institute

The European Telecommunications Standards Institute (ETSI) is responsible for producing standards for information and communications technologies, including fixed, mobile, radio, converged, broadcast, and Internet technologies in Europe. ETSI was created by the European Conference of Postal and Telecommunications Administrations (CEPT) in 1988.

In Europe, radio frequency use is managed by CEPT, which develops guidelines and provides national administrations with tools for coordinated European radio frequency spectrum management.

Radio Frequency Characteristics

To have a basic understanding of how radio frequency is used for data communications, it is important to become familiar with the RF characteristics that encompass the technology. In the following sections, you will learn about the four basic RF characteristics:

- Wavelength
- Frequency
- Amplitude
- Phase

Wavelength

The *wavelength* is the distance of one complete cycle or one oscillation of an AC signal. Wavelength is typically identified by the Greek symbol lambda λ, which is used in formulas for calculations. This distance is usually measured in centimeters or inches. Figure 3.5 shows an example of the wavelength for an IEEE 802.11 wireless LAN operating in the 2.4 GHz ISM band on channel 6.

FIGURE 3.5 The wavelength is the distance of one complete cycle, measured in centimeters or inches.

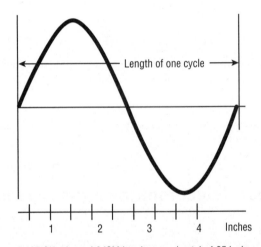

2.437 GHz channel 6 ISM band, approximately 4.85 inches

IEEE 802.11 wireless LANs use both the 2.4 GHz and 5 GHz unlicensed frequency ranges for communication. The IEEE 802.11-2012 standard also specifies some additional frequency ranges in which wireless LANs can operate. Although these do not fall under the "unlicensed" category, 4.9 GHz public safety and 3.650 GHz (IEEE 802.11y amendment) can also be used for IEEE 802.11 wireless LAN communications. Table 3.2 lists some examples of wavelengths for IEEE 802.11 wireless LANs using unlicensed frequencies.

TABLE 3.2 Typical radio transmission wavelengths for IEEE 802.11 WLAN

RF Channel	Frequency (GHz)	Length (in)	Length (cm)
6	2.437 GHz	4.85 in	12.31 cm
40	5.200 GHz	2.27 in	5.77 cm
153	5.765 GHz	2.05 in	5.20 cm

Frequency

Frequency is defined as the number of complete cycles in 1 second. Low frequencies correspond to long radio waves and high frequencies to short radio waves, so the higher the frequency, the shorter the wavelength (range). In formulas, frequency is identified by the lowercase letter *f*.

Figure 3.6 shows an example of frequency.

FIGURE 3.6 Frequency is the number of complete cycles in 1 second.

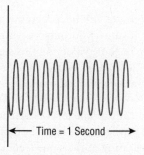

How Far Can a Signal Travel?

A few years back, 900 MHz cordless telephones were very popular. Cordless telephones were introduced in the 1970s, and in 1990 the FCC opened the 900 MHz range for these telephones. With a 900 MHz phone, you could potentially be up to 500 feet away from its base station before losing the signal and no longer being able to make a phone call. In the past few years, higher-frequency phones have increased in popularity. After upgrading to a 2.4 GHz phone, you may have noticed that you can get only about 250 feet away (half the distance compared to a 900 MHz phone) in the same environment before losing the signal. This is because the 2.4 GHz wavelength is about half the distance of a 900 MHz wavelength, assuming both phones are operating at the same output power. With 5.8 GHz cordless telephones, the range will be still less, assuming the same amount of RF transmit power.

IEEE 802.11 wireless LANs operate in several unlicensed frequency ranges. The unlicensed ranges used for wireless LANs are 2.4 GHz to 2.5 GHz and 5.15 GHz to 5.875 GHz. There are some areas in the 5.15 GHz to 5.875 GHz range that are not used for standards-based wireless networking. Cellular phones operate in several different licensed frequency ranges depending on the carrier. These ranges include but are not limited to 800 MHz, 1.7 GHz, and 1.9 GHz for voice communications.

Amplitude

From a wireless communication perspective, the *amplitude* is the signal strength which is the amount of power of an RF signal. The amplitude is calculated from the height (in a two-dimensional view), measured on the y-axis of the sine wave. This represents the voltage level of the RF signal. A basic sine wave is a change in voltage over a period of time. Using a formula, the voltage at the peak of the signal can be used to calculate the amount of RF power.

An increase in amplitude is equal to an increase in RF power. This increase in RF power is also known as *gain*.

Conversely, any decrease in amplitude will be a decrease in power. A decrease in RF power is also known as *loss*. If a transmitter outputs a certain amount of RF power—for example, 50 mW—it has amplitude of some value. As this signal travels through an RF cable, it will have a specific level of loss based on the cable in use, resulting in attenuation. Therefore, the result will be less signal or smaller amplitude at the end of the cable due to the loss value of the cable. The same effect occurs when an RF signal travels through open air. The air will diffuse the RF and cause some level of attenuation. In over-the-air communication technology, the free space produces the largest amount of signal loss.

Figure 3.7 shows two signals operating at the same frequency with different amplitudes. The signal with the higher amplitude (Signal A) is more powerful than the signal with the lower amplitude (Signal B).

FIGURE 3.7 Two signals at the same frequency with different amplitudes

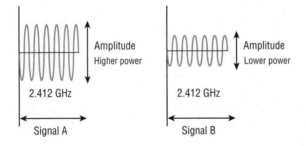

Phase

Phase is the difference in degrees at a particular point in the time of a cycle, measured from some arbitrary zero and expressed as an angle. For example, based on an x-axis scale of from 0° to 360°, if a second sine wave starts a quarter of a wavelength after the first sine wave, it is considered to be 90° out of phase with the first sine wave. Figure 3.8 shows an example of the phase relationship between two AC signals. Two radio waves that have the same frequency but start at different times are known to have a phase difference and are considered out of phase with one another. The amount of the phase difference is measured in degrees, ranging anywhere from 0° to 360°.

FIGURE 3.8 Phase is the difference in degrees between two signals.

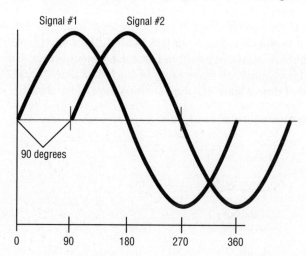

Given that a single radio receiver is not capable of distinguishing the difference between radio waves that arrive at a receiver out of phase, the received signal will experience some level of distortion. The receiver will combine these out-of-phase (reflected) signals, which will cause some corruption of the received signal. This is known as *multipath*. Reflection and multipath are discussed later in this chapter. The difference in time of arrival of the main signal and an out-of-phase, reflected signal that causes the multipath problem is called the *delay spread*. If two waves were to arrive at a receiver 180° out of phase, this will usually result in a cancellation effect, which nullifies the two signals. Conversely, two waves that arrive in phase are additive, and this will result in an increase in signal strength, also known as *upfade*. Keep in mind, however, that the amplitude of the waves that experience the upfade effect will never be higher than the original wave that was transmitted.

Bandwidth

Depending on the context in which the term *bandwidth* is used, it may have several different meanings. With respect to radio frequency, *bandwidth* can be thought of as a range of frequencies used for a specific application. For example, the operating frequency of an IEEE 802.11 access point in the 5 GHz U-NII band using channel 36 is from 5,170 MHz to 5,190 MHz. This is a difference of 20 MHz. Therefore the bandwidth of channel 36 is 20 MHz wide. Another example is the amount of information or computer data that can potentially be sent over a specific period of time, such as 1,000 Mbps (1 Gbps). This is equal to 1 billion bits per second, which is the bandwidth. The number in both examples is the total maximum capacity of the radio frequency channel or computer network infrastructure.

One simple way to analogize this concept is using a garden hose connected to a water faucet. If you have a ½-inch-diameter 10-foot garden hose, it may take 10 minutes to fill up a bucket with water. If you switched to a 1-inch-diameter garden hose and everything else remained the same, it would take only 5 minutes to fill the bucket with water. Therefore, the wider diameter hose (1 inch) can pass twice as much water over the same period of time, just as a 40 MHz channel can move twice as much data as a 20 MHz wide channel over the same period of time. Figure 3.9 shows this concept of bandwidth.

FIGURE 3.9   Representation of bandwidth

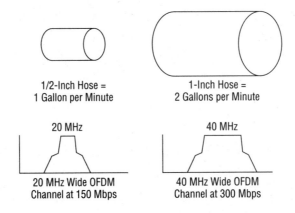

Modulation

Wireless communications utilizes radio frequency to send and receive information. This is possible because all wireless devices have transmitter and/or receiver capability. Unlike a television or FM radio you may have in your home, both of which are only receivers, other wireless communication devices such as a cellular phone or Wi-Fi–capable computer have the ability to act as either a transmitter or a receiver (also known as a transceiver).

For a transmitter and a receiver to communicate, the information that is sent over the air using radio frequency must be modulated. Depending on the type of communication taking place, *modulation* is varying one or more of the RF characteristics, such as wavelength, frequency, amplitude, or phase. For example, the modulation process will add digital data to the AC signal and prepare it to be sent over the air using radio frequency. This modulation concept is analogous to speaking a specific language. For two people to communicate, they must both be able to speak and understand the same language.

Now we will explore how modulation works as it relates to wireless computer networking and using radio transmitters and receivers (transceivers) to send and receive computer data:

Radio Transmitter   With wireless communications, a radio transmitter combines binary digital computer data (all of the 1s and 0s) with high-frequency alternating current (AC) signals to prepare it to be sent across the air. The connected antenna then transforms this signal into radio waves and propagates them through the air. The frequency of the signal depends on the technology in use. With IEEE 802.11 standards–based wireless networking, there are a few select frequency ranges. This signal that is sent over the air now contains the computer data, and the characteristics of the RF signal will vary based on the modulation type used.

Radio Receiver   A radio receiver collects the propagated signal from the air using an antenna and reverses the process by transforming the received signal back into an alternating current signal. Through the use of a demodulation process, the digital data is recovered from the received signal. The modulation/demodulation technology used depends on the wireless communication technology in use. For example, an IEEE 802.11g wireless LAN will use 64 QAM (quadrature amplitude modulation) when transmitting data at 54 Mbps. Figure 3.10 illustrates this entire process. In this figure, the access point is the receiver and the client device is the transmitter.

FIGURE 3.10   A wireless access point (transmitter) and wireless client device (receiver) with computer data traversing the air using radio frequency

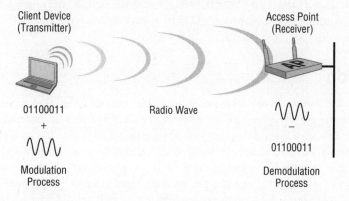

Specific modulation types will be discussed in more detail in Chapter 5.

Radio Frequency Range and Speed

How far and fast a radio frequency signal can travel depends on a variety of factors, including interference and the types of materials in the environment. In this part of the chapter you will learn about these concepts.

The *range* for wireless communications is based on the wavelength or distance of a single cycle. The frequency of an RF signal is based on how many cycles occur during a specific time frame. The higher the frequency, the shorter the range of the signal. The lower the frequency, the longer the range of the signal. A 2.4 GHz signal will travel almost twice as far as a 5 GHz signal at the same RF output power level. For example, with an IEEE

802.11 wireless network, if the network design calls for dual-band access points, range will need to be considered to ensure proper coverage for both the 2.4 GHz ISM and 5 GHz U-NII bands.

The speed of the communication link is based on the quality of the received signal strength at the receiver. Therefore, a receiver that is operating in the 2.4 GHz band and is in very close proximity to the transmitter will be receiving a stronger, better-quality signal. With regard to wireless networking, this means the receiver can use a more sophisticated modulation technology, which will result in a higher data rate. So the range of the signal may have an impact on the speed of the data transfer due to the quality of the signal at different distances.

A wireless LAN site survey will help determine the usable range an access point will produce. A wireless site survey can involve physically walking around the proposed space to determine infrastructure device placement and/or a predictive modeling process using one of many software programs. This process is discussed further in Chapter 8, "Site Survey, Capacity Planning, and Wireless Design."

There are a lot of factors that can affect RF speed and range. The first thing to understand is the concept of attenuation, which is a decrease in signal intensity. There are a host of environmental factors that can affect the behavior of RF signals, causing attenuation and other problems. Another consideration is radio frequency interference from other signals. In the following sections, we'll explore all of these factors.

Attenuation

With respect to RF, *attenuation* is a decrease in the intensity of a signal or a decrease in amplitude. Radio frequency *attenuation* can be either intentional or unintentional.

Intentional attenuation would be the result of inserting a device called an attenuator within an RF circuit. An attenuator is a passive device that does not require an external power source and is installed in series in the RF circuit. The signal will decrease based on the rating of the attenuator. For example, an attenuator rated at 3 dB would decrease the signal by 50%. (Relative RF power expressions such as dB are discussed later in the chapter.) The reason for this type of device would be to purposely cause a decrease in signal strength or amplitude of the signal without distorting or corrupting the RF signal. An attenuator is the opposite of an amplifier, which will cause an increase in amplitude or signal strength. An amplifier is an active device that does require an external power source.

Unintentional attenuation can be caused by a variety of items, including environmental conditions. RF signal attenuation can be caused by the following:

- RF signal traveling through free space
- Wood
- Sheet rock or drywall walls

- Plaster walls
- Glass
- Concrete
- Metal
- Items that contain a liquid content

This is a short list of some of the items that will cause an RF signal to be attenuated. It is important to take these and other factors into consideration when designing a wireless system and while troubleshooting or supporting a wireless system.

Environmental Factors, Including Building Materials

The interaction between RF and the surrounding environment can seriously affect the performance of IEEE 802.11 wireless networks. RF behavior is affected by environmental conditions that exist:

- Reflection
- Refraction
- Diffraction
- Scattering
- Absorption
- Diffusion

Keep in mind that the environment in which wireless devices are communicating will have an effect on how successful the communications will be. All of the environmental conditions will have an effect on RF propagation, and one or more of these behaviors will be present every place wireless is used. Now let's take a closer look at each of these behaviors, and then we'll examine the Faraday cage, the use of which can result in some of these behaviors.

Reflection

Reflection occurs when an RF signal bounces off a smooth, nonabsorbing surface, resulting in a change of direction of the RF. A table top is an example of a surface that will produce this type of behavior. Metal storage racking in a warehouse is another example. Reflections can affect indoor wireless installations fairly significantly in certain cases. Depending on the interior of a building—such as the type of walls, floors, and furnishings—there could be a large number of reflected signals. If not properly handled, reflections could cause poor network performance and a decrease in throughput or other communication indicators. Figure 3.11 illustrates reflection.

FIGURE 3.11 Radio frequency reflection

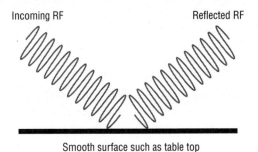

 Think of a Ping-Pong game when it comes to reflection. When a Ping-Pong ball is served or hit, it comes in contact with the table—a smooth, hard surface—and bounces off in a different direction. This is similar to how reflection works with radio frequency. Once the ball hits the table, it will slow down in speed because of the impact with the table. The same concept is true with radio frequency. An RF signal that experiences reflection will change directions, resulting in attenuation of the original signal.

Refraction

When an RF signal passes between mediums of different densities, it may change speeds and change direction. This behavior of RF is called *refraction*. Glass is an example of a material that may cause refraction. When an RF signal comes in contact with an obstacle such as glass, the signal is refracted by changing direction as it passes through and is attenuated. The amount of attenuation depends on the type of glass, its thickness, and other properties. Figure 3.12 shows an example of RF refraction.

FIGURE 3.12 Radio frequency refraction

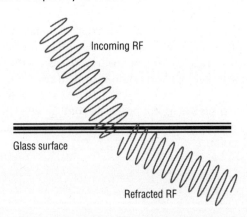

Diffraction

When an RF signal comes in contact with an obstacle, the wave may change direction by bending around the obstacle. This RF behavior is called *diffraction*. A building or other tall structure could cause diffraction, as could a building support column in a large open area or conference hall. Figure 3.13 shows an example of diffraction. When the signal bends around a column, building, or other obstacle, the signal weakens, resulting in some level of attenuation.

FIGURE 3.13   Radio frequency diffraction

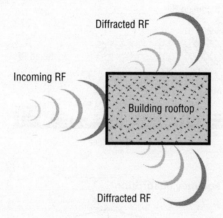

 You can demonstrate diffraction by using a pond of still water. Place a large object such as a two-by-four piece of lumber in the pond. After the water settles, try to drop a pebble or small rock off to the side of the piece of lumber. Watch closely and you will see the ripple of the water bend around the lumber, resulting in a diffraction effect.

Scattering

When an RF signal strikes an uneven surface, wavefronts of the signal will reflect off the uneven surface in several directions. This is known as *scattering*. Scattering, illustrated in Figure 3.14, is another RF behavior that may severely degrade a signal, therefore causing attenuation of this signal. Scattering, if significant enough, can render the received signal unusable.

Absorption

When materials absorb an RF signal, little if any of the signal will penetrate through the material. This may cause severe attenuation of the original RF signal. Many different types of building materials can cause *absorption* in addition to other items within an environment. Absorption can be caused by objects that contain a high content of liquid. This includes areas where products that contain liquid such as water and paint are stored. Another cause of absorption is the human body. The human body contains a high liquid content and will absorb some RF signals. This type of absorption can be a problem for wireless network deployments in certain environments. For densely populated areas such as stadiums, airports, and conference halls, absorption needs to be considered when designing for a wireless system. Figure 3.15 shows an example of absorption.

FIGURE 3.14 Radio frequency scattering

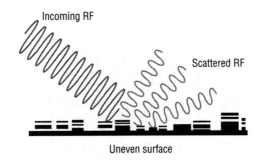

FIGURE 3.15 Radio frequency absorption

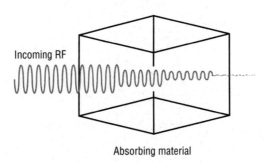

Diffusion and Free Space Path Loss

RF *diffusion* occurs when a transmitted RF signal naturally widens as it leaves an antenna element. Figure 3.16 shows this widening effect. As a result, the transmitted radio frequency signal will decrease in amplitude and be less powerful at any distance from the antenna. This is known as *free space path loss (FSPL)*. FSPL is the greatest form of attenuation in a radio frequency link, resulting in a decrease of the signal strength. The antenna on the receiver side is able to acquire a much smaller amount of the transmitted signal because of this widening effect of the diffused signal as it propagates through the air. Any signal that is not received by the intended receiving device is considered RF loss.

FIGURE 3.16 Radio frequency diffusion

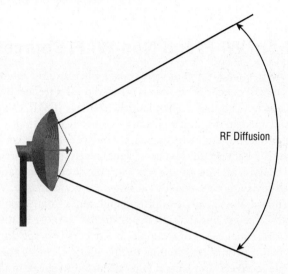

You can calculate the FSPL with a mathematical formula using the variables of frequency and distance.

FSPL (dB) = $20\log_{10}(d) + 20\log_{10}(f) + 32.45$

f = frequency in MHz and d = distance in km

Faraday Cage

Michael Faraday was an English scientist who invented the Faraday cage in 1836. The Faraday cage, also called a Faraday shield, is an enclosure of some sort that will block external static and nonstatic electric fields. This enclosure is made from or coated with a conductive material and usually consists of a metal mesh material. The shape is not important and is based on the use of the cage.

Although a Faraday cage can be created for electrical testing purposes, other structures or enclosures can have the same effect. The design of an enclosure could have Faraday cage–like properties to shield it from or to retain the electromagnetic radiation. Anything that may have the capability to either keep signals from entering or keep them from leaving the enclosure will have Faraday cage characteristics. This may include but is not limited to the following:

- Building materials
- Microwave ovens
- Elevators
- Wire mesh or screens
- Coaxial cable
- Magnetic resonance imaging (MRI) areas
- Tempest building or room

When you think about it, the possibilities are endless as to what type of enclosures can act as Faraday cages for both intentional and unintentional purposes.

Interference from Wi-Fi and Non-Wi-Fi Sources

Interference from a radio frequency point of view occurs when a radio receiver hears unintended signals on the same frequency as the one it is tuned to or a frequency that's close. This type of interference, also known as electromagnetic interference (EMI), causes a received RF signal to be distorted and can result in a partial or complete disturbance of the intended signal. In wireless communications systems such as a wireless computer network, the interference may have a severe impact on the quality of signal received by a wireless device. The distorted or corrupted signal will decrease the efficiency of the communication or the amount of data a device can effectively receive, thereby causing poor performance. This poor performance will result in dropped phone calls, less throughput, and similar problems.

A wireless LAN radio receiver has characteristics similar to those of the human ear. Both can hear a range of frequencies. For example, if one person is speaking to a group at a conference and a number of people are listening to this speaker, this is analogous to a single transmitter and multiple receivers. If a second person started to speak in the area at the same time, people listening might not be able to understand either speaker. In a sense, they are experiencing interference.

An IEEE 802.11 wireless network may use the unlicensed 2.4 GHz ISM band. The following list is a small sample of what can cause RF interference to wireless communication systems that use unlicensed frequency bands:

- Cordless phones
- Microwave ovens
- Medical devices
- Industrial devices
- Baby monitors
- Wireless cameras
- Two-way radios
- Other IEEE 802.11 wireless networks

Because these devices also use radio frequency to operate, and the frequency is in the same unlicensed band as IEEE 802.11 wireless networks, they have the potential to interfere with one another. Although they may coexist in the same RF space, the interference factor needs to be taken into consideration. This can be done as part of the site survey process. RF interference with regards to wireless networking site survey and design concepts will be discussed further in Chapter 8.

Measurement Units of Radio Frequency

If a person were given a dollar bill, they would be 1 dollar richer. If this person were given 100 cents, they would still be 1 dollar richer. From this example, you can see that 1 dollar = 100 cents and 1 cent = 1/100th of a dollar. One dollar and 100 cents are the same net amount, but a cent and a dollar are different units of currency.

The same is true for radio frequency measures of power. The basic unit of measure for radio frequency power is the watt. For example, a wireless LAN access point may be set to an RF output power of 30 mW (milliwatts). A milliwatt is 1/1000 of a watt. Just as cents and dollars are both denominations of currency, watts and milliwatts are measurements of RF power. Other units of measurement for RF power are decibel milliwatt (dBm), decibel (dB), decibel isotropic (dBi), and decibel dipole (dBd).

Absolute Measurements of Radio Frequency Power

The amount of RF power leaving a wireless access point is one example of an *absolute* measure of power. This is an actual power measurement and not a ratio or relative value. In other words, this is a measurable amount of power, and it can be determined with the proper calibrated measurement device, such as a watt meter. A typical maximum amount of transmit output power from an enterprise-grade access point in the 2.4 GHz ISM band is 100 mW.

The measure of AC power can be calculated using a basic formula. The formula is as follows:

$$P = E \times I$$

That is, power (P) equals voltage (E) multiplied by current (I).

A simple example would be to calculate the power from 1 volt and 1 amp. Using the given variables, here's the formula:

$$P = 1 \text{ volt} \times 1 \text{ amp}$$

The answer would be power = 1 watt.

Watt (W)

The *watt* (W) is a basic unit of power measurement. It is a measurable, absolute value. Most wireless computer networks and mobile wireless devices function in the milliwatt range; however, some cellular phones may approach 1 W or more. An RF power level in watts is a common measurement in long distance point-to-point and point-to-multipoint connections.

Milliwatt (mW)

One *milliwatt* (mW) is 1/1000 of a watt. This is a common value used in RF work with cellular networks and IEEE 802.11 wireless networks. The output power of an enterprise-grade wireless access point typically ranges from 1 mW to 100 mW. Most of these access points will allow you to change the RF output power. Most small office, home office (SOHO) access points have a fixed output power, typically 30 mW. The milliwatt is another example of an absolute unit of power measurement.

Decibel Relative to a Milliwatt (dBm)

dBm is the power level of a decibel compared to 1 milliwatt. This is based on a logarithmic function. A good rule of thumb to remember is that 0 dBm = 1 mW. This value is considered as absolute zero because it is a logarithmic expression. Using a formula or basic RF calculation rules, you can easily convert any milliwatt value to decibels: 100 mW = 20 dBm, for example.

The *dBm* is also an absolute unit of power measurement. A decibel (dB) is an example of a change in power or relative measurement of power, where dBm is measured power referenced to 1 milliwatt, or an absolute measure of power. The next section discusses relative measurements of power.

Absolute values are measurable values of power such as watt, milliwatt, and decibel milliwatt.

Relative Measurements of Radio Frequency Power

Changes in radio frequency power are known as *relative* values. Decibel (dB) and decibel isotropic (dBi) are relative measurements of RF power. For example, if the input RF power to an amplifier is 10 mW and the output RF power is 100 mW, the gain of the amplifier is 10 dB—a change in power. This is because the RF power was increased from 10 mW to 100 mW, or a times 10 increase.

If the input power to an antenna is 100 mW and the output power is 200 mW, the gain of the antenna is 3 dBi—a change in power. This is because the power was increased from 100 mW to 200 mW, which is two times the power. Both 10 mW to 100 mW and 100 mW to 200 mW are examples of changes in power and are known as relative expressions of RF power.

Figure 3.17 shows absolute and relative RF power with respect to a wireless LAN access point.

FIGURE 3.17 Absolute and relative RF power measurements

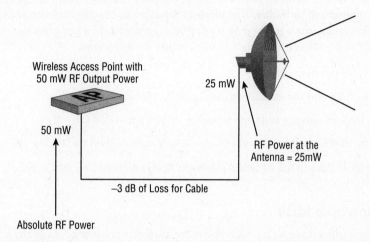

The figure shows that the wireless access point has an RF output power of 50 mW. This is an absolute expression because it is a measurable value at that point. The cable introduces a loss of −3dB, resulting in 25 mW of power at the antenna element. Because the cable resulted in half of the original power at the antenna element, and there was a change from 50 mW to 25 mW and is therefore a relative RF power expression.

Decibel (dB)

The decibel (*dB*) is a ratio of two different power levels caused by either a positive change in power (gain) or a negative change in power (loss). Figure 3.18 shows how an amplifier will provide an increase or change in power.

FIGURE 3.18 Output doubled in power from 100 mW to 200 mW from amplifier with a gain or change in power of +3 dB

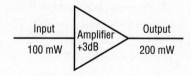

> **Basic RF Math: The 3s and 10s Rule**
>
> There is a simple way to perform any RF math calculation without having to use logarithms and mathematical formulas. This method is known as the 3s and 10s rule (or sometimes referred to as the 10s and 3s rule). If you remember five basic steps, you can perform any RF math calculation. The five basic steps are as follows:
>
> - 0 dBm = 1 mW (starting point)
> - Increase by 3 dB and the absolute power in mW doubles (or × 2)
> - Decrease by 3 dB and the absolute power in mW is cut in half (or ÷ 2)
> - Increase by 10 dB and the absolute power in mW is multiplied by (10 or × 10)
> - Decrease by 10 dB and the absolute power in mW is divided by (10 or ÷ 10)

Decibel Isotropic (dBi)

Decibel isotropic (*dBi*) is the unit of RF power measurement that represents the gain or increase in signal strength of an antenna. The term *isotropic* in the RF world means energy broadcast equally in all directions in a spherical manner. An imaginary, perfect antenna would be considered an *isotropic radiator*. This is a theoretical concept and is used as a reference and in calculations. dBi is discussed in more detail later in this chapter.

 Remember, relative measurements in RF power are changes in RF power from one value to another value. An example is 100 mW to 50 mw. dB, dBi, and dBd are relative RF power measurements.

Decibel Dipole (dBd)

The gain of some antennas may be measured in decibel dipole (*dBd*). This unit of measurement refers to the antenna gain with respect to a reference dipole antenna. The gain of most antennas used in wireless LANs is measured in decibel isotropic (dBi); however, some manufacturers may reference the gain of an antenna in dBd. The following simple formula derives the dBi value from the dBd value:

dBi = dBd + 2.14

This formula converts from dBi to dBd:

dBd = dBi − 2.14

> **Real World Scenario**
>
> ### dBd vs. dBi
>
> You are a procurement agent working for a manufacturing company. An engineer orders some antennas to be used in a wireless LAN deployment. The part number you received from the engineer on the bill of materials is for antennas that are currently out of stock at your normal supplier. The order has to be placed as soon as possible, but technical support for the vendor is gone for the day and you are not able to get any assistance.
>
> You found what appears to be a reasonable alternative for the requested antennas. However, the gain of the antennas does not exactly match what the engineer documented on the bill of materials. The engineer requested omnidirectional antennas with a gain of 6 dBi. You found what appears to be a comparable alternate with a gain of 6 dBd. It will be necessary for you to determine whether these antennas will work. Not quite understanding the difference, you do some research to determine the difference between dBd and dBi. After searching various websites, you find a formula to convert the two different units:
>
> dBi = dBd + 2.14
>
> Using your calculator, you enter the value from the specification sheet for the alternate antennas:
>
> 6 dBd + 2.14 = 8.14 dBi
>
> Unfortunately, the antennas found will not be a good alternative in this example. Back to the drawing board!

Radio Frequency Antenna Concepts

It is important to understand some of the basic theory, characteristics, and terminology associated with antennas prior to learning how they operate. Becoming familiar with this will help in making decisions when it comes to purchases and support of antennas and wireless LAN systems. This list includes some of the terms for characteristics of antennas explored in the following sections:

- **Radio frequency lobes:** Shape of the radiation patterns
- **Beamwidth:** Horizontal and vertical angles

- **Antenna charts:** Azimuth and elevation
- **Gain:** Changing the radio frequency coverage pattern (beamwidths)
- **Polarization:** Horizontal or vertical orientation

Radio Frequency Lobes

The term *lobe* has many meanings, depending on the context in which it is used. Typically it is used to define the projecting part of an object. In anatomical terms, an example would be part of the human ear known as the earlobe. In botanical terms, a lobe is the divided part of a leaf. As a radio frequency technology term, *lobe* refers to the shape of the RF energy emitted from an antenna element.

RF lobes are determined by the physical design of the antenna. The antenna design also determines how the lobes project from an antenna element. The effect of antenna design, particularly on the shape of the RF lobes, is one reason choosing the correct antenna is a significant part of a wireless design.

Antennas may project many lobes of RF signal, some of which are not intended to be usable areas of coverage but nevertheless can be in some cases. The RF lobes that are not part of the main lobe coverage—that is, the back or rear and side lobes—contain usable RF but in most cases are not intended to be used to provide coverage for the wireless cell. They are for the most part unintentional coverage areas and, depending on the environment, may not be part of a good wireless design or planned coverage area.

The type of antenna utilized—omnidirectional, semidirectional, or highly directional—will determine the usable lobes. These antennas, as well as the RF radiation patterns they project, will be discussed in more detail later in this chapter.

Figure 3.19 shows an example of RF lobes emitted from an antenna element. The "main signal" is the lobe intended to be used.

FIGURE 3.19 Radio frequency lobes' shape and coverage area are affected by the type and design of an antenna.

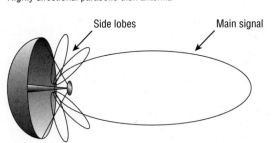

Antenna Beamwidth

The design of an antenna will determine how and in which specific patterns the RF energy propagates. The patterns of energy emitted from an antenna are known as lobes. For antennas, the *beamwidth* is the angle of measurement of the main RF lobe measured at what is called the half-power point, or the –3 dB point. *Beamwidth* is measured both horizontally and vertically, in degrees. It is important to understand that antennas shape the RF coverage or isotropic energy that radiates from the antenna element. Changing types or remaining with the same type of antenna but changing the gain will also change the coverage area provided by the wireless system.

Documents for antenna specifications are available to illustrate the horizontal and vertical beamwidths. Azimuth and elevation charts available from the antenna manufacturer will show the beamwidth angles.

The *azimuth* refers to the horizontal RF coverage pattern, and the *elevation* is the vertical RF coverage pattern. The azimuth is the view from above, or the "bird's-eye view" of the RF pattern; in some cases it will be 360°.

Think of the elevation as a side view. If you were to look at a mountain from the side view, it would have a certain height, or elevation, measured in feet or meters. For example, Pikes Peak, a mountain in the front range of the Rocky Mountains, has an elevation of 14,115 feet (4,302 meters). If you were to look at Pikes Peak from ground level (a side view), you would be able to visualize this impressive elevation.

Figure 3.20 shows a representation of horizontal and vertical beamwidths for a parabolic dish antenna.

Some wireless predictive modeling site survey software programs will allow the wireless designer to adjust the azimuth and elevation of the antennas used in the predictive modeling design to more closely depict the real-world coverage of the wireless system. Wireless site surveys will be discussed in more detail in Chapter 8, "Site Survey, Capacity Planning, and Wireless Design."

FIGURE 3.20 Horizontal (azimuth) and vertical (elevation) beamwidths measured at the half power, or –3 dB point

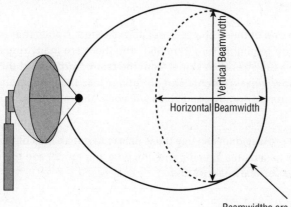

Reading Azimuth and Elevation Charts

Understanding how to read an azimuth and elevation chart is useful from a technical sales, design, and integration perspective. Knowing these patterns will help when making hardware recommendations for customers based on needed coverage and device use. Azimuth and elevation charts show the angles of radio frequency propagation from both the azimuth (horizontal or looking down, top view) and the elevation (vertical, side view). They give a general idea of the shape of the RF propagation lobe based on antenna design.

Antenna manufacturers test antenna designs in a laboratory. Using the correct instruments, an engineer is able to create the azimuth and elevation charts. The charts show only the approximate coverage area based on the readings taken during laboratory testing and do not take into consideration any environmental conditions such as obstacles or interference. The following image shows an example of an azimuth and elevation chart for a semidirectional antenna.

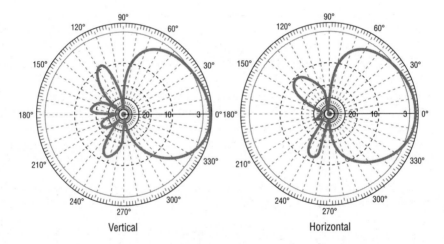

Understanding how to read one of these charts is not complicated. Notice that each chart is a circular pattern with readings from 0° to 360° and there are many rings within it. The outermost ring shows the strongest signal from the testing process of this antenna. The inner rings show measurements and dB ratings less than the strongest measured signal from the outside ring. A good chart will show the most accurate readings from the testing process.

A sales or technical support professional can use these charts to get an idea of how the radiation pattern would look based on a specific antenna type and model and to determine both the horizontal and vertical beamwidths.

Antenna Gain

The *gain* of an antenna provides a change in coverage that is a result of the antenna focusing the area of radio frequency propagation. The gain is produced from the physical design of the antenna element.

One of the characteristics of radio frequency is amplitude, which is defined as the height (voltage level) and is the amount of power of an RF wave. The amplitude is created by varying voltage over a period of time and is measured at the peaks of the signal from top to bottom. Amplification of an RF signal will result in gain. An antenna is a device that can change the RF coverage area, thus propagating an RF signal further.

Antenna gain is measured in decibels isotropic (dBi), which is a change in power as a result of reshaping the isotropic energy. Isotropic energy is defined as energy emitted equally in all directions. The sun is a good example of isotropic energy, emitting energy in a spherical manner equally in all directions. Figure 3.21 illustrates energy being emitted from an isotropic radiator.

FIGURE 3.21 A perfect isotropic radiator emits energy equally in all directions.

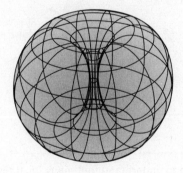

Passive Gain

It's quite intriguing how an antenna can provide *passive gain*, a change in coverage without the use of an external power source. Because of how antennas are designed, they focus isotropic energy into a specific radiation pattern. Focusing this energy increases coverage in a particular direction. A common example used to describe passive gain is a magnifying glass. If a person is standing outside on a beautiful sunny day, the sun's energy is not intense because it is being diffused across the entire earth's hemisphere. Thus, there is not enough concentrated energy to cause any harm or damage in a short period of time. However, if this person were to take a magnifying glass and point one side of it toward the sun and the other side toward a piece of paper, more than likely the paper would start to heat quickly. This is because the convex shape of the magnifying glass focuses, or concentrates, the sun's energy into a specific area, thus increasing the amount of heat to that area.

Antennas are designed to function in the same way by focusing the energy they receive from a signal source into a specific RF radiation pattern. Depending on the design of the antenna element, as the gain of an antenna increases, both the horizontal and vertical radiation patterns (beamwidths) will decrease, or create narrower beamwidths. Conversely, as the gain of an antenna decreases, the beamwidths will increase, making a larger radiation pattern.

One exception to this behavior is the omnidirectional antenna. This type of antenna has a horizontal beamwidth of 360°. When the gain is increased or decreased, the beamwidth will remain 360° but the size of this coverage area will increase or decrease depending on the change in the gain. Omnidirectional antennas are discussed in more detail later in this chapter.

Figure 3.22 shows a drawing of a wireless network with 100 mW of RF power at the antenna. Because of passive gain, the antenna has the effect of emitting 200 mW of RF power.

FIGURE 3.22 A wireless LAN access point supplying 100 mW of RF power and an antenna with a gain of 3 dBi for an output at the antenna of effectively 200 mW of RF power

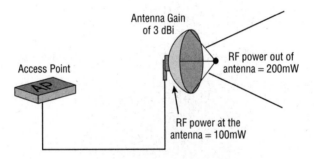

 It is important to understand that many local radio frequency regulatory domains or agencies restrict the amount of RF power that can be emitted from an unlicensed RF system. This "system" includes all the components certified by the local regulatory agency and may include the transmitter (access point), the connectors, and the antenna. Changing and increasing the gain of an antenna will increase the amount of effective RF energy leaving the antenna and may violate the regulations set forth by the local regulatory agency and void the certification. The Federal Communications Commission (FCC) has modified what it allows several times, to the point of much confusion for installers trying to remain compliant with the regulations. Additionally, altering the original design in any way may require that the entire system be recertified based on the laws in each RF agency and that agency's interpretation of the term *licensed system*.

Exercise 3.1 is a simple way to demonstrate the concept of passive gain.

Radio Frequency Antenna Concepts

EXERCISE 3.1

Demonstrating Passive Gain

You can demonstrate passive gain by using a standard 8.5" × 11.0" piece of notebook paper or cardstock.

1. Roll a piece of paper into a cone or funnel shape.
2. Speak at your normal volume and notice the sound of your voice as it propagates through the air.
3. Hold the cone-shaped paper in front of your mouth.
4. Speak at the same volume.
5. Notice that the sound of your voice is louder. This occurs because the sound is now focused into a specific area or radiation pattern and passive gain occurs.

Active Gain

Active gain will also provide an increase in signal strength. In an RF circuit, *active gain* is accomplished by providing an external power source to an installed device. An example of such a device is an amplifier. An amplifier is placed in series in the wireless system and increases the signal strength based on how much gain it provides.

To elaborate, if an amplifier is used in a wireless system, certain regulatory domains require that the amplifier be certified as part of the system. It is best to carefully consider whether an amplifier is necessary before using such a device in an IEEE 802.11 wireless LAN system. Using an amplifier may nullify the system's certification and potentially cause the system to exceed the allowed RF limit for the frequency band in use.

Antenna Polarization

Antenna *polarization* describes how a radio wave is emitted from an antenna and the orientation of the electrical component or electric field of the waveform. The electromagnetic field that propagates away from an antenna consists of two fields:

- E-plane, which is the electrical field
- H-plane, which is the magnetic field

These two fields leave the antenna element at a 90° angle and represent the polarization of an antenna. Together they create the energy that allows for wireless devices to communicate.

To maximize radio frequency signals, the transmitting and receiving antennas should be polarized in the same direction or as closely as possible. Antennas polarized in the same way ensure the best possible RF signal. If the polarization of the transmitter and receiver are different, the power of the RF signal will decrease depending how different the polarization is.

Figure 3.23 shows an example of horizontally and vertically polarized antennas.

FIGURE 3.23 Horizontally and vertically polarized antennas

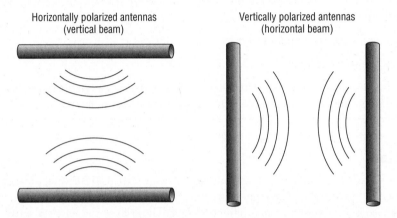

With the large number of wireless devices available, it is a challenging task to accomplish the same polarization for all devices within a wireless system. When you're working with wireless computer networks, performing a wireless LAN site survey will show signal strength based on several factors, including polarization of access point antennas. This survey will help determine the received signal strength of the wireless LAN devices.

Exercise 3.2 shows how you can observe antenna polarization activity.

EXERCISE 3.2

Antenna Polarization Example

It is fairly simple to demonstrate antenna polarization with a notebook computer or other wireless LAN device and either a wireless network adapter client utility or other third-party software that shows signal strength and/or signal-to-noise ratio. One such utility is InSSIDer, a network scanner for various operating systems from MetaGeek. You can download the InSSIDer program from www.metageek.net. You can visualize polarization by performing the following steps. This experiment should be performed using a notebook computer within close proximity to an access point.

1. Verify that you have a supported wireless network adapter.
2. Install and launch the InSSIDer program or other utility that shows signal strength.

3. Monitor the received signal strength indicator (RSSI) value.
4. While monitoring the RSSI value, change the orientation of the notebook computer.
5. Notice the change in the RSSI value (either an increase or decrease) when the orientation of the computer changes with respect to the access point.

This demonstrates how polarity can affect the received signal of a device.

Antenna Types

The type of antenna that is best for a particular installation or application will depend on the desired radio frequency coverage pattern. Making the correct choice is part of a good wireless design. Using the wrong type of antenna can cause undesirable results, such as interference to neighboring systems, poor signal strength, or incorrect coverage pattern for your design.

There are three common types of antennas for use with wireless systems:

- Omnidirectional/dipole antennas
- Semidirectional antennas
- Highly directional antennas

The following sections describe each type of antenna in more detail and provide specifications and installation or configuration information about them.

Omnidirectional Antennas

Omnidirectional antennas are used in a variety of wireless system applications and are the most common type used with wireless LAN access points of either SOHO or enterprise grade.

An *omnidirectional antenna* has a horizontal beamwidth (azimuth) of 360°. This means that when the antenna is vertically polarized (perpendicular to the earth's surface), the horizontal radiation pattern is 360° and will propagate RF energy in every direction horizontally.

The vertical beamwidth (elevation) will vary depending on the antenna's gain. As the gain of the antenna increases, the horizontal radiation pattern will increase, providing more horizontal coverage. Keep in mind that the beamwidth is still 360°, but it will be a larger 360° area that is covered because of the higher gain of the antenna. However, the vertical radiation pattern will decrease, thus providing less vertical coverage.

The shape of the radiation pattern from an omnidirectional antenna is known as a torus. Figure 3.24 shows an example of the toroidal radiation pattern of an omnidirectional antenna.

FIGURE 3.24 The omnidirectional radiation pattern has a toroidal shape.

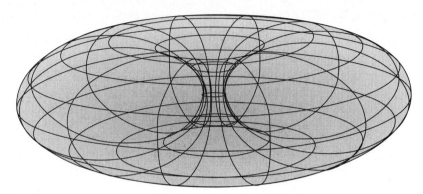

Omnidirectional antennas are one of the most common types of antenna for indoor wireless deployments of any type. Most wireless LAN access points use omnidirectional antennas to provide the needed RF coverage. Indoor cellular devices may also use omnidirectional antennas for coverage. Wireless LAN access points come with fixed, removable, or integrated antennas. If the antenna is removable, the installer can replace it with one of different gain. Enterprise-grade access points typically have removable antennas that are sold separately.

Some regulatory domains require the use of proprietary connectors with respect to antennas. These connectors limit wireless access points to the specific antennas tested with the system. Therefore, it is best to consult with the manufacturer of the access point or other wireless LAN transmitting device to determine which antennas may be used with the system.

The most common type of omnidirectional antenna used indoors with wireless LAN systems is known as the rubber duck antenna. This type of antenna typically has a low gain of 2 dBi to 3 dBi and connects directly to an access point. Rubber duck antennas usually have a pivot point so the polarization can be adjusted vertically or horizontally regardless of how the access point is mounted.

Some antennas for wireless LANs will operate in both the 2.4 GHz ISM band and the 5 GHz U-NII band and can thus work with a multiband wireless device.

Figure 3.25 shows a rubber duck omnidirectional antenna.

Omnidirectional Antenna Specifications

In addition to the beamwidth and gain, omnidirectional antennas have various other specifications to be considered:

- Frequency range
- Voltage standing wave ratio (VSWR)
- Polarization
- Attached cable length

- Dimensions
- Mounting requirements

FIGURE 3.25 2.4–2.5 GHz and 5.1–5.8 GHz dual-band rubber duck antenna

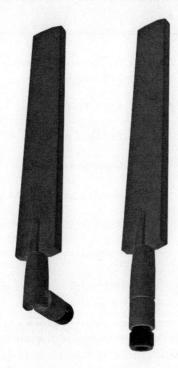

Image provided by www.L-com.com.

Table 3.3 is an example of a specification sheet for a rubber duck omnidirectional antenna.

TABLE 3.3 Omnidirectional 2.4–2.5 GHz and 5.1–5.8 GHz dual-band rubber duck antenna

Electrical Specifications	Ranges
Frequency ranges and gain	2400-2483 MHz @ 3 dBi
	5150-5900 MHz @ 5 dBi
Horizontal beamwidth	360°
Impedance	50 ohm

TABLE 3.3 Omnidirectional 2.4–2.5 GHz and 5.1–5.8 GHz dual-band rubber duck antenna *(continued)*

Electrical Specifications	Ranges
VSWR	<2:1
Mechanical Specifications	
Weight	0.18 lbs. (.07 kg)
Length	10.1" (256 mm)
Base diameter	1.6" (40.6 mm)
Finish	Matte black
Connector	Reverse polarity SMA plug
Operating temperature	–30°C to 60°C (–22°F to 140°F)
Polarization	Vertical
RoHS compliant	Yes

A physical representation of the antenna is also helpful for sales and integration professionals. Figure 3.26 shows the physical specifications diagram for a rubber duck omnidirectional antenna. Keep in mind that this is an example from only one specific antenna manufacturer and the physical design will vary.

FIGURE 3.26 Omnidirectional 2.4–2.5 GHz and 5.1–5.8 GHz dual-band rubber duck antenna physical specifications

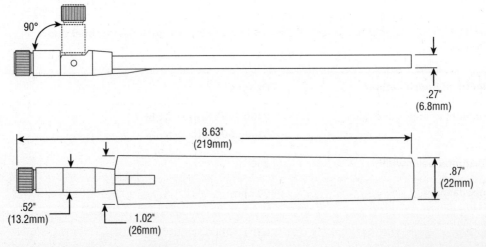

Image provided by www.L-com.com.

Azimuth and elevation charts are usually available to allow visualization of the radio frequency radiation pattern emitted from the antenna. This can help a wireless network professional determine the approximate RF propagation pattern. The purpose of these charts, and how to read them, were explained in the sidebar "Reading Azimuth and Elevation Charts" earlier in this chapter. Figure 3.27 shows the charts for a rubber duck omnidirectional antenna.

FIGURE 3.27 Vertical (elevation) and horizontal (azimuth) charts for omnidirectional antenna

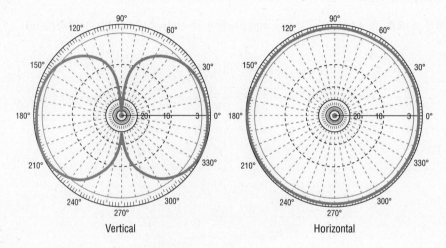

Image provided by www.L-com.com.

Semidirectional Antennas

Semidirectional antennas take radio frequency power from the transmitting system and focus the energy into a more specific pattern than an omnidirectional antenna offers. Semidirectional antennas are available in various types, including patch, panel, sector, and yagi. These antennas are manufactured for either indoor or outdoor use and are designed to provide more specific coverage by focusing the horizontal radiation pattern to a value of less than 360°. A semidirectional antenna will allow the wireless system designer to provide RF coverage to a specific area within a deployment. The coverage area may consist of rooms or areas in which an omnidirectional antenna may not be the perfect solution. For indoor installations, such areas include rectangular rooms or offices, hallways, and long corridors. For outdoor deployments, they include point-to-point and point-to-multipoint installations.

Patch/Panel Antennas

In the wireless LAN world, the terms *patch* and *panel* are commonly used to describe the same type of antenna. The intended use will affect the choice of patch/panel antenna to be used in a specific application. Choosing the correct patch/panel antenna will require knowing the dimensions of the physical area to be covered as well as the amount of gain required. A *patch/panel antenna* can have a horizontal beamwidth of as high as 180°, but usually the horizontal beamwidth is between 35° and 60°. The vertical beamwidth usually ranges between 30° and 80°. Figure 3.28 shows a 2.4 GHz 11 dBi dual polarization diversity/MIMO/802.11n flat panel antenna. At least two antennas are used for MIMO diversity. The difference between this and simple diversity is that with MIMO diversity, each antenna is connected to a separate radio chain. Simple diversity uses a single radio with multiple antennas and a diversity switch technology. Sector antennas are a type of semidirectional antenna that can be configured in an array to provide omnidirectional coverage. Sector antennas are covered in more detail later in this chapter.

FIGURE 3.28 2.4 GHz 11 dBi dual polarization diversity/MIMO/802.11n flat panel antenna

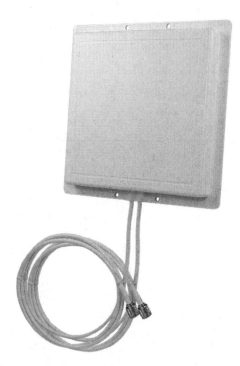

Image provided by www.L-com.com.

 Real World Scenario

Appropriate Use of a Semidirectional Antenna

A small business consultant is tasked with providing wireless LAN access to several offices in a multitenant building. The client wants to provide adequate coverage for the offices it leases but would like to minimize the number of access points. The client wishes to use access points and antennas that are aesthetically pleasing because these offices allow public access. The areas to be covered are rectangular, as shown here.

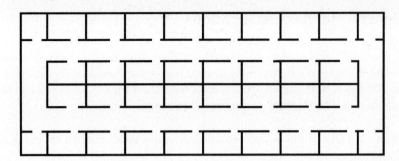

One solution would be to provide several access points using low-gain omnidirectional antennas. The following image illustrates how several access points could be used to provide coverage to this area.

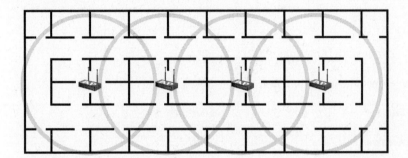

However, the consultant believes that if low-gain rubber duck omnidirectional antennas are used, an access point with significant output power would be required to cover the length of the rooms. In addition, the client wants to minimize the number of access points and make the installation aesthetically pleasing.

An alternate solution is to use a patch antenna on both sides of the office, thereby providing adequate coverage and minimizing the use of access points. The following image shows patch antennas mounted at both ends of the office area as well as the projected coverage area of both antennas.

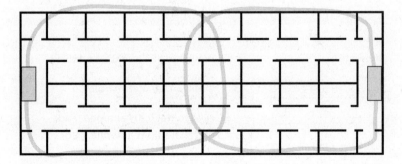

Patch/Panel Antenna Specifications

The specifications for semidirectional antennas such as patch or panel vary based on the design of the antenna. Semidirectional antennas are available in single- or multi-band capability. Semidirectional antennas may be used indoors or outdoors depending on the application. Table 3.4 is an example of a specification sheet for a 2.4 GHz 11 dBi dual-polarization diversity/MIMO/802.11n flat panel antenna.

TABLE 3.4 2.4 GHz 11 dBi dual-polarization diversity/MIMO/802.11n flat panel antenna specifications

Electrical Specifications	
Frequency ranges	2400–2500 MHz
Gain	11 dBi
Horizontal beamwidth (Antennas 1 & 2)	60°
Vertical beamwidth (Antennas 1 & 2)	30°
Impedance	50 ohm
Maximum power	25 W
VSWR	<1.5:1 avg
Mechanical Specifications	
Weight	0.95 lb. (0.43 Kg)
Dimensions	8.5" x 8.5" x 1" (216 x 216 x 26 mm)
Radome material	UV-inhibited polymer
Connector	RP SMA
Operating temperature	−40°C to 85°C (−40°F to 185°F)
Mounting	Four 1/4" (6.3 mm) holes
Polarization	Horizontal (Left antenna lead)
	Vertical (Right antenna lead)

Electrical Specifications

Flame rating	UL 94HB
RoHS-compliant	Yes
Wind survival	>150 mph (241 kph)

A radome cover will protect an antenna from outdoor elements and certain weather conditions. Attenuation from the materials that the radome covers are constructed of will be minimal. They mainly protect the antenna from the collection of elements such as snow and hail.

Azimuth and elevation charts are also available for patch/panel antennas. Figure 3.29 shows the charts for the 2.4 GHz 11 dBi dual polarization diversity/MIMO/802.11n flat panel antenna.

FIGURE 3.29 Vertical (elevation) and horizontal (azimuth) charts for the 2.4 GHz 11 dBi dual polarization diversity/MIMO/802.11n flat panel antenna

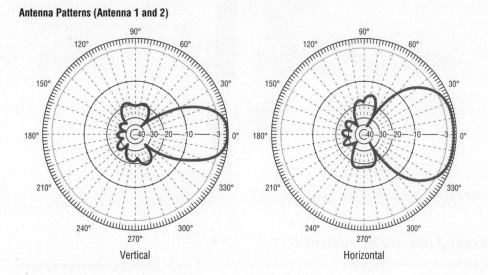

Image provided by www.L-com.com.

Sector Antennas

Sector antennas can be used to create omnidirectional radiation patterns using semidirectional antennas. These antennas are often used for base station connectivity for point-to-multipoint connectivity. *Sector antennas* usually have an azimuth that varies from 90°

to 180°. They are typically configured to offer a total azimuth of 360° when installed in groups. For example, using sector antennas with an azimuth of 120° each would require three antennas in order to get omnidirectional or 360° coverage. This is a common configuration used with cellular phone technology. Figure 3.30 shows a dual-feed, dual-band sector panel antenna.

FIGURE 3.30   2.4 GHz and 4.9 to 5.8 GHz dual-feed dual-band 90° sector panel antenna

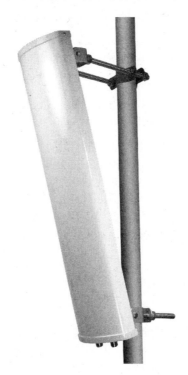

Image provided by www.L-com.com.

Sector Antenna Specifications

Sector antennas are commonly configured in an array to allow semidirectional antennas to provide omnidirectional coverage. This is useful in a campus environment or community arrangement to provide wireless access to a local network or the Internet. Sector antennas will usually have wide horizontal beamwidth (azimuth) and a narrow vertical beamwidth (elevation). Table 3.5 shows an example of a specification sheet for a 2.4 GHz and 4.9 to 5.8 GHz dual-feed dual band 90° sector panel antenna.

TABLE 3.5 2.4 GHz and 4.9 to 5.8 GHz dual-feed dual-band 90° sector panel antenna specifications

Electrical Specifications	
Frequency ranges	2400–2500 MHz
	4900–5900 MHz
Gain	14 dBi
Horizontal beamwidth	90°
Vertical beamwidth	16° (2400–2500 MHz)
	8° (4900–5900 MHz)
Impedance	50 ohm
Maximum input power	50 W
VSWR	<1.5:1 avg
Front-to-back ratio	>21 dB
Lightning protection	DC ground
Mechanical Specifications	
Weight	4.4 lbs. (2 kg)
Dimensions	24 × 6.3" × 2.3" (620 × 160 × 60 mm)
Radome material	UV-inhibited fiberglass
Connector	(2) Integral N-female
Operating temperature	−40°C to 60°C (−40°F to 140°F)
Mounting	1.5–2 in (40–53 mm) dia. mast max.
Polarization	Vertical
RoHS compliant	Yes
Rated wind	>130 mph (210 Km/h)

TABLE 3.5 2.4 GHz and 4.9 to 5.8 GHz dual-feed dual-band 90° sector panel antenna specifications *(continued)*

Electrical Specifications

Wind Loading Data

Wind speed (mph)	Loading
100	34 lb.
125	54 lb.

Figure 3.31 shows the charts for the 2.4 GHz and 4.9 to 5.8 GHz dual-feed dual band 90° sector antenna.

FIGURE 3.31 Vertical (elevation) and horizontal (azimuth) charts for 2.4 GHz and 4.9 to 5.8 GHz dual-feed dual-band 90° sector panel antenna

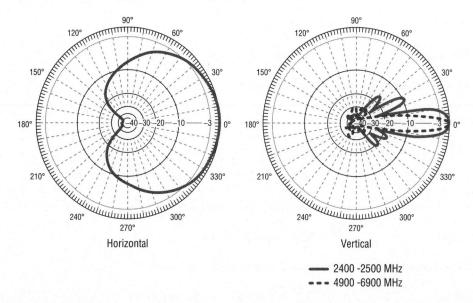

Image provided by www.L-com.com.

Yagi Antennas

Yagi antennas are designed to be used indoors in long hallways and corridors or outdoors for short-range bridging (typically less than two miles). *Yagi antennas* have vertical and horizontal beamwidths ranging from 25° to 65°. The radiation pattern may look like a funnel or a cone. As the signal propagates away from the antenna, the RF coverage naturally widens (diffusion). The aperture of the receiving antenna is much narrower than the signal at that point. This is a result of diffusion, which is the biggest form of loss in an RF link. Figure 3.32 shows a yagi antenna.

FIGURE 3.32 2.4 GHz 15 dBi yagi antenna

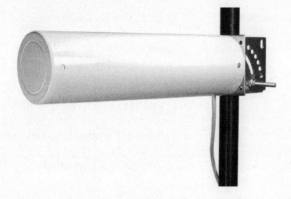

Image provided by www.L-com.com.

Yagi Antenna Specifications

Table 3.6 is an example of a specification sheet for a 2.4 GHz 15 dBi yagi WLAN antenna.

TABLE 3.6 2.4 GHz 15 dBi yagi antenna specifications

Electrical Specifications	
Frequency ranges	2400–2500 MHz
Gain	14.5 dBi
–3 dB beamwidth	30°
Impedance	50 ohm
Maximum power	50 W
VSWR	<1.5:1 avg
Lightning protection	DC short

Mechanical Specifications

TABLE 3.6 2.4 GHz 15 dBi yagi antenna specifications *(continued)*

Mechanical Specifications	
Weight	1.8 lbs. (.81 kg)
Dimensions – Length × diameter	18.2" × 3" (462 × 76 mm)
Radome material	UV-inhibited polymer
Connector	12" N-female
Operating temperature	−40°C to 85°C (−40°F to 185°F)
Mounting	1-1/4" (32 mm) to 2" (51 mm) diameter masts
Polarization	Vertical and horizontal
Flame rating	UL 94HB
RoHS compliant	Yes
Wind survival	>150 mph (241 kph)

Wind Speed (mph)	Loading
100	12 lb.
125	19 lb.

Figure 3.33 shows the charts for the 2.4 GHz 15 dBi yagi antenna.

A yagi antenna may be encased within a weatherproof enclosure. This is not required, but it may be useful in outdoor installations. The weatherproof enclosure will prevent collection of certain elements such as snow and ice. Radome covers are available for parabolic dish antennas for the same purpose.

FIGURE 3.33 Vertical (elevation) and horizontal (azimuth) charts for 2.4 GHz 15 dBi yagi antenna

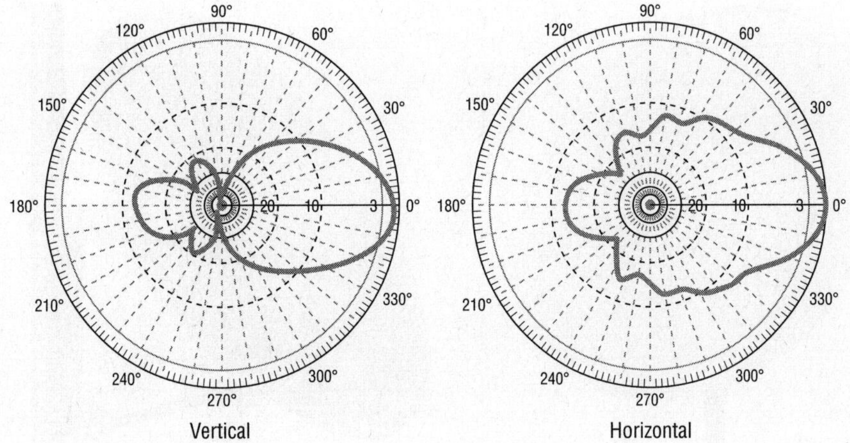

Image provided by www.L-com.com.

Highly Directional Antennas

Highly directional antennas are typically *parabolic dish antennas* used for long-range point-to-point bridge connections. These antennas are available with a solid reflector or a grid. Some manufacturers of parabolic dish antennas used with wireless networking advertise ranges of 25 miles or more depending on the gain, RF power of the transmitter, and environmental conditions. Parabolic dish antennas have very narrow horizontal and vertical beamwidths. This beamwidth can range from 3° to 15° and has a radiation pattern similar to that of a yagi, with the appearance of a funnel. The beamwidth starts very narrow at the antenna element and naturally widens because of diffusion.

Because these antennas are designed for outdoor use, they will need to be manufactured to withstand certain environmental conditions, including a wind rating and appropriate mounting. Grid antennas can provide similar coverage and are less susceptible to wind loading. Figure 3.34 shows a parabolic dish antenna.

FIGURE 3.34 Front and back views of a 5.8 GHz 29 dBi ISM / U-NII band solid parabolic dish antenna

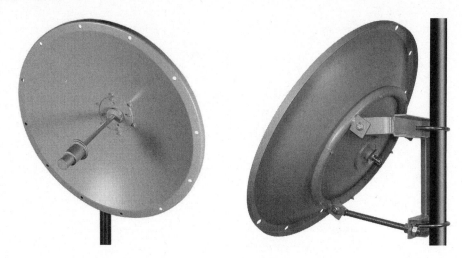

Image provided by www.L-com.com.

Highly Directional Antenna Specifications

Table 3.7 is an example of a specification sheet for a 5.8 GHz 29 dBi ISM/U-NII band solid parabolic dish antenna. Notice that the vertical and horizontal beamwidths of this antenna are 6°, very narrow compared to other antenna types discussed.

TABLE 3.7 5.8 GHz 29 dBi ISM/U-NII band solid parabolic dish antenna specifications

Electrical Specifications	
Frequency ranges	5725–5850 MHz
Gain	29 dBi
Horizontal beamwidth	6°
Vertical beamwidth	6°
Impedance	50 ohm
Maximum power	100 W
VSWR	<1.5:1 avg

Mechanical Specifications

Weight	13.2 lbs. (6 kg)
Dimensions	23.6" diameter (600 mm)
Grid material	Galvanized steel
Operating temperature	−40° C to 85° C (−40° F to 185° F)
Mounting	1.5" (38 mm) to 3" (76 mm) dia. masts
Lightning protection	DC Short

Wind Speed (mph)	Loading	With Radome
100	113 lb.	75 lb
125	177 lb.	116 lb

Figure 3.35 shows the charts for the 5.8 GHz 29 dBi ISM/U-NII band solid parabolic dish antenna.

FIGURE 3.35 Vertical (elevation) and horizontal (azimuth) charts for the 5.8 GHz 29 dBi ISM/U-NII band solid parabolic dish antenna

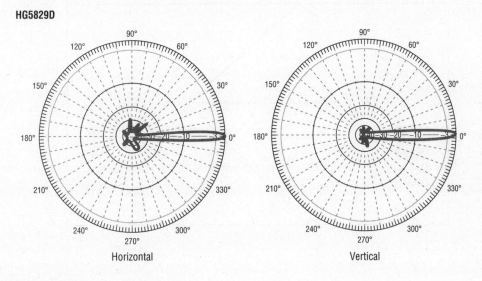

Image provided by www.L-com.com.

One thing to consider regarding the sale and procurement of a highly directional parabolic dish antenna is the size and shipping weight. Since these antennas are much larger and heavier than other antennas used in wireless LANs, shipping cost may be a factor. Some specification sheets will detail shipping information for this reason.

Minimizing the Effects of Multipath Using Antenna Diversity

Reflection is caused by an RF signal bouncing off a smooth, nonabsorbing surface and changing direction. Indoor environments are areas that are prone to reflections. Reflections are caused by the RF signal bouncing off walls, ceilings, floors, and furniture; thus some installations will suffer from reflection more than others.

The effect of reflection will be a decrease in signal strength due to a phenomenon called *multipath*. Multipath is the result of several wavefronts of the same transmission signal received out of phase at slightly different times. This can cause the receiver to be confused about the received signals. The result is corrupted signal and less overall throughput. Figure 3.36 illustrates the effects of multipath.

FIGURE 3.36 Effects of multipath

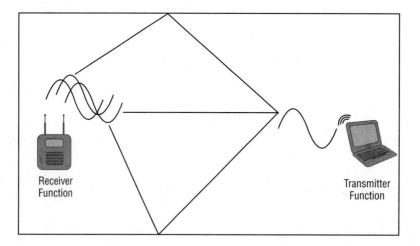

Think of multipath as an echo. If you were to stand near a canyon and speak to somebody at a high volume some distance away, the other person would notice an echo. The echo is due to the fact that the sound of your voice is reflecting off the canyon walls. Therefore, the other person is hearing variations of your voice at slightly different times—as with RF multipath, several wavefronts of the same signal are arriving out of phase.

Antenna diversity is one way to help reduce the effects of multipath. *Antenna diversity* is a technology used in IEEE 802.11 a/b/g wireless LANs; a station (access point or client device) will utilize two antennas combined with one radio to decrease the effects of multipath. Using multiple antennas and some additional electronic intelligence, the receiver will be able to determine which antenna will receive and send the best signal. In diversity systems, two antennas are spaced at least one wavelength apart. This allows the receiver to use the antenna with the best signal to transmit and receive. With respect to radio frequency diversity, the antennas are required to be of the same design, frequency, gain, and so on. Newer IEEE 802.11 technologies (802.11n and 802.11ac, for example) are capable of high throughput (HT) multiple-input, multiple-output (MIMO), which uses radio chains to transmit and receive radio signals. IEEE 802.11a/b/g devices used a single radio to transmit and receive radio signals, which is known as single-input, single-output (SISO) technology. SISO systems are subject to multipath. This is a problem for IEEE 802.11a/b/g systems, whereas MIMO actually uses the reflections to help enhance the performance and throughput using several radio chains in newer wireless LAN technology. You can read more about MIMO in Chapter 4, "Standards and Certifications for Wireless Technology," in the section "The IEEE 802.11n Amendment."

When you are using a diversity system such as an access point, it is important to have both antennas oriented the same way. They cannot be used to cover different areas. Using diversity antennas in an attempt to provide coverage for different areas will defeat the purpose of the diversity design.

Distributed Antenna System (DAS)

Because of the close proximity of wireless devices indoors, a visual line of sight is usually not required for indoor wireless communications. Although attenuated, RF is capable of propagating through walls, doors, windows, and floors within a structure.

However, in some cases RF signals such as cellular phone communications that are received from outside a building are already weak or barely above the desired receive signal threshold. Bringing that signal in from the outside and allowing it to exist at an acceptable level in a building can be accomplished using technology known as a distributed antenna system (DAS). This antenna technology may use several antennas to create small cells to provide wireless coverage throughout an entire area or building. Figure 3.37 shows an example of a DAS installation.

FIGURE 3.37 Example of a distributed antenna system (DAS)

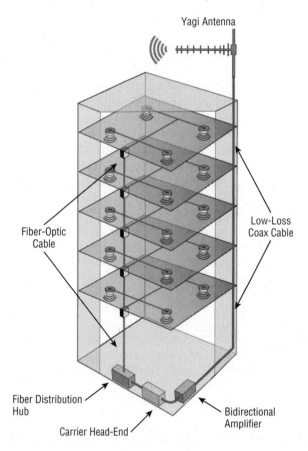

Image provided by www.L-com.com.

As you can see in the figure, a single yagi antenna captures the RF signal from the outside of the building. This is known as the donor signal. The signal is then distributed throughout the building using several antennas by way of a bidirectional amplifier, service provider/carrier head-end, and fiber distribution hub. This is becoming a common application for DAS in structures that have weak cellular signals within a building, especially in difficult areas like basements and parking garages where low or no signal is received. There are two different types of DAS solutions:

Passive DAS   Just as antennas do not require any external power source to provide passive gain, a passive DAS deployment does not require an external power source to amplify the RF signal. The system consists only of coax cabling, RF connectors, RF splitters, and DAS-capable antennas. The benefit in this type of deployment is that it has less cost and is not as complex. However, passive solutions may not be as scalable and are intended for smaller installations. Passive DAS works well when the carrier signal is strong and needs assistance with propagation throughout the entire building.

Active DAS   This type of DAS deployment requires external power sources to operate. In addition to the coax cables, RF connectors and DAS-capable antenna amplification are used to help improve the RF signal throughout the entire DAS installation. This type of DAS is scalable and intended to be used in larger installations. Active DAS works well in buildings that have a weak initial signal. Fiber optics with hubs are components of an active DAS. This allows for much longer distances and the ability to cover larger areas such as airports, sports venues, conference halls, tunnels, and university campuses.

The DAS installation does not need to be tied to one specific wireless carrier. Different carriers can share a DAS. The carriers will need to provide their head-end equipment and be able to bring in the signal from outside the building to provide the indoor service.

Leaky Coax DAS

Another type of DAS you may come across is something known as leaky coax, or leaky cable. Basically, there are cuts throughout the length of a special slotted coax cable. For the most part, this type of DAS has the same effect as an antenna with the RF propagating from the slots, allowing an RF signal to be sent and received. That is where the name *leaky coax* comes from. Since the signal is not amplified, the coverage area is small and usually less than 25 feet from the cable. This is one solution in hard-to-reach areas like basements and tunnel-like long corridors, and it's used with two-way radio (walkie-talkie) type communications and some cellular applications.

DAS and IEEE 802.11 Wireless Networking

DAS is often used with wireless communication applications such as cellular phones, two-way radios, paging, and medical applications.

There are mixed views about using DAS with IEEE 802.11 wireless LAN technology. Some equipment manufacturers do not certify, endorse or support any DAS vendors/solutions, while others may offer specialty solutions that operate with DAS.

One manufacturer of IEEE 802.11 wireless LAN equipment, Ruckus Wireless (www.ruckuswireless.com), is marketing an access point for Wi-Fi on coax distributed antenna systems (DASs). This product is the ZoneFlex 7441. The 7441 is a bit of a unique product for Ruckus Wireless. According to its website, with this unit it is now possible to deploy Wi-Fi over an existing DAS network, enabling operators of DASs to create and market Wi-Fi services for property management, building operations, guest access, and even wholesale Wi-Fi hotspot services across a building.

Incorporating DAS with wireless computer networks (IEEE 802.11) has its share of challenges and may not be a good solution for many scenarios. Because DAS is basically a shared antenna system, it can create problems with wireless computer networks that include IEEE 802.11 devices that use newer MIMO technology. DAS itself requires a system design, and you will need to have a special skill set or knowledge base in order execute a successful deployment. Before you incorporate IEEE 802.11 wireless networking within a DAS, I highly recommend performing the research to verify that it is the best solution based on the circumstances and will satisfy the requirements of the installation.

DAS Antennas

If you plan to use a DAS, you will need to use antennas designed to work with DAS technology. These antennas must be capable of operating at the correct frequency ranges for the wireless carriers used. Also, any specific physical attributes required for proper installation need to be considered.

Figure 3.38 shows an example of a multiband 698 - 960/1710 - 2700 MHz ¾ dBi ceiling mount DAS antenna with a standard N type female connector.

FIGURE 3.38 Multiband DAS antenna, front and back views

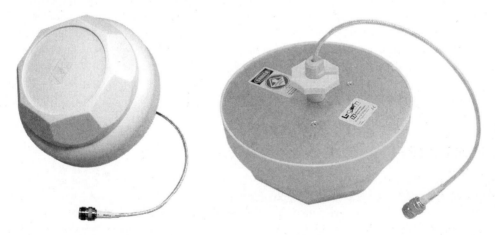

Image provided by www.L-com.com.

Notice that this is a broadband/multiband DAS antenna that is designed to operate at different frequencies that will allow it to be used for different wireless technologies, eliminating the need to purchase different antennas for each frequency. The standard N-type connector is compatible with a variety of implementations.

DAS Antenna Specifications

Table 3.8 is an example of a specification sheet for multiband DAS antenna.

TABLE 3.8 Multiband 698-960/1710-2700 MHz 3/4 dBi ceiling mount DAS antenna

Electrical Specifications	Frequency 1	Frequency 2
Frequency range	698–960 MHz	1710–2700 MHz
Gain	3 dBi	4 dBi
Polarization	Vertical	
Horizontal beamwidth	360°	
Vertical beamwidth	75°	40°
Impedance	50 ohm	
Maximum power	50W	
VSWR	< 1.6	
Lightning protection	DC ground	
Mechanical Specifications		
Cable length	12 in. (300 mm)	
Weight	0.77 lbs. (.35 Kg)	
Dimensions	7.2 Dia. x 3.4 in. (184 Dia. x 85 mm)	
Radome material	UV Resistant ABS	
Radome color	White	
Operating temperature	-40°C to 60°C (-40° F to 140° F)	
Mounting	0.687" (17.4 mm) diameter hole	
RoHS compliant	Yes	

Figure 3.39 shows the azimuth and elevation charts for the various frequency responses of the 698 to 960 MHz and 1710 to 2700 MHz ¾ dBi ceiling mount DAS antenna.

FIGURE 3.39 698 to 960 MHz and 1710 to 2700 MHz 3/4 dBi ceiling mount DAS antenna azimuth and elevation charts

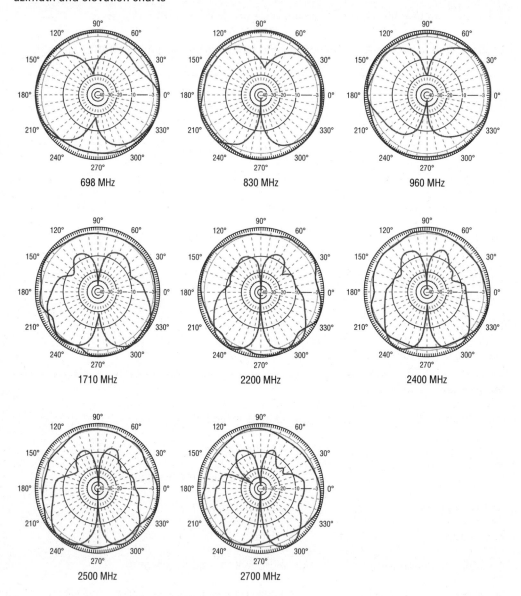

Image provided by www.L-com.com.

Radio Frequency Cables and Connectors

Radio frequency cables play a role in various wireless systems. For example, cables may be used to connect wireless access points and client devices to antennas or to connect other devices that may used in wireless systems. Several factors need to be taken into consideration when using cables in a wireless system:

- Type of cable
- Length of cable
- Cost of cable
- Impedance rating

Choosing the correct cable for use in wireless LAN systems is an important part of a successful wireless LAN deployment. The right cable for the right job will help ensure that *signal loss*—a decrease in signal strength—is minimized and performance is maximized.

Impedance and Voltage Standing Wave Ratio

Impedance is the measurement of alternating current (AC) resistance. It is normal to have some level of impedance mismatch in a wireless system, but the impedance of all components should be matched as closely as possible in order to optimize performance of the system. Impedance mismatches can result in what is called *voltage standing wave ratio (VSWR)*. A large impedance mismatch can cause a high level of VSWR and will have an impact on the wireless system and transmitted or received signal.

Electrical resistance is measured in ohms. IEEE 802.11 wireless LAN devices and cellular systems have an impedance of 50 ohms.

RF Cable Types

Many wireless systems use cables to extend from the wireless device such as an access point or a cell signal booster to an antenna located outside of a building. It is important to choose the correct type of cable in order to optimize the performance of the wireless system.

Cables vary in diameter, and the application will determine the type of cable to use. For example, connecting a wireless LAN adapter on a notebook computer to an external antenna requires a specific type of cable that should be short and flexible. Thick, rigid cables are best used for longer runs.

The radio frequency range in which the cable will be used also is important to consider. Where the cable is used will determine its radio frequency rating. For example, wireless LANs and cellular systems use 50 ohm cable, whereas television (such as satellite and cable) will use 75 ohm cable.

Using cable with the correct rating will minimize VSWR.

Figure 3.40 shows a spool of high-quality 50 ohm cable.

FIGURE 3.40 L-com spool of low-loss 400-series coaxial bulk cable

Image provided by www.L-com.com.

RF Cable Length

The length of an RF cable used in a wireless system is another factor to consider. A cable of even a very short length will have some level of attenuation, causing signal loss, which is a decrease in signal strength. This decrease in signal strength means less overall performance and throughput for users of the wireless LAN.

Professionally manufactured cables typically are available in many standard lengths. Best practices recommend using the correct length and minimizing connections. For example, if a run from an access point to an external antenna is 27 feet, it would be best to use a single cable as close to that length as possible. Connecting two or more pieces of cable together will increase the loss to the system. One might be tempted to use a longer piece, such as 50 feet, but this is not recommended because the extra length will add loss to the system.

Figure 3.41 shows a short length of cable known as a pigtail used to connect a standard cable to a proprietary cable. If an RF cable is used or extended, the attenuation that is introduced can be offset with the use of an amplifier or with a higher-gain antenna. An amplifier will provide active gain and an antenna will provide passive gain. Keep in mind that using an incorrect amplifier may void the wireless system certification and that using a higher-gain antenna may exceed the rules set by the local RF regulatory agency.

RF Cable Cost

RF cable cost may also play a role in the type of cable to be used. The old saying "You get what you pay for" is true with cables as well. I recommend using high-quality, name-brand RF cables to optimize the performance of your system. Premium cables may come at a higher price, but the benefit of a better-quality signal is the main advantage. Keep in mind that RF coax cables are rated differently based on the quality. I recommend reviewing the specifications of cables to verify that they will meet your requirements prior to making any purchase.

FIGURE 3.41   Short pigtail adapter cable

Image provided by www.L-com.com.

RF Cable Connectors

In a wireless system, radio frequency connectors are used to join devices, allowing the RF signal to transfer between the devices. The devices may connect from access point to antenna, antenna to cable, cable to cable, or in various other combinations. Standard RF connectors may be used in wireless systems to connect devices that are not part of the point connecting to the antenna. For example, an access point connecting to a length of cable that is then connected to an amplifier could use a standard RF connector. The cable connecting the amplifier to the antenna would require a proprietary connector. Figure 3.42 shows examples of common RF connectors.

FIGURE 3.42   Several common RF connectors used with wireless LANs

Image provided by www.L-com.com.

RF connectors also cause an impedance mismatch to some degree and increase the level of VSWR. To minimize the effects of VSWR, best practices suggest keeping the use of connectors to a minimum. Using connectors can also result in *insertion loss*. Insertion loss is usually minor by itself, but it can contribute to overall loss in a system, thereby resulting in less RF signal and less throughput.

> **Using Proprietary Connectors for Regulatory Domain Compliance**
>
> Some regulatory domains require the use of *proprietary connectors* on antennas and antenna connections in wireless LAN systems. Proprietary connectors prevent an installer or integrator from unintentionally using an antenna that might exceed the maximum amount of power allowed for the transmission system. Although these connectors are considered proprietary, many manufacturers share proprietary connectors:
>
> - MC connectors are used by Dell, Buffalo, IBM, Toshiba, and Proxim-ORiNOCO.
> - MMCX connectors are used by 3Com, Cisco, Proxim, Samsung, and Motorola.
> - MCX connectors are used by Apple and SMC devices.
> - RP-MMCX connectors are used by SMC devices.

Antenna Installation Considerations

Several factors are important to consider when you are planning to install a wireless network. These include earth curvature, multipath, and radio frequency line of sight. The following sections include information about how to take these factors into account when planning a wireless system installation.

Maintaining Clear Communications

Factors that affect whether two wireless devices can communicate with each other include line of sight (both visual and RF) and Fresnel zone. Indoor wireless installations use a low amount of radio frequency transmit power, usually around 30 mW to 50 mW, and will be able to communicate effectively even if the wireless client device does not have a line of sight with an access point. This is because the RF will be able to penetrate obstacles such as walls, windows, and doors. Outdoor installations usually use a much higher output transmit power and will require an RF line of sight for effective communication.

Line of Sight in Wireless Communications

Radio frequency communication between devices using wireless technology involves different types of line of sight. There are two types of line of sight to take into account when planning, designing, and installing wireless networks and using wireless technology:

- Visual line of sight
- Direct link radio frequency (RF) line of sight

Visual Line of Sight

Visual *line of sight (LoS)* is defined as the capability of two points to have an unobstructed view of one another. With IEEE 802.11 wireless LAN systems, a visual LoS is implied with the RF LoS. If a wireless LAN engineer was planning to connect two buildings using wireless LAN technology, one of the first things the engineer would do is to verify that there is a clear, unobstructed view between the planned locations in order to provide an RF LoS. In this case a radio transmitter and radio receiver can "see" each other. For wireless networking direct link communication to be successful in an outdoor wireless connection, there should be a clear, unobstructed view between the transmitter and receiver. An unobstructed line of sight means no obstacles are blocking the RF signal between these devices.

Radio Frequency Line of Sight

For two devices to successfully communicate at a distance via radio frequency, including point-to-point or point-to-multipoint connections, a clear path for the RF energy to travel between the two points is necessary. This clear path is called RF LoS. This RF LoS is the premise on which the Fresnel zone is based.

Fresnel Zone

The *Fresnel zone* for an RF signal is the area of radio frequency coverage surrounding the visual LoS. The Fresnel zone consists of a number of concentric ellipsoidal volumes that surround the direct RF line of sight between two points, such as an RF transmitter and receiver. One example is a point-to-point connection between two radio towers.

The width or area of the Fresnel zone will depend on the specific radio frequency used as well as the length or distance of the signal path. There is a mathematical formula used to calculate the width of the Fresnel zone at the widest point.

In an outdoor point-to-point or point-to-multipoint installation, it is important for the Fresnel zone to be clear of obstructions for successful communications to take place between a radio frequency transmitter and receiver. Best practices recommend maintaining an obstruction-free clearance of at least 60 percent for the Fresnel zone in order to have acceptable RF LoS. Maintaining a clear RF LoS becomes more difficult as the distance between two points increases. Obstructions such as these can cause the Fresnel zone to be blocked enough for communications to suffer between a transmitter and receiver:

- Trees
- Buildings or other structures
- Earth curvature
- Natural elements such as hills and mountains

Figure 3.43 illustrates the Fresnel zone between two highly directional antennas.

FIGURE 3.43 Visualization of Fresnel zone

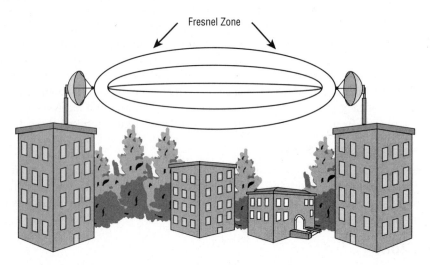

To stay clear of obstructions, carefully plan your antenna placement and antenna height. Keep in mind that a wireless link may cross public areas in which an integrator or installer will have no control over the environment. There is a possibility, depending on the environmental conditions, that a wireless link may not be a feasible solution due to the inability to maintain an RF LoS. You should perform an outdoor site survey prior to the procurement and installation of wireless system hardware to ensure that the installation and operation of the wireless LAN will be successful.

One way to analogize line of sight is by picturing two people looking at each other. If two people about the same height and standing at some distance apart are making direct eye contact, they have a good visual line of sight. In addition to being able to see directly in front of them, people have peripheral vision. Peripheral vision gives people the ability to see movement and objects outside of their direct line of sight or direct eye contact. This peripheral vision or side vision is analogous to the Fresnel zone theory.

Exercise 3.3 shows a simple way to demonstrate the concept of a blocked Fresnel zone.

EXERCISE 3.3

Demonstrating Fresnel Zone and Blockage

Here is one way to demonstrate the Fresnel zone. Focus your eyes at a location on a wall. Make sure there are obstacles or movement off to both left and right sides of your view. Hold your hands down to your sides. Continue to focus your eyes for a minute or so, and then take your right, left, or both hands and slowly raise them from your sides toward the side of your head while blocking your peripheral vision. You'll notice as your hands get closer to the side of your head the view of the objects or movement to the sides will be blocked by your hands. This is an example of a blocked Fresnel zone.

Sixty percent of the total area of the Fresnel zone must be clear of obstacles in order to have a good RF line of sight.

Because of the low transmit and receive power and a short range, a visual line of sight is not required for most indoor wireless deployments. An indoor wireless LAN access point may cover areas that are divided by walls and other obstacles. With this short range, and if the radio frequency is able to penetrate the obstacles, wireless communication between a transmitter and a receiver may be successful even when the devices do not have a visual line of sight.

Earth Curvature Effects on Propagation

Beyond a distance of seven miles, the curvature of the earth will have an impact on point-to-point or point-to-multipoint wireless connections. Therefore, it is important to add height to the antenna to compensate for the *earth curvature*, sometimes referred to as earth bulge. A mathematical formula is used to calculate the additional height of antennas when a link exceeds seven miles (11.2654 kilometers).

Antenna Placement

The installation location and placement of antennas depend on the type of antenna and the application in which it will be used. When installing antennas, consider the placement based on the design of the wireless system and the intended use of the antenna. When antennas are used outdoors, lightning arrestors, grounding, and adherence to local codes, laws, and government regulations must be followed as well as good RF design. Increasingly, local ordinances dictate how or if outdoor antennas can be mounted for looks as well as safety. Lightning arrestors and grounding methods are discussed in the upcoming section "Wind and Lightning."

Omnidirectional Antenna Placement

Placement of an omnidirectional antenna will depend on the intended use. Some omnidirectional antennas can be connected directly to a wireless access point or may be integrated within the access point. In this configuration, the installation is fairly straightforward; it involves simply attaching the antenna to the access point or using the integrated antenna.

Omnidirectional antennas are usually placed in the center of the intended coverage area. High-gain omnidirectional antennas are typically used in outdoor installations for point-to-multipoint configurations. This configuration is more complex because more than likely it requires mast or tower mounting. The exact placement depends on the intended coverage area as well as the gain of the antenna.

Semidirectional Antenna Placement

Semidirectional antennas may be used for either outdoor or indoor installations. When mounted indoors, a patch/panel antenna typically will be mounted flat on a wall with the connector facing upward for connections to a cable or directly to an access point. A template with the hole placement may be included for ease of installation. Usually four mounting holes (one in each corner) will be used to securely fasten the antenna to the wall.

Yagi antennas can also be mounted either indoors or outdoors. The most common installation is outdoors for short-range point-to-point or point-to-multipoint bridging solutions. This will require a mounting bracket such as a tilt and swivel for wall mounting or U-bolts and a plate for mast or pole mounting.

Highly Directional Antenna Placement

Highly directional antennas such as a parabolic dish are almost always used exclusively in outdoor installations. This type of antenna is used mostly for long-range point-to-point bridging links and will require installation on building rooftops or antenna towers. Alignment for long-range links is critical for reliable communications. Software and hardware tools are available for the installer to use for accurate alignment. As with other outdoor installations, secure mounting is essential to maintain safety and link reliability.

Wind and Lightning

Weather conditions such as rain, snow, and sleet typically do not affect wireless LAN communications unless the conditions are extreme or snow and sleet collect on antenna elements. However, wind and lightning *can* affect wireless communications.

Most outdoor antennas that can be affected by wind will have wind-loading data in the specification sheet. *Wind loading* is the result of wind blowing at high speeds and causing the antenna to move.

Lightning can destroy components connected to a network if the antenna takes either a direct or an indirect lightning strike. A properly grounded lightning arrestor will help protect the wireless system and other networking equipment from indirect lightning strikes.

Lightning Arrestors

Potentially damaging transient or induced electrical currents are the result of an indirect lightning strike in the area of a wireless antenna system. *A lightning arrestor* is an in-series device installed after the antenna is installed and prior to the installation of a transmitter/receiver. Although this device will not provide protection from a direct lightning strike, it will help protect against an indirect lightning strike, which can damage electronics at distances away from the source of the strike. When the induced electrical currents from a lightning strike travel to the antenna, a lightning arrestor will shunt the excess current to ground, protecting the system from damage. Figure 3.44 shows a lightning arrestor.

FIGURE 3.44   L-com AL6 series 0-6 GHz coaxial lightning and surge protector

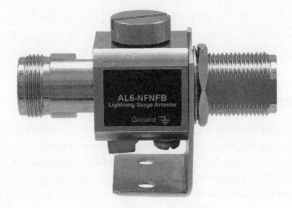

Image provided by www.L-com.com.

Grounding Rods

A *grounding rod* is a metal shaft used for grounding a device such as an antenna used in a wireless system. The rod should be driven into the ground at least 8 feet deep. Grounding rods are available in various types of steel, including stainless, galvanized, and copper clad. They are also available in a variety of diameters and lengths. Depending on the local electrical code, the grounding system should measure resistance between 5 and 25 ohms. Local code should also be consulted regarding material, diameter, and length of grounding rods. You should not share grounding rods with other equipment because interference or damage may occur.

 WARNING   It is imperative to install a grounding rod properly to ensure correct operation. If installing a grounding rod and other lightning protection equipment is beyond the knowledge level of the wireless engineer or installer, it is best to have a professional contractor perform the job.

Installation Safety

Professional contractors should be considered in the event you are not comfortable with performing the installation of a wireless system antenna yourself. Bonded or certified technicians may be required to install antennas. Be sure to check local building codes prior to performing any installation of a wireless system antenna. Never underestimate safety when installing or mounting antennas. All safety precautions must be adhered to while performing an installation. The following are some general guidelines and precautions to be considered for a wireless system antenna installation:

- Read the installation manual from the manufacturer.
- Always avoid power lines. Contact with power lines can result in death.
- Always use the correct safety equipment when working at heights.
- Correctly install and use grounding rods when appropriate.
- Comply with regulations for use of antennas in the area and for use of towers as well.

Antenna Mounting

In addition to choosing the correct antenna to be used with a wireless system, you must take into account the antenna mounting. The required antenna mounting fixture will depend on the antenna type, whether it will be used indoors or outdoors, and whether it will be used for device/client access or bridging solutions such as point-to-point or point-to-multipoint. It is best to consult with the antenna or device manufacturer to determine which mounting fixture is appropriate for use based on the intended deployment scenario. The following are several mounting types that may be used for a wireless system antenna solution:

- Internal and external (to the AP) antennas
- Pole/mast mount
- Ceiling mount
- Wall mount

Pole and Mast Mounting

Pole/mast mounts typically consist of a mounting bracket and U-bolt mounting hardware. The mounting bracket is commonly L-shaped. One side of the bracket has a hole to mount an omnidirectional or similar antenna. The other side of the bracket has predrilled holes for fastening the bracket to a pole using U-bolts. Figure 3.45 shows an example of a stationary antenna mast mounting kit.

FIGURE 3.45 Stationary antenna mast mounting kit with U-bolts

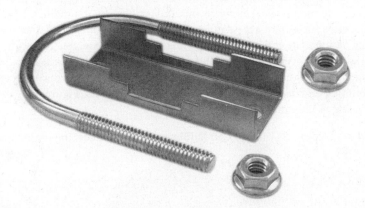

Image provided by www.L-com.com.

Exercise 3.4 describes the basic steps for installing an omnidirectional antenna using a mast mount adapter.

EXERCISE 3.4

Installing a Pole/Mast Mount

This exercise explains the process for installing an omnidirectional antenna to a mast using a mounting bracket designed specifically designed for this purpose.

1. Attach the mounting bracket to the mast using the supplied hardware.
2. Remove the antenna mounting bolt and washer from the base of the antenna.
3. Insert the antenna into the hole in the top of the mounting bracket. Without overtightening, securely fasten the antenna to the mounting bracket using the washer and antenna mounting bolt.
4. For outdoor installations, remember to use the proper sealant for weatherproofing when connecting the cable to the antenna.
5. Lightning arrestors and proper earth grounding must also be considered if you are in an area that could be affected by this type of weather condition.

Ceiling Mounting

It may be necessary to mount to a ceiling certain types of external antennas or wireless infrastructure devices such as an access point with attached or integrated antennas. Many antennas can be mounted directly to a hard ceiling made from concrete, drywall, or similar material. Another possibility is a drop ceiling with acoustic tiles. Regardless of the type of ceiling in question, follow the manufacturer's instructions for the appropriate fixture to be used for mounting and detailed instructions for ceiling mounts. Figure 3.46 shows an example of a ceiling mount antenna.

FIGURE 3.46 L-com 2.3 GHz to 6 GHz 3 dBi omnidirectional ceiling mount antenna

Image provided by www.L-com.com.

Wall Mounting

Antennas or access points with attached antennas may need to be mounted to a wall based on the use or site survey results. Just as with a ceiling mount, follow the manufacturer's instructions for the appropriate fixture for wall mounting.

When mounting an antenna to the wall, consider the polarization of the antenna. Keep in mind that some antennas are designed to be mounted on the ceiling; these types should not be mounted on a wall. This is especially true for access points or other wireless infrastructure devices with integrated antennas. It is best to try to match the polarization of the wireless infrastructure device and the wireless client devices. In other words, if the infrastructure device antennas are vertically polarized, the wireless client devices should be polarized in the same manner to promote better connectivity. However, with the wide variety of newer wireless client devices available, this is getting harder to achieve.

Choosing the correct antenna and mounting position is typically part of a wireless site survey.

Summary

This chapter explored radio frequency basics and the essential role radio frequency (RF) plays in the world of wireless computer and cellular network communications. You learned the definition of RF as it pertains to wireless networking and examined its basic characteristics or properties, such as wavelength, frequency, amplitude, and phase.

I then described devices such as transmitters and receivers and how they communicate. In wireless LAN technology, an example of a transmitter and receiver is an access point and client device. I also discussed the unlicensed RF bands and channels used in the 2.4 GHz ISM and 5 GHz U-NII ranges for wireless LAN communications as well as other frequency ranges that may be allowed for use with IEEE 80211 wireless networking and cellular phone communications.

You learned about RF behaviors such as reflection, refraction, absorption, and others and the impact of building materials and the effect they have on the propagation of radio waves. Also covered were RF units of measure (both absolute and relative), including the watt, milliwatt, dB, and dBi.

Antennas are a critical component in a successful operation of a wireless communications system. You learned about radio frequency signal characteristics and basic RF antenna concepts such as radio frequency lobes, beamwidth, gain, and polarization.

If you understand these characteristics and concepts, a sales engineer, integrator, or other wireless systems professional can help you choose the best antenna for a specific use.

Understanding the radio frequency propagation patterns of various antenna types as well as the recommended use of an antenna will assist in deciding which antenna is best suited for the desired application. Omnidirectional antennas are one of the most common types of antenna used for indoor applications of wireless computer networks and other wireless systems. Omnidirectional antennas provide a horizontal radiation pattern of 360°. Other antennas such as patch/panel, yagi, and parabolic dish can also be used. You learned about the radiation patterns of each of these types of antennas as well as how each may be used.

A proper mounting fixture is required to ensure safety and correct operation of the antenna and wireless network. In this chapter we looked at various methods for mounting antennas, including integrated and external (to the access point), pole/mast mount, ceiling mount, and wall mount.

Finally, you learned about other factors to be considered when choosing and installing an antenna for use with wireless LANs. These include line of sight and Fresnel zone.

Understanding the concepts covered in this chapter will help you achieve more successful deployment, operation, and use of antennas in wireless communication systems.

Chapter Essentials

Know the basic characteristics or properties of radio frequency.   Understand the characteristics of radio frequency such as wavelength, phase, frequency, and amplitude.

Be familiar with the organizations that are responsible for radio frequency management.   The ITU-R, FCC, and ETSI are some of the organizations that manage and regulate radio frequency. Understand that both licensed and unlicensed frequency bands are used for wireless communication systems.

Know what RF factors will affect the range and speed of wireless networks.   Understand the effects of interference and the devices that cause interference. Be familiar with the environmental conditions that cause reflection, refraction, diffraction, scattering, and absorption. Understand their impact on the propagation of RF signals.

Identify basic RF units of measurement.   Understand the difference between absolute and relative measures of RF power. Define W, mW, dB, dBm, and dBi.

Understand RF signal characteristics and basic RF concepts used with antennas.   Know the difference between passive and active gain. Understand that antennas use passive gain to change or focus the radio frequency radiation pattern. Understand the difference between beamwidth and polarization.

Know the different types of antennas used in wireless networking.   Be familiar with different types of antennas used with wireless networking, including omnidirectional antennas, semidirectional antennas, and highly directional antennas. Understand the various radiation patterns each of these antennas is capable of.

Identify various RF cables, connectors, and accessories used in wireless LANs.   Understand that, depending on the local regulatory body, proprietary connectors may be required for use with antennas. Know that cables will induce some level of loss in a wireless LAN system. Be familiar with the types of connectors available.

Identify the mounting options of antennas used in wireless communication systems.   Antennas may be integrated or external to a wireless access point. Identify different types of antenna mounts, including internal and external (to the access point), pole/mast, ceiling, and wall mounts.

Understand additional concepts regarding RF propagation.   Understand and know some of the additional concepts when choosing and installing antennas used with wireless LANs. These concepts include visual line of sight, radio frequency line of sight, and Fresnel zone.

Chapter 4

Standards and Certifications for Wireless Technology

TOPICS COVERED IN THIS CHAPTER:

- ✓ The IEEE Standards for Wireless Communication
- ✓ Wireless Networking and Communications
- ✓ Interoperability Certifications
- ✓ The Wi-Fi Alliance
- ✓ Common IEEE 802.11 Deployment Scenarios

IEEE standards-based wireless technology continues to take mobile device and computer communications to a new level. This communication technology combines computer networking and radio frequency (RF) technology, giving users the opportunity to access and share information in ways that would have seemed unattainable not too many years ago. In this chapter, I'll explain the role of the organization that is responsible for creating and managing wireless communication standards. We'll take an in-depth look at the IEEE 802.11 standard and its amendments, including the communications and functional aspects. We will look briefly at two other IEEE wireless standards, IEEE 802.15 and IEEE 802.16. You will also see interoperability certifications for IEEE 802.11 wireless networks for communications, quality of service, security, and voice that are available from the Wi-Fi Alliance. It is also important to understand the various ways in which wireless technology fits with computer networking in general and the various physical applications in which it is used. In this chapter, we will take a look at how wireless technology allows users to connect to and use resources in a wireless technology environment. Finally, we will examine various common ways in which wireless local networks are used and deployed.

The IEEE

The *IEEE* (originally known as the Institute of Electrical and Electronics Engineers), commonly pronounced "eye triple E," is a nonprofit organization responsible for generating a variety of technology standards, including those related to information technology. According to the mission statement on the IEEE's website, its core purpose is to foster technological innovation and excellence for the benefit of humanity. The IEEE is the world's largest technical professional society.

Since 1997 a series of standards related to wireless technology have been released by the IEEE. These standards are all named IEEE 802 followed by combinations of numbers and letters to define specific standards and amendments.

The IEEE consists of working groups that are made up of individual members and experts in the field. With respect to wireless technology the IEEE working groups are as follows:

- IEEE 802.11 – Wireless Local Area Networking
- IEEE 802.15 – Wireless Personal Area Networking

- IEEE 802.16 – Broadband Wireless Metropolitan Area Networks
- IEEE 802.20 – Mobile Broadband Wireless Access
- IEEE 802.22 – Wireless Regional Area Networks

For additional information about the IEEE, visit www.ieee.org.

IEEE Standards for Wireless Communication

The IEEE wireless networking standard and its amendments that identify advancements in the technology define the power, range, and speed of the radio frequency and WLAN technology. The IEEE 802.11 standard specifies the maximum amount of radio frequency (RF) transmit power, the allowed radio frequency spectrum (which is related to the range), and the allowed data rates, or speed. At the time of this writing, all ratified IEEE 802.11 amendments have been incorporated into the IEEE 802.11-2012 standard with the exception of the most recent amendments including IEEE 802.11ac, IEEE 802.11ad and a few others.

Even though some of these are now legacy amendments, it is important to understand the foundation and advancements in standards-based wireless technology that have developed into our current system.

The IEEE 802.11 Standard

The *IEEE 802.11* standard, released in 1997, is what initially defined the wireless LAN communication standards. The data rates used in this original standard (1 and 2 Mbps) are considered very slow compared to today's technology.

As of this writing, the IEEE 802.11-2012 standard is the most current ratified IEEE 802.11 standard. This latest version combines the IEEE 802.11-2007 standard and outstanding amendments at the time of ratification, such as IEEE 802.11k/n/p/r/s/u/v/w/y/z, into one document. However, many in the industry still refer to the original names of the amendments: 802.11b, 802.11a, 802.11g, 802.11n, and so on. The IEEE considers all of the previously published amendments and revisions retired as a result of the release of the newest IEEE 802.11-2012 standard.

Here are the frequency range, spread spectrum/Physical layer (PHY) technology, and data rates for the original IEEE 802.11-1997 standard:

- 2.4 GHz ISM band
- Frequency hopping spread spectrum (FHSS)
- Direct sequence spread spectrum (DSSS)
- Infrared (IR)
- 1 and 2 Mbps

Frequency hopping spread spectrum (FHSS) is considered legacy technology with regard to IEEE 802.11 wireless networking. However, it is still used in other wireless technologies, such as IEEE 802.15 Bluetooth devices and wireless, cordless public-switched telephone network (PSTN) telephones.

You will learn more about these Physical layer technologies in Chapter 5, "IEEE 802.11 Terminology and Technology."

To see the most up-to-date status of the IEEE 802.11 standard and amendments, visit the Official IEEE 802.11 Working Group Project Timelines web page at www.ieee802.org/11/Reports/802.11_Timelines.htm.

The IEEE 802.11a Amendment

This amendment to the IEEE 802.11 standard is what defines operation in the 5 GHz U-NII band. (U-NII stands for Unlicensed National Information Infrastructure.) Released in 1999, this amendment originally defines only three frequency ranges in three bands in the 5 GHz frequency range—U-NII-1, U-NII-2, and U-NII-3. The U-NII-1 band is intended for indoor use only, the U-NII-2 band is for indoor or outdoor use, and the U-NII-3 band may be used indoors or outdoors but is most commonly used outdoors. The data rates for IEEE 802.11a are up to 54 Mbps using orthogonal frequency division multiplexing (OFDM). OFDM and other Physical layer technologies will be discussed in Chapter 5. Although this amendment was released in 1999, devices were not widely available until 2001.

Here are the frequency range, PHY technology, and data rates as specified in the original IEEE 802.11a amendment (keep in mind this has since changed because of relaxed local regulatory requirements):

- 5GHz U-NII band
  - 5.150–5.250 GHz U-NII-1
  - 5.250–5.350 GHz U-NII-2
  - 5.725–5.825 GHz U-NII-3

- Orthogonal frequency division multiplexing (OFDM)
- 6, 12, and 24 Mbps OFDM required data rates
- 9, 18, 36, 48, and 54 Mbps OFDM data rates supported but not required

Figure 4.1 shows the data rates available on a wireless LAN controller for IEEE 802.11a.

FIGURE 4.1 Cisco wireless LAN controller IEEE 802.11a data rates

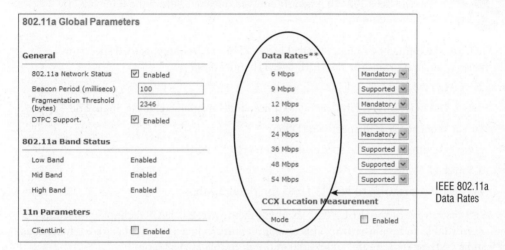

A benefit to using the 5 GHz U-NII band is that more bandwidth is available compared to the 2.4 GHz ISM band and there is less interference because not all wireless devices support operation in this band. The 5 GHz U-NII band supports up to 23 nonoverlapping 20 MHz wide channels compared to only three nonoverlapping channels in the 2.4 GHz ISM band, therefore providing more available and usable 5 GHz frequency. Currently, fewer devices use the 5 GHz U-NII license-free band than use the 2.4 GHz ISM license-free band, including non-802.11 devices. However, the number of wireless devices that operate in the 5 GHz band is always increasing. If there are fewer devices that utilize the band, it means less interference, which allows for increased performance and reliability. With the introduction of the IEEE 802.11ac amendment, the number of devices using the 5 GHz band will continue to grow because this is the only band supported for that technology. IEEE 802.11ac is discussed later.

 Since the IEEE 802.11a amendment was ratified, some changes have been implemented and are now addressed in the IEEE 802.11-2012 standard and previously in the 802.11-2007 standard. They include a new frequency range (5.470–5.725 GHz) that is allowed by some local regulatory agencies and is known as the U-NII-2e band. This extra frequency space allows for 11 additional 20 MHz-wide nonoverlapping channels. The new IEEE 802.11ac amendment added one more channel to the 5 GHz U-NII band which now provides a total of 24 nonoverlapping channels.

The IEEE 802.11b Amendment

The IEEE 802.11b amendment to the 802.11 standard specifies technology that works only in the 2.4–2.5 GHz ISM band. This amendment, released in 1999, specifies what is known as high-rate direct sequence spread spectrum (HR/DSSS) for 5.5 and 11 Mbps.

> The IEEE 802.11b amendment was released before the IEEE 802.11a amendment.

Here are the frequency range, Physical layer (PHY) technology, spread spectrum technology, and data rates for the IEEE 802.11b amendment:

- 2.4 GHz ISM band
- 2.4–2.4835 GHz in North America, China, and Europe (excluding Spain and France)
- Direct sequence spread spectrum (DSSS)
- High-rate direct sequence spread spectrum (HR/DSSS)
- 5.5 and 11 Mbps
- Backward compatible to 802.11 DSSS for 1 and 2 Mbps

With the release of the IEEE 802.11b amendment, wireless LAN technology became more affordable and mainstream. This amendment introduced two higher-rate data speeds, 5.5 and 11 Mbps, making the technology more desirable at that time.

Today wireless infrastructure device manufacturers still support IEEE 802.11b wireless technology; however, it is unlikely you would be able to purchase any "new" devices that support only IEEE 802.11b technology. As wireless technologies continue to evolve, there are very few if any IEEE 802.11b–only networks, and many organizations are no longer supporting this legacy technology at all and are disabling the capability within their systems. Devices sold today that operate in the 2.4 GHz ISM band support IEEE 802.11b/g/n and may be marketed as such or as IEEE 802.11g/n, which implies support for IEEE 802.11b.

The IEEE 802.11g Amendment

This amendment to the IEEE 802.11 standard was released in 2003. Like 802.11 and 802.11b, it operates in the 2.4 GHz ISM band. This amendment addresses extended data rates with orthogonal frequency division multiplexing (OFDM) technology and is backward compatible with 802.11 and 802.11b.

Here are the frequency range, PHY technology, spread spectrum technology, and data rates for the IEEE 802.11g amendment:

- 2.4 GHz ISM band
- 2.4–2.4835 GHz in North America, China, and Europe (excluding Spain and France)

IEEE Standards for Wireless Communication

- Direct sequence spread spectrum (DSSS)
- High-rate direct sequence spread spectrum (HR/DSSS)
- Extended rate physical orthogonal frequency division multiplexing (ERP-OFDM)
- Packet binary convolutional code (PBCC; optional)
- 1 and 2 Mbps (compatible with DSSS)
- 5.5 and 11 Mbps complementary code keying (CCK; compatible with HR/DSSS)
- 6, 12, and 24 Mbps OFDM required data rates
- 9, 18, 36, 48, and 54 Mbps OFDM data rates supported but not required

Figure 4.2 shows the data rates available on a wireless LAN controller for IEEE 802.11g.

FIGURE 4.2 Cisco wireless LAN controller IEEE 802.11g data rates

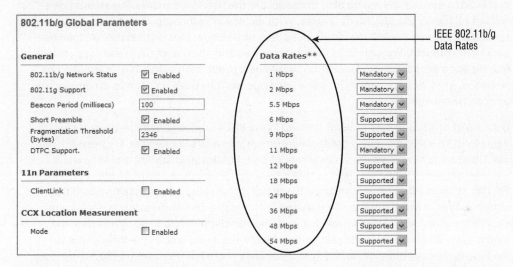

IEEE 802.11g is backward compatible with 802.11 and 802.11b because it operates in the same 2.4 GHz ISM license-free band and supports the same access methods or technology as 802.11b. One benefit of IEEE 802.11g compatibility is that many established network infrastructures and devices have used 802.11 and 802.11b for years. This allowed them to continue to operate as normal with upgrades or replacements as appropriate or necessary.

 To allow the slower DSSS and HR/DSSS data rates of 1, 2, 5.5, and 11 Mbps to operate in an IEEE 802.11g network, the amendment addresses the use of protection mechanisms. These protection mechanisms will degrade the performance of IEEE 802.11g clients to some degree when IEEE 802.11b radios are present in the basic service area (BSA).

Many organizations have dropped support for IEEE 802.11b devices altogether. One way to accomplish this is to disable the 802.11b data rates on the infrastructure devices—access points, controllers, and so on. This is done to minimize performance issues such as lower throughput when IEEE 802.11b devices are present in the IEEE 802.11g BSA. In some deployment scenarios, such as public wireless hotspots or areas with wireless guest networks, it would be difficult to disallow the use of IEEE 802.11b devices. However, this is strictly a decision to be made by the organization and information technology staff.

Maximizing Throughput in an IEEE 802.11g Network

In certain cases the only way to maximize the throughput of an 802.11g network is to set the data rates of the access points to support 802.11g data rates only. The trade-off is that 802.11b devices will not be able to connect to the network because the access point will not recognize the 802.11b data rates. With the newer technology available today, turning off support for 802.11b is becoming less of an issue. This would work well where backward compatibility with 802.11b is not required and all equipment in use supports 802.11g or newer capability. An analogy would be a group of individuals all speaking one language. They all understand the same language, so they have no need to accommodate a second language.

Because of protection mechanisms defined in the 802.11g amendment, throughput will degrade in an 802.11b/g mixed mode environment when 802.11b devices are present. The 802.11b devices have a maximum data rate of only 11 Mbps (HR/DSSS), and they share the medium with the 802.11g devices that have a maximum data rate of 54 Mbps (ERP-OFDM). Think of the language analogy. If a group of individuals are speaking two different languages, a translator may be required for complete communication. A discussion among the group would take longer because the translator would need to translate the languages. Likewise, protection mechanisms will have an impact on the throughput for the 802.11g devices because the 2.4 GHz medium is shared. If there are no 802.11b devices in the radio range of an access point in an 802.11b/g mixed mode environment, protection mechanisms should not affect throughput because the access point will not have to share the medium with the two different technologies, ERP-OFDM and HR/DSSS.

If you do not have any 802.11b devices on your network, you can set your access point to 802.11g–only mode by disabling the 802.11b data rates. In this configuration, your 802.11g devices will perform better because protection mechanisms will not be enabled. However, if there are any 802.11b devices that don't belong to your network in the "listening" range of the access point, data collisions will increase at the access point. This is because 802.11b and 802.11g operate in the same RF range and the 802.11g (ERP-OFDM) access point would stop listening to the 802.11b (HR/DSSS) transmissions. (It would simply see them as RF noise.) In this configuration, overall throughput will still exceed that of an access point set to 802.11b/g mixed mode in the presence of 802.11b devices. The access point will hear the 802.11b transmissions, but they will not be serviced since they are only seen as RF noise. Thus they will have less impact on throughput.

The IEEE 802.11n Amendment

After several years of drafts, the IEEE 802.11n amendment was finally approved in September 2009. The release of this document opened the doors for manufacturers of IEEE 802.11 wireless LAN equipment, giving them the opportunity to move forward with new technology that allows for better performance, higher throughput, and several other benefits. Wi-Fi–certified devices under 802.11n draft 2.0 were available for several years prior to the ratification of IEEE 802.11n. Most if not all enterprise manufacturers had at least one wireless infrastructure device certified under draft 2.0 by the Wi-Fi Alliance prior to the release of the new amendment.

Here are the frequency range, PHY technology, data rates, and other details for the IEEE 802.11n amendment:

- 2.4 GHz ISM band
- 5 GHz U-NII band
- Multiple-input, multiple-output (MIMO technology)
- HT-OFDM
- Physical layer (PHY) layer enhancements
- Data Link layer (MAC) layer enhancements
- Data rates up to 600 Mbps

Figure 4.3 shows the modulation and coding scheme (MCS) and data rates available on a wireless LAN controller for IEEE 802.11n operating on the 5 GHz band.

FIGURE 4.3 Cisco wireless LAN controller IEEE 802.11n MCS and data rates on the 5 GHz band

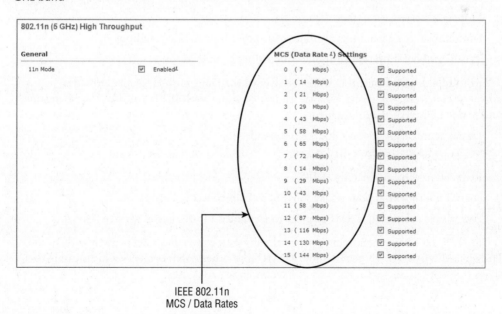

IEEE 802.11n devices are capable of operating in unlicensed frequency bands, the 2.4 GHz ISM band, and the 5 GHz U-NII band. This means that they must be backward compatible with previous technologies, such as IEEE 802.11b/g devices that operate in the 2.4 GHz ISM band and IEEE 802.11a devices that operate in the 5 GHz U-NII band.

Multiple-input, multiple-output (MIMO) is a big part of what makes IEEE 802.11n such an amazing technology. Prior to 802.11n, IEEE 802.11a/b/g devices used a single radio to transmit and receive radio signals. This is known as single-input, single-output (SISO) technology. MIMO uses multiple radios or "radio chains" to transmit and receive radio signals. SISO systems were subject to multipath, in which several wavefronts of a signal would be received out of phase because of reflections. This is a problem for IEEE 802.11a/b/g systems, whereas MIMO actually uses the reflections to help enhance the performance and throughput using several radio chains in 802.11n. With the use of MIMO technology, a device can get better signal than with previous technologies at some distance from a wireless access point, which will in turn provide a higher data rate at that distance. Therefore, in a sense the device will potentially get better usable range with 802.11n technology. MIMO consists of several types of new technologies:

- Transmit beamforming (TxBF)
- Maximal ratio combining (MRC)
- Spatial multiplexing (SM)
- Space time block coding (STBC)

It is best to check with the specific manufacturers of the wireless equipment to determine how they implement MIMO technology. IEEE 802.11n provides many enhancements to the Physical layer 1 (PHY), including the following:

- 40 MHz channels through the use of channel bonding
- More subcarriers for higher data rates
- Optional short guard intervals to provide more potential throughput
- Varying modulation types for data rates of up to 600 Mbps

The Media Access Control (MAC) sublayer of the Data Link layer also provides enhancements to improve performance and throughput with IEEE 802.11n. These include but are not limited to the following:

- Frame aggregation for less 802.11 overhead
- Block acknowledgments (block ACKs)
- Reduced interframe spacing (RIFS)
- Spatial multiplexing power save (SMPS) to help conserve battery life
- Power save multi-poll (PSMP) for devices enabled for quality of service (QoS)

The IEEE 802.11n amendment opened the doors for the continued growth with standards-based wireless LAN technology. MIMO and other enhancements will continue to play a role with newer innovation as this technology progresses.

The IEEE 802.11ac Amendment

The IEEE 802.11ac amendment is an exciting step forward with standards-based wireless LAN technology. This new gigabit wireless networking technology is used with wireless local area networking for both home and enterprise applications. As technology continues to advance, the users of the connected devices want more data, they want it faster, and they want it available everywhere. Like the IEEE 802.11a amendment to the standard, 802.11ac technology operates in the 5 GHz band only and is a Physical layer extension of IEEE 802.11n, providing backward compatibility with that technology in the 5 GHz frequency band. IEEE 802.11ac brings the following enhancements:

Data Rates up to 6.93 Gbps IEEE 802.11ac, Very High Throughput (VHT) provides the capability of up to 6.93 Gbps of aggregate throughput. The actual amount of throughput will vary and can be affected by several factors, including the number of radio chains/spatial streams and antennas.

256-QAM (Quadrature Amplitude Modulation) The Modulation and Coding Scheme (MCS) concept was introduced with the IEEE 802.11n amendment. Basically, MCS is a way to represent the data transfer rates with the newer, more sophisticated technology. The MCS index will vary based on the technology in place, such as the number of spatial streams, the channel width, and the short guard interval. IEEE 802.11ac allows for 256-QAM up from 64-QAM in 802.11n.

Wider Channels Bandwidth with respect to radio frequency can be considered to be a range of frequencies used for a specific application. IEEE 802.11n increased the RF channel width from 20 MHz to 40 MHz. This newest amendment to the standard allows for 80 MHz and even 160 MHz wide channels. Wider channels mean more bandwidth per channel and therefore higher throughput.

More Spatial Streams The IEEE 802.11n amendment allowed for a possible four spatial streams maximum; however, most manufacturers utilized only three. IEEE 802.11ac will allow for up to eight spatial streams. More spatial streams means a higher data rate and this will parlay into higher throughput.

Multi-user MIMO (MU-MIMO) Until now, IEEE 802.11 technology was based on unicast communication between two stations, such as a client device and an access point, or broadcast/multicast communication, such as an access point to many or to all connected stations. IEEE 802.11ac provides multi-user MIMO (MU-MIMO) for communication. MU-MIMO will allow different stations (client devices) to send or receive independent data streams simultaneously using multiple antennas / radio chains.

IEEE 802.11ac is backward compatible with devices operating in the 5 GHz band and will be beneficial for certain types of applications, such as streaming video and/or areas that may need additional capacity, and for minimizing issues in high-density deployments. One thing that is often overlooked is the actual physical wired infrastructure in which the wireless access points are connected. It is important to be certain that the wired infrastructure capacity is available to handle these faster speeds.

The IEEE 802.11ad amendment (Very High Throughput 60 GHz) was ratified in December 2012. This technology operates in the unlicensed 60 GHz frequency band and is not backward compatible with IEEE 802.11a/b/g/n/ac technology. Because 802.11ad operates at such a high frequency, it has many limitations and does not propagate well, and communications may be limited to a very small physical space, such as a single room. There are a limited number of IEEE 802.11ad chipsets available that will provide this technology. However, they currently are not big within the wireless LAN market or widely available. Some manufacturers will produce network adapters that will work within the 2.4 GHz, 5 GHz, and 60 GHz frequency range, but it is important to understand that even though these adapters will work in all three frequency ranges, the 802.11ad technology is not backward compatible with 802.11a/b/g/n/ac. The use of 802.11ad wireless technology includes peer-to-peer or ad hoc communications for streaming video and other multimedia type uses within a smaller physical area.

It is beneficial to understand all the frequencies, PHY technology, spread spectrum technologies, and data rates for all the IEEE 802.11 standards and amendments mentioned in this chapter. This will help with understanding the technology. Table 4.1 provides a summary and comparison of the IEEE 802.11 communication standards and amendments.

TABLE 4.1 Summary of 802.11 communication standards and amendments

Details	802.11	802.11a	802.11b	802.11g	802.11n	802.11ac
2.4 GHz ISM band	✓		✓	✓	✓	
5 GHz U-NII bands		✓			✓	✓
FHSS	✓					
DSSS	✓		✓	✓	✓	
HR/DSSS			✓	✓	✓	
OFDM		✓			✓	✓
ERP-OFDM				✓	✓	
HT-OFDM					✓	✓
VHT-OFDM						✓
1 and 2 Mbps	✓		✓	✓	✓	

Details	802.11	802.11a	802.11b	802.11g	802.11n	802.11ac
5.5 and 11 Mbps			✓	✓	✓	
6, 9, 12, 18, 24, 36, 48, 54 Mbps		✓		✓	✓	✓
Up to 600 Mbps					✓	✓
Up to 6.93 Gbps						✓

Additional IEEE 802.11 Amendments

In addition to communications, the IEEE creates amendments defining specific technical functionality, including QoS and security. We will look at those functions next.

The IEEE 802.11e Amendment

The original IEEE 802.11 standard lacked Quality of Service (QoS) functionality features. In the original IEEE 802.11 standard, Point Coordination Function (PCF) mode provided some level of QoS. PCF mode is a function of the access point and allows for polling of connected client devices. This creates a contention-free period for data transmissions and provides QoS-like functionality. However, few if any vendors implemented this mode of operation.

The IEEE 802.11e amendment defines enhancements for QoS in wireless LANs. 802.11e introduces a new coordination function, the hybrid coordination function (HCF). HCF defines traffic classes and assigns a priority to the information to be transmitted. For example, voice traffic is given a higher priority than data traffic, such as information being sent to a printer. The IEEE 802.11e amendment was incorporated into the IEEE 802.11-2007 standard and is now part of the IEEE 802.11-2012 standard. The Wi-Fi Alliance created a proactive interoperability certification for 802.11e called Wi-Fi Multimedia (WMM). The Wi-Fi Alliance and interoperability certifications are discussed later in this chapter.

The IEEE 802.11i Amendment

The IEEE 802.11i amendment addresses advanced security solutions for wireless LANs because the original IEEE 802.11 standard was known for security weaknesses. Manufacturers of IEEE 802.11 WLAN equipment addressed the following:

- Wired Equivalent Privacy (WEP)
- Service set identifier (SSID) hiding
- Media access control (MAC) address filtering

Wired Equivalent Privacy (WEP), defined by the IEEE 802.11 standard, was intended to prevent casual eavesdropping. WEP was compromised early on, making wireless LANs vulnerable to intrusion and providing little if any security. This issue was addressed by stronger security mechanisms (mainly Counter Mode with Cipher-Block Chaining Message Authentication Code Protocol/Advanced Encryption Standard, or CCMP/AES) that became available with the introduction of the IEEE 802.11i amendment to the standard.

Service set identifier (SSID) hiding and media access control (MAC) address filtering are both manufacturer-implemented features that may be used by some for pseudo-security. It is important to understand that neither of these provides any kind of security for an IEEE 802.11 wireless network.

WEP, SSID hiding, and MAC filtering all have known security vulnerabilities, allowing for security weaknesses in IEEE 802.11 wireless LANs. The IEEE 802.11i amendment addressed security weaknesses with wireless LANs by including several enhancements. The IEEE 802.11i amendment was incorporated into the IEEE 802.11-2007 standard and is now part of the IEEE 802.11-2012 standard.

The IEEE 802.11r Amendment

The IEEE 802.11r amendment was approved in May 2008. The 802.11r amendment specifies fast basic service set (BSS) transition (FT) technology. The original 1997 IEEE 802.11 standard did not address standards-based transition mechanisms, so manufacturers used proprietary methods. The IEEE attempted to standardize transition techniques for wireless LAN technology with the ratification of a recommended practice, IEEE 802.11F. This recommended practice was never implemented by many (if any) manufacturers and was eventually withdrawn by the IEEE.

The main goal of IEEE 802.11r was to provide fast basic service set (BSS) transition for Voice over IP (VoIP) with wireless LAN technology. Although this amendment has been ratified for some time and most enterprise equipment manufacturers support it, they still rely on the use of proprietary methods for fast transition. This is partly because until recently there was no interoperability certification by the Wi-Fi Alliance. The IEEE 802.11r amendment was incorporated into the IEEE 802.11-2012 standard.

The IEEE 802.11k Amendment

IEEE 802.11k is the amendment to the IEEE 802.11 standard that addresses radio resource management. This amendment was approved in May 2008, the same day as the IEEE 802.11r amendment. 802.11k and 802.11r work together to form fast, secure basic service set transition for mobile devices. IEEE 802.11k aids the wireless device in locating the best access point to transition to by defining the technology to be used to manage the radio frequency. The IEEE 802.11k amendment was incorporated into the IEEE 802.11-2012 standard.

The IEEE 802.11w Amendment

Wireless LAN management frames, such as the 802.11 authentication frames and 802.11 association frames used in IEEE 802.11 wireless LANs, are susceptible to intrusion and can cause security issues. This is because the IEEE 802.11 standard did not provide any protection for management frame information that traverses the air. With some basic knowledge of the technology and the correct software tools, an intruder can perform a denial of service (DoS) or hijacking attack. When implemented, technology specified in the IEEE 802.11w amendment helps to mitigate these types of attacks or security issues. The IEEE 802.11w amendment was incorporated into the IEEE 802.11-2012 standard.

The IEEE 802.11s Amendment

The IEEE 802.11s amendment specifies wireless mesh networking. Mesh networking with wired networking has been available for many years. Wireless mesh networking started with military deployments but has evolved into the public sector. Mesh networking allows infrastructure devices such as wireless access points or mesh routers to create a self-forming, self-healing, and intelligent network infrastructure. Most manufacturers of enterprise wireless equipment have been using mesh technology for years with proprietary protocols. Although the IEEE has ratified the standard for this technology, most manufacturers still use proprietary methods. The IEEE 802.11s amendment was incorporated into the IEEE 802.11-2012 standard.

A wireless network has sometimes been referred to as a wireless distribution system (WDS). However, the newest 802.11-2012 standard states that "this standard specifies such a frame format and its use only for a mesh basic service set (MBSS). Because of this, the term *WDS* is obsolete and subject to removal in a subsequent revision of this standard."

Wireless Networking and Communications

Wireless networks come in a variety of types and sizes and include the following wireless topologies:

- Wireless personal area network (WPAN)
- Wireless local area network (WLAN)
- Wireless metropolitan area network (WMAN)
- Wireless wide area network (WWAN)

In Chapter 1, "Computer Network Types, Topologies, and the OSI Model," you learned about the different types of computer networks and some basic networking concepts. You will now look at some of these network types from a wireless communication perspective.

> There are also wireless campus area networks (WCANs). CANs are described in Chapter 1, and some considerations regarding WCANs are discussed in the section "Educational Institution Deployments" later in this chapter.

Wireless Personal Area Networks and the IEEE 802.15 Standard

A personal area network (PAN) is a network that connects devices within the immediate area of individual people. PANs may consist of either wired or wireless connections or both. Here we will look at wireless PANs. This type of network allows users to connect various devices wirelessly to their own personal area network, including but not limited to computer keyboards, mice, and headsets.

Even though the 802.15 Working Group addresses the PAN in various forms, we will examine IEEE 802.15 Task Group 1, which addresses Bluetooth technology. Bluetooth technology is becoming the most popular technology for WPANs and uses frequency hopping spread spectrum (FHSS) for communications. Bluetooth falls under the IEEE 802.15 standard, which specifies the WPAN standards. Bluetooth devices operate in the unlicensed 2.4 GHz industrial, scientific, and medical (ISM) band, as do wireless local area networks.

WPANs may also use infrared technology, which uses near-visible light in the 850 nm to 950 nm range for communications. Infrared technology was specified in the original 802.11 standard, but regarding the infrared (IR) specification, the IEEE 802.11-2012 standard states that "the mechanisms described in this clause are obsolete. Consequently, this clause may be removed in a later revision of the standard. This clause is no longer maintained and may not be compatible with all features of this standard."

Wireless Local Area Networks

Local area networks (LANs) can be defined as a group of computers connected by a physical medium in a specific arrangement called a topology. LANs are contained in the same physical area and usually are bounded by the perimeter of a building or a group of buildings.

Wireless local area networks (WLANs) fall under the same description as LANs but no longer require a physical wire to connect devices together. Wireless LANs have been in existence for many years, even prior to IEEE 802.11 standards–based technology, and mostly include proprietary technology or are government deployments. Since the IEEE released the 802.11 standard in 1997, WLAN technology has continued to excel and is becoming a major component of every computer network.

WLANs may operate in either the licensed or unlicensed radio frequency spectrum. The most commonly used frequency spectrum for WLANs are the unlicensed 2.4 GHz ISM band and the unlicensed 5 GHz U-NII band. The frequency bands used with IEEE 802.11 wireless networking were discussed in Chapter 3, "Radio Frequency and Antenna Technology Fundamentals."

Wireless Broadband, Metropolitan Area Networks, and the IEEE 802.16 Standard

Wireless broadband allows for high-speed wireless Internet access and wireless computer networking using a variety of technology devices. The performance and data rates will vary based on the type of technology used. Metropolitan area networks (MANs) consist of networks that may span from several blocks of buildings to entire cities and interconnect devices for access to computer resources in an area larger than that covered by LANs but smaller than the areas covered by WANs.

You can expand on this technology and add much flexibility to MANs by incorporating wireless technology and creating the wireless metropolitan area network (WMAN). The IEEE 802.16 standard was developed to define the wireless metropolitan area network (WMAN) standard for broadband wireless networks. This technology may fall under the Worldwide Interoperability for Microwave Access (WiMAX) category and addresses different technologies. WiMAX is covered in more detail in Chapter 7, "Cellular Communication Technology." The WMAN may include a combination of public and private entities that encompass town services such as police, fire, and public utility access.

IEEE 802.16 is actually a series of standards that specifies wireless broadband communications. IEEE 802.16-2012 is the most current version of the standard and is a roll-up of some of the previous versions. Like the IEEE 802.11 standard, 802.16 specifies the Physical layer (PHY) and the Medium Access Control (MAC) sublayer of Layer 2, the Data Link layer, for fixed and mobile point-to-multipoint broadband wireless access systems.

Wireless Wide Area Networks

A WAN consists of point-to-point or point-to-multipoint connections between two or more LANs. WANs have the capability of extending very long distances through the use of fiber-optic connections or leased lines from telecommunications providers. When it comes to the wireless wide area network (WWAN), this extends beyond the point of connecting LANs together. The WWAN will encompass very large geographical areas and may include different wireless technologies, including cellular.

The WWAN also provides wireless broadband communications for Internet access through the use of special external adapters or even adapters built into notebook computers and other mobile devices, including smartphones. Because of the technology that is used, performance such as data rates may be lower than that expected and realized with IEEE 802.11 wireless networking.

Interoperability Certifications

By creating standards, the IEEE is encouraging technological progress. Manufacturers often implement wireless devices and networks in a proprietary manner, within or outside the standard. The proprietary approach often leads to a lack of interoperability among devices. In the wireless community, such practices are not widely accepted. Users want all of their devices to function well together. The combination of proprietary implementations and user dissatisfaction fostered the creation of interoperability testing and certifications.

The following sections will cover vendor interoperability certifications related to IEEE 802.11 standards–based wireless LAN equipment. These certifications address communications, QoS, and security.

The Wi-Fi Alliance

The IEEE is responsible for creating the standards for wireless networking and technology. However, equipment manufacturers are not required to provide proof that their equipment is compliant with the standards. Starting with the release of the IEEE 802.11b amendment, several early WLAN equipment manufacturers—including Symbol Systems, Aironet, and Lucent—formed an organization known as Wireless Ethernet Compatibility Alliance (WECA) to promote the technology and to provide interoperability testing of wireless LAN equipment manufactured by these and other companies. In 2000, WECA was renamed the *Wi-Fi Alliance*. The term *Wi-Fi* represents a certification and is often misused by people in the industry. Wi-Fi is a registered trademark, originally registered in 1999 by WECA and now registered to the Wi-Fi Alliance. People often use the term Wi-Fi synonymously with wireless LAN technology; in fact it means wireless technology certified to be interoperable.

 For additional information about the Wi-Fi Alliance, visit www.wi-fi.org.

Figure 4.4 shows an example of a Wi-Fi Certified logo, showing that the device has met the interoperability testing criteria.

Wi-Fi Protected Access Certification

The *Wi-Fi Protected Access (WPA)* certification was developed because security in the original IEEE 802.11 standard was weak and had many vulnerabilities. This certification was designed as an interim solution until an amendment to the IEEE 802.11 standard addressing security improvements was released. The IEEE 802.11i amendment addressed security for the IEEE 802.11 family of standards. The bottom line is that WPA is a pre-802.11i certification, introducing more advanced security solutions such as Temporal Key Integrity Protocol (TKIP), passphrase, and 802.1X/EAP.

FIGURE 4.4   Wi-Fi certified logo for devices that are Wi-Fi certified

This pre-802.11i certification addressed two options for wireless LAN security: personal mode and enterprise mode. Personal mode is intended for small office, home office (SOHO) and home users. Enterprise mode is intended for larger deployments. Personal mode allowed for a user to enter an 8- to 63-character passphrase (password) on both the access point and all of the devices that connected to the access point. Enterprise mode provides user-based authentication utilizing 802.1X/EAP. Both personal and enterprise modes are discussed in more detail in Chapter 16, "Device Authentication and Data Encryption."

Wi-Fi Protected Access 2 Certification

The WPA certification by the Wi-Fi Alliance worked out so well that the alliance decided to certify wireless LAN hardware after the IEEE 802.11i amendment was released. This new certification, known as *Wi-Fi Protected Access 2 (WPA 2.0)*, is a post-802.11i certification. Like WPA, WPA 2.0 addresses two options for wireless LAN security: personal mode and enterprise mode. This certification addresses more advanced security solutions and is backward compatible with WPA. The following is a preview of its key points; we will look at both WPA and WPA 2.0 in more detail in Chapter 16.

- The personal mode security mechanism uses a passphrase for authentication, which is intended for SOHO and personal use. The use of a passphrase to generate a 256-bit pre-shared key (PSK) provides strong security. Personal mode may also be identified as pre-shared key.
- The enterprise mode security mechanism uses 802.1X/EAP for user-based authentication, which is port-based authentication and is designed for enterprise implementations. 802.1X/EAP provides strong security using external authentication and Extensible Authentication Protocol (EAP). 802.1X/EAP uses an authentication server for the user authentication. Remote Authentication Dial-In User Service (RADIUS) is a common authentication server. This works well as a replacement for legacy IEEE 802.11 security solutions.

Table 4.2 provides a high-level description of the WPA and WPA 2.0 certifications.

TABLE 4.2 Details of the WPA and WPA 2.0 certifications

Wi-Fi Alliance Security Mechanism	Authentication Mechanism	Encryption Mechanism/Cipher
WPA - Personal	Passphrase	TKIP/RC4
WPA - Enterprise	802.1X/EAP	TKIP/RC4
WPA 2.0 - Personal	Passphrase	CCMP/AES or TKIP/RC4
WPA 2.0 - Enterprise	802.1X/EAP	CCMP/AES or TKIP/RC4

Temporal Key Integrity Protocol (TKIP)
Rivest Cipher 4 (RC4), named after Ron Rivest of RSA Security
Counter Mode with Cipher-Block Chaining Message Authentication Code Protocol (CCMP)
Advanced Encryption Standard (AES)

Encryption mechanisms and ciphers will be discussed further in Chapter 16.

Wi-Fi Protected Setup Certification

Wi-Fi Protected Setup (WPS) was defined because SOHO users wanted a simple way to provide the best security possible for their installations without the need for extensive technical knowledge of wireless networking. Wi-Fi Protected Setup provides strong out-of-the-box setup adequate for many SOHO implementations.

The Wi-Fi Protected Setup certification requires support for two types of authentication that enable users to automatically configure network names and strong WPA2 data encryption and authentication:

- Push-button configuration (PBC) allows for quick setup for consumer-grade Wi-Fi equipment. Typically a hardware button on the router is pushed and within two minutes a software "button" on the client device is pushed. The intent is to provide easy, secure setup for the home wireless network.

- PIN-based configuration is based on a personal identification number. It is similar to PBC, but with this method a PIN is entered on all devices that you wish to connect together on the same wireless network.

Support for both PIN and PBC configurations are required for access points; client devices at a minimum must support PIN. A third, optional method, near field communication (NFC) tokens, is also supported. With NFC, if the client device is within a very close proximity to the wireless access point, it will allow for radio communications. NFC will also allow for the exchange of information such as photos or contacts between mobile devices in a peer-to-peer or ad hoc environment. NFC evolved from radio-frequency identification (RFID) technology that provided radio communications using either passive or active tags.

In December 2011, a security flaw was reported with WPS. This allegedly allowed an intruder to recover the personal identification number (PIN) used to create the 256-bit pre-shared key. Acquiring the PIN would allow access to the wireless network. Wherever possible, it is recommended that users disable certain features in the wireless router or access point that allow this to happen. A firmware update may also be available to provide adequate protection. Keep in mind that this solution to the issue may be possible only with newer-model wireless routers. You should check with the manufacturer to determine if a solution (either a software setting or firmware upgrade) is available for a specific device. Otherwise, consider a different method, such as WPA or WPA 2.0, to secure the wireless router. Upgrading to a newer wireless router is another possible solution.

Wi-Fi Multimedia Certification

The *Wi-Fi Multimedia (WMM)* certification was designed as a proactive certification for the IEEE 802.11e amendment to the 802.11 standard. The 802.11e amendment addresses QoS in wireless LANs. The WMM certification verifies the validity of features of the 802.11e amendment and allows for a vendor-neutral approach to quality of service.

Quality of service is needed to ensure delivery of information for time-sensitive, time-bounded applications such as voice and streaming video. If a wireless network user were to send a file to a printer or save a file to a server, it is unlikely they would notice any minor delay, or latency. However, in an application that is tuned to human senses such as hearing or eyesight, latency would more likely be noticeable.

Wi-Fi Multimedia Power Save Certification

Wi-Fi Multimedia Power Save (WMM-PS) is designed for mobile devices and specific uses of wireless LAN technology that require advanced power-save mechanisms for extended battery life. Here are some of these devices and technology that benefit from WMM-PS:

- Voice over IP (VoIP) phones
- Notebook computers
- Tablet devices

Power-save mechanisms allow devices to conserve battery power by "dozing" for short periods of time. Depending on the application, performance could suffer to some degree with power-save features enabled. WMM-PS consumes less power by allowing devices to spend more time in a "dozing" state—an improvement over legacy power save mode that at the same time improves performance by minimizing transmission latency.

The Wi-Fi Voice-Enterprise Certification

The Wi-Fi Voice-Enterprise certification is designed to provide certification for IEEE 802.11 wireless LAN client and infrastructure devices used for voice applications in a variety of enterprise deployment types. These include carpeted office deployments, educational institution deployments, industrial deployments, healthcare deployments, and similar deployments. These deployment types are discussed later in this chapter. The Wi-Fi Voice-Enterprise certification is based on protocol adherence for three amendments to the IEEE 802.11 standard:

- IEEE 802.11k—Radio resource measurement
- IEEE 802.11r—Fast BSS transition
- IEEE 802.11v—Wireless network management

The performance metrics testing includes latency, jitter, and packet loss. The Wi-Fi Alliance released this certification in 2012.

Common IEEE 802.11 Deployment Scenarios

The availability and technology enhancements of IEEE 802.11 wireless networking have increased while the cost continues to decrease, making wireless LANs a viable solution for many business models, including personal use, home offices, small offices, and enterprise organizations. In the following sections, we'll look at various scenarios in which this type of wireless networking is used. We'll explore the following common deployment scenarios that utilize wireless local area networks (WLANs):

- Small office, home office (SOHO)
- Enterprise deployments: corporate data access and end-user mobility
- Extension of existing networks into remote locations
- Public wireless hotspots
- Carpeted office deployments

- Educational institution deployments
- Industrial deployments
- Healthcare deployments
- Last-mile data delivery: wireless ISP
- High-density deployments
- Other deployments, including municipal, law enforcement, and transportation networks

Small Office, Home Office

Many small office, home office (SOHO) businesses have the same needs as those of larger businesses with regard to technology, computer networking, and communication. These common needs, regardless of network size, include access to a common infrastructure for resources such as computer data (files), printers, databases, other networks, and the Internet. Computer networking technology is common regardless of the size of the business. Whether there are one or 100 employees, many are categorized as small businesses. Wireless LANs now play a major role in small businesses. Many small business locations have a high-speed Internet connection such as digital subscriber line (DSL) or cable modem for access outside the local area network.

With the number of work-at-home professionals continuing to grow at a very high rate, the need for wireless networking in this environment is also continuing to grow. The same goes for the small office environment. Deployments such as these typically involve a smaller number of users. Therefore, the equipment used may be consumer models sold in consumer electronics department stores and online retailers.

In addition, many companies and organizations now allow for employees to work remotely part or full time. In these cases, the company network is now extended to the remote location, which, whether it is a home office or other location, may be considered a branch office of the company's corporate network. When wireless LAN technology is used at a remote location, new concerns arise, such as data security and network availability.

Depending on the size of the small office or home office and the number of potential users and devices, a WLAN RF site survey may be required. A site survey will help determine areas of radio frequency (RF) coverage and interference as well as the number and placement of access points. Even if the small office or home office will require only a single access point, it is still beneficial to know if there are other wireless networks or devices in the same coverage area that may cause radio frequency interference.

Figure 4.5 shows a SOHO configuration with a wireless LAN router connected to an Internet service provider allowing access to the necessary network/Internet resources.

FIGURE 4.5 Example of a SOHO wireless LAN configuration

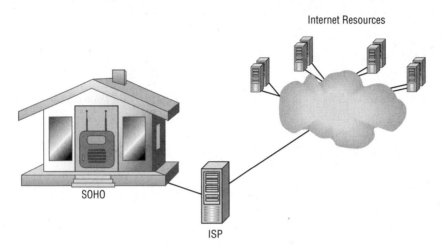

Enterprise Deployments: Corporate Data Access and End-User Mobility

Enterprise organizations have used wired local area networking technology for decades. With the increased need for mobility, wireless LANs within enterprise organizations have also increased in popularity. In earlier years, due to lack of interoperability and security features, many enterprise organizations limited wireless LAN deployments to extensions of networks where wired connectivity was either not feasible or too costly. Because of advancements in wireless LAN technology over the recent years, IEEE 802.11 deployments in enterprise organizations are continuing to grow at a rapid pace.

Wireless LANs in the enterprise are used with—but not limited to—client workstation connectivity (desktop, notebook, and tablet devices), printers, barcode scanners, voice handsets, and location services. The cost of this technology has decreased, whereas capabilities, performance, speed, and security have increased, making wireless an attractive solution for many enterprise organizations. The cost savings over hardwired solutions such as Ethernet are enormous, adding to the attractiveness of this option. Finally, wireless connectivity is the only option in some cases, such as mobile Voice over Wi-Fi handsets for voice communications.

Figure 4.6 shows a floor plan of an office area that may include a wireless deployment. Each individual or shared office would contain one or two networked desktop computers and phones, and many would also have laptops. Printers might be placed in centrally located common areas accessible to the individuals who have permissions to use them. The conference room might contain a videoconferencing system and an access point depending on the number of available seats, and the reception area might have wireless guest access for vendors or other visitors not belonging to the company or organization. Connecting all these networked devices to each other and the outside world are the wireless access point and other WLAN infrastructure devices discussed in Chapter 6, "Computer Network Infrastructure Devices." They will be located throughout the facility based on the wireless network design to provide coverage and capacity for all wireless devices.

FIGURE 4.6 Floor plan of a typical office area that may use IEEE 802.11 wireless LAN technology

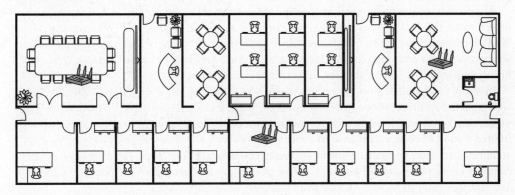

Extension of Existing Networks into Remote Locations

In its early days, wireless networking technology was typically deployed as an extension of an existing wired network infrastructure. For example, some users who required access to the computer network were farther than the physical limit of 100 meters that the IEEE 802.3 Ethernet standard allows for a copper-wired connection, so other solutions were needed to provide this connectivity. Other wired technology, such as fiber optics and leased lines, was sometimes cost prohibitive or not logistically feasible. Wireless local area networks were an excellent alternative.

Now, IEEE 802.11 wireless LANs are a major part of almost every network, including home, corporate, and branch/remote locations. Remote network locations may include the SOHO locations, branch office locations, public wireless hotspots, and wireless Internet service providers (WISPs). When a user connects to a corporate office network from any of these scenarios, the network is basically being extended to a remote location. This extension should be treated as such with regard to network, security, availability, and performance.

Public Wireless Hotspots

Portability and mobility are major benefits of wireless networking. Portability allows users to access information from a variety of locations, either public or private. Mobility allows the continuous connection to a wireless network while a device is on the move. One example of portability is the wireless hotspot. In today's world, it is rare to visit any public location, whether a restaurant, hotel, coffee shop, or airport, and not be able to find a public wireless hotspot.

A *wireless hotspot* is defined as a location that offers wireless network connectivity for free or as for-profit public or patron services. It allows a variety of mobile devices (computers, tablets, smartphones, and so on) to connect to and access public Internet and private network resources. Many users work from remote locations and require Internet access as part of their job. This can include access from a wireless hotspot.

A typical wireless hotspot will be configured with at least one wireless LAN router connected to an ISP. In some cases, this setup could be as simple as a location offering free Wi-Fi Internet access for its customers. More sophisticated hotspots will have several wireless access points or a complete wireless infrastructure and will be connected to a remote billing server that is responsible for collecting revenue from the user.

In many cases, when a user connects to the hotspot router, they will be prompted with a web page for authentication. At this point they might be asked to enter information such as an account number, username and password, or a credit card number to allow usage for a limited period of time. In the case of a free hotspot, typically this web page lists terms and conditions the user agrees to prior to accessing the Internet. This type of web page configuration is known as a *captive portal*. Captive portals are discussed in more detail in Chapter 10, "Introduction to Mobile Device Management."

Wireless hotspots can raise security concerns for the user. Without a secure connection, all information is passed in cleartext through the air via radio frequency, potentially allowing an intruder to capture usernames, passwords, credit card numbers, or other information that could lead to identity theft. Most hotspots do not have the capability to provide a secure wireless connection from the user's computer or wireless device to the wireless router or network. The secure connection then becomes the responsibility of the user. Since many corporations allow employees to work remotely from wireless hotspot connections, extra security measures need to be explored and implemented. In this case, usually a *virtual private network (VPN)* is used to ensure security. A VPN creates a secure tunnel between the user and the corporate network, allowing for a secure encrypted connection for the user from the wireless hotspot to their corporate network over the Internet or public network.

For users who connect to wireless hotspots, it is important for their wireless devices to be secured with the appropriate antivirus software, firewall software, and up-to-date operating system patches or service packs. Following these guidelines can help protect the user from attacks when they are connected to and using a wireless hotspot.

Figure 4.7 shows a simple wireless hotspot implementation.

Carpeted Office Deployments

Computer networking in traditional office space, or "carpeted offices," now relies on wireless technology to a large extent. This is for several reasons:

- Cost
- Portability

- Mobility
- Convenience

FIGURE 4.7 Wireless hotspot allows users to connect to the Internet from remote locations.

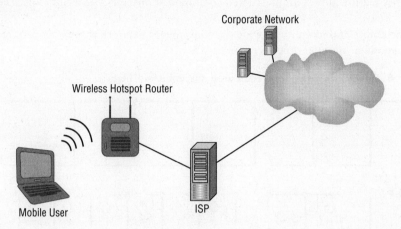

Many offices have an existing wired network infrastructure, which is not likely to go away anytime soon. In most cases copper wire for Ethernet connectivity is already in place and is adequate for the intended use. However, the cost to upgrade the copper wiring or install new wired network drops can be expensive. Therefore, wireless LAN technology is an attractive alternative to wired networks in many office deployment scenarios. We are now seeing a new era of wireless networks by default and wired networks only as needed based on the type of devices connected such as wired and wireless infrastructure devices. These include, servers, printers, wireless access points and others. Depending on the use of the network—that is, the types of software applications and the number of devices requiring connectivity—wireless may be the best solution simply because of the cost.

Without trying to set firm limits, it is common to connect as many as 20 to 25 users/devices to a single wireless access point. However, the maximum size depends on the software applications and the number of devices connected. A major benefit of IEEE 802.11 wireless LAN technology is that an access point will require only a single Ethernet drop to support all the devices or users. Of course, don't forget that an access point is part of a shared medium for everything that connects and performance and throughput can be an issue if proper design practices are not used.

It is important to understand the difference between portability and mobility. Portability allows users to access information from a variety of set locations, and mobility allows a continuous connection to a wireless network while the device is moving. In carpeted space offices, there may be a need to provide support for both portability and mobility.

For the user who moves from an office cubicle to a conference room to attend a meeting, portability will be sufficient. In this situation they will probably shut down their mobile device, such as a laptop computer, and carry it to a conference room for the meeting. Restarting the computer will then require the device to reconnect to the wireless network, hence portability.

Devices such as mobile phones using Voice over IP (VoIP) or tablet devices usually require continuous connectivity to the wireless network while the user/device is in motion. This mobility feature allows uninterrupted communications and a pleasant experience for the user.

Both portability and mobility provide the convenience network that people desire. Figure 4.8 shows a common office scenario.

FIGURE 4.8 Office with conference area and cubicle offices

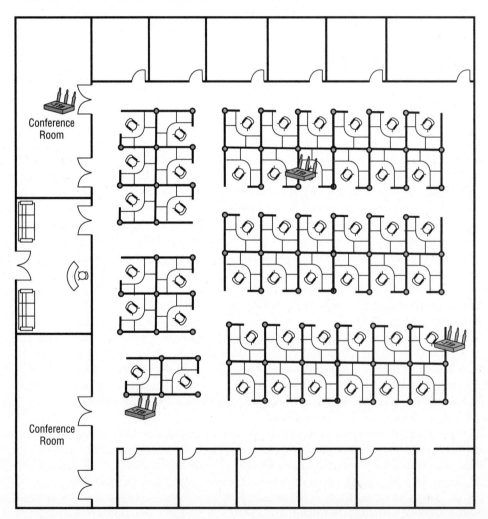

Educational Institution Deployments

Educational institutions can benefit from wireless networking in many ways. Wireless LAN deployments are common in both public and private elementary and high schools. Universities deploy campus-wide wireless LANs with thousands of access points servicing tens of thousands of users on a single campus.

Wireless LAN technology allows for increased mobility in the educational environment, providing huge cost savings when technology needs to be refreshed. Mobile carts with notebook computers are one example. A high school can deploy wireless infrastructure devices such as access points in classrooms and purchase several mobile carts with notebook or tablet computers to be used when and where needed. This is beneficial in terms of cost savings because all classrooms won't need to be supplied with computers or devices when continuous need for them may be low.

Some school buildings may be older or historic buildings in which installing cabling is impossible or cost prohibitive. Wireless provides the solution. The architecture of many school buildings may also pose concerns that need to be addressed with many wireless network deployments. Among the concerns are building materials, such as these:

- Brick and concrete walls
- Lath and plaster walls
- Inconsistent materials due to building additions

These materials can cause issues because, depending on their density and composition, the radio frequency may not propagate well. There is thus the potential for additional wireless access points and extra design considerations.

In addition, there may be modular or temporary classrooms; the issue with these is the density of devices and users, which will affect the wireless network capacity and may result in performance issues. Also, the location and distance from the main building should be taken into consideration because it may be necessary to install a point-to-multipoint connection and line of sight needs be taken into account.

Some educational institutions are implementing a "one-to-one" initiative—in other words, the goal is to have one Internet-accessible device for every one student. This type of initiative will introduce density concerns because of the potentially high number of students in a single classroom. Educational institutions, whether elementary schools, high schools, or college campuses, should always consider starting with an RF wireless site survey and follow best practices from the equipment manufacturer to ensure a successful deployment.

Figure 4.9 shows a typical small school environment.

Industrial Deployments

Some industrial organizations have been using wireless LAN technology for many years, even prior to the development solutions based on IEEE 802.11 wireless standards. Examples of these deployments include bar code and scanning solutions for manufacturing, warehousing, inventory, and retail. Although this type of deployment may not be very dense, coverage is important. Many businesses of this type include the following characteristics:

- High ceilings
- Tall storage racks
- Large inventory of product
- Forklifts

FIGURE 4.9 Classrooms for wireless LAN deployment

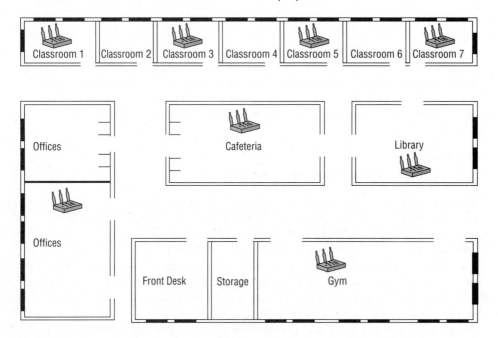

These characteristics can cause issues with wireless networks because of the way radio frequency propagates. With high ceilings, various antennas will need to be tested and coverage will need to be verified throughout the facility. Tall storage racks may have varying levels of inventory or product, resulting in poor propagation. Depending on what the products are made of, this will have a direct impact on the radio frequency behavior. For example, a high density of water products or paper products will absorb radio frequency. In many cases, forklifts will be outfitted with wireless bar code scanners or other mobile devices that require fast and secure transition capabilities.

In this type of environment, it is important to understand that radio frequency will behave in ways that could impact the performance of the wireless network. The behaviors of radio frequency are discussed in Chapter 3. These behaviors can lead to coverage issues for the devices in use. Careful evaluation of this type of environment is essential, and an RF site survey is highly recommended to ensure proper RF coverage.

The physical characteristics of this type of environment are fairly static, although racks or shelving may occasionally be added. However, product inventory is dynamic and may change constantly. Moreover, forklifts and other product-moving equipment are constantly moving and in different locations. These are some of the factors that must be taken into consideration when deploying wireless networking in an industrial environment.

Figure 4.10 shows a typical warehouse facility with 35-foot-high ceilings.

FIGURE 4.10 Warehouse facility with high ceilings and storage shelving inside

Healthcare Deployments

The growth of wireless LAN deployments in the healthcare industry is quite impressive. Today, healthcare is one of the fastest-growing sectors of the US economy. Healthcare environments pose many challenges for the design, deployment, and support of wireless networking.

Hospitals in most cases run nonstop 365 days a year. Wireless LANs have numerous applications in hospitals, including these:

- Patient registration
- Patient charting
- Prescription automation
- Treatment verification
- Inventory tracking
- Electronic medical records

- Location services
- Electronic imaging

One of the obstacles to take into consideration for wireless networking is interference. Hospitals use many devices that operate in the unlicensed ISM RF band. This can create challenges for design and reliability of the wireless network. You should also be aware of these other potential issues for healthcare environment deployments:

- Building materials that can hinder RF propagation, such as lead-lined walls used in radiology areas to protect people from X-rays
- Identical floor layouts above and below, which leads to stacking access point issues
- Limited accessibility to areas such as surgery and patient care rooms
- Aesthetics of the installed equipment

Compliance with legislation such as the Health Insurance Portability and Accountability Act of 1996 (HIPAA) also needs to be taken into consideration when designing wireless installations for healthcare institutions.

Figure 4.11 illustrates a common medical office that uses wireless LAN technology.

FIGURE 4.11 Medical offices often use wireless LAN technology

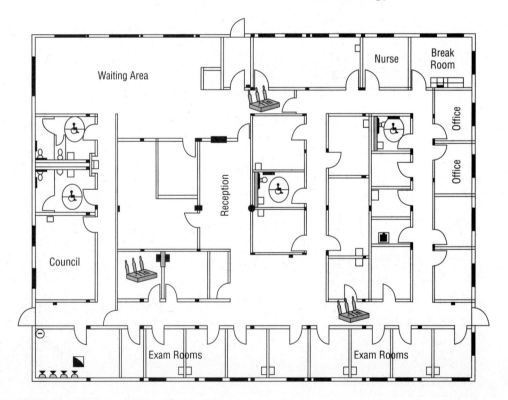

Last-Mile Data Delivery: Wireless ISP

Last-mile data delivery is a term commonly used in telecommunications to describe the connection from a provider to an endpoint such as a home or business. (*Last-mile* does not necessarily indicate a mile in distance.) This can be a costly solution in many applications because each endpoint needs a separate physical connection. Wireless technology provides a more cost-effective solution for last-mile data delivery.

Some communication technologies, such as DSL, have physical limitations that prohibit connections in some cases. It may not be cost effective for telecommunication service providers to supply connections in rural or semirural areas. Wireless LANs can service areas that may not be part of a last-mile run. Providing Internet access from a wireless ISP is one application. Things to consider for feasibility are line of sight, obstacles, and RF interference. Figure 4.12 shows an example of wireless last-mile data delivery.

FIGURE 4.12 Wireless last-mile data delivery

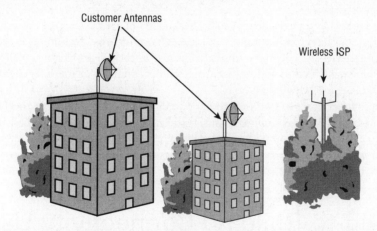

In December 2009, the 802.11 task group TGaf was formed to define the 802.11af amendment (not to be confused with IEEE 802.3af for Power over Ethernet). This amendment to the 802.11 standard addresses the use of TV white space frequency bands for use with wireless networking technology. These lower frequencies propagate well over longer distances and may be ideal to provide Internet connectivity for rural or semirural areas as well as other types of wireless technology innovation.

High-Density Deployments

What does the term *high-density Wi-Fi deployment* really mean? People have differing opinions because it can be subjective. Some industry experts claim that in the next few years the number of installed wireless devices will exceed the number of installed wired devices. When you think about it, this projection may be realistic. Take a moment and count the number of wireless devices that you have in your possession, at your home, the classroom, and the office. This includes notebook computers, smartphones, tablets, and broadband Internet devices. The average person may have between one and five separate wireless devices, many of which include IEEE 802.11 wireless technology. This gives you an idea of how the density of wireless LAN devices in all environments—home, office, education, and industrial—will continue to increase in the coming years.

Some high-density deployments have already been discussed in this chapter, including educational institutions and healthcare deployments such as in hospital environments. Issues to consider in this type of environment are the frequency band to use, co-channel interference, cell sizing, and access point capacity.

Municipal, Law Enforcement, and Transportation Networks

Wireless LANs are valuable in the industrial, municipal, and law enforcement fields and in transportation networks.

Federal and local law enforcement agencies frequently maintain state-of-the-art technology utilizing computer forensics and wireless LAN technology. Technologies that use 19.2 Kbps connectivity are becoming obsolete because of their slower data transfer rates. Municipal deployments that include police, fire, utilities, and city or town services are often all connected to a common wireless LAN.

Transportation networks are no exception. Wireless LAN installations are becoming more common in places like commuter buses, trains, airplanes, and automobiles. Users can connect for free or by paying a nominal fee. This type of connectivity now allows a user to better employ idle time. This is especially helpful to the mobile user or "road warrior" who needs to make the best use of available time.

Building-to-Building Connectivity Using Wireless LAN Technology

Connecting two or more wired LANs together over some distance is often necessary in computer networking. Depending on the topology, this can be an expensive and time-consuming task. Wireless LAN technology is often used as an alternative to copper cable, fiber optics, or leased line connectivity between buildings. When connecting two or more locations, point-to-point or point-to-multipoint links can be a quick and cost-effective solution for building-to-building connectivity.

When at least two wired LANs are connected, it's known as a *point-to-point connection*. Wireless point-to-point connections can provide long-range coverage depending on terrain and other local conditions. These links can serve both wired and wireless users on the connected local area networks. Wireless point-to-point connections typically call for semidirectional or highly directional antennas. Correct antenna selection is important and will be discussed in more detail in Chapter 3. Figure 4.13 shows a wireless point-to-point connection.

 With some regulatory agencies, when an omnidirectional antenna is used in a point-to-point configuration, it is automatically considered a point-to-multipoint connection.

FIGURE 4.13 A wireless point-to-point connection using directional antennas

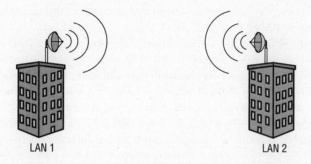

A network connecting more than two LANs is known as a point-to-multipoint connection. With wireless networking, this configuration usually consists of one omnidirectional antenna and multiple semi- or highly directional antennas (see Figure 4.14). Point-to-multipoint connections are often used in campus-style deployments where connections to multiple buildings or locations may be required.

FIGURE 4.14 A typical point-to-multipoint connection using omnidirectional and directional antennas

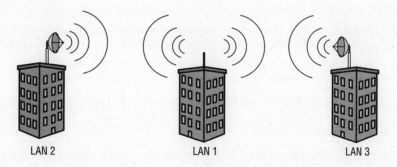

Wireless point-to-point and point-to-multipoint connections are becoming more common with many IEEE 802.11 wireless LAN deployments, thanks to the low cost of the equipment and the ease of installation. The installation time for a wireless point-to-point connection can be as little as a few hours.

Summary

The IEEE is an organization that creates standards and amendments used for wireless communication technologies. This chapter described the released communication standards for wireless LANs, wireless PANs, and wireless MANs.

Amendments to the IEEE 802.11 standard that addressed quality of service, security, fast transition, radio resource management, and management frame protection were also discussed. The Wi-Fi Alliance is an organization that addresses interoperability testing for equipment manufactured to the IEEE standards. The testing results in a variety of certifications for communication, quality of service, security, and voice enterprise.

We looked at many scenarios in which IEEE wireless LANs are currently used, from small office, home office and corporate deployments to last-mile connectivity. The use of standards-based wireless deployments continues to grow at a fast pace as new installations are implemented or proprietary and legacy-based installations are updated.

Chapter Essentials

Understand the function and roles of organizations that are responsible for the development of WLAN technology. The IEEE and the Wi-Fi Alliance play important roles with wireless technology. Know the function and role of each organization.

Know the details for IEEE 802.11 wireless LAN communication standards. Understand the details of the 802.11, 802.11b, 802.11a, 802.11g, 802.11n, and 802.11ac standard and amendments. It is important to know the operating radio frequency, supported data rates, and PHY layer technology of each.

Know the purpose of IEEE-specific function amendments. Be familiar with the details of 802.11e- and 802.11i-specific function amendments. Know that 802.11e is for quality of service and 802.11i addresses security.

Be familiar with IEEE 802.15 and IEEE 802.16 standards. IEEE 802.15 specifies personal area networking (PAN) and includes WPAN. IEEE 802.16 defines wireless metropolitan area network (MAN) for broadband wireless network technology.

Understand the differences among interoperability certifications from the Wi-Fi Alliance. Know the purpose of the WPA, WPA 2.0, WMM, and WMM-PS Wi-Fi Alliance certifications. Understand which address security, quality of service, and power-save features.

Understand details of common wireless deployment scenarios. These common WLAN applications can include small office, home office (SOHO), corporate data access, end-user mobility, and building-to-building connectivity.

Chapter 5

IEEE 802.11 Terminology and Technology

TOPICS COVERED IN THIS CHAPTER:

- ✓ Wireless LAN Operating Frequencies and RF Channels
- ✓ Wireless LAN Modes of Operation and the SSID
- ✓ IEEE 802.11 Physical Layer Technology Types
- ✓ Wireless LAN Coverage and Capacity
- ✓ Radio Frequency Signal Measurements
- ✓ Connecting to an IEEE 802.11 Wireless Network
- ✓ The Distribution System
- ✓ Data Rates
- ✓ Throughput
- ✓ Wireless LAN Transition or Roaming
- ✓ IEEE 802.11 Power Save Operations
- ✓ IEEE 802.11 Protection Modes and Mechanisms

This chapter looks at the technology used in IEEE 802.11 wireless networking, including ad hoc and infrastructure models. You will learn about RF channels and the frequencies of the unlicensed radio frequency bands. We will explore spread spectrum and other Physical layer (PHY) technologies. We will look at RF signal measurements, including received signal strength indicator (RSSI) and signal-to-noise ratio (SNR). You will also learn about IEEE 802.11 passive and active scanning as well as the authentication and association processes.

The distribution system allows access points to communicate with each other. It gives wireless devices the capability to transition or move between access points and maintain consistent connectivity across the wireless network.

This chapter will discuss the differences between data rates and throughput. Finally, we will look at IEEE 802.11 protection mechanisms. Some of the topics we will see in this chapter have been briefly touched on in earlier chapters. One of the goals of this chapter is to tie the terminology and topics together.

Wireless LAN Operating Frequencies and RF Channels

As discussed in Chapter 3, "Radio Frequency and Antenna Technology Fundamentals," RF spectrum is governed by local regulatory agencies. The country where the RF is used determines the regulations, such as frequency use and maximum power. Table 5.1 illustrates some examples of local RF regulations.

TABLE 5.1: Local RF regulations

Location	Regulation
Canada	Industry of Canada (ISC) RSS-210
China	RRL/MIC Notice 2003-13
Europe	ETS 300.328 ETS 301.893
Israel	Ministry of Communications (MOC)

Location	Regulation
Japan	TELEC 33B TELEC ARIB STD-T71
Singapore	IDA/TS SSS Issue 1
Taiwan	PDT
United States	Federal Communications Commission (FCC) (47 CFR) Part 15C, Section 15.247; FCC (47 CFR) Part 15C, Section 15.407

US (FCC) Unlicensed Frequency Bands

In the United States, the Federal Communications Commission (FCC) is the local regulatory agency responsible for regulating licensed and unlicensed radio spectrum. Listed here are the unlicensed RF bands available in the United States for use with IEEE 802.11 wireless LAN communications:

- ISM (industrial, scientific, and medical)
    - 902–928 MHz (not specified for use with standards-based IEEE 802.11 wireless networks)
    - 2.400–2.4835 GHz
    - 5.725–5.875 GHz
- U-NII (Unlicensed National Information Infrastructure)
    - 5.15–5.25 GHz: U-NII-1, lower
    - 5.25–5.35 GHz: U-NII-2, lower middle
    - 5.470–5.725 GHz: U-NII-2e, upper middle
    - 5.725–5.825 GHz: U-NII-3, upper

The 5.725–5.875 GHz ISM band is used in the United States and a few other areas for a single IEEE 802.11 channel (channel 165). The IEEE 802.11-2012 standard states that "the OFDM PHY shall not operate in frequency bands not allocated by a regulatory body in its operational region. Regulatory requirements for a given frequency band are set by the regulatory authority responsible for spectrum management in a given geographic region or domain." The FCC in the United States allows this frequency for IEEE 802.11 wireless networking.

The IEEE 802.11 standard addresses the 2.4 GHz ISM band and the 5 GHz U-NII bands. The 5 GHz U-NII band consists of four bands utilizing four frequency ranges: U-NII-1, the lower band; U-NII-2 and U-NII-2e, the middle bands; and U-NII-3, the upper band. Table 5.2 shows a summary of the unlicensed frequency bands and number of channels for each band used with IEEE 802.11 wireless LAN technology. Keep in mind that the 5.725–5.875 GHz ISM band is used only where allowed by the local regulatory agency.

TABLE 5.2: IEEE 802.11 frequency usage and channel allocations

Band	Frequency Range	Number of RF channels
ISM	2.400–2.4835 GHz	14
U-NII-1	5.150–5.250 GHz	4
U-NII-2	5.250–5.350 GHz	4
U-NII-2e	5.470–5.725 GHz	12
U-NII-3	5.725–5.825 GHz	4
ISM	5.725–5.8750 GHz	1

A chart of the United States frequency allocations is available from the National Telecommunications and Information Administration. To view this chart, visit www.ntia.doc.gov/osmhome/allochrt.pdf. The Federal Communications Commission (FCC) now allows the use of channel 144 in the U.S. based on the ratification of the IEEE 802.11ac amendment which increased the number of available channels in the U-NII-2e band from 11 to 12.

Radio Frequency Channels

As you have seen, radio frequency is divided into bands. These bands can be further separated into *channels*. A channel is a smaller allocation of the radio frequency band.

One familiar application in which this is accomplished is television. Until over-the-air television became available in digital format, television was allocated certain frequency ranges. Common television channels operated in the very high frequency (VHF) band—for example, channels 2 through 13 operated from 54 through 216 MHz. This frequency range was divided into 12 channels, allowing optimal use of the frequency range for the application, in this case television signals. A viewer can change channels on a television to watch different programs running simultaneously. However, only one program can be viewed at any one time depending on which channel is currently selected. (Picture-in-picture televisions can show two or more channels at once on the screen, but each picture is still being received on a different channel.)

Wireless LANs use channels in the same way. Certain frequency ranges are allocated for wireless networking, and those frequency ranges are subdivided into channels. For a transmitter and receiver to communicate with one another, they must be on the same RF channel. The 2.4 GHz ISM band has a total of 14 channels available for IEEE 802.11 wireless networking. The locale where they are used will determine which of the 14 channels can be legally used for this technology. In the United States, IEEE 802.11b/g/n wireless networks use 11 of the 14 channels available in the 2.4 GHz ISM band. Each of these 11 channels for DSSS or HR/DSSS is 22 MHz wide, and for OFDM it is 20 MHz wide. Understand that these channels are further defined by their center frequency; for example, channel 1 in the 2.4 GHz ISM band has the center frequency at 2.412 GHz. Simple mathematics show there will be some overlap in order to accommodate all of the 20 MHz or 22 MHz wide channels in this frequency range. Table 5.3 shows the 14 available channels in the 2.4 GHz range with examples of which channels are used in certain local regulatory domains.

TABLE 5.3: Radio frequency channels in the 2.4 GHz ISM band

Channel Number	Frequency in GHz	United States	Europe	Israel*	China	Japan
1	2.412	✓	✓	✓	✓	✓
2	2.417	✓	✓	✓	✓	✓
3	2.422	✓	✓	✓	✓	✓
4	2.427	✓	✓	✓	✓	✓
5	2.432	✓	✓	✓	✓	✓
6	2.437	✓	✓	✓	✓	✓
7	2.442	✓	✓	✓	✓	✓
8	2.447	✓	✓	✓	✓	✓
9	2.452	✓	✓	✓	✓	✓
10	2.457	✓	✓	✓	✓	✓
11	2.462	✓	✓	✓	✓	✓
12	2.467		✓	✓		✓
13	2.472		✓	✓		✓
14	2.484					✓

*Israel allows only channels 5–13 outdoors but 1–13 indoors.

Figure 5.1 shows the 14 available channels and the amount of overlap in the 2.4 GHz ISM band.

FIGURE 5.1 The 2.4 GHz ISM band allows for 14 RF channels.

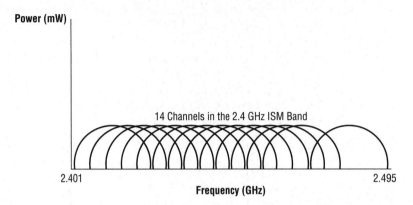

Of these 14 channels, mathematically there are only three adjacent nonoverlapping channels, with the exception of channel 14. According to the IEEE 802.11-2012 standard, "channel 14 shall be designated specifically for operation in Japan." Channel 14 is separated by 12 MHz on the center frequency from channel 13, whereas channels 1 through 13 are separated by 5 MHz on the center frequency of each channel. There are 3 MHz of separation where the radio frequency of one channel ends and the next adjacent nonoverlapping channel begins. For example, channel 1 and channel 6 are adjacent nonoverlapping channels. Channel 1 ends at 2.423 GHz and channel 6 begins at 2.426 GHz. Mathematically this is a separation of 3 MHz. This means that three access points can be colocated in the same physical space without experiencing overlapping channel interference. However, there is still theoretically a small amount of overlapping RF, or harmonics, between these two channels. This small level of overlap is not large enough to cause any real interference issues. Figure 5.2 illustrates 3 of the first 11 channels that do not overlap in the 2.4 GHz ISM band.

FIGURE 5.2 Three nonoverlapping channels possible in the 2.4 GHz ISM band

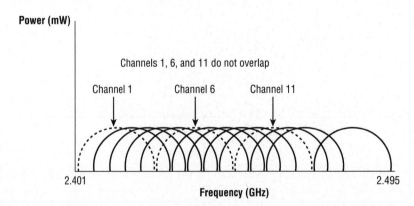

Each DSSS channel is 22 MHz wide. Using spread spectrum technology, a 22 MHz wide channel helps add resiliency to interference for data transmissions and gives the capability to move large amounts of data with a small amount of power. Some early IEEE 802.11 devices included bar code scanners, and they worked with limited battery life. Using a spreading technology instead of narrowband technology helped to conserve battery life and increased the use of IEEE 802.11 devices as a whole.

The 5 GHz U-NII band that is used with IEEE 802.11 wireless networking is divided into smaller sections of frequency ranges and then into RF channels. The 5 GHz U-NII band consists of four smaller frequency ranges —lower, lower middle, upper middle, and upper. Since there are fewer channels in the same amount of frequency space compared to the 2.4 GHz ISM band, channels in the U-NII band do not overlap. In the 5 GHz U-NII band, channels are 20 MHz wide. Table 5.4 shows the 5 GHz U-NII band for the FCC and ETSI locales.

TABLE 5.4 Channels in the 5 GHz bands

Locale	Frequency	Number of channels
Americas/EMEA	U-NII-1 band (5.15–5.25)	4
Americas/EMEA	U-NII-2 band (5.25–5.35)	4
Americas/EMEA	U-NII-2e band (5.470–5.725)	12
Americas/EMEA (with restrictions)	U-NII-3 band (5.725–5.825)	4
Americas	ISM (5.725–5.850)	1

Radio Frequency Range

The range a wireless LAN signal will travel is based on the wavelength or distance of a single cycle. The higher the frequency, the shorter the range of the RF signal; the lower the frequency, the longer the range of the RF signal. At the same output power level, a 2.4 GHz signal will travel almost twice as far as a 5 GHz signal.

If a wireless network design is going to have dual-band access points, RF range will need to be considered to ensure proper coverage for both the 2.4 GHz ISM and 5 GHz U-NII bands. A wireless site survey will help determine the usable range an access point will produce. A survey can involve physically walking around the proposed space and/or predictive modeling using one of many software programs. This process is discussed further in Chapter 8, "Site Survey, Capacity Planning and Wireless Design."

Wireless LAN Modes of Operation

IEEE 802.11 wireless LANs can be configured to operate in different modes for device and user access: ad hoc mode and infrastructure mode. These two modes can be broken down into four different service set configurations:

- Independent basic service set (IBSS)
- Basic service set (BSS)
- Extended service set (ESS)
- Mesh basic service set (MBSS)

The application/deployment scenario for a wireless LAN is the determining factor for the best mode to use.

IBSS configuration does not require the use of a wireless access point, and unless this configuration is specifically justified, it is not used in enterprise wireless LAN deployments. In addition, if not properly implemented, the IBSS can introduce security vulnerabilities, such as potentially bridging a wired network infrastructure to an unsecured wireless network.

The most common configuration for an IEEE 802.11 wireless LAN is infrastructure mode, which uses at least one access point. Infrastructure mode requires a minimum of one access point but can consist of up to thousands. The access points are connected by a common medium known as the distribution system. BSS, ESS, and MBSS are all configurations that use wireless infrastructure mode and access points. These operation modes are discussed in more detail later in this chapter.

The basic service area (BSA) is the area of radio frequency coverage, or the RF cell that encompasses a wireless LAN access point and its associated stations. A wireless client device will be contained in the basic service area as long as it has enough required signal strength to maintain an association state with the wireless access point.

Three parameters must be set on the wireless devices that wish to participate in a wireless LAN:

- The service set identifier (SSID)
- The radio frequency channel
- Security configuration

These parameters must be the same on all the devices in order for them to effectively communicate with one another. These wireless LAN configuration parameters are part of every type of wireless LAN installation.

We will now look at each of the configurations and the details of how they operate. The discussion of parameters will be explored in the greatest detail in the next section, "The Independent Basic Service Set (IBSS)," where it is particularly important to understand the issues related to the parameters.

The Independent Basic Service Set (IBSS)

It is important to understand what the *independent basic service set (IBSS)* is, how it works, and its potential uses, advantages, and disadvantages. This wireless LAN operation configuration uses no access points (APs) and consists of only wireless mobile devices or client computers. This is the most basic type of IEEE 802.11 service set, consisting of a minimum of two stations (STAs), of which neither is an AP station. Communication occurs only among the wireless devices that are part of the same IBSS. Unlike when using an access point, this configuration has no centralized control or manageable security or accounting features. Figure 5.3 shows devices in an IBSS.

FIGURE 5.3 Example of an independent basic service set (IBSS)

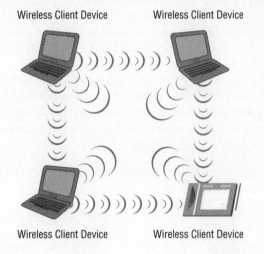

IBSS Terminology

The wireless LAN industry uses several different terms to identify an IBSS. The term used is up to the manufacturer or depends on a specific implementation. An IBSS is usually identified by one of three terms:

- Independent basic service set (IBSS)
- Ad hoc
- Peer-to-peer

Regardless of the terminology used—*IBSS*, *ad hoc*, or *peer-to-peer*—it comes down to wireless LAN devices connecting to each other without the use of an access point or other wireless infrastructure device connected to a distribution system. All devices in an IBSS network work independently of one another, and there is no centralized management or administration capability. This type of connection may be useful in homes or small offices for ease of installation but is rarely if at all used with enterprise or corporate wireless networks.

The Service Set Identifier (SSID)

The *service set identifier (SSID)* is a common parameter used in all wireless LAN operation configurations. The SSID is the logical name of the service set used to easily identify the wireless network using wireless network discovery utilities. The SSID is used by wireless devices to select a wireless network to join. This is accomplished through processes that are known collectively as the discovery phase and include passive and active scanning, both of which will be discussed later in this chapter.

In some cases, choosing a name for a wireless network can be a tough decision. Organizations that deploy a wireless network may already have a naming convention in place for such scenarios. If not, a decision will need to be made regarding the wireless network names (SSIDs) used for access points and other devices to identify the wireless LAN.

Every device that wishes to be part of the same wireless LAN IBSS, BSS, or ESS will use a common network name, the SSID. (See Figure 5.4 for an IBSS example.) For infrastructure devices such as wireless LAN access points, the SSID parameter is manually set on the access point. From the client access side, the SSID is a user-configurable parameter that can be set manually in the wireless client software utility or received automatically from networks that broadcast this information element.

FIGURE 5.4 IBSS, ad hoc, or peer-to-peer network using common configuration parameters

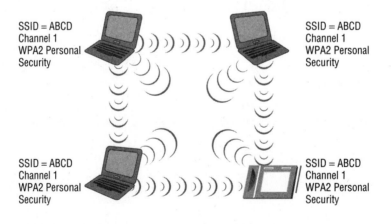

The SSID should be unique and should not divulge who you are or the location of the wireless LAN devices, unless you are trying to create a wireless hotspot or a publicly accessible IEEE 802.11 wireless network. For example, if a fictitious bank by the name of ABC Bank used an SSID like ABC_Bank, it would identify exactly where the wireless network is and could be a potential security threat because financial institutions are common targets. However, it is important to understand if the proper wireless LAN security is enabled, the SSID should not be an issue, but there is no point in broadcasting certain types of information. Rather than using the bank name in the SSID, consider a unique name that does not describe the type of business where it is used or exact location.

The SSID assigned to a wireless LAN service set is case sensitive and must include a minimum of one character and a maximum of 32 characters or, as specified in the IEEE 802.11 standard, 32 octets because an ASCII character consists of 8 bits. An octet is 8 bits in length.

SSID Hiding

Most manufacturers of small office, home office (SOHO) and enterprise-grade access points allow the SSID to be hidden from view for devices attempting to locate a wireless network. In this case, a client device would need to know and specify the SSID in the client device software utility profile in order to connect to the network. Even though this is not an effective way to secure a wireless network and should not be used to do so, it is a practice some choose to use for various reasons, including to prevent unauthorized devices from attempting to access a network in which they do not have the proper authority or credentials. Some disadvantages to not broadcasting the SSID are that it may cause an increase in transition times in an enterprise deployment, and in a SOHO deployment, it may cause neighbors to deploy on the same channel that you are using because they are unable to see your network. I have also seen wireless devices that are unable to connect to the network if the SSID is hidden, even if it is specified in the client device software profile. A network on which SSID hiding is used is also known as a closed network.

Figure 5.5 shows an example of entering the SSID in the Microsoft Windows 7 AutoConfig wireless configuration client utility for an ad hoc network. First you select "Set up a wireless ad hoc (computer-to-computer) network" in the Set Up A Connection Or Network dialog box; then you enter the SSID on the "Give your network a name and choose security options" properties page.

FIGURE 5.5 Entering the SSID and other parameters in the Microsoft Windows 7 wireless configuration client utility

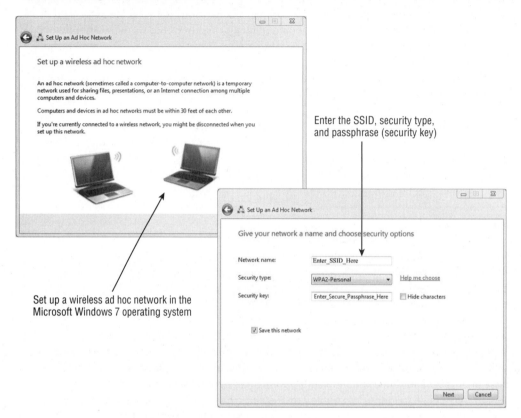

The Radio Frequency Channel

The IBSS wireless network configuration requires a user to set the specific radio frequency channel that will be used by all devices that are part of the same IBSS network. This is accomplished in the client software utility for the wireless network adapter. Some client software utilities set this automatically, in which case the IBSS will use the channel automatically specified. You may also be able to specify the RF channel in the advanced properties of the wireless network adapter device driver properties.

It is important to understand that all wireless devices in any common IBSS must be communicating on the same radio frequency channel. If the client utility does allow a channel to be set, the channel chosen is up to the user but based on the local regulatory domain in which the network is used. Additional devices wishing to join the IBSS must do so by scanning, either passively or by use of active scanning. Figure 5.6 shows an example of setting the RF channel on a notebook computer.

FIGURE 5.6 Setting the RF channel for an IBSS, ad hoc wireless network in the Intel 5100 IEEE 802.11a/g/n wireless network adapter driver advanced settings page

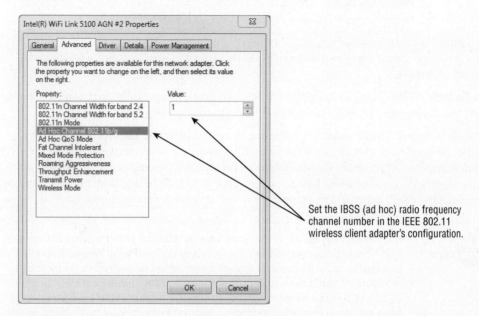

Set the IBSS (ad hoc) radio frequency channel number in the IEEE 802.11 wireless client adapter's configuration.

IBSS Security

With IBSS networks, there is no centralized control and no security management features. Security is left up to the individual user or wireless client device. If a user inadvertently shares a resource, it could expose sensitive information and pose security threats to the organization where it is used. This can be a concern for many enterprise installations, and therefore the use of an IBSS is against corporate security policy in most enterprise wireless deployments.

Advantages and Disadvantages of an IBSS

The advantages and disadvantages of an IBSS network will vary depending on the application.

Some of the advantages of IBSS are as follows:

Easy to Configure   To create an IBSS, the user only needs to specify an SSID, set the radio frequency channel, and enable the security settings.

No Investment in Wireless Infrastructure Equipment   An IBSS can be created with the IEEE 802.11 wireless LAN adapter that is built into the wireless computer or other device. No infrastructure device such as an access point is required to connect the wireless LAN devices together.

Here are some disadvantages of IBSS:

Limited Radio Frequency Range   Because radio communications is two-way, all devices need to be in a mutual communication range in order to operate effectively.

No Centralized Administration Capability   In many large or enterprise deployments, IBSS connectivity is against corporate security policy because it is impossible to manage such networks centrally.

Not Scalable   There is no set maximum number of devices that can be part of an IBSS network, but the capacity of such networks is low compared to other types of networks.

Difficult to Secure   Some computer operating systems have made the setup of an IBSS wireless network very easy for any type of user. Users may inadvertently share or allow access to sensitive or proprietary information. This security threat is worse if an IBSS user is also physically connected to a wired network and provides a bridge from an unsecured or unmanaged wireless network to a company's wired network infrastructure.

A wireless LAN client device such as a notebook computer with the wireless networking capabilities configured to be in an IBSS network can be a potential security threat if it is also connected to a wired network infrastructure. It could provide a bridge for unsecured wireless access to the company's wired network. In this configuration, potential intruders would have access to information from a corporate network by connecting to the unsecured ad hoc network. For this reason, this type of configuration is against the corporate security policies of most companies. Many organizations use wireless intrusion prevention systems to detect and shut down wireless IBSS (ad hoc) networks. It is important to inform visitors and contractors who may be physically connected to the company's infrastructure when this type of wireless network is against the corporate security policy to prevent potential security issues.

Setting up an IBSS network is similar to setting up a workgroup for a computer operating system such as Microsoft Windows. All devices with the same workgroup name will be able to communicate with each other, sharing resources such as files, printers, and so on. Although this configuration is against most corporate security policies and not recommended in corporate or enterprise networks, its use may be justified in some instances. One such example includes the home or SOHO wireless device market for use with streaming video or other multimedia applications that include IEEE 802.11ad technology.

The Basic Service Set (BSS)

The *basic service set* (BSS) is the foundation of the wireless network. This mode consists of an access point connected to a network infrastructure and its associated wireless devices.

This is considered the foundation because it may be one of many access points that form a wireless network. With a BSS setup, each access point is connected to a network infrastructure, also known as the distribution system (DS), which allows connected wireless LAN client devices to access network resources based on the appropriate permissions the wireless device or user has access to. The radio frequency area of coverage depends on several factors, such as the antenna gain and RF output power settings; this area of RF coverage is known as the *basic service area (BSA)*. Any IEEE 802.11 wireless device in radio range and part of the BSA with the correct configuration parameters, including the SSID and security settings, will be able to successfully connect to the access point. Figure 5.7 shows an example of a BSS.

FIGURE 5.7 Basic service set consisting of a single access point connected to a distribution system and its associated stations

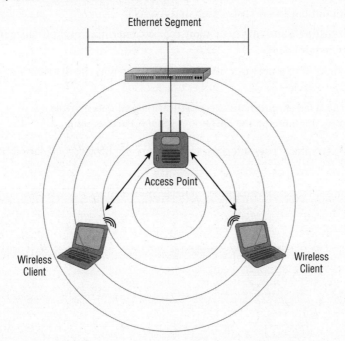

Infrastructure mode consists of a wireless access point connected to a distribution system. The BSS consisting of one access point is a common implementation in many homes, SOHO, or small to medium businesses (SMBs). The decision to use a single access point depends on several factors, among them the size of the location, how the wireless network is used, and how many wireless devices will be connected.

Just as in an IBSS configuration, several parameters must be configured for a BSS. These include the SSID or name of the network, the radio frequency channel to be used, and any security parameters that are set on the BSS. The access point will broadcast these and other parameters about the wireless network to devices that want to connect to the BSS, thus requiring minimal configuration on the wireless client side. Unlike the independent basic service set (IBSS), in a BSS the radio frequency channel is configured on the access point and not on the wireless client devices.

A BSS has many advantages and disadvantages. Some of the advantages are as follows:

- Uses intelligent devices with a large feature set to provide users with consistent, reliable, and secure communications to a wireless network.
- Useful in a variety of situations: homes, SOHO, and small to large businesses.
- Very scalable; you can increase the coverage and capacity of a wireless network by adding additional access points.
- Centralized administration and control.
- Security parameters and specific access can be set centrally.

Some of the disadvantages of a BSS are listed here:

- Incurs additional hardware costs compared to IBSS.
- Usually will require a site survey of some type to determine radio frequency coverage and capacity requirements.
- Must be connected to a network infrastructure known as the distribution system, either wired or wireless.
- Additional knowledge required for configuration and deployment.

Figure 5.8 shows configuring the SSID on an SOHO access point.

FIGURE 5.8 Graphical user interface to configure the SSID for a D-Link DGL-5500 Gaming router

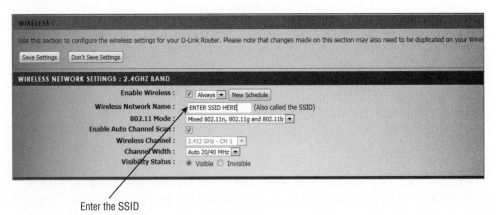

The Extended Service Set (ESS)

As stated in the IEEE 802.11-2012 standard, an *extended service set (ESS)* is defined as a "set of one or more interconnected basic service sets (BSSs) that appears as a single BSS to the logical link control (LLC) layer at any station (STA) associated with one of those BSSs." In basic terms, this can be one or more access points connected to a common distribution system. An ESS is a common configuration in most wireless LAN deployments for small to medium businesses as well as large enterprise organizations.

In most cases, an ESS would be used to provide consistent and complete coverage across an entire organization. An ESS can be thought of as several basic service sets (BSSs) that must have matching parameters, such as SSID and security settings. If the SSIDs of two access points do not match, then they are considered separate basic service sets, even though they are connected by a common network infrastructure. It is the distribution system connecting these together and a common network name (SSID) that makes up the ESS. In most cases, the basic service area for each BSS will overlap to allow transition (roaming) from one BSS to another. Figure 5.9 shows an example of an extended service set (ESS).

FIGURE 5.9 Two basic service sets connected by a common distribution system, making an extended service set

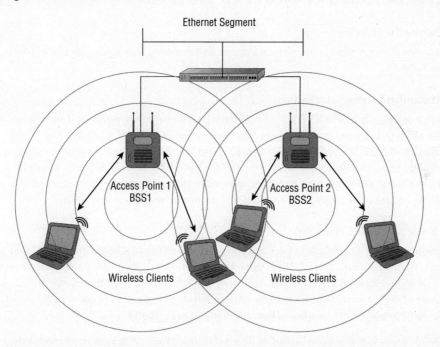

Transitioning (roaming) between access points is a critical component of wireless LAN technology in almost every wireless network deployment. This is because the wireless LAN is a major part of every corporate computer network. Many envision a complete wireless network for all communications, including data, voice, and video, and we are now closer to reaching that goal. Transition is so important that the IEEE added the 802.11r amendment to the standard to provide a standardized methodology for a fast secure transition within the enterprise wireless LAN.

The Mesh Basic Service Set (MBSS)

The *mesh basic service set (MBSS)* by definition in the IEEE 802.11-2012 standard is a basic service set (BSS) that forms a self-contained network of mesh stations (STAs) that use the same mesh profile. An MBSS contains zero or more mesh gates and can be formed from mesh STAs that are not in direct communication. In layman's terms, this is a self-forming intelligent network that allows redundant connections to multiple wireless infrastructure devices to ensure reliable communications. Mesh networking is not new technology. Mesh has been available for years with various computer networking models and continues to become a part of many wireless LAN deployment scenarios.

Know Your Abbreviations: SSID vs. ESSID vs. BSSID

It is easy to confuse the abbreviations (acronyms) for several of the wireless LAN terms. Here is a summary of the differences of some of these terms.

SSID (Service Set Identifier)

SSID (*service set identifier*) is the network name and provides some segmentation of the wireless network.

ESSID (Extended Service Set Identifier)

Although not defined by the IEEE 802.11 standard or amendments, *extended service set identifier* (*ESSID*) is a term that some manufacturers use in place of *SSID*. For the most part, *ESSID* and *SSID* are synonymous terms for the name or segmentation of a wireless network. The term used will vary among manufacturers. The term *ESSID* was adopted by some manufacturers because it implies that more than one access point is using the same SSID and security settings connected to a common distribution system.

BSSID (Basic Service Set Identifier)

It is sometimes easy to confuse the *basic service set identifier (BSSID)* with the SSID, or the name, of the network. The BSSID is defined as the unique identifier, the Media Access Control (MAC) address of the basic service set. It is important to note that some manufacturers may allow for several BSSIDs to be connected to a single access point radio or for a single common BSSID to be shared among many access points.

The MAC address is the unique identifier of a network adapter; it is known as the hardware address. The radio in an access point is also for the most part a network adapter. The difference between a wired and a wireless network adapter is simply that no Ethernet jack is available on a wireless adapter. Instead, a radio is used for Layer 1 communications.

The MAC address is a 48-bit IEEE 802 format address that uniquely identifies the network interface adapter—or in this case, radio. The format of the BSSID is *xx:xx:xx:yy:yy:yy*, where *x* is the number assigned to a manufacturer and *y* is the unique hardware address of the device.

Although the BSSID uniquely identifies the access point's radio using a MAC address, the SSID is broadcast as the name of the network in order to allow devices to connect. Therefore, the SSID is a logical name mapping to the BSSID. Some devices allow for multiple SSIDs, which use multiple BSSIDs for a single radio. This lets a single access point connected to a wired infrastructure provide multiple WLANs using a single radio.

In an ad hoc or IBSS network, there is no access point for centralized communication. Instead, wireless LAN devices communicate directly with each other. Because there is no access point in this configuration, the BSSID is a randomly generated number that has the same format as the 802 MAC address and is generated by the first ad hoc wireless device at startup. Any subsequent wireless devices that join the IBSS will use the BSSID that was generated by the wireless station that started the IBSS.

IEEE 802.11 Physical Layer Technology Types

In the following sections, you will learn about the different Physical layer (PHY) technologies used with IEEE 802.11 wireless networking:

- Spread spectrum technology
- Extended rate physical (ERP)
- Orthogonal frequency division multiplexing (OFDM)

The type of PHY technology used will vary based on the communication methods, modulation type, and several other factors. That will be explained in more detail in the sections that follow.

Spread Spectrum Technology

Two types of *spread spectrum* technology were specified in the original IEEE 802.11 wireless LAN standard ratified in 1997:

- Frequency hopping spread spectrum (FHSS)
- Direct sequence spread spectrum (DSSS)

These spread spectrum technologies communicate in the 2.4 GHz ISM frequency range. There are advantages and disadvantages to each of these spread spectrum types.

Spread spectrum technologies take the digital information generated by a computer (ones and zeros) and, through the use of modulation technologies, send it across the air between devices using radio frequency (RF).

In order for devices to communicate effectively and understand one another, they must be using the same spread spectrum and modulation technology. This would be analogous to two people trying to talk with each other. If the two people don't know the same language, they will not be able to understand each other and a conversation could not take place.

Frequency Hopping Spread Spectrum

Frequency hopping spread spectrum (FHSS) is used in a variety of devices in computer technology and other communications. FHSS was used by many early adopters of IEEE 802.11 wireless networking, including with computers, bar code scanners, and other handheld or portable devices. Although defined in the original IEEE 802.11 standard, this technology is considered legacy (out-of-date) technology in IEEE 802.11 wireless networking. However, FHSS is still found today in many devices such as cordless telephones and IEEE 802.15 wireless personal area networks (WPANs), including Bluetooth mice, cameras, phones, wireless headsets, and some older wireless LAN technology devices. Bluetooth technology using FHSS is comparatively slower than newer IEEE 802.11 wireless communications.

FHSS operates by sending small amounts of information such as digital data across the entire 2.4 GHz ISM band. As the name implies, this technology changes the frequency ("hops") constantly in a specific sequence or hopping pattern and remains on a frequency for a specified amount of time known as the *dwell time*. The dwell time value will depend on the local regulatory domain where the device is used. In the United States, for example, the Federal Communications Commission allows a maximum dwell time of 400 milliseconds. A transmitter and receiver will be synchronized with the same hopping sequence, therefore allowing the devices to communicate.

The data rate for IEEE 802.11 FHSS is only 1 and 2 Mbps, which is considered slow by modern computer applications. However, that data rate was more than adequate for some applications when the technology became available. For example, various wireless devices in retail and manufacturing used FHSS in handheld scanners and other portable technology for many years. The cost of upgrading these devices to support higher data rates was cost prohibitive and unnecessary in many cases.

Figure 5.10 illustrates what FHSS would look like if you could see the RF signal hopping through the entire 2.4 GHz ISM band.

802.11 Direct Sequence Spread Spectrum

Direct sequence spread spectrum (DSSS) is spread spectrum technology used with wireless LANs and defined by the original IEEE 802.11 standard. Like FHSS, DSSS supports data rates of 1 and 2 Mbps and is considered to be very slow by today's computer networking requirements.

DSSS uses special techniques to transmit the digital data (ones and zeros) across the air using radio frequency (RF). This is accomplished by modulating or modifying the RF characteristics such as phase, amplitude, and frequency (see Chapter 3.)

In addition to modulation, DSSS uses technology known as a *spreading code* to provide redundancy of the digital data as it traverses through the air. The spreading code transmits information on multiple *subcarriers*, and this redundancy helps the receiver detect transmission errors due to radio frequency interference. Subcarriers are smaller segments of the radio frequency channel that is in use. The spreading of information across the 22 MHz wide channel is what helps makes DSSS resilient when it comes to some types of interference. This spreading code technology allows the receiver to determine if a single bit of digital data (symbol) received is a binary 0 or binary 1. Depending on the data rate, the transmitter and receiver understand the spreading code in use and therefore are able to communicate.

FIGURE 5.10 FHSS hops the entire 2.4 GHz ISM band.

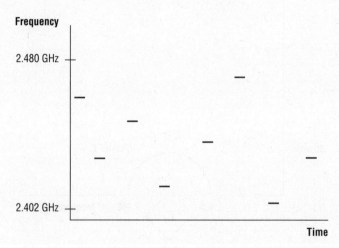

An example of a coding technique is Barker code. Barker code is used as the spreading code for DSSS at the data rates of 1 and 2 Mbps. IEEE 802.11 wireless LANs can use this 11 "chip" spreading code for communications. Each digital data bit (binary 1 or 0) is combined with a set Barker code through what is called an exclusive OR (XOR) process. XOR is a way of combining binary data bits in digital electronics. The result then spreads the binary 0 or 1 over a 22 MHz wide channel, helping to make it resilient to radio frequency interference. Since both the transmitter and receiver understand the same code, they would be able to determine the information that was sent over the air.

 A chip is part of a string of binary digits that are used with a Boolean exclusive OR (XOR) process to create an 11-bit symbol and prepare the digital data to be sent across the air.

DSSS operates within a range of RF frequency also known as a *channel*. The channel in a wireless LAN is defined by its center frequency; that is, channel 1 is 2.412 GHz on center, channel 2 is 2.417 GHz on center, and so on. Each channel in the 2.4 GHz ISM band is separated by 5 MHz on center with the exception of channel 14, which is 12 MHz on center with respect to channel 13. Unlike with narrowband communication, which operates on a single narrow frequency, a DSSS channel is 22 MHz wide and is one of 14 channels in the 2.4 GHz to 2.5 GHz ISM band. The country and location of the device will determine which of the 14 channels are available for use in that specific area.

Figure 5.11 shows that channel 6 is 22 MHz wide for DSSS in the ISM unlicensed RF band.

FIGURE 5.11 DSSS is limited to a 22 MHz wide channel in the 2.4 GHz ISM band. Each channel for DSSS is separated by 5 MHz on the center frequency.

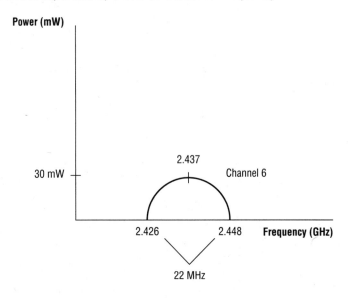

 FHSS and DSSS operate in the same frequency range. If devices that use both technologies are occupying the same physical area, the devices may encounter some level of interference. Therefore, RF interference may occur from either IEEE 802.11 wireless LAN devices using the same RF channel or non-IEEE 802.11 wireless LAN devices that are using the same RF channel. This includes anything wireless that does not use IEEE 802.11 technology but is operating in the same frequency range, such as 2.4 GHz wireless cameras, two-way radios or radio monitors, and even microwave ovens.

802.11b High-Rate Direct Sequence Spread Spectrum

High-rate direct sequence spread spectrum (HR/DSSS) is defined in the *IEEE 802.11b* amendment to the IEEE 802.11 standard. HR/DSSS (802.11b) introduced higher data rates of 5.5 and 11 Mbps. At the time the 802.11b amendment was released, because of the higher data rates, this technology helped fuel the acceleration of IEEE standards based on wireless LAN technology. The desire for 802.11 wireless LANs grew as the availability became greater and the cost decreased.

Like DSSS, HR/DSSS uses one of fourteen 22 MHz wide channels to transmit and receive digital computer data. The main difference between these two technologies is that HR/DSSS supports higher data rates of 5.5 Mbps and 11 Mbps.

HR/DSSS (802.11b) also uses a different spreading code or encoding technique than DSSS. HR/DSSS uses complementary code keying (CCK) for transmitting data at 5.5 and 11 Mbps. The detailed operation of CCK is beyond the scope of this book.

It is important to note that IEEE 802.11b technology is becoming less relevant in larger wireless networks and many administrators are disabling the data rates that support DSSS and HR/DSSS technology due to ERP protection mechanisms. This is true with both public guest and private networks and will help with the overall performance of the wireless network by optimizing the communications for newer technologies.

IEEE 802.11g Extended Rate Physical

The IEEE 802.11g amendment was released in 2003 and introduced technology that allowed for higher data rates for devices and operation in the 2.4 GHz ISM band. The objective of this amendment was to allow for these higher data rates (up to 54 Mbps) using orthogonal frequency division multiplexing (OFDM) and still maintain backward compatibility with existing 802.11b technology and devices. This technology, known as extended rate physical (ERP), builds on the data rates of 1 and 2 Mbps DSSS (802.11) and 5.5 and 11 Mbps HR/DSSS (802.11b). The 802.11g amendment addresses several compatibility operation modes:

- ERP-DSSS/CCK
- ERP-OFDM
- ERP-PBCC (optional)
- DSSS-OFDM (optional)

The *802.11g* amendment required support for ERP-DSSS/CCK and ERP-OFDM. This allowed for both the 802.11b data rates of 1, 2, 5.5, and 11 Mbps and the new OFDM data rates of 6, 9, 12, 18, 24, 36, 48, and 54 Mbps with a 20 MHz wide channel. Manufacturers of wireless LAN equipment implement this in various ways. In a graphical user interface, there may be a drop-down menu that allows a user to select a specific operation mode such as mixed mode, b/g mode, b-only mode, and so on. Another possibility is to select the individual data rates using radio buttons. For manufacturers that provide a command-line interface (CLI) option, the appropriate commands would need to be executed in order to enable or disable the desired data rates. Packet Binary Convolution Code (PBCC) is an optional feature defined in the IEEE 802.11g amendment. IEEE 802.11b provided data rates of 1, 2, 5.5, and 11 Mbps. PBCC extended these to allow for potential data rates of 22 and 33 Mbps. However, to achieve these extended data rates, the technology must be enabled on both the transmitter and the receiver, making this more a proprietary feature that was found only in SOHO wireless LAN equipment. Figure 5.12 illustrates an example of how manufacturers may allow a user to select the ERP operation mode on an access point.

FIGURE 5.12 Selecting an operation mode on a D-Link DGL-5500

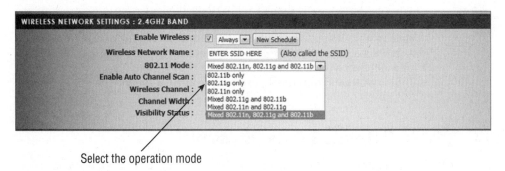

IEEE 802.11n High Throughput

The IEEE 802.11n amendment was ratified in September 2009. This new high throughput (HT) Physical layer (PHY) technology is based on the OFDM (PHY) in Clause 17 (802.11a) PHY. 802.11n HT allows extensibility of up to four spatial streams, using a channel width of 20 MHz. Also, transmission using one to four spatial streams is defined for operation in 20/40 MHz channel width mode. This technology is capable of supporting data rates up to 600 Mbps using four spatial streams with a 20/40 MHz channel. IEEE 802.11n HT provides features that can support a throughput of 100 Mbps and greater. The following optional features are included on both the transmitter and receiver sides:

- HT-Greenfield format
- Short guard interval (GI), 400 ns
- Transmit beamforming (TxBF)
- Space–time block coding (STBC)

The *802.11n* amendment allows for operation in both the 2.4 GHz ISM and 5 GHz U-NII bands with either 20 MHz or 40 MHz wide channels. Although 40 MHz wide channels are allowed in the 2.4 GHz ISM band, best practices recommend against it. Using a 40 MHz channel in this band would equate to only a single channel without any channel overlap.

It is important to understand the difference between data rate and throughput. Data rates are the rates at which a station is capable of exchanging information, whereas throughput is the rate at which the information is actually moving. Data rate and throughput are compared in more detail later in this chapter.

IEEE 802.11a, 802.11g, 802.11n, and 802.11ac Orthogonal Frequency Division Multiplexing (OFDM)

OFDM is used by the IEEE 802.11a (OFDM), IEEE 802.11g Extended Rate Physical Orthogonal Frequency Division Multiplexing (ERP-OFDM), IEEE 802.11n High Throughput (HT-OFDM), and IEEE 802.11ac Very High Throughput (VHT-OFDM) amendments to the IEEE 802.11 standard. OFDM allows for much higher data rate transfers than DSSS and HR/DSSS, up to 54 Mbps for 802.11a and 802.11g, potentially up to 600 Mbps for 802.11n, and 6.93 Gbps aggregate for 802.11ac.

Orthogonal frequency division multiplexing (OFDM) is a technology designed to transmit many signals simultaneously over one transmission path in a shared medium and is used in wireless and other transmission systems. Every signal travels within its own unique frequency subcarrier (a separate signal carried on a main RF transmission). *802.11a* and 802.11g OFDM distribute computer data over 52 subcarriers equally spaced apart, and 4 of the 52 subcarriers do not carry data and are used as pilot channels. 802.11n allows for 56 subcarriers, of which 52 are usable for data with a 20 MHz wide channel, and 114 subcarriers, of which 108 are usable for data with a 40 MHz wide channel. Having many subcarriers allows for high data rates in wireless LAN IEEE 802.11a and IEEE 802.11g devices. 802.11n devices (HT-OFDM) may use a MIMO technology known as spatial multiplexing (SM), which uses several radio chains to transmit different pieces of the same information simultaneously, greatly increasing throughput. In addition to high data rates, OFDM helps provide resiliency to interference from other wireless devices.

IEEE 802.11a, 802.11g, 802.11n, and 802.11ac OFDM Channels

OFDM functions in either the 2.4 GHz ISM or the 5 GHz U-NII band. The channel width is smaller than with DSSS or HR/DSSS. The width of an OFDM channel is only 20 MHz compared to 22 MHz for DSSS. Figure 5.13 shows a representation of a 20 MHz wide OFDM channel.

Like DSSS, when OFDM is used in the 2.4 GHz ISM band, there are only three nonoverlapping adjacent channels for use. This will limit the use of bonded channels (20/40 MHz wide channels) in IEEE 802.11n (HT-OFDM) deployments that are located in the same radio frequency physical area. In the 5 GHz U-NII bands, the channel spacing is such that there is no overlap. The frequency range used will determine how many nonoverlapping channels are available for use. In the lower and upper U-NII bands, 4 nonoverlapping channels are available. The middle U-NII band has 15 nonoverlapping channels available. All U-NII band channels are 20 MHz wide and separated by 20 MHz from the center frequencies of each channel. Certain regulatory domains, including the United States Federal Communication Commission (FCC) and the European Telecommunications Standards Institute (ETSI), require the use of dynamic frequency selection (DFS) support for wireless devices such as access points that operate in the middle 5 GHz (5.250–5.725 GHz) U-NII band. DFS will allow an access point to change the radio frequency channel it is operating on in order to avoid interfering with certain types of radar systems. Table 5.5 shows the 23 available channels, center frequency, and channel number in the 5 GHz U-NII band. Also displayed is a single 5 GHz ISM channel that is available from some regulatory agencies for use with wireless networking.

FIGURE 5.13 OFDM transmit spectral mask for 20 MHz transmission

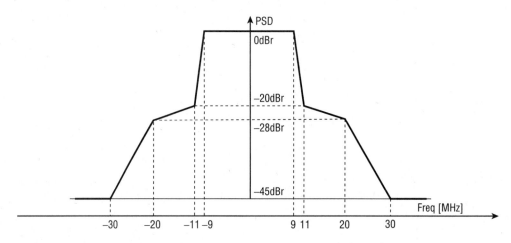

TABLE 5.5 Examples of RF channels in the 5 GHz U-NII band

Channel Number	Frequency in GHz	United States	Europe	Israel*	China	Japan
U-NII-1 Band - Lower						
36	5.180	✓	✓	✓	✓	✓
40	5.200	✓	✓	✓	✓	✓
44	5.220	✓	✓	✓	✓	✓
48	5.240	✓	✓	✓	✓	✓
U-NII-2 Band - Lower						
52	5.260	✓	✓	✓	✓	✓
56	5.280	✓	✓	✓	✓	✓

Channel Number	Frequency in GHz	United States	Europe	Israel*	China	Japan
60	5.300	✓	✓	✓	✓	✓
64	5.320	✓	✓	✓	✓	✓
U-NII-2 Band - Upper						
100	5.500	✓	✓			
104	5.520	✓	✓			
108	5.540	✓	✓			
112	5.560	✓	✓			
116	5.580	✓	✓			
120	5.600	✓	✓			
124	5.620	✓	✓			
128	5.640	✓	✓			
132	5.660	✓	✓			
136	5.680	✓	✓			
140	5.700	✓	✓	✓	✓	✓
144	5.720	✓				
U-NII-3 Band						
149	5.745	✓	✓	✓	✓	✓
153	5.765	✓	✓	✓	✓	✓
157	5.785	✓	✓	✓		✓
161	5.805	✓	✓			
5 GHz ISM Band						
165	5.825	✓	✓	✓		✓

The 5 GHz channel numbers may vary based on the IEEE 802.11 technology standard in use such as IEEE 802.11n or IEEE 802.11ac. The channel width will also vary. For example 20 MHz, 40 MHz, 80 MHz and 160 MHz wide chanels may be used with IEEE 802.11ac technology vs. 20 MHz and 40 MHz channels used with IEEE 802.11n technology. The Federal Communications Commission (FCC) now allows the use of channel 144 in the U.S. based on the ratification of the IEEE 802.11ac amendment.

Wireless LAN Coverage and Capacity

Coverage and capacity are two key factors to take into consideration when designing and implementing an IEEE 802.11 wireless LAN. During the design phase of an IEEE 802.3 wired network, the design engineer will take capacity into account, verifying and validating that there are enough capacity switches, ports, required infrastructure speeds, and so on for the user base of the network. The same is true for a wireless network. The number of devices/users connected to an access point is something that needs to be carefully considered. The fact that wireless networks use a shared medium is an issue because the more devices connected to an access point, the lower the performance may be, depending on what the devices are doing. This capacity consideration will ensure satisfied end users and excellent network performance—proof of a successful wireless network design and deployment.

In wireless networks, coverage also needs to be considered. Coverage is determined by the RF cell size. In IEEE 802.11 wireless networks, a cell is the area of RF coverage of the transmitter, in most cases an access point. Depending on implementation, wide coverage or large cell size may not be the best solution. A large space covered by a single access point could result in less than adequate network performance based on factors such as the users' distance from the access point. The farther away from an access point, the less throughput a device or user will experience. If users will be scattered throughout a large space, it may be best to have several access points covering the space to allow for optimal performance.

The term *cell* has several different meanings depending on the context. In the world of IEEE 802.11 wireless networks, a cell is the radio coverage area for a transmitter such as an access point or a client device.

Wireless LAN Coverage

The term *coverage* has different meanings depending on the context in which it used. For example, if you buy a gallon of paint, the label will specify the approximate coverage area in square feet. If one gallon of paint covers 300 square feet and the room you wish to paint is 900 square feet, simple math shows at least three gallons of paint would be needed to effectively cover the room.

The concept is similar in IEEE 802.11 wireless networking. However, unlike with paint, there is no simple rule that determines how much space an access point will cover with the RF energy it is transmitting. This coverage will depend on many factors, some of which are listed here:

- Physical size of the area
- Bandwidth-intensive software applications or hardware applications in use that may negatively impact the performance, therefore requiring smaller RF coverage cells

- Obstacles, including building materials and propagation (the way radio waves spread through an area)
- Radio frequency range
- WLAN hardware in use (affects coverage because higher frequencies, such as 5 GHz, do not travel as far as lower frequencies, such as 2.4 GHz)
- Transmitter output power

You might initially assume that you want the RF signal to propagate over the largest area possible. But this may not be the best solution. A very large cell may allow too many devices to connect to a single access point, causing a decrease in overall performance. For those client devices connected at a greater distance, the performance will be lower than for stations closer to the access point. Figure 5.14 shows a large coverage area, approximately 11,250 square feet (1,046 square miles) covered with a single access point. This is an example of an area that's too large for a single access point.

FIGURE 5.14 Wide coverage with only a single access point is not recommended.

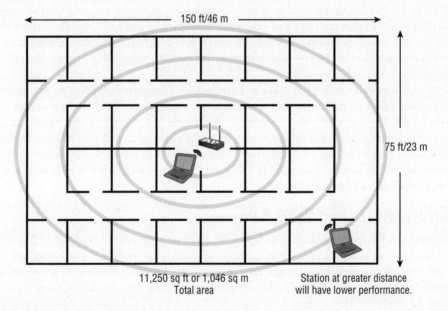

Physical Size of the Area

Rarely, if at all, will a manufacturer of enterprise-grade IEEE 802.11 wireless LAN hardware commit to the amount of area an access point will cover. There are too many variables to take into consideration, which makes it difficult to specify an exact number. However, some manufacturers may estimate the effective range of the device or access point. A site survey of the area will help determine the coverage area of an access point.

A manual survey will allow for testing to verify the distance a signal will travel. A predictive site survey will model the environment and determine the signal propagation. This concept will be discussed further in Chapter 8, "Site Survey, Capacity Planning and Wireless Design."

Applications in Use

The application types in use—either software or hardware—can affect the bandwidth of an access point. If the devices connected to an access point use bandwidth-intensive applications such as a computer-aided design/computer-aided manufacturing (CAD/CAM) application, it could result in poor throughput for all devices or users connected to that access point. This is another example of where more access points, with each covering a smaller area, could be a better solution than a single access point covering a large area. Multiple access points could allow the high-bandwidth users to be separated from other parts of the network, increasing overall performance of the network.

Obstacles, Building Materials, Propagation, and RF Range

Obstacles in an area (building materials such as walls, doors, windows, and furnishings) as well as the physical properties of these obstacles (thickness of the walls and doors, density of the windows, and type of furnishings) can also affect coverage. The radio frequency used—either 2.4 GHz or 5 GHz—will determine how well a signal will propagate and handle an obstacle.

For example, a wall made from sheetrock or drywall materials may have an attenuation value of about 3 dB to 4 dB, whereas a wall made of concrete may have an attenuation value of about 12 dB. Therefore, the sheetrock wall would have less impact on the RF propagation than the concrete wall. Partitions, walls, and other obstacles will also determine the coverage pattern of an access point because of the way RF behaves as it travels through the air. Behaviors of RF were be discussed in Chapter 3 in the section titled "Environmental Factors, Including Building Materials."

WLAN Hardware and Output Power

The wireless LAN hardware in use can also have an impact on the coverage area. Examples include the antenna type, antenna orientation, and gain of the antenna. The higher the gain of an antenna, the greater the coverage area; conversely, the lower the gain of an antenna, the smaller the coverage area. The polarization of an antenna (horizontal vs. vertical) will also have an effect on the coverage area because of the different shapes of the electromagnetic radiation patterns. The output power of the transmitter or access point will also have an effect on coverage. The higher the output power, the greater distance a signal will propagate. A higher power signal will provide more coverage. Most enterprise-grade access points provide the capability to control or adjust the output power.

Wireless LAN Capacity

One definition of *capacity* is the maximum amount that can be received or contained. For example, an elevator will typically have a maximum number of people or amount of weight it can hold; this is usually stated on a panel within the elevator. To ensure safety, the elevator may have a safety mechanism to prevent overloading. Likewise, a restaurant has a certain number of chairs for customers; therefore, it would have a maximum capacity of customers who can be served at any one time. Does this mean that when a restaurant fills its seats to capacity, the doors close and no other customers can enter the building? Not necessarily. In some cases, a restaurant could have customers standing and waiting to be seated.

Just as a limited number of people can be accommodated comfortably in an elevator or a restaurant, a limited number of devices can be handled by wireless access points, and that's known as capacity. The capacity of an access point is how many devices or users the AP can service effectively, offering the best performance based on the available bandwidth. Capacity depends on several factors:

- Software and hardware applications in use
- Desired throughput, or performance
- Number of devices/users

The following sections discuss how these factors affect the capacity of an access point.

 Real World Scenario

What Happens When an Access Point Is Overloaded?

If the capacity of a single access point has exceeded the maximum number of users or devices based on the performance metrics, access points may need to be added. If a wireless network is installed correctly, an access point will not be overloaded with an excessive number of users or devices. An overloaded access point will result in poor performance and therefore unhappy users.

To understand why, look back at the restaurant example. If a restaurant seats 20 customers and all 20 seats are taken, the restaurant has reached its seating capacity. Let's say the restaurant is short-staffed because two servers did not show up for work. The servers who did show up will have to work extra hard to handle the customers. This may cause delays in service because the servers need to handle more than their normal number of tables. The delays may result in unhappy customers.

The same is true for wireless access points. If a wireless access point has reached its capacity (the devices it can handle efficiently), it could get overloaded. This would result in it taking longer to handle any individual request for access. The delays may result in unhappy users. Therefore, this situation could justify adding another access point in the area to handle the additional users. Just as a restaurant will not close its doors when all seats are taken, an access point will continue to accept users trying to connect unless restrictions such as load balancing are implemented.

Software and Hardware Applications in Use

The software and hardware applications in use may affect the capacity of an access point. Some applications are more bandwidth intensive than others. For example, word processing applications may not require much bandwidth, but database or CAD/CAM applications may require much more than other applications. If high-bandwidth applications are in use, the contention among the connected users will increase because they are using a shared medium (air and RF). Therefore, performance will potentially be reduced for all users connected to the access point. The access point is providing the same amount of bandwidth, but the overall performance has been decreased for the connected users because the software applications are all using a lot of bandwidth.

The use of Voice over Wireless LAN (VoWLAN) technology is also increasing steadily in many wireless network deployments. This is an example of a hardware application. Voice technology on wireless networks is subject to latency. Therefore, the network must be carefully planned to take into account the number of voice client devices connecting to an access point. Capacity planning and quality of service (QoS) features are important when it comes to deploying voice technology on wireless networks.

Desired Throughput or Performance

The desired *throughput* or performance can also affect capacity. A large number of users connected to an access point using a bandwidth-intensive application will cause poor performance. Therefore, it may be necessary to limit the capacity to a certain number of devices to give the connected users the best performance possible. Any software application that is bandwidth intensive, such as CAD/CAM, streaming video, or File Transfer Protocol (FTP) downloads, can have an effect on overall performance. One way to help resolve this would be to use load balancing to limit the number of users that can connect to an access point. Another way would be to create RF cells with smaller coverage and add more access points.

Number of Devices/Users

The number of devices or users in an area will also affect the access point capacity. A single access point covering a large area will potentially allow for a large number of devices connecting to the access point. For example, an office of 8,000 square feet may consist of 100 people, each with their own wireless device. This is an example of wide coverage and large capacity. The software applications in use on the wireless network will have an impact on the overall performance. If all 100 devices connected are using a CAD/CAM application, which is a bandwidth-intensive application, the overall performance will be poor because this type of application requires a lot of resources. Therefore, more access points, each covering less space and having less capacity, would parlay into better overall performance for all of the users.

Wide coverage in a densely populated area may allow too many devices to connect to a single access point, resulting in poor performance overall. Wireless LANs use what is known as a shared medium. In other words, all devices connected to an access point will share the available bandwidth. Too many devices using powerful applications will overload the access point, adding to the poor performance issues. This scenario is considered a capacity issue. In this situation, a better solution would be more access points that each cover a smaller area and a lower number of devices, or have less capacity.

Many organizations now have to deal with the bring your own device (BYOD) situation, which needs to be evaluated closely. In addition to potential problems with technical support and security, wireless LAN capacity is also a major concern. The BYOD expansion of devices that are Wi-Fi capable is causing a wireless client device density issue within the enterprise market. If a company's corporate network policy allows employees to bring their own wireless-capable devices, then wireless LAN density and capacity need to be carefully evaluated to address potential issues.

Radio Frequency Channel Reuse and Device Colocation

The 2.4 GHz ISM band has a total of three nonoverlapping channels. In the US FCC implementation of this band, the three *nonoverlapping channels* are 1, 6, and 11. This means there must be a separation of five channels in order for them to be considered nonoverlapping. In the 2.4 GHz ISM band, channels are separated by 5 MHz on the center frequency. Taking this into consideration, channels must be separated by 25 MHz or greater on the center frequency in order to be considered nonoverlapping (IEEE 802.11-2012, Clause 18). This is calculated from five channels of separation multiplied by 5 MHz on the center frequency ($5 \times 5 = 25$). With deployments larger than a few access points, a channel plan may be necessary. A channel plan will minimize the chance of interference caused by two transmitters (access points) set to the same or adjacent overlapping channels.

IEEE 802.11-2012 Clause 16 (formerly Clause 15 in the IEEE 802.11-1007 standard) specifies 30 MHz or greater of separation to be considered nonoverlapping: "Adjacent channel rejection is defined between any two channels with ≥ 30 MHz separation in each channel group defined in 16.4.6.3." Keep in mind that this specification is for 802.11 DSSS. For HR/DSSS (IEEE 802.11-2012, Clause 18), channels must be separated by 25 MHz or greater in order to be considered nonoverlapping.

Figure 5.15 illustrates a 2.4 GHz deployment with no channel planning. Users in the areas where the circles overlap will experience co-channel interference. This interference will result in lower overall throughput for the connected users because of the Physical layer

(PHY) technologies that wireless LANs use. It basically has the same effect as collisions in an Ethernet network, resulting in retransmissions of data.

FIGURE 5.15 Users of these access points will experience co-channel interference in a multichannel architecture because they are all set to the same RF channel.

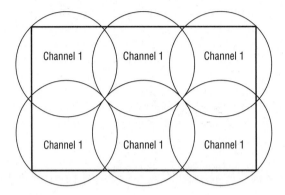

A correct channel plan will implement channel reuse and ensure that overlapping cells will not use overlapping channels. Channel reuse is using nonoverlapping channels—for example 1, 6, and 11 in the 2.4 GHz range—in such a way that the overlapping cells are on different RF channels. Figure 5.16 shows a 2.4 GHz deployment utilizing proper channel reuse. Channel reuse may be accomplished manually by mapping out the access points on a floor plan and minimizing the chances that the RF cells propagated by the access points do not overlap on the same RF channels. Modern enterprise wireless infrastructure devices have the capability to manage the RF automatically by selecting the best RF channels to operate on. This is done electronically by evaluating the surrounding RF environment. Channel planning can de done manually or with the help of wireless site survey software applications.

FIGURE 5.16 Colocation of access points with proper channel reuse. Overlapping areas use different channels in a multichannel architecture to help minimize interference.

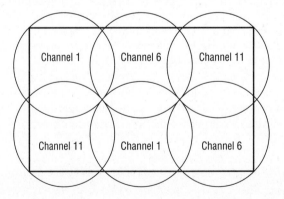

Radio Frequency Signal Measurements

It is important to understand the various signal measurements of radio frequency used in wireless LAN technology. Using tools like a wireless adapter client utility or a *spectrum analyzer* will allow you to view different statistics that pertain to a wireless network:

- Receive sensitivity
- Radio frequency noise
- Received signal strength indicator (RSSI)
- Signal-to-noise ratio (SNR)

Receive Sensitivity

The basic definition of *receive sensitivity* is the measurable amount of radio frequency signal usable by a receiver. This is also determined by how much radio frequency noise is in the area of the radio receiver. Figure 5.17 shows a wireless adapter client utility that displays statistics, including the strength of signal received.

FIGURE 5.17 The Orinoco 8494-US IEEE 802.11a/b/g/n USB adapter shows the amount of signal received.

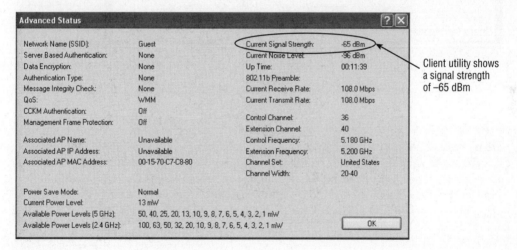

Radio Frequency Noise

Radio frequency noise is the term for RF signals from sources other than the transmitter and receiver that are in communication. Here is an analogy to help explain. You and a guest are in a crowded open-space restaurant for dinner. There are many unrelated conversations occurring at the same time at the various tables throughout the restaurant. If you and your dinner guest momentarily paused in your conversation, you would hear these other conversations as well as the noise from equipment, telephones, and tables that are being cleared. This would be the restaurant equivalent of radio frequency noise, as shown in Figure 5.18.

FIGURE 5.18 Restaurant analogy example of radio frequency noise

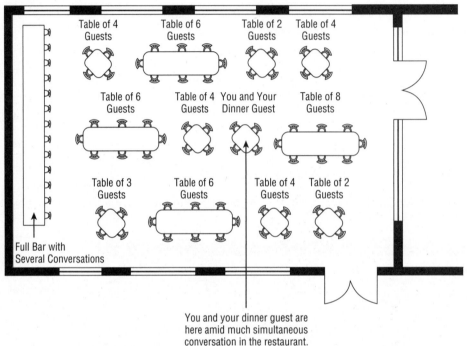

In a wireless LAN environment, several RF devices may be operating in the same physical space as the wireless transmitter and receiver. Depending on the level of this radio frequency noise, it may be difficult for the transmitter and receiver to understand each other.

In Figure 5.19, a screen capture from a noise analyzer utility shows a wireless basic service set on channel 40 in the 5 GHz U-NII-1 band. Also shown is the radio frequency noise floor of about –95 dBm.

Received Signal Strength Indicator (RSSI)

Received signal strength indicator (RSSI) is an arbitrary number assigned by the radio chipset or device manufacturer. There is no standard for this value, and it will not be comparable between devices from different manufacturers. The calculation of the RSSI value is done in a proprietary manner and a wireless device from one manufacturer may indicate different signal strength than that indicated by another, even though they both are receiving the exact same signal and at the same actual amount of radio frequency power.

This value is a key determinant of how well the wireless LAN device will perform. How the device is used with the network will determine the required levels of signal for optimal connectivity. Most wireless client device manufacturers allow their chipsets to access the higher data rates as long as they are getting a −70 dBm signal or stronger. Wireless VoIP manufacturers recommend deploying so that the client devices can receive a −67 dBm or better signal from the access point, a strength that is double the −70 dBm required for higher data rate use due to the need for better signaling in QoS communications. Most deployments are targeting −67 dBm or better for both data and voice applications.

FIGURE 5.19 The MetaGeek Chanalyzer software utility shows a noise floor for the tested site of about −95 dBm.

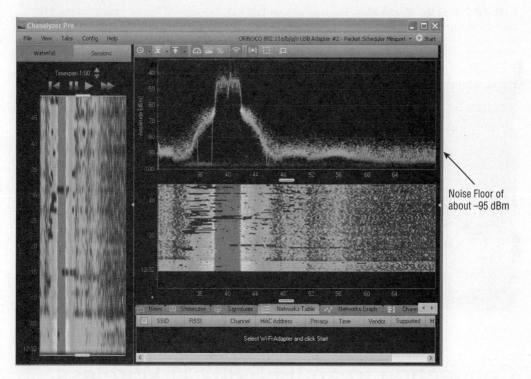

Signal-to-Noise Ratio (SNR)

The *signal-to-noise ratio (SNR)* is the difference between the amount of received signal and the noise floor. Looking back at the restaurant analogy, if you were to continue your conversation and the tables surrounding yours were all speaking at higher volumes, you might not be able to hear your dinner guest very well because of the amount of noise created in the open area of your table. Looking at this from a wireless LAN perspective, if a client device records a received signal of –85 dBm and the noise floor is –95 dBm, the signal-to-noise ratio will be 10 dB. This value is calculated by subtracting the received signal from the noise, in this case –85 dBm – (–95 dBm) = 10 dB. This would not be an adequate signal-to-noise ratio because the receiver would have a difficult time determining the difference between the wanted RF signal and the surrounding RF noise. On the other hand, if the received signal is –65 dBm and the noise floor is –95 dBm, then the signal-to-noise ratio will be 30 dB. This value again is calculated by subtracting the received signal from the noise, in this case –65 dBm – (–95 dBm) = 30 dB. This would be an excellent signal-to-noise ratio because the receiver would easily be able to determine the intended RF signal from the surrounding RF noise. Figure 5.20 illustrates the SNR as seen in a spectrum analyzer tool.

FIGURE 5.20 Graph showing the received signal strength vs. noise floor and the SNR using a wireless LAN spectrum analyzer program

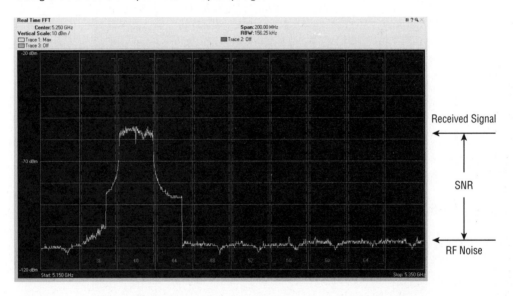

Connecting to an IEEE 802.11 Wireless Network

For a device to successfully connect to a wireless network, several different frame exchanges must take place. Various frame types allow for specific functions to occur. The details of the specific functions of each of these frame types are beyond the scope of this book, but some of these frames need to be briefly introduced in order to explain how connecting to a wireless network works. After looking at frame types, we will explore wireless network discovery, the authentication and association process, reserving the medium, exchanging data, and power save functions.

IEEE 802.11 Frame Types

Wireless LAN devices communicate by sending radio frequency waves to each other through the air. The RF waves carry the digital data from one wireless device to another. At this stage, the information traveling through the air is organized into what are known as frames. There are several different types of frames, and the frame types play various roles depending on the information being sent. Wireless LANs use three different frame types: management frames, control frames, and data frames.

Management Frames

Management frames are used to manage the wireless network. They assist wireless LAN devices in finding and connecting to a wireless network. This includes advertising the capabilities of the WLAN and allowing connections by the authentication and association process. Management frames are exchanged only between immediate wireless devices such as an access point and client device and never cross the Data-Link layer (Layer 2) of the OSI model. It is important to understand that management frames are always transmitted at the lowest mandatory data rate of the service set so that all stations on the same radio frequency channel in the basic service area can understand them. The following are examples of management frames:

- Beacon
- Probe request
- Probe response
- Authentication
- Association request
- Association response

Control Frames

Control frames are used to control access to the wireless medium by allowing devices to reserve the medium and acknowledge data. In addition, some control frames are used to request data from the access point after returning from a power save state and are used with IEEE 802.11 protection mechanisms to allow wireless device coexistence. Here are some examples of control frames:

- RTS
- CTS

- CTS-to-Self
- PS-Poll
- ACK

Data Frames

As their name implies, data frames are used to carry data payload or Layer 3 information between wireless devices. The following list includes examples of data frames:

- Data
- QoS data
- Null data

The null data or null function frame is a special type of data frame that helps implement power save features and is not used to carry any data payload. There is also a variant of the null frame called the QoS null frame, which is used with wireless LAN quality of service functions.

Wireless Network Discovery

Wireless LAN discovery is the process of a client device looking for wireless networks and identifying the parameters of the network, including but not limited to the SSID, supported data rates, and security settings. Wireless network discovery prepares a wireless client device to perform an IEEE 802.11 authentication and association, which will allow a wireless device access to the network. The wireless network discovery phase consists of the passive scanning and active scanning processes.

Passive Scanning

The first part of the wireless network discovery phase in IEEE 802.11 wireless networking is known as *passive scanning*. This process allows wireless LAN devices to "listen" for information about wireless networks in the radio receiving area of the wireless network or the basic service area (BSA). During the passive scanning process, wireless LAN devices will listen for specific information to make them aware of networks in the area. An analogy to this process would be using an FM radio tuner to scan through the entire band, listening for a station to tune in to. The radio will scan through the band listening for different stations. The person using the tuner can stop on the desired station once it is heard.

Management frames assist wireless LAN devices in finding and connecting to a wireless network. An example of a management frame that works in the discovery phase or passive scanning is a beacon frame. This frame for the most part is an advertisement of the wireless network. It carries specific information about the access point or basic service set such as the SSID, the radio frequency channel that it is operating on, the available data rates it is configured for, the security parameters, and much more. During the passive scanning phase, wireless devices listen for beacons advertising the details about the wireless networks in the area or radio range of the client device. Wireless LAN devices are constantly listening for beacon frames. Figure 5.21 shows a wireless LAN client passively scanning and listening for an access point to connect with.

FIGURE 5.21 An example of passive scanning with a wireless LAN client listening for access points in the basic service area

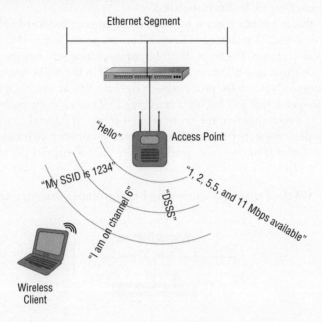

By default, most manufacturers set beacons to broadcast at about 10 times a second. This value is actually 1,024 microseconds and is identified as the target beacon transmission time (TBTT). Although this interval can be changed, it is best to do so only if necessary or recommended by the manufacturer. In some cases, manufacturers may suggest specific timing intervals for such frames as beacons.

Figure 5.22 shows a packet analyzer capturing beacon frames generated from an access point.

FIGURE 5.22 Packet analyzer capture of beacon frames

00:14:A8:53:5F:C0	Ethernet Broadcast	00:14:A8:53:5F:C0	47%	802.11 Beacon
00:14:A8:53:5F:C0	Ethernet Broadcast	00:14:A8:53:5F:C0	45%	802.11 Beacon
00:14:A8:53:5F:C0	Ethernet Broadcast	00:14:A8:53:5F:C0	44%	802.11 Beacon
00:14:A8:53:5F:C0	Ethernet Broadcast	00:14:A8:53:5F:C0	44%	802.11 Beacon
00:14:A8:53:5F:C0	Ethernet Broadcast	00:14:A8:53:5F:C0	44%	802.11 Beacon
00:14:A8:53:5F:C0	Ethernet Broadcast	00:14:A8:53:5F:C0	44%	802.11 Beacon
00:14:A8:53:5F:C0	Ethernet Broadcast	00:14:A8:53:5F:C0	45%	802.11 Beacon
00:14:A8:53:5F:C0	Ethernet Broadcast	00:14:A8:53:5F:C0	47%	802.11 Beacon

Active Scanning

Active scanning is the second part of the wireless LAN discovery phase. In *active scanning*, wireless LAN devices wishing to connect to a network send out a management frame known as a probe request. The function of this management frame is to find a specific wireless access point to connect with. Depending on the wireless client utility software used, if an SSID is specified in the client utility software active profile, the device will join only a network with the matching SSID that is specified.

An exception to this is a probe request frame that contains a "wildcard SSID" or a "null SSID." The IEEE 802.11 standard requires all access points to respond to a "null" or broadcast probe request frame. This type of probe request frame will not specify an SSID value and will rely on the access points to provide the SSID in the probe response frame.

Access points constantly listen for probe request frames. Any access point within hearing range of the wireless device and having a matching SSID sends out a probe response frame to the wireless device that sent the probe request frame. If more than one access point responds, the device selects the "best" access point to connect with based on certain factors such as signal strength and signal quality. Figure 5.23 illustrates the active scanning process.

FIGURE 5.23 Wireless client device sending a probe request frame to access points in radio range

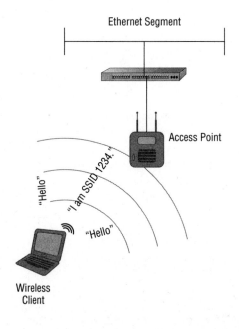

> **Frames Used for Active Scanning**
>
> During the active scanning process, two frames are exchanged between the client device and the access point.
>
> 1. The wireless LAN client device sends a broadcast probe request frame to all devices, including access points within radio range.
> 2. The access point(s) send a probe response frame to the device so it can identify the parameters of the network before joining.

Figure 5.24 shows a packet analyzer capturing frames of the active scanning process.

FIGURE 5.24 Packet analyzer capture of probe request and probe response frames

00:19:7E:43:4E:E8	Ethernet Broadcast	Ethernet Broadcast	74%	802.11 Probe Req
00:14:A8:53:5F:C0	00:19:7E:43:4E:E8	00:14:A8:53:5F:C0	48%	802.11 Probe Rsp
00:19:7E:43:4E:E8	00:14:A8:53:5F:C0		75%	802.11 Ack

The IEEE 802.11 standard requires access points to respond to devices that are sending a null or blank SSID in a probe request frame. The standard refers to this as a wildcard SSID. It is important not to confuse this with disabling the SSID broadcast on an access point. Most wireless equipment manufacturers provide the capability to set the access point not to respond to a probe request frame with a null or wildcard SSID. If the AP is set not to respond to such probe requests, the wireless device is required to have the SSID specified in the client utility in order to connect to the BSS.

IEEE 802.11 Authentication

Authentication in general is defined as verifying or confirming an identity. We use a variety of authentication mechanisms in our daily lives, such as when logging onto a computer or network at home or at the office, accessing secure sites on the Internet, using an ATM machine, or showing an identification badge to get access to a building.

IEEE 802.11 devices must use an authentication process to access network resources. This IEEE 802.11 authentication process differs from conventional authentication methods such as providing credentials like a username and password to gain access to a network. The authentication discussed here is wireless device or *IEEE 802.11 authentication*, required for the device to become part of the wireless network and participate in exchanging data frames. (Providing credentials such as a username and password or a preshared key is a different type of authentication, to be discussed in Chapter 16, " Device Authentication and Data Encryption.") The IEEE 802.11 standard addresses two types of IEEE authentication methods: open system and shared key.

IEEE 802.11 Open System Authentication

This 802.11 authentication method is defined by the IEEE 802.11 standard as a null authentication algorithm. IEEE 802.11 open system authentication is the only valid IEEE 802.11 authentication process allowed with newer wireless LAN security amendments and interoperability certifications for the network to be considered a robust security network (RSN).

IEEE 802.11 open system authentication is a two-step authentication process. Two management frames are exchanged between the wireless client device and the access point during open system authentication. It is a simple process. A wireless LAN device will ask an access point, "Can I be a part of this network?" and the access point will respond, "Sure, come join the party." So there really is no validation of identity. *IEEE 802.11 open system authentication* is considered a two-way frame exchange because two authentication frames are sent during this process. It is not a request and response situation; it is authentication and success.

Figure 5.25 illustrates open system authentication.

FIGURE 5.25 A wireless client authenticating to an access point using open system authentication

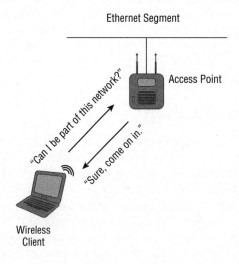

Open system authentication does not provide any type of data encryption. With open system authentication, Wired Equivalent Privacy (WEP) is optional and can be used for data encryption if desired. However, because of the inherit security vulnerabilities with WEP, using it as a security solution for wireless networks is highly discouraged. WEP will be discussed in more detail in Chapter 16.

These are the two steps used for open system authentication. One management frame is sent in each step.

1. The wireless LAN client device wanting to connect sends an authentication frame to the access point. This frame is acknowledged by the access point.
2. The access point accepting the connection sends a successful authentication frame back to the client device. This frame is acknowledged by the authenticating device.

Figure 5.26 shows a packet capture of the two-way open system authentication frame exchange.

FIGURE 5.26 Packet capture of open system authentication

00:19:7E:43:4E:E8	00:14:A8:53:5F:C0	00:14:A8:53:5F:C0	75%	802.11 Auth
00:14:A8:53:5F:C0	*00:19:7E:43:4E:E8*		47%	802.11 Ack
00:14:A8:53:5F:C0	00:19:7E:43:4E:E8	00:14:A8:53:5F:C0	50%	802.11 Auth
00:19:7E:43:4E:E8	*00:14:A8:53:5F:C0*		75%	802.11 Ack

For the most part, open system authentication cannot fail unless other security measures such as MAC filtering are put in place that will prevent the device from accessing the network. It may also fail because of a corrupt network adapter driver, although this is not very common.

Keep in mind that IEEE 802.11 open system authentication always exists, even with the most secure wireless LANs. It is used to allow the wireless station to connect to the access point and then after association use additional credentials such as a passphrase or username and password pair for authentication. If the wireless station did not perform an open system authentication and association first, the station would not have a Layer 2 connection and there would be no way to use additional security mechanisms.

IEEE 802.11 Shared-Key Authentication

Shared-key is another authentication method defined by the IEEE 802.11 standard. It is a little more complex than open system authentication. This IEEE 802.11 authentication method is a four-way frame exchange. During *IEEE 802.11 shared-key authentication*, four management frames are sent between the wireless client device wanting to join the wireless network and the access point. Shared-key authentication differs from open system authentication in that shared-key authentication is used for both IEEE 802.11 authentication and for data encryption.

IEEE 802.11 shared-key authentication is considered flawed from a security perspective. It requires the use of Wired Equivalent Privacy (WEP) for both wireless client device authentication and data encryption. Because WEP is mandatory, a hacker could potentially identify the WEP key used for the network by capturing the authentication process using a wireless packet analyzer. If a hacker was to capture the four wireless frames used during the shared-key authentication process, they would be able to use this information with the appropriate software to extract the security (WEP) key to gain access to the network and see the encrypted data. Shared-key authentication should be avoided whenever possible and is not allowed when using newer IEEE 802.11i, WPA, and WPA2 security methods.

Some manufacturers have removed the option to set shared-key authentication, both in infrastructure devices such as access points and bridges and in client software utilities. Although very unlikely, there is a slight chance that some legacy client devices may still use IEEE 802.11 shared-key authentication as the only authentication option. If this is the case, steps need to be taken to protect the integrity of the network and also to identify an appropriate upgrade path for the devices using IEEE 802.11 shared-key authentication.

It is important not to confuse legacy IEEE 802.11 shared-key authentication with modern WPA pre-shared key. WPA and WPA2 pre-shared key is used with personal mode authentication and passphrases. When you enter a passphrase into an access point or wireless client device an algorithm defined in the IEEE 802.11i amendment will create a 256-bit pre-shared key.

Figure 5.27 illustrates the four frames exchanged during the shared-key authentication process.

Because WEP is mandatory with shared-key authentication, it makes a system vulnerable to intrusion. Therefore, open system authentication is considered more secure than shared-key authentication when WEP is used with the IEEE 802.11 open system authentication. This is because WEP is used to encrypt the data only and not used for the IEEE 802.11 authentication process. WEP was designed as a way to protect wireless networking users from casual eavesdropping, and it does just that.

FIGURE 5.27 Shared-key authentication uses a four-way frame exchange

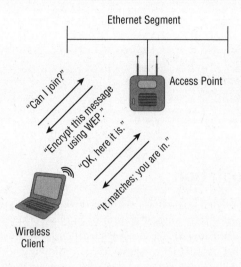

Frames Used for Shared-Key Authentication

An 802.11 wireless device must perform an IEEE 802.11 authentication to the wireless network prior to associating according to the 802.11 standard. The following steps show the four-way frame exchange used for shared-key authentication. This method should not be used but is shown here to illustrate the process.

1. The wireless LAN device wanting to authenticate sends an authentication frame to the access point. This frame is acknowledged by the access point.

2. The access point sends a frame back to the WLAN device that contains a challenge text. This frame is acknowledged by the WLAN device.

3. The WLAN device sends a frame back to the access point containing an encrypted response to the challenge text. The response is encrypted using the device's WEP key. This frame is acknowledged by the access point.

4. After verifying the encrypted response, the access point accepts the authentication and sends a "successful authentication" frame back to the device. This final frame is acknowledged by the device.

Figure 5.28 shows the four authentication frames used in shared-key authentication.

FIGURE 5.28 Packet capture of four-frame exchange 802.11 shared-key authentication

00:19:7E:43:4E:E8	00:14:A8:53:5F:C0	00:14:A8:53:5F:C0	88%	802.11 Auth
00:14:A8:53:5F:C0	00:19:7E:43:4E:E8		44%	802.11 Ack
00:14:A8:53:5F:C0	00:19:7E:43:4E:E8	00:14:A8:53:5F:C0	45%	802.11 Auth
00:19:7E:43:4E:E8	00:14:A8:53:5F:C0		84%	802.11 Ack
00:19:7E:43:4E:E8	00:14:A8:53:5F:C0	00:14:A8:53:5F:C0	88%	802.11 Auth
00:14:A8:53:5F:C0	00:19:7E:43:4E:E8		42%	802.11 Ack
00:14:A8:53:5F:C0	00:19:7E:43:4E:E8	00:14:A8:53:5F:C0	44%	802.11 Auth
00:19:7E:43:4E:E8	00:14:A8:53:5F:C0		87%	802.11 Ack

IEEE 802.11 Association

IEEE 802.11 association takes place after a wireless device has been successfully 802.11 authenticated either by open system authentication or by shared-key authentication. In the association state, the authenticated device can pass traffic across the access point to the network infrastructure or other associated wireless devices, allowing access to resources that the device or user has permissions to access. After a device is authenticated and associated, it is considered to be part of the basic service set. Figure 5.29 illustrates the association process, and Figure 5.30 shows the frames used during the association process.

FIGURE 5.29 Authentication and association

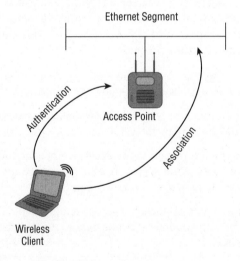

FIGURE 5.30 Packet capture of the association request and association response process

00:19:7E:43:4E:E8	00:14:A8:53:5F:C0	00:14:A8:53:5F:C0	78%	802.11 Assoc Req
00:14:A8:53:5F:C0	00:19:7E:43:4E:E8		44%	802.11 Ack
00:14:A8:53:5F:C0	00:19:7E:43:4E:E8	00:14:A8:53:5F:C0	50%	802.11 Assoc Rsp
00:19:7E:43:4E:E8	00:14:A8:53:5F:C0		71%	802.11 Ack

After successful association, the IEEE 802.11 authentication and association process is complete. Keep in mind that this is very basic access to the network using either open system authentication or WEP for authentication and encryption. After this process is complete, more sophisticated authentication mechanisms such as IEEE 802.1X/EAP (which provides user-based authentication) or pre-shared key (passphrase) can be used to secure the wireless network. These and other security components will be discussed in more detail in Chapter 16.

Frames Used for IEEE 802.11 Association

After a successful 802.11 authentication, the association process will begin. Association means the device has successfully connected to the wireless network and is allowed to send information across the access point to the network infrastructure.

1. Wireless LAN device sends an association request frame to the access point. This frame is acknowledged by the access point.
2. The access point sends an association response frame to the device. This frame is acknowledged by the associating device.

IEEE 802.11 Deauthentication and Disassociation

The opposite of authentication and association can occur in a wireless LAN. These events are known as *deauthentication* and *disassociation*. Deauthentication occurs when an existing authentication is no longer valid. This can be caused by a wireless LAN device logging off from the current connection or roaming to a different BSS. A disassociation occurs when an association to an access point is terminated. This may occur when the associated wireless LAN device roams from one BSS to another.

Both deauthentication and disassociation frames are notifications and not requests. Since neither can be refused by either side, they are both considered automatically successful from the sender's perspective. Unless IEEE 802.11w is implemented, deauthentication can be a security issue. These frames can be used for denial of service attacks or to hijack a wireless device.

Both deauthentication and disassociation frames are management frames. Figure 5.31 shows how disassociation and deauthentication frames would look on a packet analyzer.

FIGURE 5.31 Wireless packet capture of disassociation and deauthentication frames

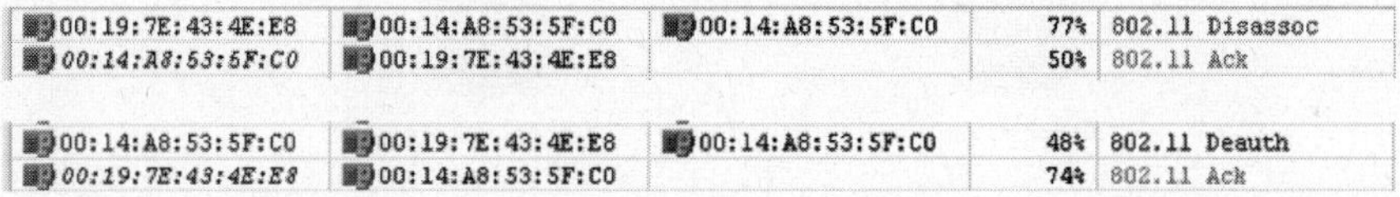

00:19:7E:43:4E:E8	00:14:A8:53:5F:C0	00:14:A8:53:5F:C0	77%	802.11 Disassoc
00:14:A8:53:5F:C0	00:19:7E:43:4E:E8		50%	802.11 Ack
00:14:A8:53:5F:C0	00:19:7E:43:4E:E8	00:14:A8:53:5F:C0	48%	802.11 Deauth
00:19:7E:43:4E:E8	00:14:A8:53:5F:C0		74%	802.11 Ack

Hacking tools are available that will continuously send deauthentication frames to a device. Use of this type of tool is considered a denial of service (DoS) attack. A hacker may also use a deauthentication frame to force a device to reauthenticate to an access point, causing the device to be hijacked. The 802.11w amendment provides enhancements to the IEEE 802.11 standard that enables data integrity, data origin authenticity, replay protection, and data confidentiality for selected IEEE 802.11 management frames that include authentication.

The Distribution System

In wireless LAN technology, the *distribution system (DS)* is the common network infrastructure to which wireless access points are connected. In most cases this would be an Ethernet segment. In this capacity, the access point acts like a Layer 2 translational bridge. A *translational bridge* is defined as a device used to connect two or more dissimilar types of LANs together, such as wireless (IEEE 802.11) and Ethernet (IEEE 802.3). From a receiver's perspective, this allows an access point to take information from the air (the communication medium in wireless networking) and make a decision either to send it back out to the same wireless radio or to forward it across to the distribution system. An access point can do this because it has enough intelligence to determine if a data frame is destined to be sent to the distribution system or if it should stay on the originating wireless side of the network. This is possible because the access point knows whether a device is part of the wireless LAN side through the authentication and association methods mentioned earlier. Figure 5.32 shows an example of a distribution system.

FIGURE 5.32 Two access points connected to a common distribution system, in this case IEEE 802.3, Ethernet

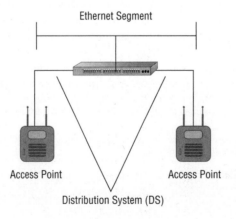

The distribution system is a network segment that consists of one or more connected basic service sets. According to the original IEEE 802.11 standard, one or more interconnected basic service sets make up an extended service set. The distribution system allows wireless LAN devices to communicate with resources on a wired network infrastructure or to communicate with each other through the wireless medium. Either way, all wireless frame transmissions will traverse through an access point.

Data Rates

The speed at which wireless devices are designed to exchange information is known as the *data rate*. Data rates will differ depending on the wireless standard, amendment to the standard, spread spectrum type, or Physical layer technology in use. Table 5.6 shows data rates for the various WLAN technologies. Data rates do not accurately represent the amount of information that is actually being transferred between devices and a wireless network. Figure 5.33 shows an 802.11a/g/n wireless LAN adapter in a notebook computer reading a data rate of 300 Mbps. To learn more about the actual amount of information transferred, see the next section, "Throughput."

FIGURE 5.33 Windows 7 wireless configuration utility showing a data rate of 300 Mbps for an IEEE 802.11n wireless LAN adapter

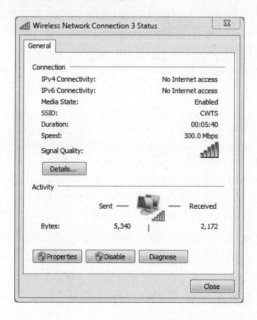

TABLE 5.6 Data rates based on spread spectrum/PHY type

Standard/Amendment	PHY Technology	Data Rates
802.11	FHSS	1 and 2 Mbps
802.11	DSSS	1 and 2 Mbps
802.11b	HR/DSSS	5.5 and 11 Mbps; 1 and 2 Mbps from DSSS
802.11a	OFDM	6, 9, 12, 18, 24, 36, 48, and 54 Mbps
802.11g	ERP-OFDM	6, 9, 12, 18, 24, 36, 48. and 54 Mbps
802.11n	HT-OFDM	Up to 600 Mbps
802.11ac	VHT-OFDM	Up to 6.93 Gbps aggregate

Throughput

Unlike the data rate (the maximum amount of information theoretically capable of being sent), *throughput* is the amount of information actually being correctly received or transmitted. Many variables affect the throughput of information being sent:

- Spread spectrum or Physical layer technology type in use
- Radio frequency interference
- Number of wireless devices connected to an access point (contention)

For example, an 802.11g wireless access point has a maximum data rate of 54 Mbps. With one user connected to this access point, chances are the best throughput that could be expected is less than 50 percent of the maximum, or about 20 Mbps. If more users connect to the same access point, the throughput for each user would be even less, because of the contention between users sharing the same wireless medium. Figure 5.34 shows an example of actual throughput for an 802.11a/g/n wireless LAN adapter.

FIGURE 5.34 Actual throughput of an IEEE 802.11a/g/n 300 Mbps wireless LAN adapter

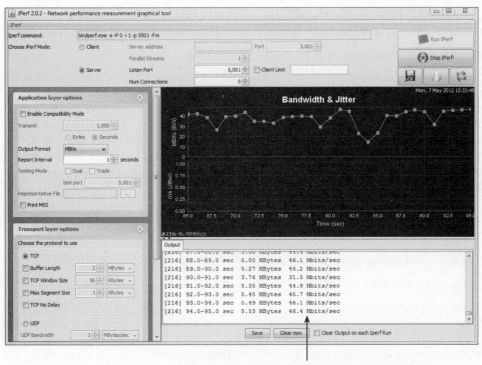

JPerf performance test utility showing an average throughput of about 44 Mbps using an IEEE 802.11a/g/n wireless adapter

 Real World Scenario

Packing and Shipping Data: A Throughput Analogy

Packing and shipping an item in a cardboard box is a way of looking at data rate versus throughput. You have a cardboard box that is rated to have a maximum capacity of two cubic feet. You want to send a fragile item such as a vase to somebody. The vase, if measured, would really only take about one cubic foot of space. However, this is a very fragile item, and you want to make sure it gets to the destination without any damage. So rather than just putting the vase by itself in a box with a capacity of one cubic foot, you want to protect it with some packing material such as Bubble Wrap. That will take an additional one cubic foot of space.

The data rate is analogous to the box capable of holding two cubic feet of material. The one-cubic-foot vase is analogous to the actual data being sent. The packing material is analogous to the contention management and other overhead that causes the throughput to be less than the theoretical capacity of the WLAN device.

In Exercise 5.1, you will measure the actual throughput of your own wireless network.

EXERCISE 5.1

Measuring Throughput of a Wireless Network

In this activity, you will measure throughput of a wireless network. If you have the proper equipment, it is not too difficult. If you already have an existing wireless network set up with a computer connected to the wired side or distribution system, you have a good part of the setup done. This exercise uses the JPerf software program for Microsoft Windows. JPerf is a graphical front-end program for IPerf from SourceForge. JPerf does require that Java be installed on the computer. The following step-by-step instructions assume a wireless access point already configured with TCP/IP settings as well as SSID. To perform this exercise, you will need the following equipment:

- Two computers
- Java installed on both computers
- One wireless access point
- One Ethernet cable
- One wireless network adapter
- JPerf software (jperf-2.0.2.zip)

The JPerf software is available from code.google.com/p/xjperf. This link will take you to the project home page. From there you can click the Downloads tab and then select the desired file to complete the download.

Complete the following steps to measure throughput:

1. Connect the required equipment as shown here.

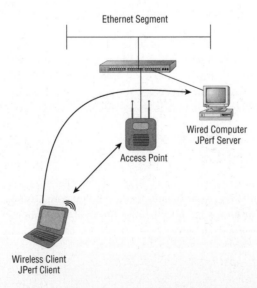

2. Create a folder named Jperf on the C:\ drive on both computers and extract the contents of the jperf-2.0.2.zip file you downloaded to the Jperf folder you created. This folder needs to be created at the root or C:\ in order for the remaining steps to work as written.

3. On the computer connected to the wired distribution system, open a command prompt. This will vary based on the operating system in use. For example, if you are using Windows 7, select Start ➢ All Programs ➢ Accessories ➢ Command Prompt.

4. In the command prompt window, type the command **ipconfig** at the C:\ prompt and note the IP address of this computer.

5. This computer will act as the JPerf server. In the open command window, type **cd\ Jperf** at the C:\ prompt and press the Enter key. This will put you in the proper location of the JPerf program you copied to this computer in step 2.

6. Enter the following command to start the JPerf server: **jperf.bat**. After a few seconds, the JPerf 2.0.2 – Network Performance Measurement Graphical Tool window will appear.

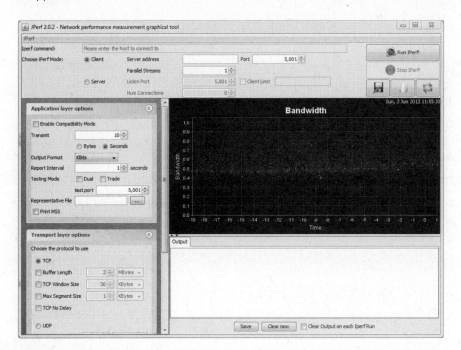

EXERCISE 5.1 (continued)

7. In the Application Layer Options section, click the Output Format drop-down box and select the MBits option. This will show the results in megabits per second.
8. Click the Server radio button.
9. Click the Run IPerf! button in the upper-right corner of the window.

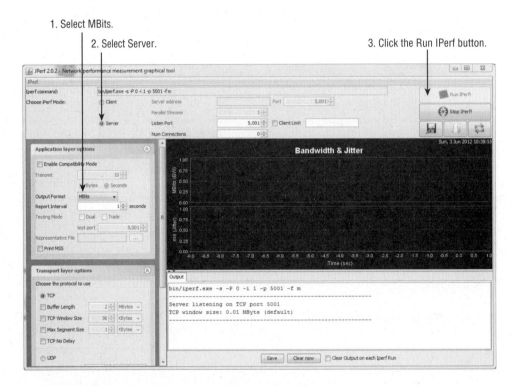

10. The JPerf server is now ready for throughput testing.
11. On the computer with a wireless network adapter, connect to your access point using the wireless network adapter. This computer will act as the JPerf client for throughput testing.
12. On this same computer, open a command prompt.
13. In the command prompt window, type the command **ipconfig** at the C:\ prompt and verify the IP address of this computer.

14. Verify connectivity to the JPerf server by typing the following command: **ping** *{IP address}*. You will need to replace *{IP address}* with the server address you noted in step 4. You should see several replies if you are correctly connected to the server through the access point.

15. This computer will act as the JPerf client. In the open command window, type the command **cd\jperf** at the C:\ prompt and press Enter. This will put you in the proper location of the JPerf program you copied to this computer in step 2.

16. In the command prompt window, type the following command to launch the JPerf graphical program: **jperf.bat**. After a few seconds, the JPerf 2.0.2 – Network Performance Measurement Graphical Tool window will appear.

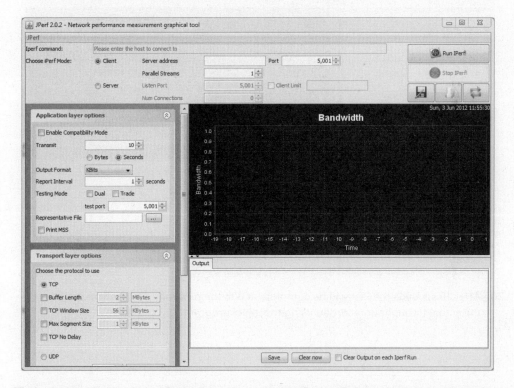

17. In the Application Layer Options section, click the Output Format drop-down box and select the MBits option. This will show the results in megabits per second.

18. Click the Client radio button, and in the Server Address field enter the IP address of the server that you noted in step 4.

19. Click the Run IPerf! button in the upper-right corner.

EXERCISE 5.1 *(continued)*

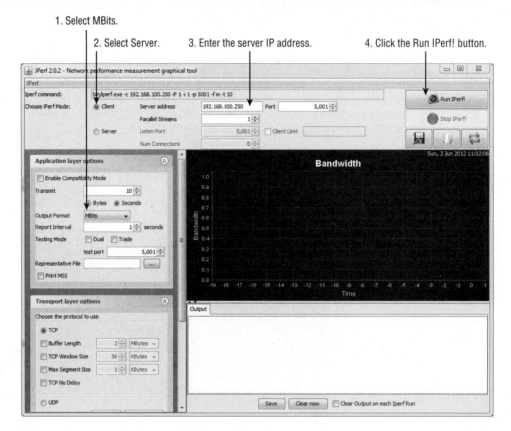

20. After 10 seconds the test will be complete and in the program window you will see the actual throughput recorded using the JPerf program.

EXERCISE 5.1

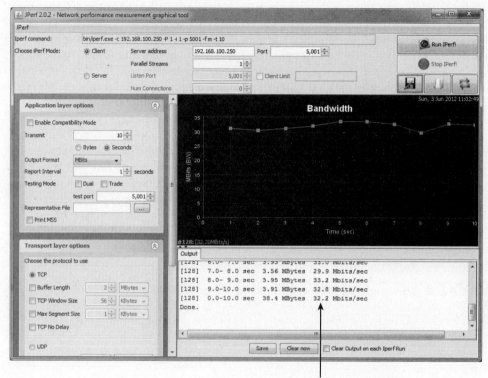

Test shows 32 Mbps of data throughput

21. Close the JPerf program window on both computers.

When you are finished, you can delete the JPerf program and folder you created in step 2.

Dynamic Rate Switching

When a wireless device moves through the basic service area (BSA) or as the distance from the access point increases, the data rate will decrease. Conversely, as the wireless device moves closer to the access point, the data rate can increase. This is called *dynamic rate switching (DRS)*, also known as dynamic rate shifting and even dynamic rate selection. This process allows an associated wireless device to adapt to the radio frequency in a particular location of the BSA. DRS is typically accomplished through proprietary mechanisms set by the manufacturer of the wireless device. The main goal of dynamic rate switching is to improve performance for the wireless device connected to an access point. As a wireless device moves away from an access point, the amount of received signal will decrease because of the free space path loss. When this occurs, the modulation type will change because the radio frequency signal quality is less and thus a lower data rate will be realized.

Different data rates use different modulation technologies. Using a less complex modulation type at a lower data rate will provide better overall performance as the station moves away from the access point. Figure 5.35 illustrates how dynamic rate switching works. As the wireless device moves away from the access point, the data rate will decrease. Keep in mind the opposite is true as well. As a wireless device moves closer to an access point, the data rate will increase.

FIGURE 5.35 A graphical representation of dynamic rate switching

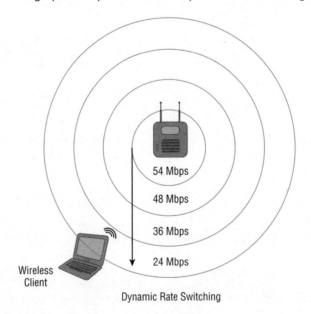

Wireless LAN Transition or Roaming

In wireless LAN technology, *transition* is the term for what happens when a wireless device moves from one basic service set or access point to another. People in the wireless LAN industry commonly refer to this process as *roaming* (this term was derived from the cellular world).

Wireless transition was not addressed in the original IEEE 802.11 standard. The process is typically accomplished in a proprietary manner based on how the manufacturer chooses to implement it. Manufacturers use different criteria to initiate transitioning from one access point to another. There was an amendment to the IEEE 802.11 standard (IEEE 802.11F, Inter-Access Point Protocol) that was ratified in June 2003 as a recommended practice intended to address multivendor access point interoperability. However, this recommended practice was implemented by few if any manufacturers and it was withdrawn by the IEEE 802 Executive Committee in February 2006. The next attempt to standardize transition between access points was IEEE 802.11r. The driving force behind this amendment to the standard was to allow for fast secure transition with wireless voice devices.

When a wireless LAN device moves through a BSA and receives a signal from another access point, it needs to make a decision whether to stay associated with the current access point or to reassociate with the new access point. The decision when to roam is proprietary and based on specific manufacturer criteria. Manufacturers use the following criteria:

- Signal strength
- Signal-to-noise ratio
- Error rate
- Number of currently associated devices

When a wireless LAN device chooses to reassociate to a new access point, the original access point will hand off the association to the new access point as requested from the new access point. Keep in mind it is the wireless client device that initiates the move to a new access point. This move is done over the wired network or distribution system based on how the manufacturer implemented the roaming criteria. Figure 5.36 illustrates a notebook computer roaming (transitioning) from one access point to a new access point.

FIGURE 5.36 The roaming (transition) process for a wireless LAN

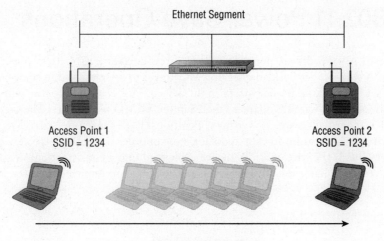

> **Frames Used for Reassociation (Roaming)**
>
> When a wireless client device transitions, or "roams," to a new access point, it needs to associate to the new access point. Because the device is already associated, in order to connect to the new access point it must complete a reassociation process.
>
> 1. A wireless LAN device sends a reassociation request frame to the new access point. This frame is acknowledged by the new access point.
>
> 2. The new access point sends a reassociation response frame to the wireless client device after handoff across the distribution system from the original access point has occurred. This frame is acknowledged by the reassociating wireless device.

Figure 5.37 shows reassociation request and response frames in a packet analyzer.

FIGURE 5.37   Packet capture of the reassociation process

00:19:7E:43:4E:E8	00:14:A8:53:5F:C0	00:14:A8:53:5F:C0	77%	802.11 Reassoc Req
00:14:A8:53:5F:C0	00:19:7E:43:4E:E8		50%	802.11 Ack
00:14:A8:53:5F:C0	00:19:7E:43:4E:E8	00:14:A8:53:5F:C0	45%	802.11 Reassoc Rsp
00:19:7E:43:4E:E8	00:14:A8:53:5F:C0		74%	802.11 Ack

The 802.11r amendment to the standard was ratified in 2008 and is now part of the IEEE 802.11-2012 standard. This amendment is for fast basic service set (BSS) transition (FT) and allows for fast secure transition (roaming) for devices between basic service sets. The main objective of this amendment is to support Voice over IP (VoIP) technology.

IEEE 802.11 Power Save Operations

Many wireless LAN devices are portable and use direct current (DC) battery power to some degree. A wireless network adapter uses DC power to operate, and in some cases this could be a significant drain on the battery in the device. This is especially true with newer IEEE 802.11n or newer wireless adapters that support MIMO technology. The original IEEE 802.11 standard addresses power save operations. Power save operations are designed to allow a wireless LAN radio to enter a dozing state in order to conserve DC power and extend battery life. If the wireless LAN device is plugged into a consistent power source such as an alternating current (AC) outlet, there is no reason to implement power save features. However, portable devices that are mobile and may not have access to an AC power source should consider using power save operations if necessary.

The original IEEE 802.11 standard addressed two different power save modes: active mode (AM) and power save (PS) mode. In some cases, power save (PS) mode is considered legacy because the IEEE 802.11e amendment for quality of service addresses new, more efficient power save mechanisms. Although the original PS mode may be considered legacy, it may still be used in some devices.

As mentioned earlier in this chapter in the section "IEEE 802.11 Frame Types," a data frame known as a *null function frame* is used with power management; it does not carry any data but is used to inform the access point of a change in power state.

Active Mode

In *active mode (AM)*, a wireless LAN client device or station (STA) may receive frames at any time and is always in an "awake" state. In this case, the wireless LAN device is not relying on battery power; thus, there is no reason for the device to assume a low power state, and it will never doze. Some manufacturers refer to active mode as continuous aware mode (CAM).

Power Save Mode

In *power save (PS) mode*, the radio in the wireless LAN client device or station (STA) will doze or enter a low power state for very short periods of time. At specific time intervals, the device will "listen" for selected beacons and determine if any data is waiting for it (buffered) at the access point. The beacon frame contains information for associated devices regarding power save. When a wireless LAN device associates to an access point, the device receives what is known as an *association ID (AID)* in the association response frame. The association ID is a value that will represent that device in various functions, including power save mode. The beacon frame contains an indicator for each AID associated device to let wireless devices know whether they have data waiting for them or buffered at the access point. If it is determined that the access point does have data buffered for a specific device, the device will send a control frame message (PS-Poll frame) to the access point to request the buffered data. Figure 5.38 shows where power save mode can be set in the advanced settings of the wireless adapter device driver.

Power save mode will cause some level of overhead for the wireless LAN device, and there is a trade-off in performance. With power save mode enabled, the battery life will be extended; however, performance will suffer to some degree because the device will not be available to receive data continuously. The device will only be able to receive buffered data during the "awake" state. Power save mode is common in applications where battery conservation is important, such as bar code scanners, voice over Wi-Fi phones, and other handheld devices.

FIGURE 5.38 The driver settings for an Intel 5100 IEEE 802.11a/g/n wireless adapter and power save mode setting

Real World Scenario

Use of Power Save Mode in Bar Code Scanners

Organizations such as retail, manufacturing, and warehousing have been using 802.11 wireless LAN technologies for many years. Many of these businesses use wireless LAN devices such as bar code scanners in addition to notebook computers and other portable devices. Bar code scanners are used heavily for inventory and asset tracking purposes. These devices must run for many hours at a time, typically in 8- or 10-hour shifts for individuals who may be using them. Applications such as this greatly benefit from using IEEE 802.11 power save features and extending battery life of wireless LAN devices. This minimizes downtime because batteries in these devices will not have to be changed or recharged as often during a work shift.

Automatic Power Save Delivery

The IEEE 802.11e Quality of Service amendment to the standard fueled the need for more efficient power save mechanisms in wireless networking. Depending on the implementation and requirements, legacy power save modes may not be efficient enough to work with applications that use QoS, such as voice and video. *Automatic power save delivery (APSD)* differs from the original power save mode in that a trigger frame will wake a device in order to receive data. APSD is a more efficient way of performing power save functions. Because of this trigger-and-delivery method, it works well with time-bound applications that are subject to latency, such as voice and video.

IEEE 802.11 Protection Modes and Mechanisms

To allow newer, faster wireless LAN technology such as 802.11g and 802.11n devices to communicate with older, slower wireless devices, technology called protection mechanisms was designed to allow for backward compatibility. The mechanisms available depend on which amendment to the standard is used. Protection mechanisms will provide the backward compatibility needed to allow different technologies to coexist in the same radio frequency space.

There are two broad categories of protection mechanism:

- Extended rate physical (ERP) protection mechanism for IEEE 802.11g networks
- High throughput (HT) protection mechanism for IEEE 802.11n networks

Each category includes several modes for specific situations.

IEEE 802.11g Extended Rate Physical Protection Mechanisms

For IEEE 802.11g and IEEE 802.11b devices to coexist in the same basic service area, the wireless access point must use *extended rate physical (ERP) protection*. Most manufacturers of IEEE 802.11 wireless LAN equipment will provide options when it comes to coexistence. These options usually include the capability to set an access point to one of three modes:

- IEEE 802.11bonly mode: DSSS and HR/DSSS
- IEEE 802.11gonly mode: ERP-OFDM
- IEEE 802.11b/g mixed mode: DSSS, HR/DSSS, and ERP-OFDM

IEEE 802.11b-Only Mode

An access point must be set to operate in 802.11b-only mode. This involves disabling all the IEEE 802.11g ERP-OFDM data rates of 6, 9, 12, 18, 24, 36, 48, and 54 Mbps and allowing only DSSS data rates of 1 and 2 Mbps and HR/DSSS rates of 5.5 and 11 Mbps. Enabling this mode limits the maximum data rate to 11 Mbps. Setting an access point to this mode has limited applications, such as using legacy devices capable of IEEE 802.11b-only, for example.

IEEE 802.11g-Only Mode

This mode is the opposite of 802.11b-only mode. It disables all of the IEEE 802.11b DSSS and HR/DSSS data rates of 1, 2, 5.5, and 11 Mbps, and it allows the IEEE 802.11g ERP-OFDM data rates of 6, 9, 12, 18, 24, 36, 48, and 54 Mbps. This operation mode is useful in an environment where backward compatibility to 802.11b is not required (such as an environment where all devices connecting have IEEE 802.11g capability) and the throughput needs to be maximized; thus there are no IEEE 802.11b devices in use.

IEEE 802.11b/g Mixed Mode

Most deployments in the 2.4 GHz ISM band use this mode for communications. It allows devices that support the IEEE 802.11g amendment and IEEE 802.11b devices to operate together in the same BSA and associate to the same access point. Because of this protection mode, throughput will decrease when IEEE 802.11b devices and IEEE 802.11g devices are both associated to the same access point.

Extended rate physical (ERP) mixed mode uses control frames to reserve the wireless medium. Two options are available:

Request to Send/Clear to Send (RTS/CTS)   One option of control frames that are used as a protection mechanism to reserve the RF medium.

Clear to Send (CTS) to Self   A single frame used as a protection mechanism. This is a common implementation used by wireless LAN equipment manufacturers. A benefit of using this frame is less overhead than the RTS/CTS process.

Both RTS/CTS and CTS-to-Self control frames allow wireless devices using different Physical layer technologies to share the wireless medium and help to avoid collisions. These control frames specify how much time is needed for a frame exchange between the transmitter and a receiver to complete. The time value is processed by all devices in the basic service area that are not part of the frame exchange. Once this time has expired, the wireless medium is considered clear.

IEEE 802.11n High-Throughput Protection Mechanisms

IEEE 802.11n devices operate in either the 2.4 GHz or the 5 GHz band. Backward compatibility for IEEE 802.11a/b/g devices needs to be taken into consideration. The IEEE 802.11n amendment identifies several different operation modes for *high-throughput (HT) protection* mechanisms. The mechanisms are known as HT protection operation modes and are a set of rules that devices and access points will use for backward compatibility:

- No protection mode
- Nonmember protection mode

- 20 MHz protection mode
- Non-HT mixed mode

These operation modes are constantly changing based on the radio frequency environment and associated wireless devices. The goal with IEEE 802.11n wireless networks is to eventually get to no protection mode. The correct encoding bit (0, 1, 2 or 3) will be set in the HT Protection field within the wireless frame header and show what the actual operating mode is. With today's wireless networks and WLAN technology, we are more than likely at non-HT mixed mode or possibly even one of the other two modes in most cases.

No Protection Mode

No protection mode, allows high-throughput (HT) devices only. These HT devices must also share operational functionality, and they must match; for example, they must all support 20 MHz or 20/40 MHz channels only. If an IEEE 802.11n (HT) access point is set to 20/40 MHz channel width and a client capable of only 20 MHz wide channels associates, the connection is considered to be in no protection mode. This operation mode does not allow IEEE 802.11a/b/g devices using the same RF channel. IEEE 802.11a/b/g devices will not be able to communicate with an access point in no protection mode. Transmissions from these devices will cause collisions at the access point, causing some degradation in throughput because it is seen by the HT system as radio frequency interference. No protection mode is what we as wireless network designers and administrators are working toward achieving, but it may be some time before we are there because of backward compatibility and legacy wireless devices.

Nonmember Protection Mode

All devices in this operation mode, or *nonmember protection mode*, must be HT capable. When a non-HT device—that is, an IEEE 802.11a/b/g access point or wireless client device—is within hearing range of the HT access point and on the same 20 MHz channel or one of the 20/40 MHz wide channels, this protection mode will be activated.

20 MHz Protection Mode

All devices in this operation mode, or *20 MHz protection mode*, must be HT capable as well. The operation of this protection mode is based on the fact that 802.11n devices can use 20 MHz or 20/40 MHz wide channels. 20 MHz protection mode means that at least one 20 MHz HT station is associated with the HT 20/40 MHz access point and that the access point provides compatibility for 20 MHz devices.

Non-HT Mixed Mode

Non-HT mixed mode, is used if one or more non-HT stations are associated in the BSS. This operation mode allows backward compatibility with non-802.11n or IEEE 802.11a/b/g wireless devices. This is the likely the most common mode for IEEE 802.11n

HT networks today because of the need for backward compatibility and the legacy IEEE 802.11 wireless devices that are still in use on most wireless networks.

Additional HT Protection Mechanisms

Two other HT protection mechanisms are also available:

- Dual CTS is a Layer 2 protection mechanism that is used for backward compatibility between IEEE 802.11n HT and IEEE 802.11a/b/g devices.
- Phased coexistence operation (PCO) is an optional BSS mode with alternating 20 MHz and 20/40 MHz phases controlled by a PCO-capable access point.

Summary

Standards-based wireless LANs can use unlicensed or licensed radio frequency bands, which are broken down further into smaller RF channels. Wireless LANs use two modes: either ad hoc mode, which means no access points are used, or infrastructure mode, where an access point provides a central point of communication. We looked at wireless LAN configurations and their parameters in this chapter.

We also looked at the spread spectrum and Physical layer (PHY) technologies used with WLANs and the differences among them. The IEEE standard and various amendments use different PHY technologies and unlicensed radio spectrum, allowing for data rates of up to 600 Mbps for 802.11n and 6.93 Gbps for 802.11ac.

We explored FHSS because it is used today in many industries in various types of wireless technologies, including IEEE 802.15 personal area networks (PANs), Bluetooth, and cordless telephones.

You learned about radio frequency coverage and capacity and various types of radio frequency signal measurements used with IEEE 802.11 wireless networking.

This chapter also looked at the processes wireless devices use to connect to and become part of a wireless LAN, including passive and active scanning (discovery) and IEEE 802.11 authentication and association. Once these processes are complete, the wireless device becomes part of the wireless network, enabling it to pass traffic across to the access point to the network infrastructure.

Additionally, you saw the components and technology that play a role with IEEE 802.11 wireless networks, including the distribution system (DS), data rate (what is advertised), and throughput (what is actual).

Dynamic rate switching—when a client is transferring more or less data depending on the proximity from an access point as well as roaming or moving through the basic service areas while being able to maintain connectivity—was also discussed in this chapter. This chapter explored the important topics of power save mode and protection mechanisms.

Last, this chapter covered IEEE 802.11 protection mechanisms and how they provide backward compatibility to older technology devices. We looked at some highlights of extended rate physical (ERP) protection for IEEE 802.11g and high throughput (HT) protection 802.11n networks.

Chapter Essentials

Know the frequencies and channels HR/DSSS and OFDM use. Understand that HR/DSSS operates in the 2.4 GHz ISM band and can use 14 channels depending on the country/location used. Know that ERP-OFDM is used for the 2.4 GHz band, OFDM is used for the 5 GHz band, and HT-OFDM is used for either band. Know the four U-NII bands OFDM uses for the 802.11a and 802.11n amendments. Understand that MIMO systems may use HT-OFDM and can operate in either the 2.4 GHz ISM band or the 5 GHz U-NII band.

Understand the different operation modes for IEEE 802.11 wireless networks. Know the difference between infrastructure and ad hoc mode as well as the use of both.

Be familiar with the different service sets used with wireless networking. Understand the differences among IBSS, BSS, ESS, MBSS, and BSA.

Identify the terminology used with IEEE 802.11 wireless networking. Understand the differences among SSID, ESSID, and BSSID. Know which one identifies the name of a network and which one identifies the physical address of an access point.

Know the differences among various Physical layer wireless technologies, such as FHSS, DSSS, HR/DSSS, OFDM, ERP-OFDM, and MIMO. The uses of Physical layer technologies vary depending on radio frequency, applications, and desired data rates. Understand the standard or amendment that defines each Physical layer technology as well as advantages and disadvantages of each, including colocation and interference.

Know the process that devices use to join a wireless LAN. Understand the process and operation of discovery, passive scanning, active scanning, IEEE 802.11 authentication, and IEEE 802.11 association.

Understand the differences between distribution systems as well as data transfer. Identify the differences as well as the functions of a wired distribution system and wireless mesh technology with respect to data flow. Know the differences between data rate, throughput, and dynamic rate switching.

Identify the power-save capabilities of IEEE 802.11 wireless networks. Know the various power save modes of both legacy IEEE 802.11 wireless technology devices and newer Wi-Fi Multimedia (WMM) technology devices, including active mode, power save mode, and APSD.

Know the various protection mechanisms available for both IEEE 802.11g and 802.11n wireless networks. Be familiar with the two types of protection mechanisms: ERP protection mechanisms and HT protection modes. Understand that these mechanisms provide coexistence for newer and legacy wireless LAN devices.

Chapter 6

Computer Network Infrastructure Devices

TOPICS COVERED IN THIS CHAPTER:

✓ Wireless Access Points

✓ Mesh

✓ Wireless Bridges

✓ Wireless Repeaters

✓ Wireless LAN Controllers and Cloud-managed Architectures

✓ Power over Ethernet (PoE) Devices

✓ Virtual Private Network Concentrators

✓ Network Gateways

✓ Proxy Servers

Choosing the correct network infrastructure devices to be installed as part of a computer network (both wired and wireless) is a critical element of a successful wireless LAN deployment. In this chapter, we will look at a variety of infrastructure devices, including wireless access points, wireless mesh devices, wireless bridges, wireless repeaters, hardware wireless LAN controllers, and cloud-managed wireless systems. Cloud-managed wireless LAN deployments are growing at a very fast pace in all markets. Where some manufacturers specialize in only cloud-managed solutions, most manufacturers of enterprise wireless equipment utilize cloud-based access points to some extent. You will learn about the features, benefits, and advantages of these and other wireless network infrastructure devices. Power over Ethernet (PoE) is an extension to the IEEE 802.3 Ethernet standard that allows direct current (DC) voltage to be supplied over Ethernet cable to wireless access points, VoIP telephones, Ethernet security cameras, and other PoE-capable devices. Finally, we will explore the basics of other network infrastructure devices, including virtual private network (VPN) concentrators, network gateways, and network proxy devices.

The Wireless Access Point

The wireless *access point (AP)* is an integral component of a wireless LAN infrastructure. Wireless access points allow a variety of wireless client devices access to any network resources that the device or user may have permissions for. The access point provides computers, voice over Wi-Fi phones, smartphones, tablets, and other wireless client devices access to a local area network, using radio frequency (RF) as the communication mechanism through free space (air) as the communication medium.

Wireless access points are available in three common types—autonomous, controller-based, and cloud-based (or "controllerless"). Autonomous access points are self-contained units and can function as independent network infrastructure devices. Controller-based access points, by contrast, function in conjunction with a hardware wireless LAN controller. Cloud-based access points provide a wireless network infrastructure without the use of a hardware controller, and are software-managed devices.

When a wireless device is connected to an access point, it is said to be in *infrastructure mode*. In this operation mode, all wireless data traffic is passed through the access point to the intended destination, whether that is a file server, a printer, the Internet, another wireless client device, or anything else capable of receiving network data.

An access point can operate as a stand-alone network device, configured independently to allow wireless devices to connect. It can also operate as part of a larger wireless network by sharing some of the same parameters, such as the service set identifier (SSID). The SSID is the logical name, or identifier, of the wireless LAN, and all wireless client devices connected to an access point will share the same SSID setting. Figure 6.1 shows an example of an access point connected to an Ethernet network with several wireless network client devices.

FIGURE 6.1 Access point connected to an Ethernet network

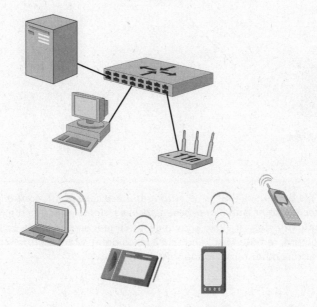

In addition to providing access through a shared medium, wireless access points are *half-duplex* devices. Half-duplex is defined as two-way communication that occurs in only one direction at a time. (By contrast, *full-duplex*, the other communication method used in computer networking, allows two-way communication to occur between devices simultaneously.) Communication only one way at a time means less data throughput for the connected device. An access point is a network infrastructure device that can connect to a distribution system (DS)—typically an Ethernet segment or Ethernet cable—and allow wireless devices to access network resources with the appropriate access permissions. According to the IEEE 802.11 standard, all wireless devices are considered stations (abbreviated STA), including access points. However, per the standard, an access point is identified as AP/STA (access point station). In a completely Ethernet-switched network, devices will communicate directly with the Ethernet switch. Figure 6.2 illustrates half-duplex communication using a wireless access point.

FIGURE 6.2 Half-duplex—communication one direction at a time

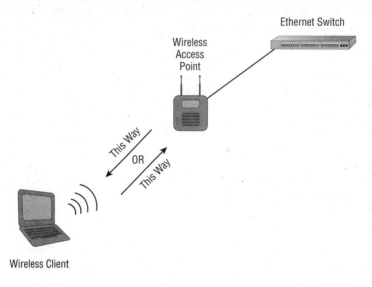

This book uses the terms *autonomous access point*, *controller-based access point*, and *cloud-based access point* to describe the devices. In the industry, they are also known as intelligent access point, split MAC architecture, remote MAC, and thin access point. Manufacturers may also use various other terms to identify them.

Autonomous Access Points

Autonomous access points are self-contained units with all the intelligence necessary to provide devices with wireless access to a wired network infrastructure, and access to the resources the devices have permission to use. There are three popular types of autonomous access points: consumer grade; small office, home office grade (SOHO grade); and enterprise grade. Not surprisingly, the enterprise type generally offers the most robust feature set. The autonomous access point is best suited for small networks with only a few access points or wireless hotspots, because they must be managed independently, and therefore the scalability is somewhat limited.

The industry terminology for access points is not clearly defined. What might be considered a consumer-grade or SOHO-grade access point from one vendor might be considered the opposite from another vendor. In fact, some access points might even be considered both consumer-grade and SOHO-grade access points.

The Consumer-Grade Access Point

Consumer-grade access points, or home broadband wireless routers, are usually equipped with an Internet port, several ports for Ethernet connections, and a wireless access point. The routers are configured through a web browser using either the HTTP or HTTPS protocol. Configuration of the devices is fairly simple for the novice user using a web browser via a built-in web server. In most cases, a broadband wireless router connects to either a cable modem or a digital subscriber line (DSL) connection available from an Internet service provider (ISP). In this configuration, a router is able to accept wired and wireless connections for computers and other devices, providing them with access to the local area network or the Internet. A wireless broadband router usually includes the following features:

- Network Address Translation (NAT)
- Dynamic Host Configuration Protocol (DHCP) server
- IP routing
- Domain Name System (DNS) services
- Built-in firewall

A wireless broadband router has many of the same features a SOHO access point has. An example of an IEEE 802.11ac dual-band Gigabit wireless broadband router is shown in Figure 6.3.

FIGURE 6.3 Netgear Model R7000 AC1900-Nighthawk Smart Wi-Fi dual-band IEEE 802.11ac Gigabit router

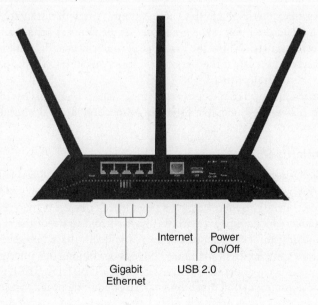

Notice in the figure that this device contains a Gigabit Ethernet switch, an Internet WAN port, a USB 2.0 port, and a wireless access point all in one self-contained unit. This is a very common configuration with the wireless broadband router.

The SOHO-Grade Access Point

Although they are powerful devices, *SOHO-grade access points* usually have a less extensive feature set than enterprise-grade access points. However, all newer-model consumer-grade, SOHO-grade, and enterprise-grade access points now support the highest standards-based security options available, including IEEE 802.11i or Wi-Fi Protected Access 2.0 (WPA 2.0) certifications. SOHO-grade access points are best used in the SOHO or home environment, and usually have a limited number of connections for computers and devices. SOHO-grade access points typically have the following features:

- IEEE 802.11 standards support
- Wi-Fi Alliance interoperability certifications
- Removable antennas
- Static output transmit power
- Advanced security options
- Wireless bridge functionality
- Wireless repeater functionality
- Dynamic Host Configuration Protocol (DHCP) server
- Configuration and settings options

Figure 6.4 shows an example of a SOHO access point. You will notice that one difference between a SOHO access point and a consumer-grade wireless broadband router is the available connection ports. Where the broadband router is usually equipped with an Internet port and several ports for Ethernet connections, the SOHO access point typically has one Ethernet port to connect to a LAN.

In the figure, notice that unlike the consumer-grade access point discussed earlier, this device has only a single PoE-capable Ethernet connection, and no Internet WAN connection or USB port.

IEEE 802.11 Standards Support

All later-model SOHO-grade access points support the current IEEE 802.11 standard. Some older devices may not have firmware updates available, which can cause implementation challenges where interoperability between newer and legacy devices is required. In this case, I recommend replacing older equipment with the most current, state-of-the art models. The 802.11 standard and amendments that are supported will vary based on several factors, including the cost and complexity of the unit. The most common consumer-grade and SOHO-grade access points support the IEEE 802.11b, IEEE 802.11g, IEEE 802.11n, and now IEEE 802.11ac communication amendments. Most equipment manufacturers do make dual-band models, but the cost is normally higher than single-band (IEEE 802.11b/g/n) access points. See Chapter 4, "Standards and Certifications for Wireless Technology," if you need to review these 802.11 amendments to the standard.

FIGURE 6.4 D-Link DAP-2590 AirPremier N dual-band PoE SOHO access point, front and end views

Wi-Fi Alliance Interoperability Certifications

Interoperability certifications from the Wi-Fi Alliance are a common feature of SOHO-grade access points. As mentioned in Chapter 4, these certifications include WPA/WPA 2.0 and WPS for security, and WMM and WMM-PS for QoS. Selecting a SOHO-grade access point that is Wi-Fi certified ensures compliance with IEEE standards, and interoperability with other IEEE 802.11 wireless devices.

 In December 2011, a security flaw was reported with WPS, and should be considered before using this feature. See the warning in the section "Wi-Fi Protected Setup Certification" in Chapter 4 and check the device manufacturer's website for more information.

Removable Antennas

Some SOHO-grade access points are equipped with removable antennas. This allows the end user to change to a larger (higher-gain) antenna, thereby allowing a radio frequency to cover a wider area. Conversely, connecting a smaller (lower-gain) antenna will decrease the coverage area. Many SOHO access points have fixed or nonremovable antennas, so you cannot add a higher-gain antenna.

 The RF coverage area of an access point can be increased by adding a higher-gain antenna to the access point. For more information on this, and other antenna-related information, see Chapter 3, "Radio Frequency and Antenna Technology Fundamentals."

Static Output Transmit Power

Occasionally an end user will have the ability to adjust the transmit output power in a SOHO-grade access point. If this is available, the settings are usually very basic, such as low, medium, and high. With enterprise access points, you can change the power in increments of a milliwatt (mW) or a decibel milliwatt (dBm). The transmit output power of the access point will determine in part the area of radio frequency coverage, also known as the RF cell or basic service area. A cell is the area of RF coverage of the transmitter, in most cases a wireless access point. The typical RF transmit power of a SOHO-grade access point is about 15 dBm, or 32 mW; however, this will vary with the manufacturer. An access point model with static output power settings cannot be adjusted, which will limit your ability to decrease or increase the size of the radio frequency cell. Changing the RF cell size will allow the access point to cover a larger area in the home or small office where the device is installed. In this case, the only way to change the RF cell size is to change the gain of the antenna in models that have the removable antenna feature. Note that replacing the antenna will also change the vertical and horizontal beamwidths, which is the electromagnetic radiation pattern that propagates away from it.

Advanced Security Options

All newer-model SOHO-grade access points support the highest security features, including the IEEE 802.11i amendment and WPA 2.0 personal and enterprise operation modes. These security features give users with limited technical knowledge the ability to provide the most up-to-date security for their wireless network. For those users who have greater technical know-how, SOHO-grade access points also provide more advanced security features, such as user-based authentication IEEE 802.1X/EAP and VPN pass-through. Users can find more information about these advanced features in most user guides provided with the access point, or online at the device manufacturers' websites.

Wireless Bridge Functionality

SOHO-grade access points can sometimes be configured in wireless bridge mode. Both point-to-point and point-to-multipoint settings are available, enabling administrators to connect two or more wired LANs together wirelessly using IEEE 802.11 equipment.

You learned about point-to-point and point-to-multipoint settings in Chapter 1, "Computer Network Types, Topologies, and the OSI Model." Wireless bridging is discussed later in this chapter.

Wireless Repeater Functionality

Some SOHO-grade access points can be configured to function as wireless repeaters. Configuring an access point as a repeater enables administrators to extend the size of the radio frequency cell so that devices not in hearing range of an access point can connect to the wireless network. However, the cost is reduced throughput for other devices accessing the network through a wireless repeater, in addition to more contention. Wireless repeaters are discussed later in this chapter.

Dynamic Host Configuration Protocol (DHCP) Server

It is also common for SOHO-grade access points to be able to act as Dynamic Host Configuration Protocol (DHCP) servers. You learned in Chapter 2, "Common Network Protocols and Ports," that a DHCP server will automatically issue an Internet Protocol (IP) address (logical address) to allow upper-layer communication between devices on the network. IP addresses are a function of Layer 3 of the OSI model, as outlined in Chapter 1. A built-in DHCP server will ease the installation and support of the access point, providing a much better overall user experience.

Configuration and Settings Options

SOHO-grade access points are configured via a web browser, using either HTTP (Hypertext Transfer Protocol) or HTTPS (Hypertext Transfer Protocol Secure). This type of browser-based configuration is an easy way for the novice administrator to make all the necessary settings based on the application in which the access point will be used. SOHO-grade access points rarely offer configuration from the command-line interface (CLI), which allows for more extensive configuration parameters. Figure 6.5 shows an example of a configuration page from a SOHO-grade access point.

For security reasons, it is best practice to configure access points from the wired side of the network infrastructure whenever possible. Configuration of an access point should only be done wirelessly if absolutely necessary. If configuring this device from the wireless side of the network is the only option, a secure connection such as HTTPS or SSH2 should be in place to prevent potential compromise of administration user credentials and unauthorized access of the wireless device.

Some manufacturers of consumer-grade and SOHO-grade wireless equipment have online emulators that allow customers to view a sample of the configuration process for a device. This allows a user to sample the configuration settings and become familiar with the device before making a purchase.

FIGURE 6.5 D-Link DAP-2590 AirPremier N dual-band PoE SOHO access point configuration screen

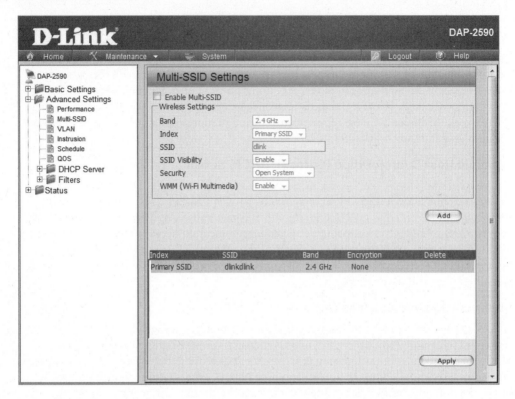

The Enterprise Access Point

Enterprise-grade access points typically have a much more extensive feature set than the previously mentioned SOHO-grade access points. Taking this into consideration, the price point can be significantly higher for enterprise-grade access points. Figure 6.6 shows an example of an enterprise-grade access point.

Enterprise-grade access points can include the following features:

- IEEE 802.11 standards support
- Wi-Fi Alliance interoperability certifications
- Removable or expandable antennas
- Adjustable output transmit power
- Advanced security options
- Multiple operation modes, including root access point, wireless bridge, wireless repeater, and mesh capabilities
- Graphical user interface (GUI) and command-line interface configurations

FIGURE 6.6 The Aruba Networks 220 series IEEE 802.11ac dual-band access point

In addition to the items listed here, enterprise-grade access points have various other features that put them a notch above consumer-grade and SOHO-grade access points. Some of these features are outdoor use, plenum ratings, industrial environment ratings, more memory, and faster processors to help handle the load and various environmental conditions.

IEEE 802.11 Standards Support

Like the other access point types discussed in this chapter, enterprise-grade access points support IEEE standards. Enterprise access points have a more extensive feature set than SOHO-grade access points, and depending on the manufacturer and model, they will support all communication standards by utilizing the 5 GHz IEEE 802.11a/n/ac and the 2.4 GHz IEEE 802.11b/g/n dual-band radios. Enterprise-grade access points can include support for some amendments to the standard not supported by SOHO-grade access points. Examples include support for IEEE 802.11e QoS, Wi-Fi multimedia certifications, IEEE 802.11r fast BSS transition (FT), and IEEE 802.11w for the security of management frames, to name a few.

Wi-Fi Alliance Interoperability Certifications

Certifications by the Wi-Fi Alliance are an important feature of enterprise-grade access points. These certifications include WPA/WPA 2.0 for security and WMM and WMM-PS for QoS. Selecting an enterprise-grade access point that is Wi-Fi certified ensures compliance with IEEE standards and interoperability with other IEEE 802.11–compliant devices.

Removable or Expandable Antennas

Many enterprise access points have removable or expandable antenna capabilities. These antenna configurations provide a lot of flexibility, because an installer can choose the appropriate antenna based on the deployment scenario. Omnidirectional, semidirectional, and highly directional antennas are all types of antennas commonly used in the enterprise environment. Enterprise-quality access points that use internal antennas can offer options for connecting external antennas, should they be required for the specific installation. Antennas were discussed in more detail in Chapter 3.

Adjustable Output Transmit Power

Unlike consumer-grade and some SOHO-grade access points, the RF output power of the radio can be adjusted with enterprise-grade access points. This feature allows an installer to select the correct amount of transmit power based on the installation needs of the access point. One benefit of having adjustable output power is that an installer can adapt to the environment in which the access point is installed. If the radio frequency dynamics of an area change, the ability to change access point settings such as output transmit power without physical intervention is beneficial.

Advanced Security Options

Compared to access points used in the consumer and SOHO environments, enterprise access points typically have more advanced security features. In addition to IEEE 802.11i, WPA/WPA 2.0, passphrase and user-based authentication, and IEEE 802.1X/EAP modes, features such as a built-in user database for local *Remote Authentication Dial-In User Service (RADIUS)* authentication are included. As discussed later in this chapter, local RADIUS authentication allows small-to medium-sized businesses to provide their own advanced authentication features without the need of external RADIUS authentication services. This reduces costs and lowers administration overhead.

RADIUS is just one example of the more advanced security features available in enterprise-level access points. Another advanced security feature that may be available is some level of a wireless intrusion prevention system (WIPS). A WIPS will help determine and have the potential to mitigate certain levels of wireless intrusions or attacks on the network. One example is the detection of a rogue (unknown) access point.

Advanced security features are discussed in more detail in Chapter 16, "Device Authentication and Data Encryption." Some of the common configuration options available for one model of enterprise access point are shown in Figure 6.7.

Multiple Operation Modes

In addition to the features just discussed, enterprise-grade access points typically have several operation modes:

Root Access Point Mode (the most common configuration)   What some in the industry refer to as *root access point mode* is typically the default operation mode in which an enterprise-grade access point is set. Root access point mode involves connecting the access point to a distribution system (DS) such as an Ethernet segment or a network infrastructure of some sort. This mode allows computers and other wireless devices to connect to the access point and use network resources based on the assigned permissions of the user, computer, or other wireless device.

Wireless Bridge Mode (for connecting LANs together)   This configuration allows an access point to be set in *bridge mode* for wireless point-to-point or point-to-multipoint configurations connecting two or more LANs. Benefits of using wireless access points to bridge

LANs together include cost savings and high data transfer rates compared to some other wired connectivity options.

Wireless Repeater Mode (to extend the radio frequency cell) An access point configured in wireless *repeater mode* can act to extend the radio frequency cell. This allows wireless client devices outside the radio hearing range of an access point to still be able to connect to the network and access network resources via the wireless repeater. This operation mode does have the downside of reduced throughput, and should be used only if justified because of no wired connectivity that allows the access point/repeater to connect to the network.

Mesh Mode (for connecting access points together wirelessly) If an access point has mesh capability, it can be configured to connect access points together without using a wired infrastructure. This is useful in areas where the physical distance from the network segment to the access points exceeds the 328-foot or 100-meter limitation. This is what was formerly known as a wireless distribution system (WDS). According to the latest version of the IEEE 802.11 standard, the term WDS is obsolete and subject to removal in a subsequent revision of the standard.

FIGURE 6.7 Motorola AP-7131 enterprise-grade access point configuration page in a web browser

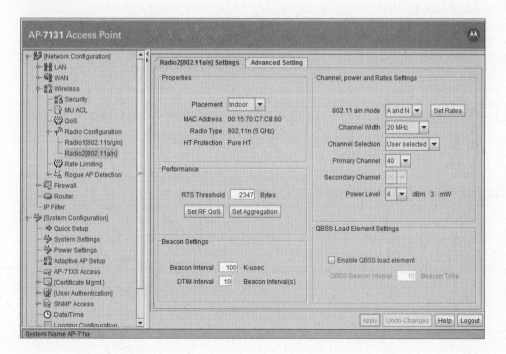

If an enterprise-grade access point is configured for an operation mode other than an access point, it is no longer considered an access point. Typically an access point can be configured to operate in only one mode at any one time. Also, if an enterprise-grade access point is dual-band capable, it may be possible to configure each band (radio) as a different operation mode.

Access Point Configuration Methods

Enterprise-grade access points can commonly be configured, or "staged," two different ways:

Graphical User Interface (GUI) Configuration   Enterprise-grade access points can be configured using a GUI configuration from a web browser using HTTP or HTTPS. This is a convenient way to configure and change settings on the access point using a common graphical interface tool, a web browser. If the access point is configured using a wireless connection, using HTTPS is recommended for security at a minimum, but an SSH or IPsec VPN connection is highly recommended.

Command-Line Interface (CLI) Configuration   Most if not all enterprise-grade access points have *command-line interface (CLI)* capabilities to allow extensive and detailed configuration of the device. In some cases, the CLI command set provides higher-level commands that allow an administrator to perform additional configuration tasks that aren't available using the web browser method. This allows consistency in configuring other network infrastructure devices, because many manufacturers share common commands among devices. CLI capabilities vary depending on the manufacturer, but most enterprise models have an extensive set of commands. Access points with CLI capabilities may have a special console port on the device, allowing a direct serial connection to a computer to access for configuration purposes.

Controller-Based Access Points

Controller-based access points differ from autonomous access points in that they are used with hardware wireless LAN controllers and not as stand-alone devices. An autonomous access point is a self-contained unit that has all the intelligence needed to provide computer and device access to a wireless network. In contrast, controller-based access points have shifted some of the intelligence to the hardware wireless LAN controller. Since a controller-based access point may contain less intelligence than an autonomous access point, the cost of a controller-based access point may be lower, depending on the local MAC capabilities.

Controller-based access points are centrally managed from the hardware wireless LAN controller. The extent of how this is accomplished depends on the type of architecture that is in use: local MAC, split-MAC, or remote-MAC. Split-MAC is the most commonly used, with the controller handling the management and control planes and the access point taking care of the real-time MAC functionality and the Physical layer communications. How manufacturers implement the technology varies. Typically, communication between the access point and the controller is handled by an Internet Protocol (IP) or proprietary tunnel that is built from the access point to the controller. Depending on the manufacturer, they may have a more extensive feature set than autonomous access points, while also including many of the features of those devices. Figure 6.8 shows front and rear views of a controller-based access point.

FIGURE 6.8   Cisco Aironet 3500e CleanAir IEEE 802.11n access point

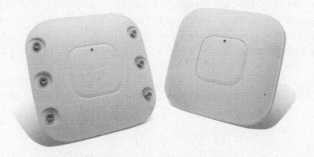

The benefits of controller-based access points are similar to those of autonomous access points, including radio frequency management, security, and quality of service. With remote MAC controller-based access points, very little or no intelligence is contained within the devices. Controller-based access points are PoE capable for ease of deployment in either mid-sized or large organizations. One thing to keep in mind is if a controller-based access point is unable to communicate with the wireless LAN controller for some reason, the access point may be able to still provide communications to the associated client devices. How this is achieved widely varies, based on how the manufacturer implements the technology.

Cloud-Based Access Points

Cloud-based access point technology provides another solution for deploying wireless local area network infrastructures. The architecture for networks that use this type of access point is sometimes referred to as a "controllerless" architecture, because the device intelligence has been pushed back out to or distributed to the access point edge, similar to that of the autonomous access point, but with much more intelligence and capabilities. The access points are managed through a "cloud" software configuration tool, eliminating the need for a hardware controller. This type of management tool can be accessed from any computer with an Internet connection, assuming the user has appropriate permissions to manage the devices. Some manufacturers also have "software appliances," eliminating the reliance on the cloud server because the software is contained locally with the organization's datacenter network. Figure 6.9 shows an example of a cloud-based access point.

FIGURE 6.9 ADTRAN/Bluesocket 2030 IEEE 802.11ac (3x3:3) indoor acess point

Cloud-based access points provide all the benefits and features of a wireless LAN controller solution without the need and extra expense of a hardware controller. This technology is scalable and performs well without relying on a "tunnel" from the access point to a controller through which some or all of the network traffic can be forwarded.

The distributed intelligence allows the cloud-based access point to make decisions about how frames traverse both the wired and wireless network without relying on the hardware controller.

Some manufacturers of controller-based solutions provide a variant of the cloud technology by allowing autonomous access points to be "adopted" by a controller in a large enterprise environment. These access points are then site survivable, meaning they will still be able to function on their own should connectivity with the controller be temporarily lost. The description of this technology includes the term *adaptive access point*.

Wireless Branch Router/Remote Access Point

A *wireless branch router* or *remote access point* can be used to extend a corporate network to a remote location such as a home, conference room, or branch office through a secure connection using a WAN or the Internet. This type of device typically has three interfaces available:

- Ethernet port(s) to connect to a LAN
- Internet port to connect to the WAN or to an Internet connection
- Wireless port to allow IEEE 802.11 computers and devices to connect to a network through a wireless connection

Wireless branch routers are usually compact and lightweight, making them easy for sales representatives and other corporate employees to travel with. They also have a more extensive feature set than wireless broadband routers, including the capability of building a Layer 3 VPN tunnel between devices in which the router on each side will act as a VPN endpoint. They can also be configured as a VPN pass-through device, allowing another device on the network such as a VPN concentrator to act as the VPN endpoint. They also include these features:

- Point-to-Point Tunneling Protocol (PPTP)
- Layer 2 Tunneling Protocol/Internet Protocol Security (L2TP/IPsec)
- SSH2
- Advanced IP networking services
- Edge router capability

Figure 6.10 shows an example of a wireless branch router.

FIGURE 6.10 Aerohive BR200 wireless branch router

Wireless Mesh

Wireless mesh networking continues to grow at a steady pace. The concept of mesh networking has been in existence for many years. In a full mesh network, all nodes connect together with at least two paths for every node. This allows for reliable communication in the event of a device or path failure.

Wireless mesh networking is popular in the outdoor market space. Most outdoor mesh infrastructure devices provide the highest levels of wireless security, and are usually inside a rugged weatherproof enclosure for protection from the elements. Wireless mesh networks are currently utilized in places such as metropolitan areas, university campuses, and amphitheaters, as well as for applications used with public safety, transportation, and government organizations.

Currently, many wireless LAN manufacturers use proprietary mechanisms and protocols for wireless mesh networking. IEEE 802.11s is an amendment to the IEEE 802.11 standard to include wireless mesh networking, and was ratified in 2011. Many enterprise-grade access points have the ability to operate in mesh mode, whereas others have a dedicated mesh function.

Wireless mesh networking for indoor deployments is starting to appear as a viable solution for some wireless LAN infrastructure deployments.

Manufacturers commonly recommend using two unlicensed RF bands for a wireless mesh operation. One common solution is to use the 2.4 GHz ISM band for wireless client device access and the 5 GHz U-NII band for mesh device infrastructure connectivity. The use of a third radio may be an option in some cases. Using two different RF bands reduces contention on a single band and increases the overall performance of the network.

Mesh can also be used in the event of Ethernet loss to an access point. Some cloud-based access points are able to automatically mesh together when they suffer an Ethernet loss. By default, they typically support clients in both bands, but can mesh in 5 GHz if an Ethernet connection fails.

Figure 6.11 illustrates mesh access points connected to a wired infrastructure.

 Some manufacturers may provide as many as four radios in a mesh access point. The fourth radio could utilize the 4.94–4.99 GHz licensed band, which is restricted to public safety use only, in the United States.

FIGURE 6.11 Mesh access points/routers connected to a common infrastructure and to the Internet

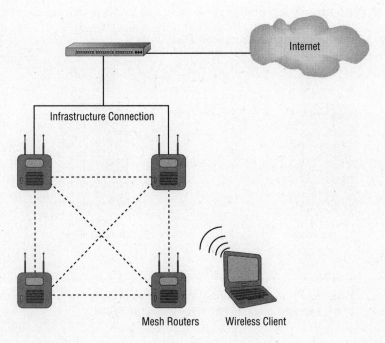

Wireless Bridges

Wireless bridges connect two or more wired LANs together. As discussed in Chapter 1, typically there are two configurations for wireless bridges: point-to-point and point-to-multipoint. A wireless bridge is a dedicated device that functions in much the same way as an access point in bridge mode. Wireless bridges have many of the same features as enterprise access points, including removable antennas and selectable power levels.

Connecting locations together using wireless bridging has many benefits, including fast installation, cost savings, and high data transfer rates. Depending on the circumstances, a wireless bridge can be installed in as little as one day or even several hours. Cost savings can be enormous compared to installing and maintaining a physical wired connection between locations, whether it is copper, fiber-optics, or a leased line from a service provider.

Most wireless bridges can work in either the 2.4 GHz ISM or 5 GHz U-NII unlicensed band. However, keep in mind that bridges using wireless technology can also consist of proprietary technology using licensed frequency bands, which may be capable of greater distances and faster speeds than wireless bridges that use IEEE 802.11 devices.

The connection can span long distances, so it is important to take security and environmental conditions into consideration, as well as the proper antenna selection.

Figure 6.12 illustrates wireless bridges connecting two LANs in two separate buildings.

FIGURE 6.12   Wireless bridges connecting two LANs

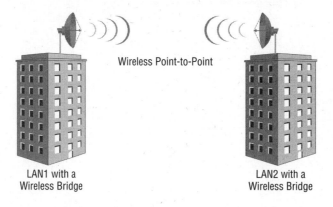

 When LANs are connected using wireless bridges, the bridges must be set to the same RF channel and have the same SSID.

Wireless Repeaters

Wireless repeaters are used to extend the radio frequency cell. In a wired Ethernet network, repeaters function at Layer 1 of the OSI model to extend the Ethernet segment. An Ethernet repeater lacks intelligence—that is, it cannot determine data traffic types, and simply passes all data traffic across the device. Since wireless infrastructure devices, including repeaters, are Layer 2 devices, they have more intelligence than Ethernet repeaters.

An Ethernet segment has a maximum distance for successful data transmission, and wireless LANs do as well. This distance depends on several factors, including the transmit power of the access point and the gain of the antenna. Like an access point, the wireless client device is also a transmitter and a receiver, and will have a radio frequency range limited by the transmit power and gain of the antenna. A wireless repeater provides the capability for computers and other devices to connect to a wireless LAN even when outside the normal hearing range of the access point connected to the network. Figure 6.13 illustrates how a wireless repeater can extend the range of a wireless network.

FIGURE 6.13 A wireless repeater extends the range of a wireless network.

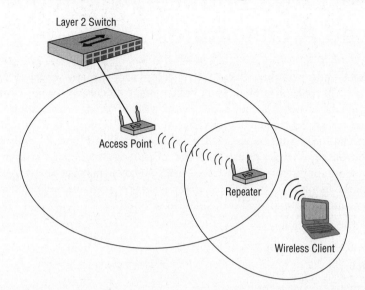

As illustrated, the wireless client device is not within hearing range of the access point, so adequate communication is not possible between these devices. In order for wireless LAN devices to communicate effectively with an access point, the transmitter must be able to hear the receiver, and the receiver must be able to hear the transmitter. It is a two-way radio communication. A wireless repeater will allow this communication to occur where the wireless client is outside the radio frequency cell or basic service area (BSA) of the access point. The wireless client will send information, or frames, to the repeater and the repeater will forward them to the access point, and vice versa. The downside of this configuration is that it will reduce the overall throughput. The wireless repeater may be named differently by the manufacturer and include the term *wireless range extender*.

 Real World Scenario

Using Wireless Repeaters Reduces Throughput

Before using a wireless repeater, consider whether it would be the best solution. Since wireless LANs are half-duplex (two-way communication but only one way at a time), data throughput will suffer when repeaters are used. When data traverses a wireless link between devices that are set to the same radio frequency channel, the data throughput can be reduced by up to 50 percent. If a physical wired connection is available, it should be used for an access point connection rather than a wireless repeater. For security purposes, the Ethernet port on a wireless repeater should be disabled because it is not connected to a wired network infrastructure.

Wireless LAN Controllers and Cloud-Managed Architectures

The *wireless LAN hardware controller* is a main component in many wireless LAN deployments. Wireless LAN controllers range from branch office models with a few controller-based access points to large-scale enterprise devices with hundreds or even thousands of controller-based access points. The branch office models are typically used in remote office installations or small/medium business (SMB) applications with a limited number of access points. The controllerless, or cloud-managed, architecture is growing at a fast pace, and provides another option for enterprise wireless LAN deployments. The following sections discuss some of the many benefits, features, and advantages available on both wireless LAN controllers and cloud-managed controllerless solutions:

- Centralized administration
- Virtual local area Networks (VLANs)
- PoE capability
- Improved mobile device transition
- Wireless LAN profiles and virtual WLANs
- Advanced security features
- Captive web portals
- Built-in RADIUS services
- Predictive modeling site survey tools
- Radio frequency spectrum management
- Firewalls
- Quality of service (QoS)
- Infrastructure device redundancy
- Wireless intrusion prevention system (WIPS)
- Direct and distributed AP connectivity
- Layer 2 and Layer 3 AP connectivity
- Distributed and Centralized Data Forwarding

Centralized Administration

Wireless LAN controllers and cloud-managed controllerless solutions provide *centralized administration,* and give an administrator complete control over the wireless network from a single physical location. Unlike autonomous access points that require intervention at each device for configuration, these solutions can be a "one-stop shop" for configuration and management of the wireless network. A wireless network management system

(WNMS) can be used as a centralized tool to manage autonomous access points that may be used in larger deployments. A WNMS can be used to help scale the autonomous access point architecture, but is not required.

Virtual Local Area Network

According to the IEEE 802.1Q standard, VLANs define broadcast domains in a Layer 2 network by inserting VLAN membership information into Ethernet frames. Layer 2 Ethernet switches can create broadcast domains based on how the switch is configured, by using VLAN technology. This allows an administrator to separate physical ports into logical networks to organize traffic according to the use of the VLAN for security profiles, QoS, or other applications. The concept of a Layer 2 wired VLAN is extended to IEEE 802.11 wireless LANs. Both hardware controller and controllerless solutions have the ability to configure broadcast domains and segregate broadcast and multicast traffic between VLANs.

Power over Ethernet (PoE) Capability

Wireless LAN controllers and cloud-managed solutions support PoE, allowing direct current voltage and computer data to be sent over the same cable. (The details of PoE are discussed later in this chapter, in the section "Power over Ethernet.") Some hardware controllers provide direct access point connectivity to the controller at the access layer and will provide the PoE. For hardware controllers that do not support direct connectivity and cloud-managed solutions, the access points will receive their PoE from the Layer 2 Ethernet switch.

Improved Mobile Device Transition

Fast, seamless Layer 2 and Layer 3 *transitioning* or *roaming* between access points is another common feature of hardware wireless controllers and cloud-managed solutions. This feature makes it possible for computers and other wireless devices connected to the wireless LAN to maintain a connection while physically moving throughout the wireless network. The IEEE 802.11r amendment specifies fast transition (FT), and the IEEE 802.11k helps with this functionality. Transitioning is more often than not an enterprise network requirement, and exists in very few SOHO deployments.

Wireless LAN Profiles and Virtual WLANs

Both the wireless LAN controller and controllerless cloud-managed solutions can give network administrators the ability to create a variety of configuration profiles. These profiles can work in conjunction with VLANs to allow or deny access based on requirements for the computer, device, or user. Profiles can be configured for various situations, including different SSIDs for guest, corporate, and voice networks; security configurations; and QoS support. Each WLAN profile will create a virtual access point with its own BSSID, and will act as though it is a separate physical device. This includes all the wireless management traffic that works with wireless LAN technology.

 Although Wired Equivalent Privacy (WEP) is not recommended to be used in any wireless LAN, it may still exist in deployments that use legacy devices such as wireless bar code scanners or wireless print servers. Using WLAN profiles, you can allow this type of legacy device to be located on a separate wireless VLAN without compromising the security of the entire network. However, if WEP is still used, I highly recommend moving away from it as soon as possible by upgrading legacy devices.

Advanced Security Features

Like autonomous access points, hardware wireless controller and cloud-managed solutions provide advanced security options. These include security options based on IEEE 802.11i and WPA/WPA 2.0, with both passphrase and enterprise configuration capabilities, allowing for the most secure mechanisms available for wireless LAN technology.

Captive Web Portals

Captive web portal capability is a common feature in hardware wireless LAN controllers and cloud-managed systems. A *captive web portal* will intercept a user's attempt to access the network by redirecting them to a web page for authorization of some sort. This web page may request account credentials, payment information from a user, or a simple agreement to terms and conditions before granting access to the wireless network. One common example of where you will see a captive portal is in a paid or free wireless hotspot. The captive portal can be hosted by an outside service provider, an autonomous access point, a hardware wireless controller, and in a cloud-based system on the access point. It is important to note that some mobile devices may experience issues while connecting to a wireless network with captive web portal technology enabled.

Built-in RADIUS Services

Another common feature of wireless LAN controllers and cloud-managed systems is RADIUS services for 802.1X/EAP authentication, which is supported by WPA and WPA 2.0. Built-in RADIUS allows a network administrator to utilize the most advanced security features available today to secure the wireless network. Built-in RADIUS servers typically limit the number of users that can be created in the user database, which means that built-in RADIUS is a good solution for SMB or remote office locations, but not for very large organizations. Larger networks can use external RADIUS services for scalability. See www.gnu.org/software/radius for more about this server.

Predictive Modeling Site Survey Tools

Predictive modeling site survey tools assist in placement of access points and other wireless infrastructure devices. These tools are sometimes a feature of a hardware wireless LAN controller. Performing a predictive modeling site survey will assist in planning to determine coverage and capacity for data and voice for both indoor and outdoor deployments. Some manufacturers of cloud-managed enterprise solutions have web-based online predictive modeling site survey tools. Aerohive and AirTight Networks are two examples of manufacturers that provide these online wireless network planning tools. Visit their websites for more information and instructions on how to access these tools:

www.aerohive.com/planner

www.airtightnetworks.com/home/products/AirTight-Planner.html

Radio Frequency Spectrum Management

Keeping an eye on the radio frequency (RF) environment is another responsibility of the wireless network administrator. RF spectrum management consists of adjusting RF parameters such as the RF channel (frequency) and the RF transmit power after deployment. This allows the network to adapt to changes in the environment and assist in the event of hardware failures.

Firewalls

An integrated stateful firewall feature helps protect a network from unauthorized Internet traffic, but still allows authorized traffic. Firewalls can be hardware based, software based, or a combination of the two. Stateful firewalls, which keep records of all connections passing through the firewall, help protect against broadcast storms, rogue DHCP server attacks, Address Resolution Protocol (ARP) poisoning, and other potential attacks against the wireless LAN.

Quality of Service (QoS)

QoS features help time-critical applications such as voice and video communications minimize latency and allow for traffic prioritization. With the continual expansion of voice and video technology in the wireless LAN arena, QoS is becoming an increasingly important component in the wireless network.

Infrastructure Device Redundancy

Infrastructure device *redundancy* allows for fault-tolerant deployments and provides uninterrupted access in the event an access point or wireless LAN controller fails. Complete redundancy will prevent a major outage caused by hardware failure for mission-critical or other deployments. Coverage is maintained by alternating access points between the redundant infrastructure devices, minimizing interruption for user access in the event of a hardware failure. Cloud-managed systems eliminate the need of redundant hardware WLAN controllers, and the technology allows for redundancy in the event of an access point failure.

Wireless Intrusion Prevention System (WIPS)

A WIPS monitors all activity across the wireless network for potential intrusion and malicious activities. A WIPS can take appropriate action to mitigate an attack based on the type of intrusion. WIPS will be discussed further in Chapter 15, "Mobile Device Security Threats and Risks."

Direct and Distributed AP Connectivity

Connecting access points that are not directly plugged into a port on the wireless LAN controller is a feature known as distributed AP connectivity, and it's beneficial in large-scale deployments. Almost all manufacturers support distributed AP connectivity. Direct AP connectivity is defined as a direct connection to ports on the switch, and is typically used with access layer hardware wireless controllers. A typical controller with distributed connectivity is shown in Figure 6.14.

FIGURE 6.14 Meru MC6000 large-scale enterprise wireless LAN controller

Layer 2 and Layer 3 AP Connectivity

Early wireless network implementations were built with dedicated Layer 2 connectivity, which meant limited wireless mobility. Layer 2 transition (roaming) occurs when a computer or other wireless client device moves out of the radio cell of the currently associated access point and connects to a different AP maintaining Layer 2 connectivity.

As wireless networking technology evolved, so did the need for Layer 3 connectivity and transition. IP addresses are logical Layer 3 addresses that identify devices on a network. All IP devices on the same network or subnet are considered to be in the same IP boundary. Layer 3 roaming occurs when a client moves to an AP that covers a different IP subnet. After a transition, the client will no longer have a valid IP address from the original subnet, and the device will be issued an IP address from the new subnet while maintaining Layer 3 connectivity.

Figure 6.15 illustrates Layer 2 and Layer 3 connectivity and wireless transition.

FIGURE 6.15 Wireless client device roaming across Layer 2 and Layer 3 boundaries

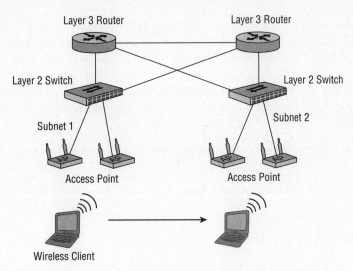

Distributed and Centralized Data Forwarding

Wireless LAN controller solutions consist of two common types of architectures: centralized and distributed. Early WLAN controller solutions supported the centralized architecture (split-MAC architecture). This design separated the intelligence from the access point and placed it into the wireless controller to allow for centralized management and control of the wireless network. The access point for the most part was just a radio and antenna, and traffic decisions were sent to the controller through an Ethernet cable. This technique is also known as *centralized data forwarding*. Depending on where the controller was placed, it could cause bottlenecks and other issues in the case of an overloaded or poorly designed network infrastructure. With the data rates possible with IEEE 802.11n and now IEEE 802.11ac, the aggregate throughput could be too much for the network to handle, resulting in poor performance.

Distributed data forwarding reduces the amount of infrastructure traffic because the controller-based access point is able to make more decisions, taking some of the load away from the wireless controller. Moving some of the intelligence back to the edge (the wireless access point) minimizes the bottlenecks and other potential issues such as latency. This is also true in a cloud-managed or controllerless architecture, thus eliminating the need for the data to be sent to the controller for handling.

Most enterprise wireless LAN equipment manufacturers now support both the centralized and distributed WLAN architectures.

Power over Ethernet

PoE sends direct current (DC) voltage and computer data over the same Ethernet cable, enabling a device to receive DC power and computer data simultaneously. This eliminates the need for an external alternating current (AC) power source to be near the Ethernet device.

An Ethernet cable has four copper wire pairs, or eight copper wires. Depending on the technology in use, either two or all four wired pairs may be used to carry data traffic. Figure 6.16 shows an example of a standard Ethernet cable pin assignment.

FIGURE 6.16 Standard Ethernet pin assignment

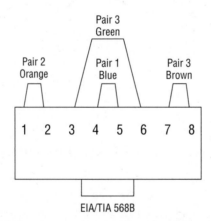

Power over Ethernet now consists of two ratified amendments to the IEEE 802.3 standard. They are defined in *802.3-2012 Clause 33*, also known as *IEEE 802.3af*, and *IEEE 802.3at*, sometimes called *PoE+*. These amendments define the specifications for devices used in wired or wireless networking to receive DC power from the Ethernet connection without the need for an external DC power source. The PoE amendments to the Ethernet standard allow electrical power to be supplied in one of two ways, either over the same wired pairs that carry computer data or over the wired pairs that do not carry data. 10BaseT and 100BaseT (Fast Ethernet) implementations use only two wired pairs (four wires) to carry data. 1000BaseT (Gigabit Ethernet) may use all four pairs (eight wires) to carry computer data. The standard defines which wire pairs are allowed to carry the DC power based on whether the network is 10BaseT, 100BaseT, or 1000BaseT, and whether the power is sourced from an endpoint or midspan injector. Both midspan and endpoint injectors are explained later in this chapter.

The nominal voltage for PoE is *48 volts of direct current (VDC)*, but the amendments allow for a range of 44 to 57 VDC at the power source. The PoE amendments address two types of devices: power sourcing equipment (PSE), the device that provides of the DC power, and the powered device (PD), the device that receives the DC power.

 Before PoE was standardized, some manufacturers used proprietary implementations. These solutions used various voltages, polarities, and pin assignments and may still be on the market today. I recommend that you verify PoE standard compliance before using this technology, to prevent potential hardware or device failures.

The IEEE 802.3-2012 Clause 33 (802.3af) amendment was released in 2003 and allocates 15.4 watts (W) of power maximum per port. This amendment has been incorporated into the IEEE 802.3-2012 standard. The IEEE 802.3at amendment, PoE+, was released in 2009 and includes changes to add to the capabilities of the IEEE 802.3-2008 standard, with higher power levels and improved power management information. IEEE 802.3at allows for 34.2W of power per port maximum, a big increase over IEEE 802.3af, which allowed for 15.4 W per port. IEEE 802.3af PoE will work with access points from all manufacturers of enterprise-grade IEEE 802.11n access points. Using IEEE 802.3af will make it easier for organizations to transition from older-model access points to the newer 802.11n technology, which will improve their client service and provide overall better performance without having to immediately upgrade their PoE infrastructure to IEEE 802.3at.

Power Sourcing Equipment

The *power sourcing equipment (PSE)* is the device that supplies the DC voltage to the end devices that receive the DC power. The DC voltage (power) can be delivered to the device in one of two ways:

- An *endpoint injector* (usually a wireless LAN controller or a Layer 2 Ethernet switch) delivers DC power directly over the same wire pairs that carry data over the unused wire pairs.
- A *midspan injector* (usually a single-port or multiple-port injector) injects DC power into the Ethernet cable over the unused wire pairs or over the data pairs, depending on the version of the standard in use.

Midspan Injectors

Midspan PoE injectors provide the required DC voltage (48 VDC) into the Ethernet cable, allowing the AP, bridge, or other powered device to receive electrical power and computer data. There are two types of midspan devices—single-port injectors and multiport injectors. A single-port injector supplies power to a single device. This is useful in an implementation that may have only a few PoE devices. A single-port injector is an in-line device that adds DC power to the Ethernet cable. A multiport injector can supply DC power to many devices simultaneously. It is an in-line device that functions like a patch panel. Two ports on this device are required to supply both DC power and computer data to a single powered device such as an access point, bridge, or IP camera. Therefore, a 24-port injector will allow connectivity for only 12 devices.

Endpoint Injectors

Endpoint PoE injectors supply DC power and computer data directly at the Ethernet port, rather than relying on an intermediate device to supply the power. Wireless LAN controllers and Ethernet switches are examples of endpoint devices. A benefit of endpoint PoE is that no intermediate adapter to inject power is necessary.

Powered Devices and Classification Signatures

The *powered device (PD)* is defined as the device receiving DC power, such as a wireless access point, wireless bridge, IP camera, IP phone, and so on. The IEEE 802.3 standard defines the maximum cable length of an Ethernet cable to be 328 feet, or 100 meters. Because of line loss, the standard specifies less maximum power than what is available at the port. Table 6.1 shows the maximum power allowed for both the PSE and the PD.

TABLE 6.1 Maximum power supplied by PSE and drawn by PD for both amendments to the IEEE 802.3 Ethernet standard

Specification	802.3-2012 Clause 33	802.3at
PSE power maximum	15.4 W	34.2 W
PD power draw maximum	12.95 W	25.5 W

Equipment manufacturers have the option of defining a *classification signature*. The classification signature determines the maximum amount of power a device requires, thereby allowing the PSE to better manage the amount of power delivered to a specific port. The PoE amendments make five classes of powered device available (class 0 through class 4). Table 6.2 shows the available classes and the amount of power in watts for each class for IEEE 802.3-2012 Clause 33 (802.3af) devices.

TABLE 6.2 Classes of powered device described in the PoE amendment to the Ethernet standard, 802.3-2012 clause 33 (802.3af)

Class	Use	PSE Power Output in Watts	PD Max Levels in Watts
0	Default	15.4 W	0.44 W to 12.95 W
1	Optional	4.0 W	0.44 W to 3.84 W
2	Optional	7.0 W	3.84 W to 6.49 W
3	Optional	15.4 W	6.49 W to 12.95 W
4	Type 2 PoE Devices	30.0 W	12.95 W to 25.5 W

Figure 6.17 shows an example of PSE and a PD.

FIGURE 6.17 PSE single-port injector and PD access port

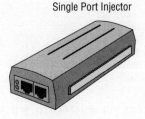
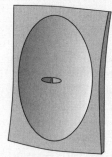

Benefits of PoE

There are many benefits to using devices that support PoE, including cost savings and convenience. The IEEE 802.3 standard (Ethernet) specifies a maximum distance of 100 meters, or 328 feet, for unshielded twisted-pair (UTP) Category 5 (CAT5) Ethernet cable. Power over Ethernet enables a PoE device to receive DC power and computer data at this distance without the need for electrical power at the point where the device is installed or located. This can amount to a big cost savings if a voltage source is not available where the device is located, because there is no need to install electrical power at that point.

Virtual Private Network (VPN) Concentrators

To understand how a VPN concentrator operates, it is important to understand the basics of VPN technology. A VPN is a network that provides private secure communications over a public unsecured network infrastructure such as the Internet. VPNs are based on Internet Protocol. They typically operate at Layer 3 of the OSI model, but some will work at Layer 2. Figure 6.18 illustrates VPN technology in relation to the OSI model.

FIGURE 6.18 OSI model representation of a Layer 3 VPN security solution

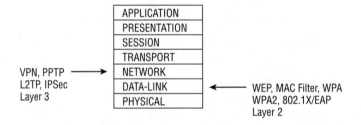

Prior to the ratification of the IEEE 802.11i amendment to the standard, wireless LAN VPN technology was prevalent in enterprise deployments as well as in remote access security solutions. Since wireless LAN Layer 2 security solutions have become stronger (mostly thanks to the 802.11i amendment and the Wi-Fi Alliance WPA and WPA2 certifications), VPN technology is not as widely used, if at all, within internal enterprise wireless LANs. However, VPNs still remain a very powerful security solution for remote access in IEEE standards–based wired and wireless networking, as well as cellular communications.

VPNs consist of two parts—tunneling and encryption. Figure 6.19 illustrates a VPN tunnel using the Internet. A stand-alone VPN tunnel does not provide data encryption, and VPN tunnels are created across Internet Protocol (IP) networks. In a very basic sense, VPNs use encapsulation methods, where one IP frame is encapsulated within a second IP frame. The encryption of VPNs is performed as a separate function.

FIGURE 6.19 Representation of a VPN tunnel using the Internet

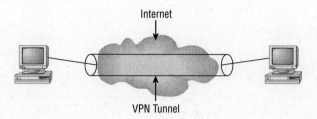

 Real World Scenario

Shipping a Crate Using VPN Technology

An analogy for the VPN process is shipping a locked crate from one location to another. You are a technical support engineer for the headquarters office of a company that has five offices in different locations around the world. You get a telephone call from a co-worker at one of the remote offices. She needs to replace an access point with a newer model at the remote office. You need to ship the preconfigured replacement access point to her using a common carrier. You want to ensure that the access point arrives at the destination location without coming into physical contact with anybody other than the intended recipient.

The access point is analogous to the IP frame. You put the access point into a crate that has a combination lock to secure it. This crate containing the access point is analogous to the second IP frame, or the one that *encapsulates* the original IP frame.

You ship the crate to the destination using a common carrier, which would be analogous to the *public infrastructure* over which the encrypted data is sent. Many other packages are shipped by this common carrier, but no one will be able to see the contents of the crate because they do not know the combination to the lock (the encryption method).

When the access point arrives at the destination, the recipient (the technical support engineer for the remote office) must know the combination of the lock on the crate in order to open it to retrieve the access point. So you tell her the combination over the telephone. This is analogous to the *encryption method*. Over a secure telephone line, only you (the sender) and she (the recipient) know the combination to the lock. The tech support engineer will be able to unlock the crate using the combination you supplied her, and she will be able to retrieve the access point.

There are two common types of VPN protocols:
- Point-to-Point Tunneling Protocol (PPTP)
- Layer 2 Tunneling Protocol (L2TP)

Point-to-Point Tunneling Protocol

Developed by a vendor consortium that included Microsoft, *Point-to-Point Tunneling Protocol (PPTP)* was very popular because of its ease of configuration, and was included in all Microsoft Windows operating systems starting with Windows 95. PPTP uses the Microsoft Point-to-Point Encryption (MPPE-128) protocol for encryption. This process provides both tunneling and encryption capabilities for the user's data.

If the PPTP configuration uses Microsoft Challenge Authentication Protocol (MS-CHAP) version 2 for user authentication on a wireless network, it can be a security issue. This authentication process can be captured using a wireless protocol analyzer or other scanning software program, and potentially allow someone to perform a dictionary attack, allowing them to acquire a user's credentials and eventually giving them the capability to log on to the network. A dictionary attack is performed by software that challenges the encrypted password against common words or phrases in a text file (dictionary). Therefore, using PPTP on a wireless network should be avoided. Keep in mind that the security vulnerability is not PPTP itself; it is that the authentication frames on a wireless LAN can be captured by an intruder, who can then acquire user credentials (username and password) and be able to gain access to the VPN.

Layer 2 Tunneling Protocol

Layer 2 Tunneling Protocol (L2TP) is the combination of two different tunneling protocols: Cisco's Layer 2 Forwarding (L2F) and Microsoft's Point-to-Point Tunneling Protocol (PPTP). L2TP defines the tunneling process, which requires some level of encryption in order to function. With L2TP, a popular choice of encryption is Internet Protocol Security (IPsec), which provides authentication and encryption for each IP packet in a data stream. Since L2TP was published in 1999 as a proposed standard, and because it is more secure than PPTP, L2TP has gained much popularity and for the most part is a replacement for PPTP. L2TP/IPsec is a common VPN solution in use today. PPTP should not be used when L2TP is available.

Components of a VPN Solution

A VPN solution consists of three components:

- Client side (endpoint)
- Network infrastructure (public or private)
- Server side/VPN concentrator (endpoint)

In many cases, both the client side and the server side (which may consist of a VPN concentrator appliance) are known as VPN endpoints. The concentrator is the association between the public network and the private network infrastructures. The public infrastructure in many cases is a public access network such as the Internet. The client-side endpoint typically consists of software, allowing it to be configured for the VPN. This software is available at a nominal cost from a variety of manufacturers. Newer Microsoft Windows operating systems include VPN client software for both PPTP and L2TP. Figure 6.20 shows a VPN client configuration screen.

FIGURE 6.20 Microsoft Windows 7 Professional built-in VPN client utility configuration

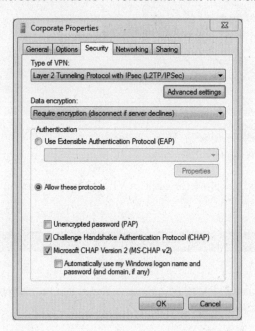

Many wireless client devices also have the capability to be a VPN client endpoint. The VPN can terminate either at an access point or across the Internet to the corporate network. Figure 6.21 shows a common example of a wireless client device connecting to a wireless hotspot to access the corporate network via a VPN concentrator hardware appliance.

FIGURE 6.21 Wireless LAN client using a wireless hotspot to connect to a corporate office VPN concentrator appliance using VPN technology

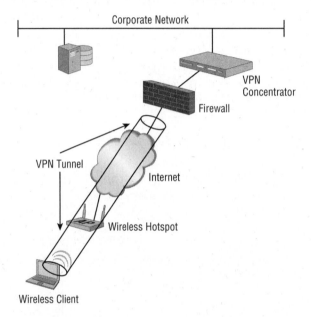

In Exercise 6.1, you will explore the built-in VPN client software utility in Windows 7 Professional.

EXERCISE 6.1

Setting Up a VPN

In this exercise, you will set up a VPN connection using the built-in VPN client utility in Microsoft Windows 7 Professional.

1. Click Start ➢ Control Panel. The Control Panel window appears.
2. Click the View Network Status And Tasks link under the Network And Internet heading in the Control Panel window. The Network and Sharing Center window appears.

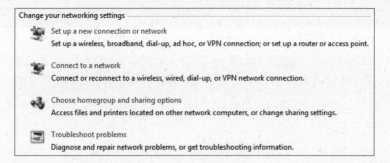

3. In the Change Your Network Settings menu, click the Set Up A New Connection Or Network link. The Set Up A Connection Or Network – Choose A Connection Option screen appears.

4. Select Connect To A Workplace – Set Up A Dial-Up Or VPN Connection To Your Workplace, and click Next. The Connect To A Workplace – How Do You Want To Connect? screen appears.

EXERCISE 6.1 *(continued)*

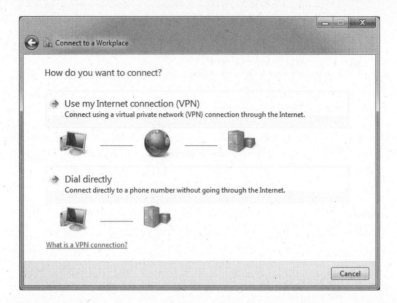

5. Click Use My Internet Connection (VPN). The Connect To A Workplace – Do You Want To Set Up An Internet Connection Before Continuing? screen appears. Click I'll Set Up An Internet Connection Later. The Connect To A Workplace – Type The Internet Address To Connect To screen appears.

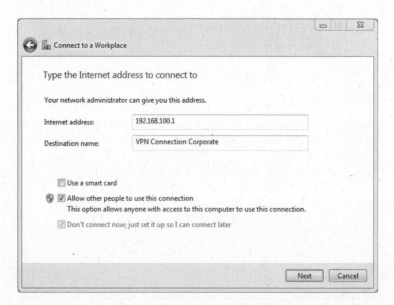

6. In the Internet Address text box, type the IP address or hostname of the remote VPN server you wish to connect to. In the Destination Name text box, type your selected name for the VPN connection and click Next. The Connect To A Workplace – Type Your User Name And Password screen appears.

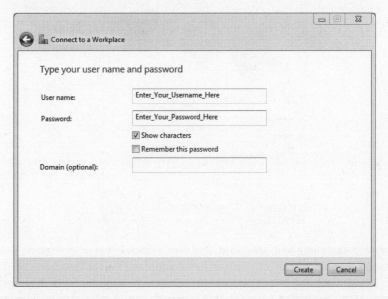

7. In the User Name and Password text boxes, enter a valid username and password and click Create.

8. The Connect To A Workplace – The Connection Is Ready To Use screen appears; click Close. The Network and Sharing Center appears.

9. To use your VPN connection, click the Connect To A Network link. The Currently Connected To screen will appear. Click the name of the VPN connection you created in step 6. The Connect VPN Connection dialog box appears, prompting for a username and password.

EXERCISE 6.1 *(continued)*

10. Enter a valid username and password, and also a domain name if required, and click Connect to connect to your VPN server. Once your credentials have been validated by the VPN server, you will have access to the network through the VPN you created.

Network Gateways

The term gateway may have different meanings in computer networking, and is sometimes misunderstood or interpreted differently. Technically, a network gateway is an interface that allows computer networks using different protocol sets to connect together and share information. Gateways can be based on software or on hardware appliances/computers. The wireless broadband router discussed earlier in this chapter is one very basic example of a network gateway. A wireless broadband router has the capability to connect IEEE 802.3 network devices and IEEE 802.11 network devices to the Internet. All three of these networks (802.3, 802.11, and the Internet) use different protocol sets, while still allowing information sharing across all of the networks. Figure 6.22 shows an example of how a network gateway may be part of a computer network.

FIGURE 6.22 A network gateway connects different types of networks with unlike protocol sets.

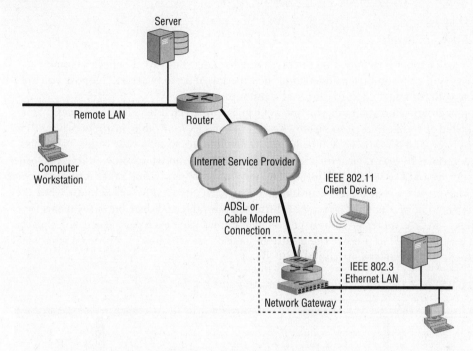

If you are using a Microsoft Windows–based computer, you may be familiar with the term default gateway. This type of gateway is the Internet Protocol (IP) address of the router port that connects the locally connected LAN segment to a remotely connected LAN segment via a network router.

Proxy Servers

In a general sense, the word *proxy* means to act or perform on behalf of another person. In computer networking terminology, a proxy is a server or hardware appliance middleman between a user or device on a private network and a destination server across another network, which can include a WAN or the Internet.

A proxy server can be used for a variety of items:

- Web page caching
- Network address obfuscation
- Bandwidth preservation
- Logging
- Content filtering

We will examine each item on the preceding list later. But first, I believe a simple analogy will help you better understand the function of a proxy server. Suppose you are taking yourself and a guest to dinner at a restaurant.

Upon entering a restaurant, you and your guest are greeted by a host or hostess. You are seated at a table and given menus. A waiter comes to your table, introduces himself as your server, and tells you he will be back in a few minutes to take your order. When the server returns to your table, you and your guest will tell him what you would like to order for your meals. The server documents the order for each person and takes it to the kitchen. The chef prepares your meals according to each order, and lets the server know when they are ready to be picked up. The server returns to your table and distributes the meals for each person. The server is the only person who knows what each person ordered, and the chef was instructed to prepare the meals but has no idea which person ordered what meals. Figure 6.23 illustrates this analogy.

FIGURE 6.23 Proxy server restaurant analogy

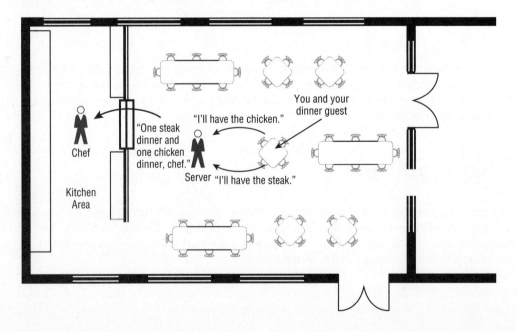

Now look at this analogy from a technology perspective. The dining room represents your office or business. You and your guest seated at the table represent users or devices on the local area network. The waiter represents your proxy server, and the kitchen is the Internet. The chef is a web server contained somewhere within the Internet. Think about the list of features earlier in this section that a proxy server can provide, and further tie that list to this restaurant analogy. The waiter is able to communicate with each person who is ordering their meal. This equates to each person being identified by a unique IP address on the LAN. The waiter is the proxy server and knows what meal should be delivered to each person. The chef, a web server on the Internet, is on a different computer subnet and has a completely different IP address than the people who ordered the meals. The waiter interfaces with each side (the chef and dinner guests), and therefore understands the addresses of both networks. The waiter is acting as the network translator, knowing where to deliver each meal. Figure 6.24 illustrates the concept of a proxy server handling a request from a client device.

FIGURE 6.24 Computer network utilizing a proxy server for content filtering

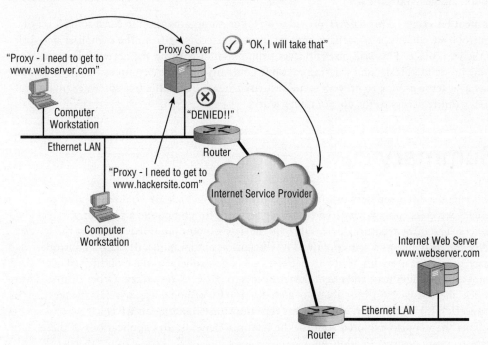

Now let's take a look at some of the features proxy servers may provide.

Web Page Caching This feature allows a proxy server to cache or temporarily store web pages that were visited by a user on the network. If another user were to make a request for the same pages, the server would be able to quickly deliver the content to the second user without having to retrieve the page(s) again from an Internet web server. In addition to improving the user's experience, it will help with the bandwidth usage.

Network Address Obfuscation   IP address obfuscation allows a proxy server to hide local internal IP addresses from outside servers and devices. This network translator, in addition to a firewall, provides some level of security from outside the local network.

Bandwidth Preservation   Because the proxy server has the capability to cache web pages based on requests from users on the internal network, this feature will save on wide area network bandwidth. If the server is able to deliver the pages directly to the requesting user, then the WAN bandwidth can be used for other internetworking tasks.

Logging   Logging any kind information about a computer network can be an invaluable tool. Proxy servers have the capability to log various types of information, including who the users are, what content the users accessed, and what sites or networks were visited. Any cleartext information is recorded and can be evaluated at a later time for forensics or troubleshooting purposes.

Content Filtering   This feature provides ways for an organization to have some level of control over what type of website content users are accessing from the computer network at the workplace. This will prevent users from potentially using the network for non-company-related tasks and wasting valuable company network resources. Content filtering may also serve as way to prevent some security threats like malicious software (malware) from gaining access to the corporate network.

Summary

This chapter discussed network infrastructure devices, which are commonly used to provide wireless connectivity to a computer network for computers and other wireless client devices. The infrastructure devices include the access point (an integral part of the wireless LAN and available as a self-contained intelligent, or autonomous, device), a controller-based device for use with hardware wireless LAN controllers, and a cloud-based access point that provides user and client device access to network resources. Other infrastructure devices include wireless broadband routers for SOHO or home use, wireless bridges for connecting LANs together, and wireless repeaters for extending an RF cell.

This chapter also explored some of the features, benefits, and applications of these infrastructure devices. In addition, we took a look at the two Power over Ethernet (PoE) amendments (IEEE 802.3-2012 Clause 33, also referred to as 802.3af, and IEEE 802.3at), components, the DC voltage and amount of DC power supplied to devices (in watts), and how the power may be delivered to a powered device. Finally, you learned about the basics of additional infrastructure devices, including virtual private network (VPN) concentrators, network gateways, and proxy servers.

Chapter Essentials

Remember the function and features of the different access point technologies. Compare and contrast the features of autonomous, controller-based, and cloud-based access points. Know that autonomous access points are self-contained units, and controller-based access points work with wireless LAN controllers. Cloud-based access points do not require a hardware controller, but use cloud-based software for configuration and management.

Understand differences between various infrastructure devices. Identify the features and applications of wireless access points, wireless bridges, wireless repeaters, and the wireless LAN controller, and how they differ from one another.

Explain the function and implementation of wireless infrastructure devices such as wireless bridges and wireless repeaters. Understand the different modes in which wireless infrastructure devices operate, as well as the uses for specific devices such as wireless bridges and wireless repeaters.

Explain the differences between Power over Ethernet devices. Know the differences between power sourcing equipment (PSE) and powered devices (PDs), and know how they are used in wireless networking. PSE devices will supply the DC voltage and PDs will receive the DC voltage.

Know the details of the IEEE 802.3-2012 Clause 33 (802.3af) and 802.3at Power over Ethernet (PoE) amendments to the Ethernet standard. Know that the nominal voltage for PoE is 48 VDC. Identify the different classifications signatures. Understand the difference between midspan and endpoint PoE solutions. IEEE 802.3at allows 34.2 W maximum power per port, while IEEE 802.3-2012 Clause 33 (802.3af) allows for 15.4 W of power maximum.

Understand how infrastructure network devices such as (VPN) concentrators operate. Understand the basic functionality of additional network infrastructure devices, including virtual private network (VPN) concentrators, network gateways, and proxy servers.

Chapter 7

Cellular Communication Technology

TOPICS COVERED IN THIS CHAPTER:

✓ The Evolution of Cellular Communications

✓ Channel Access Methods

✓ Circuit Switching vs. Packet Switching Network Technology

✓ Mobile Device Standards and Protocols

✓ Worldwide Interoperability for Microwave Access (WiMAX)

✓ Roaming between Different Network Types

In earlier chapters you learned about radio frequency and communications based on IEEE 802.11 standards. In this chapter we will explore the common communications methods used with wireless mobile devices and cellular technology. Starting with first generation (1G) in the early 1980s, cellular technology has evolved, with new advanced wireless technologies appearing about every 10 years. The most current cellular technology, fourth generation (4G), started appearing around 2008. We will take a quick look at the evolution of these different generations of cellular technology.

Networks of any type, wired or wireless, need some way of accessing the medium used for the communications. Wired Ethernet networks use Carrier Sense Multiple Access with Collision Detection (CSMA/CD) and wireless IEEE 802.11 networks use Carrier Sense Multiple Access with Collision Avoidance (CSMA/CA).

Cellular networks also require an access method of some sort to access the medium; in this case, the medium is the air. I will explain the common access methods used with the various generations of cellular technology. In addition to an access method, cellular networks must use protocols to effectively exchange information. We will examine the different protocols used with wireless cellular technology. We will also take a look at Worldwide Interoperability for Microwave Access (WiMAX), which falls under the IEEE 802.16 wireless standard. Finally, will take a look at roaming between different network types, including Wi-Fi and cellular.

The Evolution of Cellular Communications

In earlier chapters you learned about various aspects of wireless communications. This includes, but is not limited to, the following:

- Radio frequency characteristics and behaviors
- Antenna technology fundamentals
- The RF spectrum, including licensed and unlicensed bands
- IEEE 802.11 wireless LAN technology

Wireless communication technologies of various types have been in use worldwide for decades and utilize both the unlicensed and licensed radio frequency bands. Commercial, private, and government sectors use a combination of proprietary and standards-based solutions and devices for wireless connectivity. Staying connected through the use of mobile devices is becoming a major part of our daily lives. Here you will learn the basic fundamentals of wireless technologies used with cellular communications and mobile devices and how they relate to indoor network usage and management.

The first mobile wireless technology used was known as Zero Generation (0G). This preceded what we know today as wireless cellular technology. 0G was a mobile radio telephone technology, not cellular communications. This pre-cellular mobile telephone technology worked with the Mobile Telephone System (MTS) protocol. The telephones were usually large, heavy, and expensive. Some were carried in briefcases or backpacks or mounted in a vehicle, which is how the phrase *car phone* was coined. Let's now take a look at the different generations of wireless cellular technology, as shown in Figure 7.1.

FIGURE 7.1 The evolution of cellular technology

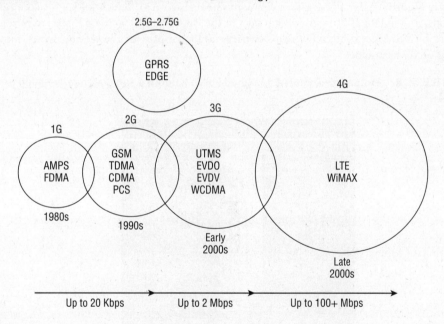

 The FDMA, TDMA, and CDMA channel access methods and other technologies mentioned in the following sections will be discussed in more detail throughout the remainder of the chapter.

First Generation (1G)

First generation (1G) cellular communications started in the early 1980s. Using analog transmission mechanisms and operating in the 800 to 900 MHz frequency range, 1G was used only for voice communications and was not very secure. The following items were key features of 1G:

- Limited to analog communications only
- Capable of download speeds of about 28.8 to 56 kilobits per second (Kbps)
- Used Advanced Mobile Phone System (AMPS) technology
- Had Frequency Division Multiple Access (FDMA) support

The speed of 1G was fairly slow by today's standards but was adequate for the technology at the time. AMPS was a cellular communication standard created by Bell Labs and implemented in the late 1970s in the United States and in the 1980s in other countries. First generation (1G) cell phones were expensive and could be heavy. One specific model is shown in Figure 7.2. First generation cellular systems used AMPS technology for communications, and in late 2000, wireless carriers in the United States were no longer required to support AMPS. Most if not all wireless carriers no longer provide the service, which meant the end of 1G technology.

FIGURE 7.2 Motorola's DynaTAC weighed about 2 pounds and cost nearly $4000 in 1983.

Second Generation (2G)

2G second generation cellular is what started the transition from analog-only communications to digital communications for cellular users. First used in 1991, second generation (2G) signified the continued evolution of digital cellular technology. Second generation (2G) cellular technology included the following features:

- Digital communications
- Varied speeds based on how it was used
- Time Division Multiple Access (TDMA) support
- Code Division Multiple Access (CDMA) support
- Global Standard for Mobile Communications (GSM) technology

New advancements in cellular wireless technology allowed for digital communications with second generation (2G) cellular. Through the use of Global Standard for Mobile Communications (GSM) and newer access methods, the speeds and performance increased, providing a better user experience. 2G also offered a greater call capacity because multiplexing and compression of the digital data is better achieved with digital communications rather than with older analog communications. This in turn allowed for more calls to take place using the same amount of radio frequency spectrum. 2G is still in use today and is popular in various parts of the world.

Third Generation (3G)

Third generation (3G) technology was approved for use in the United States and implemented by cellular wireless carriers in early 2000. 3G technology provided faster data transmission, more network capacity, and additional advanced services along with the following features:

- Speeds of up to 200 Kbps
- Wireless broadband speeds of up to about 2 megabits per second (Mbps) based on 3.5G and 3.75G technologies
- Wireless voice, Internet access, video calls, and TV capable
- Code Division Multiple Access (CDMA) support
- Universal Mobile Telecommunications System (UMTS)

Third generation (3G) technology provided better call quality, coverage, capacity, security, and capabilities.

Fourth Generation (4G)

The newest wireless cellular technology is known as fourth generation (4G). This technology first became available around 2008 in the United States but was not widely adopted until 2011. 4G actually consists of several different technologies, including LTE and WiMAX. 4G includes the following features:

- IP-based wireless mobile device support
- Data speeds that vary from 100 Mbps to 1 Gbps, depending on the implementation
- Used for digital voice and data
- Enhanced security
- Long Term Evolution (LTE) technology
- Worldwide Interoperability for Microwave Access (WiMAX) support

All newer mobile cell phones support fourth generation (4G). However, the technology is not available in all areas or countries. If a 4G service is not available, the mobile device is backward compatible and will drop to 3G. Figure 7.3 shows several new technology 4G mobile phones.

FIGURE 7.3 The Apple 4G iPhone 5s. We have come a long way.

Comparing the Features of the Generations

Table 7.1 summarizes the different features available with the generations of cellular technology.

TABLE 7.1 Comparison of feature sets

Feature	First Generation (1G)	Second Generation (2G)	Third Generation (3G)	Fourth Generation (4G)
FDMA	✓			
TDMA		✓		
CDMA		✓	✓	✓
CSD		✓		
PSD		✓	✓	✓
GSM	✓	✓		
UTMS			✓	
LTE				✓

 According to the 3rd Generation Partnership Project (3GPP), "organizational partners" are groups of telecommunications associations that initially were developed to create specifications for 3G mobile devices based on GSM. This was later expanded to include the development and maintenance of GSM, GPRS, and EDGE. For more information on the role of 3GPP, visit www.3gpp.org.

Channel Access Methods

Channel access methods give devices connected to a common transmission medium the opportunity to communicate and send information across that network medium from one device to another. Several types of channel access methods are used in mobile and cellular communications. We will explore the following types in this chapter:

- Frequency Division Multiple Access (FDMA)
- Time Division Multiple Access (TDMA)
- Code Division Multiple Access (CDMA)

Frequency Division Multiple Access

In Chapter 3, "Radio Frequency and Antenna Technology Fundamentals," you learned about the basic radio frequency characteristics, which include wavelength, frequency, amplitude, and phase. Frequency is the number of complete cycles the sine wave repeats as the result of an electrical current varying uniformly in voltage in 1 second. Low frequencies correspond to long radio waves and high frequencies to short radio waves, so the higher the frequency, the shorter the wavelength.

The Frequency Division Multiple Access (FDMA) method can be broken down as follows:

- Frequency Division: The frequency is divided into small sections to be used by several devices simultaneously.
- Multiple Access: Many devices access the transmission medium at the same time within the same radio frequency space.

As the name implies, FDMA allows for multiple connections across a specified frequency range. These 30 KHz wide RF channels provide one channel for each call and use full duplex communication. Full duplex allows for two-way communication (transmit and receive simultaneously). In addition, FDMA is not limited to only analog voice calls. It can also be used to carry digital data if configured as such. The downside to FDMA is that only one device at a time can communicate while the channel is in use, and that is not an efficient use of the radio frequency spectrum.

Figure 7.4 shows how the frequency band is divided into smaller RF channels.

FIGURE 7.4 Frequency Division Multiple Access (FDMA)

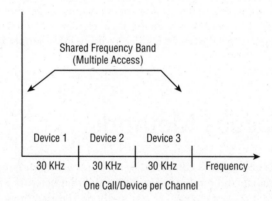

In this example, each mobile device is using a 30 KHz channel to communicate. That frequency channel can service only one device (voice call) at a time. Therefore, each device has that frequency space allocated for its use only during the duration of the call.

Time Division Multiple Access

Time Division Multiple Access (TDMA) is another channel access method that allows for network communications using a shared medium. TDMA allows several users to use the same frequency channel by dividing the radio frequency channel into different slots of time.

Time Division Multiple Access can be broken down as follows:

- Time Division: Each device is allocated a specific time slot of the channel to use for transmissions.
- Multiple Access: Many devices access the transmission medium within the same radio frequency space.

Every device that is accessing the medium will have its own time slot, and the slices of time will vary from device to device. Global Standard for Mobile Communications (GSM) is based on TDMA. Using a 200 KHz wide radio frequency band, TDMA will allow for up to eight voice transmissions on this single RF channel. GSM is discussed in more detail later in this chapter.

Figure 7.5 illustrates the TDMA process.

FIGURE 7.5 Representation of TDMA

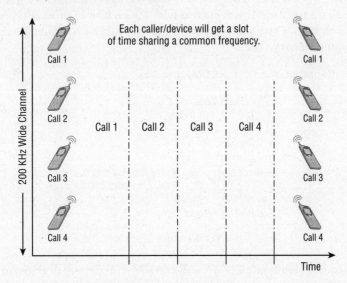

You can see in Figure 7.5 that there are several wireless mobile telephones accessing the same 200 KHz wide channel. This is different from the previously mentioned FDMA because with that technology, each device is assigned its own channel to use. In this example, four devices are sharing the same channel but are each allocated their own time interval.

Code Division Multiple Access

This access method is based on spread spectrum technology and allows for several transmitters to send information simultaneously over a single radio frequency communications channel. This will in turn allow several mobile devices to communicate on the same RF channel by uniquely identifying their own transmissions.

Code Division Multiple Access (CDMA) can be broken down as follows:

- Code Division: Devices share the same frequency channel by using a unique transmission code.
- Multiple Access: Many devices access the transmission medium within the same radio frequency space.

You learned about direct sequence spread spectrum (DSSS) in Chapter 5, "IEEE 802.11 Terminology and Technology." CDMA uses this same method by combining data bits with a common unique code known by the transmitter and its receiver. In the case of HR/DSSS, the coding is either Barker code or complementary code keying. With a Boolean algebra process known as exclusive OR (XOR), the single data stream bits are combined to create a code that is spread across the entire radio frequency channel that is in use.

Figure 7.6 shows a single digital data bit of 1 that is combined with a unique code to give the resultant data stream that was created with the exclusive OR (XOR) process. The result is what is known as a series of *chips* (which is the effect of combining a single digital data bit with a unique binary code).

FIGURE 7.6 Representation of the CDMA process

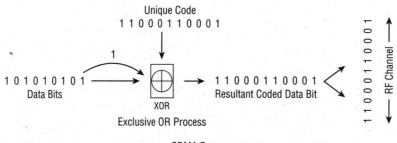

By using this coding technique, CDMA will allow several devices to transmit at the same time using a common frequency band. This is a fairly efficient use of the radio frequency spectrum and is possible because of digital multiplexing technology. As cellular technologies continue to evolve, some wireless carriers will eventually phase out CDMA technology in favor of technology advancements such as Long Term Evolution (LTE). Even though CDMA will still be supported in the foreseeable future, you can expect newer cellular phones to no longer have that capability. Of course, this depends on the part of the world in which you reside.

Circuit Switching vs. Packet Switching Network Technology

Before we explore the different telecommunications methods, protocols, and standards used with cellular networks, it will be beneficial to get a basic understanding of two common communications technologies used to get data and other information from a source node to a destination node. This could be as simple as a telephone call using an analog phone or using wireless cellular technology via a smartphone. These two methods are as follows:

- Circuit switching via Circuit Switched Data (CSD)
- Packet switching

Circuit Switched Data

Circuit Switched Data (CSD) technology requires that a dedicated physical circuit or channel be in place during the entire conversation for the information (or data) exchange. One analogy to this is the public-switched telephone network (PSTN) that is used to make a telephone call using a standard copper cable. Basically, when a call is initiated, a direct connection "circuit" is created between the two telephones across the entire network and will remain intact until the telephone conversation ends or either party hangs up the telephone to end the call (see Figure 7.7). Although reliable, this method is older technology and is expensive due the need to create the dedicated reliable circuits for each call.

Circuit Switched Data (CSD) was designed to operate with TDMA systems such as GSM (discussed later in this chapter). TDMA uses slices, or slots, of time within a radio frequency channel, allowing transmitters to send their data during that time period. Each period of time allowed is for some specific amount of data to be sent.

FIGURE 7.7 Example of Circuit Switched Data (CSD)

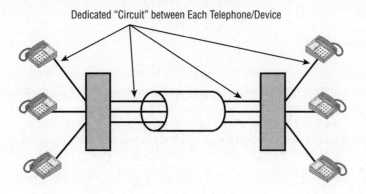

Global Standard for Mobile Communications (GSM) uses Circuit Switched Data and is discussed later in this chapter.

Packet Switching

A packet-switched network uses TCP/IP for communications. In this scenario, there is not a set path or dedicated circuit as in circuit switching. Packets (small blocks) of information may use various routes to get to the destination. A basic example of packet switching is shown in Figure 7.8. Due to the nature of this type of communications, it is not as reliable as circuit switching and can result in issues such as congestion, packet loss, and latency.

There are two major packet-switching modes. The first one is connectionless packet switching, which is also known as datagram switching. This type of packet switching is not considered guaranteed delivery. It uses connectionless protocols such as Ethernet, IP, and UDP. In this method, each packet includes complete addressing or routing information. The packets are routed individually, sometimes resulting in different paths and out-of-order delivery. The second packet switching mode is connection-oriented packet switching, also known as virtual circuit switching. It is generally considered to be a guaranteed delivery. The packets include a connection identifier rather than address information and are delivered in order.

FIGURE 7.8 Example of a packet-switching network

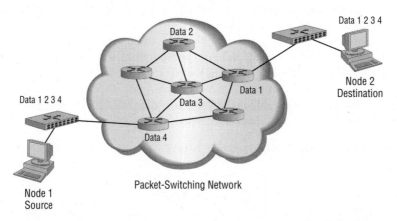

You can see in Figure 7.8 that the data stream is broken into smaller pieces at the source node and is sent across different routes (paths) on the network based on the conditions and the best route to take. Once the data is received at the destination, it is reassembled into the message that was sent by the source. This is different from a circuit-switched network, where a dedicated circuit is created between the source and destination nodes.

Mobile Device Standards and Protocols

In the following sections, we will examine the different technologies that operate with cellular networks for both voice and data communications. As this technology evolved over the past 30 years, so has the need for better and faster communications. In addition to voice, this includes data and Internet connections for use with a variety of mobile devices, including cellular telephones, tablets, notebook computers, and others. Mobile devices have become more sophisticated and are loaded with new technology features, for the most part allowing a "one-stop shop" for communications to satisfy voice and data needs. The following protocols and standards are used with cellular voice and data communications:

- Global Standard for Mobile Communications (GSM)
- General Packet Radio Service (GPRS)
- Enhanced Data Rates for GSM Evolution (EDGE)
- Universal Mobile Telecommunications System (UMTS)
- Evolution Data Optimized (EVDO)
- High Speed Downlink Packet Access (HSDPA)
- High Speed Uplink Packet Access (HSUPA)
- High Speed Packet Access (HSPA)
- Evolved High Speed Packet Access (HSPA+)
- Long Term Evolution (LTE)

Let's take a quick look at each of these technologies. This will give you perspective on each technology and help you to see how mobile device standards and protocols have continued to evolve with the need for more advanced mobile wireless devices.

Global Standard for Mobile Communications

Implemented in the early 1990s, Global Standard for Mobile Communications (GSM) is a standard developed by the European Telecommunications Standards Institute (ETSI) and was first used with second generation (2G) networks to allow digital cellular data communication for mobile devices and to replace first generation (1G) systems that were analog based. GSM then became the standard for mobile communications, with over 80 percent market share worldwide. GSM initially used Circuit Switch Data (CSD) technology, which was explained earlier in this chapter. Because CSD is used with GSM, it allows for data rates only up to 14.4 Kbps. GSM mobile devices use a Subscriber Identity Module (SIM) card. This electronic circuit card is removable and contains a unique identification number (serial number), and the user can store their unique data on the card.

General Packet Radio Service

General Packet Radio Service was first implemented in the late 1990s. Unlike GSM, which was circuit-switched technology, this technology involves a packet-switched mobile protocol used with 2G and 3G cellular systems and allows for both voice and data communications. GPRS allows for faster speeds of up to 114 Kbps and additional capabilities. GPRS was originally developed by the European Telecommunications Standards Institute (ETSI). It is now maintained by the 3rd Generation Partnership Project (3GPP). GPRS falls between second generation (2G) and third generation (3G) cellular technologies and is commonly referred to as 2.5G.

Enhanced Data Rates for GSM Evolution

The year 2003 is when Enhanced Data Rates for GSM Evolution (EDGE) was first deployed in the United States. EDGE is a pre-3G technology that, as the name implies, provides greater data transmission rates; it is a backward-compatible extension of GSM technology. Like GPRS, EDGE used packet-switching technology for data transmissions. In order for EDGE to operate on the infrastructure, wireless carriers needed to upgrade their systems to support the technology. The first wireless carrier to use EDGE technology was the company that designed it, Cingular. After some rebranding and other reorganizations, Cingular was acquired by AT&T in late 2006. Like GPRS technology, EDGE actually is an interim technology between second generation (2G) and third generation (3G) cellular technologies and is commonly referred to as 2.75G.

Universal Mobile Telecommunications System

Universal Mobile Telecommunications System (UMTS) is a 3G mobile system for cellular networks based on the GSM standard. UTMS first appeared in implementations in 2001. This technology is not backward compatible to GSM and requires new infrastructure and equipment. However, wireless carriers can provide support for multiple technologies. The data rates are up to 384 Kbps for downlink speeds and 128 Kbps uplink speeds, which is higher than the previous technologies mentioned. Keep in mind that additional technologies used with UTMS may provide much faster data speeds. A few examples are HSPA+ 42 Mbps and HSPDA 7.2 Mbps. GSM uses circuit-switching (CSD) technology, and GPRS uses packet-switching technology. UMTS differs from them in that it can support both circuit- and packet-switching technologies.

Evolution Data Optimized

Evolution Data Optimized (EVDO) is a standard for fast wireless broadband Internet access. A cable modem or digital subscriber line (DSL) will provide high-speed Internet access, and EVDO was designed to accomplish the same task. This technology can be part of the mobile wireless cellular device (cell phone) or it could be in the form of a separate external adapter that would plug into a PC Card slot (legacy) or USB port on a computer. First generation EVDO provided data rates of about 2 Mbps downlink and about 150 Kbps uplink. Enhancements to the technology over several revisions support much higher speeds of up to 14.7 Mbps. EVDO is for data-only communications and not used for voice calls. A mobile cellular phone that is EVDO capable will use other communications technologies, such as GSM, for voice phone calls. Because of advances in cellular technology such as 4G and LTE, some carriers in parts of the world will be terminating CDMA communications within the next decade. Therefore, EVDO will no longer be available in those areas.

High Speed Downlink Packet Access

With the continued evolution of wireless cellular protocols, fast speeds are a natural progression. High Speed Downlink Packet Access (HSDPA) is a packet-switching technology that allows for much higher downlink data transmission speeds. HSDPA is commonly identified as 3.5G technology because it is an enhancement to the third generation (3G) technology. Depending on the technology used, the data rates can be as high as 20 Mbps. This would be with multiple-input, multiple-output (MIMO) technology utilizing a 5 MHz wide RF channel.

High Speed Uplink Packet Access

This protocol, commonly referred to as 3.75G, is another interim technology in the third generation (3G) technology evolution. High Speed Uplink Packet Access (HSUPA) works with the previously mentioned HSPDA. The high uplink data rates will enhance the user experience for device-to-device communications for a variety of application types. Depending on the technology used, HSUPA provides uplink data rates of up to 5.8 Mbps.

High Speed Packet Access

This newer third generation (3G) protocol is a derivation of the two previously mentioned protocols, High Speed Downlink Packet Access (HSDPA) and High Speed Uplink Packet Access (HSUPA). It provides an enhancement for the performance of 3G mobile networks utilizing the WCDMA protocols. HSPA is the predecessor to Evolved High Speed Packet Access (HSPA+).

Evolved High Speed Packet Access

This new technology provides faster uplink and downlink data rates Like HSPDA, Evolved High Speed Packet Access (HSPA+) is multiple-input, multiple-output (MIMO) technology and is capable of 2x2 MIMO utilizing 64 quadrature amplitude modulation (QAM) for higher data rates. The uplink speeds can be as high as 10.8 Mbps and downlink data rates can be up to 84 Mbps based on the 5 MHz RF channel width. HPSA+ provides higher speeds for wireless broadband users.

Long Term Evolution

The most recent standard for wireless communication of high-speed data for mobile phones and other devices is known as Long Term Evolution (LTE). This technology is often promoted as fourth generation 4G LTE. It is based on the GSM, EDGE, UMTS, and HSPA network technologies, increasing the capacity and speed using a different radio interface together with core network improvements. The motivation for creating LTE technology was faster data rates and better quality of service. LTE is based on Internet Protocol (IP) and supports Internet-accessible applications such as World Wide Web (WWW) browsing, email, VoIP, and more. The support for LTE depends on your geographic area. Figure 7.9 shows a coverage map for the Sprint wireless LTE network. The places where LTE is available are shown with the solid dots.

FIGURE 7.9 Sprint's LTE coverage map of the United States

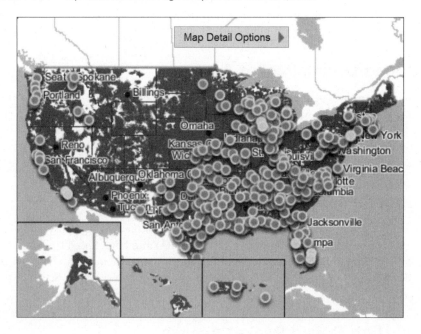

WiMAX

The Wi-Fi Alliance provides interoperability certifications for IEEE 802.11 wireless LAN devices. Worldwide Interoperability for Microwave Access (WiMAX) is the wireless broadband equivalent to the Wi-Fi Alliance. The WiMAX forum (www.wimax.org), established in

June 2001, provides interoperability certifications for WiMAX wireless broadband communications based on the IEEE 802.16 standard.

WiMAX is sometimes confused with IEEE 802.11 Wi-Fi technology. They are different technologies and are specified under different IEEE standards, but they do have similar characteristics. Both Wi-Fi and WiMAX technology operate at Layer 1 (the Physical layer) and Layer 2 (the Data Link layer) of the OSI model. One distinct difference between the two technologies is that Wi-Fi is contention based and WiMAX is connection orientated.

The data rates that WiMAX supports can be fairly high. Depending on the conditions and how it is used (for example, with an external USB adapter or with a mobile telephone), 30 to 40 Mbps is not unrealistic.

WiMAX has had its share of ups and downs. One major provider, Clearwire Inc., was a major player in the US market and was working on large scale deployments throughout the Unites States. Although they previously had part ownership, Sprint acquired 100 percent of Clearwire in June 2013. Since that time, Sprint has been shifting from the WiMAX side in favor of building its own LTE network.

Roaming between Different Network Types

When a wireless device moves from one service area to another service area, it is said to be roaming or transitioning between networks. This can happen for a variety of reasons.

In Chapter 6, "Computer Network Infrastructure Devices," we looked at roaming from an IEEE 802.11 wireless LAN perspective. You learned that the decision to roam between access points in an IEEE 802.11 wireless network is always made by the wireless client device. That decision is based on different criteria, depending on how the manufacturer of the device chooses to implement it. You also saw that IEEE 802.11 roaming is now standardized through the IEEE 802.11r amendment, but most manufacturers of wireless LAN equipment use proprietary mechanisms on the infrastructure side. Just as in wireless LAN technology, cellular systems use radio coverage areas to allow wireless mobile devices to connect to a wireless provider's network.

With the advancements in wireless mobile device technology, modern cellular phones are also IEEE 802.11 Wi-Fi capable, IEEE 802.15 Bluetooth capable, and in some cases, IEEE 802.16 WiMAX capable. This allows the user of the mobile device to use either the cellular network or another network type depending on whether it will be used for voice calls or data. One thing to keep in mind is that different technologies use different mechanisms to allow a wireless client device to connect to a wireless infrastructure. For example, if an IEEE 802.11 device wants to connect to an infrastructure access point, it must perform an IEEE 802.11 authentication and association. You learned about that process in Chapter 5. As long as the device has a matching set of capabilities (such as SSID and security parameters), it will successfully connect or roam to a comparable access point. However, for a cellular telephone to connect to a wireless carrier's service, the process is more complex.

Cellular phones are identified with unique numbers to allow them to be identified. The following kinds of identification numbers are used:

- Electronic serial number (ESN)
- Mobile equipment identifier (MEID)
- Mobile identification number (MIN)

These numbers will allow a cellular phone to connect to and use the cellular network. They are also used when a phone roams from one cellular tower to another. You can connect to and roam within IEEE 802.11 wireless LANs that are home or corporate based at no charge. One exception is a paid wireless hotspot such as in a hotel. Cellular networks charge the user of the device to pay fees, usually monthly, to use the cellular network. Therefore, these unique numbers are required for the device to be used. When wireless mobile devices are capable of connecting to different network types (such as Wi-Fi and 3G cellular) and are used within the a building, it makes perfect sense to use the best and most cost-effective way to communicate. When a smartphone that is Wi-Fi capable and connected to a Wi-Fi network is used, the voice calls still traverse the cellular network, as do any text messages. The Wi-Fi network is used for web browsing and email and for most other data applications.

In some cases, the decision of which network the mobile device will use is automatic. In other cases, it could be a manual configuration on the device itself. Unless your cellular service plan includes unlimited data usage (which is getting to be rare), it would be best to use the Wi-Fi network for data exchanges. This would eliminate the data usage charges because the cellular 3G, 4G, or LTE network is not providing that service for the mobile device at that time.

Summary

In this chapter you learned the basics of the common communication methods used with mobile devices and cellular technology. Cellular technology started this evolution in the early 1980s and continued with new technologies about every 10 years. We looked at some of the details and features of each of the generations. You learned that first generation (1G) used analog communications and voice communications only and 2G, 3G, and 4G use digital technologies that are capable of both voice and data communications, including high-speed Internet access via mobile broadband technology.

You also learned that cellular networks require an access method to access the medium, which in this case is the air. This chapter covered Frequency Division Multiple Access (FDMA), Time Division Multiple Access (TDMA), and Code Division Multiple Access (CDMA).

The access method was determined by the type of cellular protocols used and the generation in which they were used. We explored the protocols cellular networks must use to effectively exchange voice and network data:

- Global Standard for Mobile Communications (GSM)
- General Packet Radio Service (GPRS)
- Enhanced Data Rates for GSM Evolution (EDGE)
- Universal Mobile Telecommunications System (UMTS)
- Evolution Data Optimized (EVDO)
- High Speed Downlink Packet Access (HSDPA)
- High Speed Uplink Packet Access (HSUPA)
- High Speed Packet Access (HSPA)
- Evolved High Speed Packet Access (HSPA+)
- Long Term Evolution (LTE)

You also learned about the highlights of Worldwide Interoperability for Microwave Access (WiMAX) technology. WiMAX falls under metropolitan area networks (MANs), which is standardized as IEEE 802.16. Finally, we looked at wireless mobile devices roaming between different types of networks such as cellular and Wi-Fi within a physical building or structure.

Chapter Essentials

Understand the different generations of cellular technology. Starting with slower, analog-only first generation (1G) technology and evolving to digital communications with second generation (2G), third generation (3G), and fourth generation (4G), cellular technology continues to evolve and provide a better user experience.

Be familiar with the different channel access methods used with cellular technology. Understand that FDMA uses 30 KHz wide channels that can carry only one call at a time. TDMA allows for multiple devices to use the same frequency by giving each device a specific time slot to transmit. CDMA uses unique coding methods that allow multiple devices to use the same RF channel.

Understand the different ways data can traverse infrastructures. Circuit-switched networks use dedicated circuits to create connections between nodes. Packet-switched networks use TCP/IP for communications. In this situation, there is no set path or dedicated circuit as in circuit switching. Packets (small blocks) of information may use a variety of routes to get to their destination.

Know the different mobile device standards and protocols that are used with cellular networks. Understand the following cellular technologies: GSM, GPRS, EDGE, UTMS, EVDO, HSDPA, HSUPA, HSPA, HSPA+, LTE, and WiMAX.

Understand the relevant concepts for connecting to different network types. Wi-Fi-capable cellular telephones can exchange data on different types of networks, that is, cellular or IEEE 802.11. Know that cellular telephones have unique identification numbers that allow them to use the cellular services from a specific provider.

Chapter 8

Site Survey, Capacity Planning, and Wireless Design

TOPICS COVERED IN THIS CHAPTER:

- ✓ Wireless Site Surveys
- ✓ Gathering Business Requirements
- ✓ Interviewing Stakeholders
- ✓ Gathering Site-Specific Documentation
- ✓ Identifying Infrastructure Connectivity and Power Requirements
- ✓ Understanding Application Requirements
- ✓ Understanding RF Coverage and Capacity Requirements
- ✓ Client Connectivity Requirements
- ✓ Antenna Use and Testing
- ✓ The Physical Radio Frequency Site Survey Process
- ✓ Radio Frequency Spectrum Analysis
- ✓ Received Signal Strength
- ✓ Performing a Manual Radio Frequency Wireless Site Survey
- ✓ Performing a Predictive Modeling Site Survey

In this chapter you will learn about the details of wireless site surveys for both IEEE 802.11 wireless networks and indoor cellular connectivity. We will explore site survey planning and the business aspects related to a wireless network site survey, including gathering business requirements, interviewing the appropriate people, and gathering additional information about the location in which the network will be installed.

You will learn about site survey requirements such as understanding the business requirements for the intended use of the wireless LAN and asking plenty of questions by interviewing department managers and users to help determine the applications and security requirements for the proposed network. Gathering information is a major part of a site survey to ensure a successful deployment.

You will also learn about the importance of documenting every step, from the design to the installation, and the importance of validating the wireless network, a process commonly known as a post survey. It is important to have all the necessary groundwork prior to starting a physical wireless RF site survey. Having all the correct information in hand will allow for a complete and thorough survey, which in turn will result in a successful wireless network deployment. In this chapter, you will see some of the components of performing an RF wireless site survey, including determining areas of RF coverage and interference by using a spectrum analyzer. Taking into consideration both Wi-Fi and non-Wi-Fi interference sources and understanding how this interference will affect the deployment are also important parts of performing a site survey. We will also look at different types of site surveys, both manual and predictive modeling. You will learn about the two types of manual site surveys, passive and active, and see the advantages and disadvantages of both. Understanding the steps involved and the details of each site survey type will help a wireless engineer determine the best methodology to use. You will see some of the tools that may be used in a wireless site survey, including spectrum analyzers, wireless protocol analyzers, and wireless scanners. Finally, we will look at the two different types of channel architecture and best practices for hardware placement, including access points and antennas, as well as some of the limitations you may encounter during the physical site survey process.

Wireless Site Surveys

Early deployments for wireless networks in many cases required only a one-time site survey. This is because wireless networking technology was fairly static and the radio frequency dynamics of the locations in which these networks were installed did not change much. The main goal for early deployments was to provide adequate RF coverage. However, thanks to the rapid pace at which wireless technology is growing and the increasing use of other devices that use unlicensed radio frequency, site surveys can be an ongoing process for the areas where these wireless networks are installed. Although there are no set rules about when or how wireless site surveys should be performed, there are guidelines that many manufacturers suggest based on the deployment scenarios for wireless networks using their equipment. Best practices and experience also play a role in the site survey process.

The term *wireless site survey* can be subjective. A site survey may include walking an area checking for RF coverage and interference, planning for network capacity, and designing the wireless network. A wireless site survey includes the following main objectives:

- Evaluate areas for radio frequency coverage.
- Locate and mitigate any sources of radio frequency interference.
- Determine wireless network capacity requirements.
- Decide on locations for hardware infrastructure devices.
- Identify the antenna types to be used.

Site surveys vary in complexity, depending on the type of business or location in which a wireless network will be used. Many organizations are now designing wireless networks based on an average of three to four wireless mobile devices per user. The devices may include computers, smartphones, and tablets and can have a major impact on capacity planning.

As mentioned earlier, there is not a specific set of rules that must be followed. However, many manufacturers of wireless networking equipment have guidelines and suggestions when it comes to a wireless site survey and where this survey will fit into the process of the design and implementation of a wireless network.

Knowing and understanding the expectations of the client or business in regard to a wireless installation is a critical part of a successful deployment. To understand client expectations, you have to gather much information. This means you will need to participate in interviews and meetings with all those who will be affected by the installation of the wireless network, which encompasses nearly all departments of the company in most cases.

The scope of the wireless site survey, capacity planning, and wireless network design is dependent on many factors, the following among them:

- Size of the physical location
- Intended use of the wireless network
- Number of wireless mobile devices
- Wireless client device capabilities
- The environment
- Understanding the performance expectations
- Bring your own device acceptance
- Building age and construction materials
- Network infrastructure devices

Size of the Physical Location

Depending on the size of the physical location where the wireless network will be installed, a complete wireless site survey may not be necessary and the design will be straightforward. For example, a small sandwich shop wishes to offer free wireless Internet access (a hotspot) as a convenience for its patrons who choose to have a meal there. This sandwich shop is approximately 1,200 square feet, has seating for about 15 people, and is located in a small street retail mall. In this case, a single access point would more than likely be sufficient for the number of users who access the wireless network at any one time and the type of data being sent across the access point to the Internet.

Although a full-blown physical site survey determining areas of RF interference, coverage, and capacity would more than likely not be required, it would still be beneficial to visit the location and determine the best place for the access point. In a situation like this, what I like to call a "site survey lite" may be all that is necessary. This would include performing some simple tests to determine the best RF channel to use as well as access point mounting, consideration of aesthetics, and connecting to the wired network for access to the Internet.

A larger installation will require a much more extensive site survey and may include either a manual RF site survey or a predictive modeling site survey using elaborate software. Manual and predictive modeling site survey software will be discussed in more detail later in this chapter.

Intended Use of the Wireless Network

Looking at the sandwich shop scenario again, chances are the intended use of this wireless network will consist of patrons staying online for short periods of time and browsing the Internet, interacting with social media, or checking email. It is unlikely that many users would be performing any high-end or bandwidth-intensive applications on this type of connection. Therefore, the single-access point model would be sufficient for this deployment, providing adequate wireless access and capacity. With larger installations, the use of the network may include bandwidth-intensive applications such as electronic imaging, computer-aided design, graphics design, and database programs and would have an impact on the number of access points that would be required to provide adequate performance. Although not bandwidth-intensive, a Voice over Wireless LAN (VoWLAN) may also have an impact on the number of access points because of the type of wireless traffic.

Number of Mobile Devices

The number of users and their wireless mobile devices that will be accessing a wireless network is also a big factor in determining the number of access points required, which in turn will determine the scope of a site survey. Many organizations plan on three to four wireless mobile devices per user, including computers, smartphones, tablets, and others.

It has already been established that in our sandwich shop example a single access point would be sufficient based on the size of the location and the intended use of the network. However, as the number of actual wireless-enabled devices increases and because of the added capacity requirements, the need for additional access points will also increase. In a case where more than one access point is required, a more extensive wireless site survey is also required.

Wireless Client Device Capabilities

In addition to the number of users and their wireless client devices, it is important to understand the type of client devices that will be connected to the wireless network and what those device capabilities are. For example, notebook computers or handheld devices such as tablets and smartphones all have specific functionality and will need to meet certain network requirements. Some client devices are multifunctional and include capabilities such as cellular voice, 3G, 4G, WiMAX, and Wi-Fi which will need to be taken into consideration. This type of device could be performing tasks as simple as basic web browsing or checking email. Other possible device types include Wi-Fi-enabled voice handsets that may need their own SSIDs for quality of service functionality and may operate in the 5 GHz U-NII bands or even bar code scanners that will be keeping track of product inventory. With the advancements in wireless networking technology such as the 5 GHz IEEE 802.11ac amendment, the device capabilities must be closely evaluated.

Part of the wireless site survey planning process is to question the appropriate people regarding the types of wireless devices and understand what their capabilities are to ensure that expectations of the users will be met. The following device capabilities are among those that need to be considered:

- IEEE 802.11b/g/n, 2.4 GHz only
- IEEE 802.11a or dual band, 2.4 GHz/5 GHz capable
- IEEE 802.11n capable and the number of transmit and receive radio chains (e.g., 2x2 MIMO)
- IEEE 802.11ac capable and the extended features 802.11ac technology provides.
- MIMO capabilities, such as the number of spatial streams, channel width, or transmit beamforming
- Notebook computers, desktop computers, multifunction smart phones, or tablets
- Radio frequency identification (RFID) tags for location services
- Wi-Fi-enabled voice handsets and the operating frequency in which they can be used

Future mobile device types that will be used with the wireless network must also be evaluated. This includes the expansion of the wireless network, new technology types that may be introduced based on the business model, and bring your own device (BYOD) acceptance and use.

The Environment

Understanding the environment in which the wireless network will be installed and the wireless mobile devices that will be used within the environment are important factors that are sometimes overlooked. Different environments will have different concerns. For example, areas that may have excess radio frequency noise, areas with harsh environmental conditions such as temperature fluctuations, or industrial areas have their own set of challenges. Different environments may have distinct types of challenges:

- General office spaces: Walls and attenuation values
- Office cube farms: User and device density
- Healthcare environments: Wireless medical and other interfering devices
- Industrial areas: Reflections and machine shop equipment
- Educational environments: High quantity of mobile devices

These are a few examples of different environments that may use IEEE 802.11 wireless networks and the challenges that will need to be considered as part of the wireless site survey. In Chapter 4, "Standards and Certifications for Wireless Technology," you learned about some of the various deployment scenarios for common wireless network types.

Understanding the Performance Expectations

Keep in mind that IEEE 802.11 wireless networks are half-duplex and contention based and that many factors will affect the performance of a wireless network, including the number and capabilities of wireless and mobile client devices, types of software and hardware applications used, location, and the number of infrastructure devices providing access. These infrastructure devices include wireless access points and bridges. Part of a wireless site survey involves defining what the customer expects for performance of the network. A mutual understanding of the factors that affect performance as well as how they will be dealt with is imperative from the beginning of the wireless site survey process. From a cellular communications perspective, it is important to keep in mind that cellular signals are received from wireless carriers, and depending on the location, the signal may be weak in certain areas of the building. Adding technology such as a distributed antenna system (DAS) will help improve performance of cellular-enabled mobile devices.

Bring Your Own Device Acceptance

With the number of IEEE 802.11–enabled wireless mobile devices continually growing, the need for increased wireless network capacity is a major concern. *Bring your own device* (BYOD) describes the recent increase in company employees bringing their own personal IEEE 802.11–capable mobile devices such as smartphones and tablets to their place of work and using these devices to access company resources, including the Internet, printers, software applications, and file servers. Take a moment and think about the number of IEEE 802.11–enabled mobile devices you may have in your office or home. Any of the following devices would qualify:

- Notebook computers
- Smartphones
- Tablet devices
- Wireless printers
- Wireless video cameras
- Wireless household appliances
- DVD or Blu-ray players
- Wireless-capable gaming devices
- Wireless-capable televisions

These are just some of the devices that are used daily and are now IEEE 802.11 wireless capable. A family of four people may have as many as 20 wireless devices in the home, and this number will likely increase as wireless technology continues to advance. When companies or organizations allow employees to bring their personal wireless devices to the office or place of work, that opens up an entire set of potential issues, including security, technical support, and wireless infrastructure capacity concerns.

If employees are allowed to bring their own devices to their place of business, you must take this into consideration when performing an RF site survey and designing a wireless network. Because IEEE 802.11 wireless networks are contention based and use a shared medium, the added number of devices will impact the performance if the network is not properly designed. Networks that support BYOD may require smaller RF cells and additional wireless access points in order to offer adequate performance for the number of connected devices. In addition to the load factor, the types of technology used—IEEE 802.11a/b/g/n and IEEE 802.11ac—will have an impact on the need for backward compatibility. Figure 8.1 shows how a number of personal devices can have an impact on the capacity of a wireless network.

Building Age and Construction Materials

It is important to identify the building age and materials that were used for construction. RF propagation is required for adequate coverage and desired signal strength. In older buildings, for example, it may be difficult to determine what materials the walls consist of based on the building practices at the time of construction. If an RF signal does not propagate sufficiently, additional access points may be required for adequate coverage and capacity requirements. Newer buildings will have lighter and thinner yet stronger construction materials that may make it easier for an RF signal to penetrate. In addition, covering an office with open space and cubicles will be easier than covering a hotel or enclosed areas with many rooms and walls. Coverage is just one part of the equation. Network capacity must also be considered and is discussed later in this chapter.

FIGURE 8.1 Various personal devices that may be connected to a wireless network as illustrated by Aerohive.

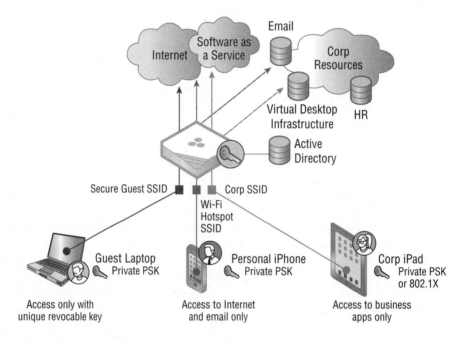

Network Infrastructure Devices

It is important to factor in network infrastructure devices such as Ethernet switches. For example, installing 500 access points in a hospital will require 500 Ethernet switch ports to connect the access points. It is unlikely that a hospital or any organization for that matter will have that many open and available Ethernet ports. Also, considering the access points will be installed throughout the building, it is important to note where the wiring closets are and how many Ethernet switches and ports will be required per wiring closet. When you're running twisted-pair Ethernet cabling, the access point must be within 100 meters (328 feet) of the wiring closet. The access points will be powered using Power over Ethernet (PoE), so choosing a proper PoE source is critical. In addition, it is important to ensure that the wiring closets have adequate supply power to support additional Ethernet switches and the PoE they will provide to the connected devices.

A wireless engineer conducting a wireless site survey will also work with the wired network engineers. For example, adding 500 access points and many additional wireless mobile client devices will increase traffic to and from the WAN and Internet. The network infrastructure devices will need the capacity to handle to additional traffic.

Gathering Business Requirements

Gathering information is typically the first step of wireless network design and implementation. The business model or type of business where the wireless network will be deployed is a major part of deciding the level of a wireless site survey. The type of business will determine the needs and use of a wireless network. Knowing the applications used—both hardware (such as voiceover wireless LAN) and software (such as accessing a database)—is a critical part of a wireless deployment because this will affect recommendations such as the number and locations of access points. Expectations can make or break a wireless deployment. The expectations of the wireless network must be discussed, evaluated, and documented up front. To completely understand what the customer expects, you will need to gather information from various areas of the business. A high-quality site survey is going to require many questions to be asked and answered, including the following:

Bandwidth and Capacity Needs How much bandwidth will be required for users of the wireless network? The types of applications in use will have an impact on this. How many wireless devices per user? Most manufacturers have wireless design guides that you can reference for various deployment scenarios.

Coverage Area In what rooms or areas of the buildings is wireless coverage expected? Is coverage limited to indoors only or will outdoor coverage be required? Many businesses want coverage over the entire location in most cases.

Applications Used What type of applications—either hardware or software—are used at the facility? How will the applications affect the performance? This may include computer-aided design, digital imaging, databases, and Voice over Wireless LAN.

Wireless Devices Used What type of wireless client devices will be used? Are the devices multifunctional? These include notebook computers, handheld scanners, wireless phones, tablets, and other devices, whether company owned or employee owned (BYOD).

Desired IEEE 802.11 Technologies What type of IEEE 802.11 infrastructure devices (IEEE 802.11a/b/g/n or IEEE 802.11ac) would be best suited to the specific environment and deployment?

IEEE 802.11 wireless networks were once primarily used as extensions of wired networks, providing access to a few users in areas exceeding the physical distance of an Ethernet or other wired medium in place. IEEE 802.11 wireless technology continues to evolve and is now a major part of every area in a business, corporation, or company's computer network infrastructure. It is difficult to find any business or organization that does not provide some type of IEEE 802.11 wireless network access. Fully understanding the *business requirements* is part of a successful wireless network site survey and deployment.

General Office/Enterprise

Office buildings and other enterprise installation locations may consist of walled offices or open spaces with many cubicles. This type of installation usually will require infrastructure devices to aesthetically fit the environment and may require antennas to be mounted to drop ceilings with the access points located out of sight. Figure 8.2 shows an example of a floor plan for a small office deployment.

FIGURE 8.2 A small office wireless network installation using omnidirectional antennas

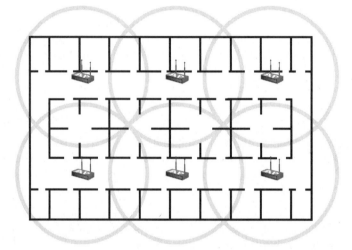

Some installations may have a high density of wireless mobile devices and therefore require more access points to handle the number of wireless devices connected to the network. The software and hardware applications used in these environments will need to be closely evaluated to ensure optimal performance for the user base. Chances are that interference sources in this type of deployment would be limited mostly to computer networking, and radio frequency challenges may be limited.

Interviewing Stakeholders

Understanding the intended use of a wireless network is a critical part of a successful deployment. Who better to explain what the wireless network will be used for than those who will be using it? Those performing a site survey may not necessarily understand the functional aspects of a certain type of business, so it is critical to get input from all who will be using the wireless network. Department managers, business unit managers, and team leaders usually know the function of their specific areas of the organization the best.

Therefore, they will also know the needs and requirements of users of the wireless network and how a successful deployment will help increase job productivity.

I recommend that you create some type of checklist or formal site survey questionnaire to use during the interview process. This will ensure that specific details about the business and proposed wireless deployment are not missed. The use of such forms helps ensure uniform, repeatable interviews. The forms can also become part of the documentation and final deliverable that will be presented to the customer. Although there are some general questions that can be asked, there will be more specific questions based on the business model of the organization or the wireless network to be installed. Figure 8.3 shows an example of notes that can be taken during the information gathering stage and the wireless site survey specifics.

FIGURE 8.3 Sample checklist showing some information collected for a wireless site survey

```
Objective
Survey dates
Testing procedure
General network description
Proposed WLAN components
Existing wireless system
RF spectrum analysis
Area-by-area analysis
Contact list


1.    Project Objective
      Site Description
2.    Wireless LAN Evaluation Process
3.    Existing Wireless LAN Configuration
4.    Ethernet Network
5.    Wireless LAN Assessment
6.    RF Coverage Test Results
7.    Recommended Network Changes
8.    Contacts
9.    Warranty
```

The following list includes some generic interview questions that will pertain to most installations:

Has a site survey ever been performed in the past? It is good to know whether a site survey has previously been performed at a location. Although the previous site survey is only as good as the person who performed it, it may be beneficial and a timesaver to have some information available. Depending on when it was performed, a previous site survey report may not be accurate because, for example, physical changes to the location may have taken place, such as additions of rooms or walls or changes to the interior design.

What blueprints, floor plans, or any other site-specific documentation is available? Blueprints, floor plans, or other documentation about the location are critical in performing an RF wireless site survey. If this information is not available, it may have to be created, which in turn would create an additional expense for the company or customer. The accuracy of these documents needs to be considered in order to provide ideal site survey results.

How many wireless mobile devices and users anticipate using the wireless network? What are the bandwidth and capacity requirements? The number of devices expected to be on the wireless network is valuable information to have. Knowing the number of devices and users will help determine the amount of infrastructure equipment, such as access points and wireless bridges, that will be required for the deployment. Discussing with department managers the number of devices and users on the network as well as the number of working shifts will help provide adequate planning. It will also be beneficial to know whether users are allowed to bring their own devices.

Will guest access be required? If guest access to the wireless network is required, that will potentially affect the number of infrastructure devices such as access points required for the deployment. Depending on what type of access, network capacity will need to be considered. In addition to the equipment, you should take into account security, the captive portal, and backward compatibility.

Is there any preference for a specific manufacturer's equipment? It is a recommended practice for a site survey to be performed with the same manufacturer's equipment that will be used in the deployment. Therefore, customer preference for equipment manufacturers must be determined at the initial phases of the site survey. This will ensure good results based on the design of the wireless network.

What is the coverage area? The intended coverage area of the facility also needs to be addressed. This helps provide a surveyor with information to accurately estimate how long a physical site survey may take and roughly estimate the amount of hardware required. Knowing the coverage area will also help determine any unexpected obstacles that may occur as part of the site survey process.

Is an existing IEEE 802.11 wireless network in place? If an existing IEEE 802.11 wireless network is in place, it needs to be addressed as part of the site survey process. The following questions need to be asked:

- What technology is currently in use, IEEE 802.11a, IEEE 802.11b/g, or IEEE 802.11n?
- How many wireless devices and users does the network currently support?
- Where are the existing access points located?
- Are the existing access points and external antennas installed correctly?
- What is the wireless network used for?

Knowing the answers to these and other questions will help determine the role, if any, that the existing IEEE 802.11 wireless network will play in the new deployment. Keep in mind that some organizations may have quite an extensive existing wireless network and may be in the process of upgrading to newer or different technology. If this is the case, it will need to be determined whether any of the existing network components can or will be used with the new deployment. Photographs of existing equipment used in a deployment are always useful. Figure 8.4 shows a photograph of an existing IEEE 802.11a/b/g access point with an incorrect installation of diversity antennas. If this was used in the new installation, the antennas would need to be moved to the proper spacing based on the frequency wavelength.

FIGURE 8.4 An IEEE 802.11a/b/g dual-band access point mounted on a ceiling 35 feet high with diversity antennas spaced incorrectly for diversity technology

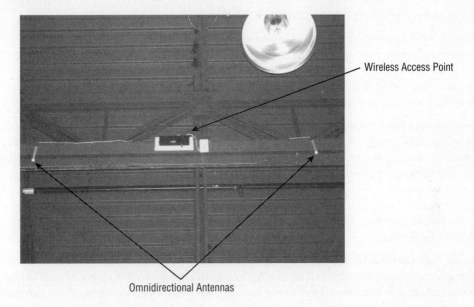

Are there any known areas of RF interference? Information regarding known areas of RF interference is useful in a site survey. It will save time if previous knowledge of RF interference is made available as part of the site survey process. Keep in mind that interference may change and can be detected only at the time of testing or when a manual site survey is performed.

Are there any known areas that may lack RF coverage? Just as previous knowledge of areas affected by RF interference is valuable, the lack of RF coverage in specific areas is also good to know. This will help a surveyor determine any special situations that may be addressed during the site survey process. Testing of various types of antennas may be required to help provide RF coverage in areas that are currently lacking. Available coverage maps may also be useful.

What type of applications will be used? It is important to know the types of applications that will be used. Applications—either software or hardware—will affect the load and number of access points or other infrastructure devices required. These applications include bandwidth-intensive software, voice over wireless LAN, and so on. The surveyor should also become familiar with any special circumstances that may be required to support these applications—for example, the frequency or IEEE 802.11 technology that may be required for the wireless devices to operate.

Will Wi-Fi voice or other real time applications that require quality of service be used? If applications (such as Wi-Fi voice handsets) are planned for the location, they will have an impact on the site survey and design of the wireless network. Because this type of application has greater requirements for signal quality and signal strength as well as fast transition, this will need to be taken into consideration during the site survey. Additional density and more access points may be required.

Real-time video communications such as a videoconference call over a wireless network connection is another application that may require quality of service. Like voice, real-time video is subject to latency and may involve special design requirements. Real-time video over a wireless connection is used in applications ranging from sports venues to security surveillance and monitoring.

Is wireless roaming (transition) required? In most cases, the answer to this question is yes. This is especially true with networks that will be using Wi-Fi-capable voice handsets. Voice handsets are among the most commonly used wireless mobile devices that require seamless roaming and secure transition capabilities. Although notebook computers and tablets may require roaming, real-time voice and video applications are subject to latency issues. Fast, secure transitions may also be required for roaming. If so, the amount of overlap between RF cells would need to be closely looked at to ensure reliable sessions for the devices connected to the network.

Is Power over Ethernet (PoE) required? In all medium to large installations, the answer to this question is yes. Understanding the PoE requirements is another essential part of the wireless site survey. The capabilities of the network devices as well as the number of devices expected to use PoE—including wireless access points, hardwired Ethernet IP phones, and security cameras—will play a role in the design and types of equipment used in the wireless network.

What are the wireless security requirements? Although security is not necessarily part of an RF site survey, as much information as possible on the security requirements is helpful with the design of the wireless network. Some security solutions may require additional hardware, such as sensors that are used with a wireless intrusion prevention system (WIPS), or software that would have to be taken into account for the network design.

Will an escort and ID badge be required? In many cases, people are not allowed to roam freely throughout a business. An escort might be and usually is needed to walk through a location with an outside contract site surveyor. In addition, the escort and surveyor will need access to areas that may be locked or secure, such as wiring closets and computer rooms. Escorts may be from various parts of the company, hospital, or organization, including facilities, information technology, medical staff, and security. Depending on the organization, proper paperwork for access may need to be submitted ahead of time to allow for adequate processing time. For example, you don't want to show up on site on a Monday morning only to find out it will take hours or even days to process the paperwork needed to get an ID badge for access.

Are there any compliance requirements? Depending on the type of business in which the wireless network will be installed, there may be legislative or other compliance requirements. For example, medical institutions may have to meet HIPAA requirements, and retail establishments may require PCI compliance. These need to be taken into consideration as part of a wireless site survey and deployment.

Are you following corporate policies? If one exists, you should use a written corporate policy as a guideline or offer suggestions for changes in the policy to accommodate other wireless requirements. The wireless site survey questioning process should bring any policy-related issues to light for discussion and/or change if needed.

Manufacturer Guidelines and Deployment Guides

The information just presented includes some types of questions that need to be addressed during the site survey process. Keep in mind that the actual questions and details depend on the business model and the implementation of the wireless network. For site survey and deployment guides, check with the manufacturer of the equipment that you will use. These guides will provide additional information that is helpful in generating a list of questions and concerns that will need to be addressed.

Gathering Site-Specific Documentation

Documentation for the location where a wireless network will be installed will make a surveyor's job much easier and result in a smoother overall deployment. Drawings and other documentation pertaining to the following items can provide valuable information:

- Floor plans
- Blueprints
- Proposed location of furnishings
- Electrical specifications
- Existing network characteristics

Floor Plans and Blueprints

Gathering any site-specific documentation that exists, such as floor plans or blueprints, is extremely helpful for a site survey. This documentation is useful to a variety of individuals who will be participating in a wireless network design and deployment. The documentation can be used during a physical or predictive RF site survey and spectrum analysis to note areas of importance and concerns. Having floor plans and blueprints available allows a surveyor to document specific parts of a site survey such as location of access points and other wireless devices. If a predictive modeling site survey is used, an electronic version of a floor plan can be imported into the software program to help streamline the surveyor's job. The standard formats for floor plan (map) files vary from product to product for most modeling software, but they typically include DWG, PDF, JPG, PNG, GIF, and TIF. Figure 8.5 shows a site survey software program that allows for various file formats to be imported.

FIGURE 8.5 Importing a floor plan map using the Ekahau Site Survey program

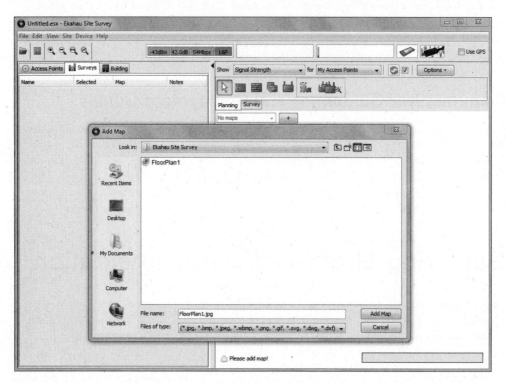

Blueprints or floor plans will also help those who install cable and mount hardware and if necessary can be provided to electricians for AC power installation.

Furnishings

The proposed types and location of furnishings or other items that may affect RF signal propagation or penetration are also good to know if that information is available. This will help during the design and site survey phase to determine access point locations and pinpoint other things that may affect RF signals, such as reflection, refraction, diffraction, and absorption. Unfortunately, most floor plan documents do not specify this kind of information because, in many cases, furnishings are not permanent fixtures and can change. Some organizations will have this information stored in asset documentation using spreadsheets or databases. Many times it will be up to the surveyor to determine where the furnishings or equipment are located and the effect it may have on the radio frequency.

Be sure to gather information about the following:

- In an office or enterprise environment, furnishings may consist of desks, cabinets, chairs, and other items.
- In warehousing and retail environments, furnishings will include storage racks and shelving as well as product inventories.
- In manufacturing environments, information should be gathered about the location of industrial equipment used in the manufacturing process and about equipment used to move product throughout the factory.
- In medical environments, furnishings or equipment will include devices that may cause interference and operate in the same frequency range as the proposed wireless network. Storage of items used within the hospital or medical environment for patients and employees may also affect RF coverage.
- In educational environments, the location of lockers and internal windows is important, as are the types of materials used in the walls, such as concrete blocks with and/or without sound dampening material inside. All of these factors will have an effect on the RF propagation.

These are just some examples of the types of furnishings and other items that may affect a wireless network deployment and are factors to consider.

Electrical Specifications

Documentation of the electrical specifications of the environment is helpful in determining whether the current electrical implementation will be sufficient to handle the proposed wireless network deployment. This will allow the site survey process to determine whether any upgrades need to be made to support devices that may be using Power over Ethernet or whether the existing infrastructure is sufficient. It is best to gather information regarding electrical power sources, electrical panels, existing wiring, and location of electrical outlets.

Documenting Existing Network Characteristics

Documentation is a major part of any business, including documentation of computer networks. In order to have a successful deployment of a wireless network, it is critical to know the details of the existing network infrastructure as well as future implementations, upgrades, and modifications. These existing infrastructures may include a wired or wireless network already in place and functioning that may be upgraded or in a new deployment. Figure 8.6 shows a floor plan with an existing wireless deployment of only two access points providing wireless access for the conference rooms. This information can be used as part of the new wireless site survey and design.

FIGURE 8.6 Floor plan with an existing wireless network deployment

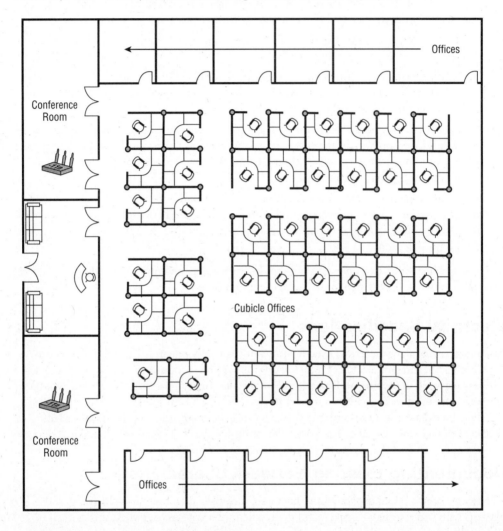

The following list includes some of the questions that you should take into consideration when performing a wireless site survey:

- What frequency range will the new wireless network operate in?
- Are there any existing IEEE 802.11 wireless networks in the same RF space?
- Will all or part of the existing wireless network be utilized in the new deployment?
- What effect will the neighboring wireless networks have on this deployment?

Ignoring existing IEEE 802.11 wireless networks may have a significant impact on how a new wireless network will operate and result in poor performance for the clients or devices that will be connecting.

Documentation of networks is usually the responsibility of the IT department. Some organizations may lack good documentation of the existing network infrastructure. If this is the case, additional work may be required prior to starting a wireless site survey.

Existing IEEE 802.11 Wireless Networks It was mentioned earlier that the scope of any existing wireless networks must be determined during the interview process. If a wireless network does exist (as it often does), it will need to be dealt with during the site survey and design procedure. The questions that are asked regarding the existing wireless network will help determine the role it is going to play. If the existing network is going to remain in place, understanding its technical details and how to work it into the design of the new or upgraded deployment will help you create a successful and productive wireless network deployment.

Existing IEEE 802.3 or Other Wired Networks In addition to knowing of any existing wireless networks, you should know about the wired network infrastructure. Existing documentation on the wired network infrastructure will help streamline the process for connecting the IEEE 802.11 wireless components of the network. The wired infrastructure is discussed in more detail in the next section of this chapter.

Existing wireless networks can play a big role in a new wireless network deployment. Understanding the current location of infrastructure and other wireless network devices is an important part of a wireless site survey. One way to see devices that are part of a wireless network is to use a protocol analyzer. Other programs will also be able to view existing wireless networks, such as the MetaGeek InSSIDer program or the Xirrus Wi-Fi Inspector. In some cases these programs may be adequate based on the complexity and size of the site survey. But they typically do not have the extensive feature set that many protocol analyzer packages have.

Identifying Infrastructure Connectivity and Power Requirements

Why are there so many wires in wireless networking? Wireless networks require some type of wired infrastructure for many reasons, including connecting access points together, allowing wireless device and user access to network resources, providing access to a wide area network, allowing Internet connectivity, and supplying electrical power.

Network infrastructure connectivity plays a big role in wireless networking. A wireless site survey will require additional information about the network infrastructure and power requirements. In a sense, a wireless site survey also requires a wired or infrastructure survey. A wireless site survey should include information about the wired network, including the location of the wiring closet, the wired infrastructure network devices in use, the connection speed between sites, and the electrical power requirement:

Location of Wiring Closets A *wiring closet* is a room (usually secured) containing electrical power and cabling for voice and data that is terminated and connected to infrastructure devices such as switches and routers. In most cases, these locations are noted on the floor plan or blueprint documents, but that may not always be the case, and you will need to find them if they are not noted. When deploying a wireless network, you should know not only the physical locations of wiring closets but also the capacity of existing infrastructure devices. This is important because wired connections such as IEEE 802.3 Ethernet have specific limits for cable lengths, and infrastructure devices such as access points and bridges will have to be placed within these physical limits of infrastructure connectivity. For example, the IEEE 802.3 Ethernet standard has a physical maximum cable length of 328 feet (or 100 meters) for unshielded twisted-pair wiring.

In addition to cable length, the feasibility of running the Ethernet cable from the wiring closet to the desired location must be considered. For example, the intended location for an access point may be on a ceiling without any access from above. In this case, lack of accessibility could pose a problem for installation of the Ethernet cabling from the wiring closet and/or AC power to the devices.

Wired Infrastructure Network Devices in Use The wired infrastructure devices in use may have an impact on a successful deployment of a wireless network. An evaluation of the infrastructure devices by a WLAN site surveyor or a network infrastructure professional may be required to determine if any additional hardware is needed or if changes must be made prior to deploying a WLAN. If the infrastructure devices, such as Layer 2 switches and routers, are not adequate to support a new wireless deployment, additional hardware may need to be purchased.

Connection Speed between Sites The connection speeds and type of connections between sites should be evaluated to determine if there are any bottlenecks that could affect the overall network performance. Placement of authentication servers and other network resources may be affected by the speed of these links.

Electrical Power Since all wireless network infrastructure devices, including access points, bridges, and wireless switches, require an AC power source, and because these devices may be supplying Power over Ethernet (PoE) to the infrastructure devices, verification of an adequate AC power supply must be performed. The AC power sources are sometimes taken for granted or not taken into consideration, which could pose a problem during the installation phase. It is also important to determine what local ordinances may apply and what they allow you to do for a wireless network installation.

Because PoE is used not only in infrastructure devices but also with IP telephones, security cameras, and user devices, it may be necessary to perform calculations and verify that the power supply to the wiring closet will be adequate to support the powered infrastructure (including the PoE devices). Electrical components of a wiring closet may need to be upgraded to support a new wireless network deployment. Newer wireless networking technologies such as 802.11ac MIMO systems may require more DC power than is available with the IEEE 802.3af PoE amendment to the Ethernet standard.

If these technologies will be used, the power requirements will need to be carefully considered to verify that enough DC power will be available to the end powered devices. Manufacturers now provide IEEE 802.3at (with new PoE capabilities) endpoint or midspan devices to provide the necessary amount of DC voltage for certain PoE-capable devices. All manufacturers of enterprise dual-band IEEE 802.11n wireless access points claim their equipment will operate with standard 802.3af power. I recommend that you read the specification data sheet for the access point, WIPS sensor, or other wireless infrastructure device to determine the type of PoE required and whether IEEE 802.3af or IEEE 802.3at will be required to provide DC power to the device.

It is important to understand that wireless infrastructure devices require some sort of wired connectivity for data and electrical power. Data access will usually come by way of an Ethernet connection from a wired network infrastructure or backbone. As stated in the IEEE 802.3 Ethernet standard, the maximum length for unshielded twisted-pair Ethernet cable is 328 feet, or 100 meters. This limitation may have an impact on how and where access points and other infrastructure devices are placed.

During the wireless site survey process, the surveyor will need access to wiring closets. This will allow the surveyor to evaluate and perform a survey of the wired network infrastructure as well.

Another consideration is electrical power requirements. Infrastructure devices need electrical power as well as data connectivity to operate. If the electrical power is decentralized and located at each device, an electrician will need to evaluate and determine the requirements.

An option for supplying electrical power to infrastructure devices is Power over Ethernet (PoE). This technology is now very common and is supported by all enterprise device manufacturers. A survey of the wired infrastructure and devices is almost always required to determine if the infrastructure will be capable of handling the new PoE devices that will be installed. Keep in mind that zoning and building regulations, electrical contractors, and labor union requirements may add to the overall cost of wireless deployment. PoE was discussed in Chapter 6, "Computer Network Infrastructure Devices."

Understanding Application Requirements

Understanding the application requirements for software applications is another important part of a successful wireless deployment. The types and number of applications used will affect the performance and the capacity of an IEEE 802.11 wireless network. Some software applications are very bandwidth intensive and will require more resources than others to perform effectively as designed. Examples of these applications are streaming video, computer-aided design (CAD), and database programs. By contrast, applications used for email, social media (if policy allows), and casual Internet browsing are not bandwidth intensive.

Because digital imaging, graphic design, CAD, and database applications can have a tremendous effect on the performance of a wireless network, they require special attention when the network is designed. The infrastructure resources that will be impacted include the processing power and memory capacity of the wireless access points and the speed of the wired infrastructure itself, such as 100 Mbps or 1,000 Mbps (Gigabit).

This in turn will affect the site survey. The wireless site survey will be impacted because the location where the wireless network is installed may require more infrastructure devices or access points. Other applications, such as Voice over Wireless LAN, may introduce another entire set of issues. This type of deployment may use only the 5 GHz band for wireless voice technology, and therefore the wireless site survey must be performed accordingly.

Understanding RF Coverage and Capacity Requirements

A major aspect of a wireless site survey is to understand and verify the RF coverage and capacity requirements based on the network design. In Chapter 5, "IEEE 802.11 Terminology and Technology," you learned about coverage versus capacity and the differences between them. A wireless network site surveyor will need to verify these requirements as part of the site survey. This can be accomplished either manually or automatically through a predictive process, both to be discussed in more detail later in this chapter.

To review, the wireless coverage and capacity requirements are going to depend on several factors:

- Physical size of the area to be covered
- Number of users or devices accessing the wireless network, including bring your own device (BYOD) users
- Software or hardware applications in use
- Obstacles and propagation factors based on the environment
- Radio frequency range of the network to be installed
- Wireless network hardware to be used
- Output RF power of the transmitters
- Receive sensitivity of the receivers

In addition to an RF coverage analysis, an *RF spectrum analysis* will be beneficial. An RF spectrum analysis allows a site surveyor to view areas of RF coverage as well as interference sources and non-IEEE 802.11 wireless devices. This topic will be discussed later in this chapter. Although a spectrum analysis is not required, it does allow a site surveyor to view sources of RF in the locations where a wireless network will be deployed.

Client Connectivity Requirements

The wireless client devices that will be connecting to the wireless network also need to be considered as part of a site survey. This includes knowing the radio type, antenna type, gain, orientation, portability, and mobility of the device. Keep in mind that a wireless client device is both a radio frequency transmitter and a receiver. The client (receiver) must be able to hear the transmitter and the transmitter must be able to hear the client. The RF cell size will vary from device to device. It is beneficial to survey with the type of devices that will be used with the wireless network whenever possible. Understanding the type and function of client devices will have an impact on the design of the wireless network. The following wireless client devices are commonly used in a wireless network:

- Notebook computers
- Tablets
- Pocket computers
- Smartphones
- Bar code scanners
- Point-of-sale devices
- Voice handsets

Other wireless devices used in various wireless applications include but are not limited to the following:

- Desktop computers
- Printers and print servers
- Manufacturing equipment
- IEEE 802.11 friendly video cameras

Many environments have both desktop and notebook computers as wireless client devices. These devices may or may not require roaming capability. Office and enterprise deployments commonly use handsets for voice communications. Although it is fairly difficult to take all potential wireless client devices into consideration, it is best to understand the type of devices that will be used. This information can be obtained through the interview process and the information gathering stage of the site survey.

Antenna Use Considerations

In Chapter 3, "Radio Frequency and Antenna Technology Fundamentals," you saw various types of antennas and accessories. The antenna used in any wireless network installation will depend on the specific scenario. As part of the site survey, various antenna types may need to be used for testing purposes to determine the best antenna for a specific application. In some cases, the customer may want a specific type of antenna, such as an omnidirectional mounted directly to the access points. Others may be using access points without external antenna capabilities.

Some businesses are concerned about aesthetics and are particular about the appearance of an antenna and the mounting location. The proper antenna selection will ensure correct coverage as intended by the design and site survey of the wireless network. The antenna type used will determine the propagation pattern of the radio frequency and is a significant part of a successful wireless deployment. The antenna used in various deployments will depend on the business model in which the network is installed:

General Office/Enterprise General office/enterprise solutions usually require complete coverage throughout the entire location. In many cases, this type of installation will require access points mounted out of sight and aesthetically pleasing antennas. This could be an omnidirectional antenna mounted to a ceiling tile or integrated within an access point.

Manufacturing Manufacturing environments are usually industrial facilities with high ceilings and various types of manufacturing and industrial equipment. These environments may use a combination of omnidirectional and semidirectional antennas due to the physical architecture of the buildings. The antennas also may need to withstand harsh environmental conditions such as extreme temperature fluctuations and dirt.

Warehousing Warehousing implementations have some characteristics in common with manufacturing. The buildings that house this type of business are in many cases large open areas with high ceilings to allow the storage of large volumes of product and equipment. Antenna mounting needs to be looked at very closely to ensure that equipment such as forklifts used to move product do not come in contact and damage antennas. Warehousing also may use a combination of omnidirectional and semidirectional antennas for proper coverage. Figure 8.7 shows a sample floor plan for a combination small office/warehouse deployment using a combination of omnidirectional and semidirectional antennas.

Retail/Point of Sale (PoS) Retail/point-of-sale installations may have to accommodate publicly accessible areas as well as warehousing and storage in the back of the buildings. In this type of installation, antennas will be a combination of omnidirectional and semidirectional. The devices may consist of computers as well as other handhelds such as wireless bar code scanners or portable wireless devices. Appropriate antenna selection and gain need to be considered to ensure that the devices have good signal connectivity to the infrastructure. In many cases, the public areas will require aesthetics to be taken into account in antenna selection. In the storage part of the building, antennas are similar to those described in warehousing.

FIGURE 8.7 Small office/warehouse floor plan showing RF coverage using different antenna types

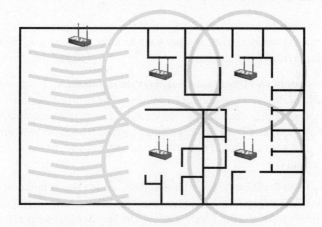

Healthcare and Medical Many healthcare and medical deployments are publicly accessible, so aesthetics and security are both important. Antenna types that fit the environment and are not accessible to the public are commonly used. Some healthcare facilities, such as hospitals, may require the use of only omnidirectional antennas mounted to the autonomous access point or integrated into an access point. Others may allow semidirectional antennas such as a yagi to cover long hallways and corridors.

Government and Military Government and military wireless environments are usually not publicly accessible and in many cases are campus based and require outdoor as well as indoor solutions. In these installations, antennas mounted directly to an access point or semidirectional antennas may well fit the environment. For outdoor point-to-point or point-to-multipoint, highly directional antennas such as parabolic dishes may be required.

Education Education deployments are typically campus based and may require the use of outdoor antennas to connect buildings together. However, for indoor solutions, ensuring that the antennas will fit the environment and not be tampered with is important. These deployments may use a combination of omnidirectional and semidirectional antenna types. Some locations may require the use of enclosures for the infrastructure devices to ensure security of the physical devices. In these cases, special connectors and antenna adapters may be required.

Public Access, Hotspot, and Hospitality Public access, hotspot, and hospitality sites are publicly accessible locations that will require aesthetics and security when it comes to antenna selection, as with some of the previous examples. These deployments can use a combination of omnidirectional, semidirectional, and highly directional antennas.

Real World Scenario

Typical Steps Used in a Wireless Site Survey

Here are some basic steps to provide an overview of the wireless site survey process. Keep in mind that these steps may vary slightly based on the location and the type of wireless network deployment.

1. **Gathering Information and Discussing Business Requirements:** Determine the need and intended use of the wireless network and interview appropriate individuals. Explain the site survey process and provide an overview of wireless networking. Doing so will help minimize unnecessary delays caused by people who are not included in the process but should have been and are unaware of the wireless network and site survey.

2. **Project Timeline and Planning:** Understand and document the extent and estimated timeline of the survey process and deployment. This will include all stages of the wireless site survey from start to finish, such as design, deployment, and verification. Because the wireless site survey may include people from many different areas, such as facilities, IT, engineering, and so on, everyone must be on the same track.

3. **Site Visit and Pre-design Audit:** It is a good idea to make a visit to the location to get an idea of any issues that may not be visible on a blueprint or floor plan. This will also allow for checking the RF propagation with a test access point if feasible. Performing a wireless site survey without a site visit—especially a predictive modeling survey—can be done, but that will come with some risks. Knowing what you are dealing with will minimize mistakes caused by not knowing the environment and therefore possibly making inaccurate assumptions.

4. **Wireless Network Design:** Determine areas of RF coverage and interference as well as potential placement of access points and other infrastructure devices. This will be done with an onsite manual site survey, a predictive modeling site survey, or a combination of both. An RF spectrum analyzer can be used to view the air to determine the areas of RF interference.

5. **RF Spectrum Analysis and RF Testing:** Perform testing of the proposed design and verify RF sources of interference. Also verify coverage and lack of coverage. Although an RF spectrum analysis is not a required part of the wireless site survey process, it is highly recommended. Knowing the potential RF interference may impact the wireless network design as a whole. It will be much easier to deal with

potential RF issues up front rather than after the wireless network is deployed. Testing specific areas or audit points with an access point will allow the surveyor to determine the coverage and how the RF propagates from the access point.

6. **Deployment of Infrastructure Devices:** Install the infrastructure devices as described by the design. This may or may not be done by the same person who performed the wireless site survey and in many cases it is not. This means that clear, concise documentation and instructions must be provided to the installer. This will reduce the risk of having to make unnecessary visits back to the location for corrections or adjustments because of mistakes that could have been avoided.

7. **Verification of RF Coverage:** Perform verification testing and spot checks of RF coverage per the design. Make necessary adjustments based on the results of RF testing. Verification testing may consist of spot checks in key areas (audit points) or could include the entire facility. It is important to be certain the wireless network will operate as designed. Sometimes cost or budget may not allow for a verification of the entire location, but I recommend it whenever possible. If the entire area cannot be checked, determining the audit points should be done carefully and should be completed in the early stages of the site survey. If needed, make adjustments to allow the wireless network to operate as designed.

8. **Support:** Provide technical support for wireless network deployment. This involves maintaining the wireless network infrastructure and includes RF spectrum monitoring and making the necessary adjustments to adapt to the dynamics of environment. This can be done manually or automatically, depending on the capabilities of the technology that is deployed.

The Physical Radio Frequency Site Survey Process

After all the up-front work is completed, such as gathering information (including the business requirements and site-specific documentation), defining physical and data security requirements, and interviewing managers and users, it is time to start the physical RF site survey process and the wireless network design. This is one of the most important parts of a successful wireless network deployment. This process includes locating areas and sources of RF interference as well as RF coverage (or lack thereof) and determining the locations of access points, bridges, sensors, and other infrastructure devices that will be used with the wireless network. Capacity planning is another important part of the process to ensure that optimal performance is achieved. The following sections detail the entire physical wireless site survey process and guidelines. The RF physical site survey is subjective, and people have different opinions on how the entire process works. In many cases, the process can be tailored based on the individual needs or requirements of the location where the wireless network will be installed. The following steps should be viewed as recommendations or guidelines:

1. Arrange a walk-through of the entire location, which may often require being escorted by an authorized person.
2. Take thorough notes and digital photographs as needed.
3. Perform an RF spectrum analysis.
4. Determine preliminary placement of infrastructure devices.
5. Perform on-site testing to verify the design.
6. Determine actual placement of infrastructure devices.
7. Install infrastructure hardware as specified.
8. Perform on-site verification testing and make adjustments to verify that the design meets the specifications.
9. Deliver the final site survey report.

In today's wireless world, an RF site survey can be considered an ongoing process. Just a few years ago, not nearly as many wireless networks existed as do today, so the interference factor was not as significant. With the RF dynamics constantly changing and more devices using RF for communications, it is up to the network engineer to take into consideration that site requirements may also be constantly changing. Therefore, the RF site survey may need to be updated periodically.

Radio Frequency Spectrum Analysis

The wireless site survey includes finding areas of RF coverage and RF interference as well as locations for hardware, such as access points, bridges, and other infrastructure devices. Although radio frequency is not visible to the human eye, some tools are available to "see" a visual representation of the RF. One common tool is the RF *spectrum analyzer*. How the analyzer will be used will determine the type of equipment you will need for the job. Radio frequency is used as a communications mechanism for so many different types of technology:

- Radio
- Television
- Computing
- Cellular voice

In Chapter 3 you learned about the fundamentals, characteristics, and behaviors of RF technology, including various antenna types. You also saw some of the common frequency bands that are used for various types of wireless technologies. Using the correct spectrum analyzer for the job is a key element to collect the data you need. Spectrum analyzers are available in various form factors, and some are designed for a specific purpose, such as wireless networking technology, while others can be used for a larger part of the RF spectrum. These tools can be PC based or dedicated instrumentation devices. In the following sections, we will explore spectrum analysis for both IEEE 802.11 wireless networks and cellular communications.

Spectrum Analysis for IEEE 802.11 Wireless Networks

A spectrum analyzer used with IEEE 802.11 wireless networking can be limited to the specific RF bands used with wireless networking. This includes the 2.4 GHz ISM band and the 5 GHz U-NII bands. Figure 8.8 shows a spectrum analyzer view of the 2.4 GHz ISM band with the access points on various channels. Spectrum analyzers will vary in cost and complexity, depending on the frequency ranges they are designed to work with.

FIGURE 8.8 MetaGeek Chanalyzer shows an RF capture of the 2.4 GHz ISM band.

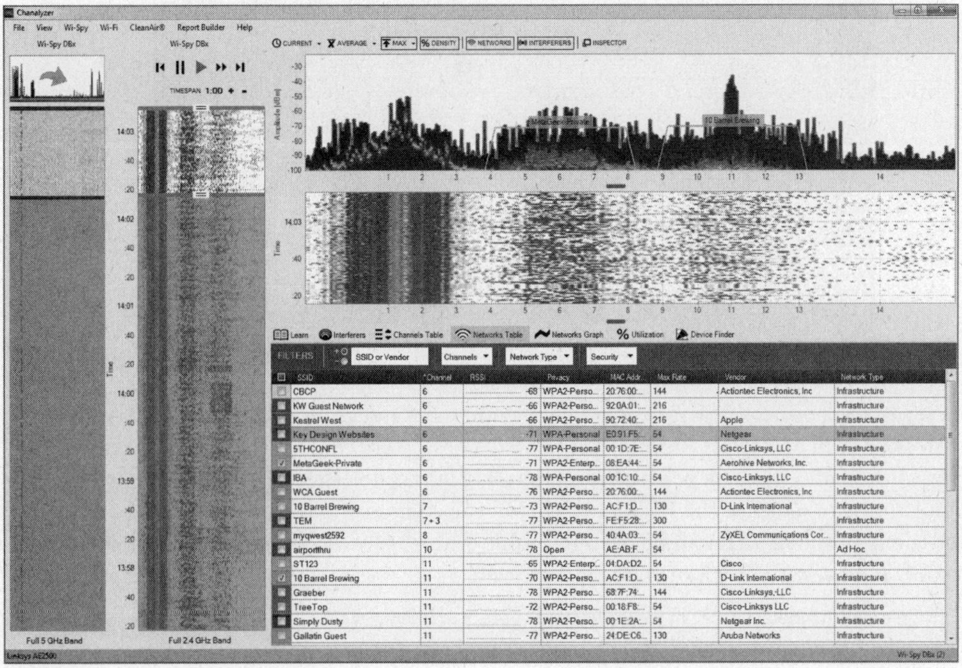

The spectrum analyzers that are designed specifically for the wireless market usually work only in the license-free radio bands and are typically less expensive because of spectrum limitations. Spectrum analyzers allow you to view the Physical layer of communication between devices used in wireless networking. Spectrum analyzers that work with IEEE 802.11 wireless networking technology are available in devices that attach to a USB port on the computer and instrumentation-style devices that can be used for applications other than wireless networking. Legacy solutions are PC card-based models, which may still be available in some markets. Figure 8.9 shows an example of a wireless network spectrum analyzer that connects to a USB port on a computer.

FIGURE 8.9 MetaGeek DBx spectrum analyzer for IEEE 802.11a/b/g/n networks operating in the 2.4 GHz and 5 GHz frequency ranges

Some manufacturers of wireless network equipment integrate spectrum analyzer tools into devices such as wireless access points or wireless LAN controllers for constant monitoring of the radio frequency. This feature is used for troubleshooting and allows a network administrator to monitor the RF in any part of the company. Some manufacturers offer stand-alone products that can be used to monitor more specific areas. Here are some manufacturers of wireless network spectrum analyzers:

- AirMagnet Spectrum XT: www.flukenetworks.com
- BumbleBee Handheld: www.bvsystems.com
- Motorola AirDefense: www.motorolasolutions.com
- Wi-Spy by MetaGeek: www.metageek.net

The ability to analyze the RF allows a site surveyor to find areas that lack coverage, also known as *dead spots*, as well as interference caused by other devices and other IEEE

802.11 wireless networks that operate in the same radio frequency range. Although a spectrum analysis is not a requirement for a wireless site survey, it is beneficial in most medium- to large-scale deployments of wireless networks. Because of the vast number of devices that use unlicensed radio bands, spectrum analysis can be considered an ongoing process. Performing regular spectrum analysis gives wireless network engineers the capability to monitor the area in which wireless devices are located for RF interference and other issues. Figure 8.10 shows a screen capture of an RF scan using a USB-based spectrum analyzer designed for IEEE 802.11 wireless networks.

FIGURE 8.10 AirMagnet Spectrum XT USB Spectrum Analyzer screen capture

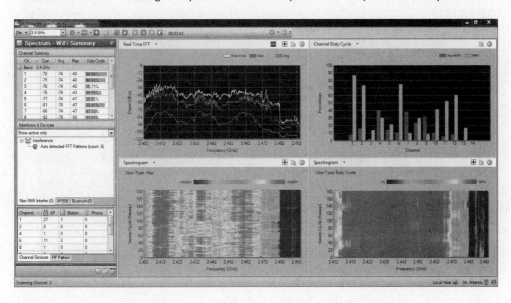

Wi-Fi and Non-Wi-Fi Interference Sources

The performance of a wireless network can be significantly affected by various types of interference sources. Interference may be either intentional or unintentional and caused both by IEEE 802.11 wireless networks and by other devices that also operate in the 2.4 GHz ISM or 5 GHz U-NII band. The use of portable and mobile devices that employ radio frequency for communication continues to rise, so interference from these devices is also more prevalent. As mentioned in previous chapters of this book, these interfering devices typically operate in the unlicensed radio spectrum 2.4 GHz ISM band. Other sources of interference from the industrial, medical, and scientific communities will also affect the amount of interference. *Non-Wi-Fi interference* in the ISM band can be caused by devices such as these:

- Microwave ovens
- Cordless telephones
- Medical equipment

- Manufacturing or industrial equipment
- Wireless video cameras
- Radar systems (5 GHz bands)

This is by no means a complete list, but it does contain some of the devices used daily that may cause some level of interference in the world of wireless networking. Figure 8.11 shows an example of interference caused by a microwave oven. Microwave ovens operate in the 2.4 GHz frequency range and, depending on the unit, may cause interference with IEEE 802.11 wireless networks.

FIGURE 8.11 MetaGeek Chanalyzer shows a microwave oven operating at maximum power. These typically create a mountain-like shape as shown in the waterfall view.

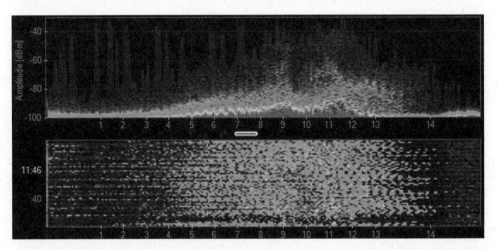

In Exercise 8.1, you will use the Chanalyzer spectrum analysis software from MetaGeek to look at some sample radio frequency captures.

EXERCISE 8.1

Using Spectrum Analysis Tools for Wi-Fi Networks

In this exercise, you will install the Chanalyzer software program by MetaGeek. This seven-day demonstration version of the software will allow you to view some predefined RF captures that are available as part of the program. To perform actual captures, you will need to purchase the Chanalyzer program and one of the Wi-Spy adapters available at www.metageek.net. To perform this exercise, you can download the Chanalyzer program from www.metageek.net or at the download page for this book, www.sybex.com/go/mobilityplus.

1. Download the Chanalyzer 5 software package; as of this writing, the file is named Chanalyzer-Installer.msi. Execute the program. Depending on the version of the Microsoft OS you are using, you may see an Open File – Security Warning dialog box. Click the Run button to continue. The setup wizard will appear.

Radio Frequency Spectrum Analysis

2. Click Next to start the installation process. The license agreement will appear.

3. Read the license agreement; then select the I Agree button and click Next. The Destination Folder screen will appear.

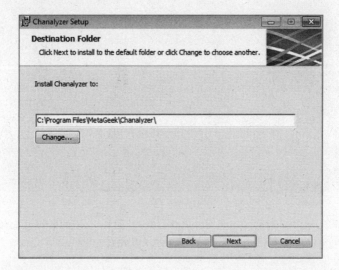

4. Click Next to accept the default installation folder. The User Experience Improvement Program screen will appear.

5. Click Next to continue the installation. The Ready To install Chanalyzer screen will appear.

EXERCISE 8.1 *(continued)*

6. Click Install on the Ready To install Chanalyzer screen to start the installation process. Depending on the version of the Microsoft OS you are using, you may see a User Account Control screen. Click Yes to continue the installation. The Chanalyzer program will now be installed on your computer. The Installation Complete screen will appear.

7. Click Finish to complete the installation process and exit the installer program. The Chanalyzer program is now installed on your computer.

8. Start the Chanalyzer program by choosing Windows Start ➤ All Programs ➤ MetaGeek ➤ Chanalyzer. The Welcome To Chanalyzer screen will appear.

9. Click the RECORDINGS tab. Scroll down to the 802.11a - 5 GHz Band recording to view the sample recording for an IEEE 802.11a network 5 GHz capture.

Radio Frequency Spectrum Analysis 353

10. Click the large white arrow to start viewing the recording. While observing the screen, let the sample recording run for a few minutes and notice the capture change as the program displays the recorded session. If desired, you can pause the capture or rewind by using the controls on the upper-left corner of the Chanalyzer window.

EXERCISE 8.1 *(continued)*

11. Select a new capture by clicking the RECORDINGS tab near the center of the screen. The additional sample captures screen will appear in the recordings window.

12. Use the scroll bar to move down and view the sample recordings. Click the large white arrow on the Hidden Security Cameras recording to view the sample recording of a hidden security cameras capture.

13. While observing the screen, let the sample recording run for a few minutes and notice the capture change as the program displays the recorded session. As with all the sample captures, you can pause, restart, or rewind the capture by using the controls on the upper-left corner of the Chanalyzer window.

14. Use the scroll bar to move down and view the other sample recordings.

15. Continue to view the other sample recordings by clicking the 802.11b and 802.11g - 2.4 GHz Band, 802.11n 40 MHz, and Cisco CleanAir AP recordings, repeating steps 10 and 11.

16. Click the LESSONS and INTERFERERS tabs and explore some of the available features of the program.

After you have completed this exercise, you can remove the Chanalyzer program from your computer. If you choose to keep and continue to use the program, before the seven-day evaluation period expires, you can click on the Buy button in the upper-right corner of the screen or visit the MetaGeek website for instructions on how to purchase the program with the Wi-Spy USB adapter.

The Wi-Spy spectrum analyzer device is available in several models. This device will plug into an available USB port on your computer. The Chanalyzer program used in this exercise is the same you would use to perform actual captures with the Wi-Spy device from MetaGeek.

Wi-Fi Interference

Wi-Fi interference is caused by IEEE 802.11 wireless network devices that operate in the 2.4 GHz ISM or 5 GHz U-NII bands. In some cases, this is the largest source of interference for wireless networks. There are two types of Wi-Fi interference: co-channel interference (other devices on the same channel) and adjacent channel interference (other devices on overlapping channels). The following network types will cause Wi-Fi interference:

- FHSS networks: 2.4 GHz (Legacy)
- DSSS networks: 2.4 GHz
- ERP-OFDM networks: 2.4 GHz
- OFDM: 5 GHz
- HT-OFDM: 2.4 GHz and 5 GHz (802.11n)
- VHT-OFDM: 5 GHz (802.11ac)

Wi-Fi devices that cause interference can be operating in either infrastructure mode or ad hoc mode. The number of devices that are part of the wireless network will also determine the extent of the interference. In Chapter 5, we discussed different types of spread spectrum technologies that wireless networks use to communicate. FHSS, DSSS, ERP-OFDM, OFDM, HT-OFDM, and VHT-OFDM are technologies used for IEEE 802.11 wireless networking in the 2.4 GHz and 5 GHz bands. However, some of these technologies are used in non-wireless-network devices as well. For example, Bluetooth devices operating in the 2.4 GHz ISM band also use FHSS technology for communication. Zigbee devices operate in the 2.4 GHz band and use DSSS technology for communication. Depending on the manufacturer, 2.4 GHz cordless phones may use either FHSS or DSSS. Performing an RF spectrum analysis is one way you can determine the source of this type of interference. If an instrumentation spectrum analyzer (a calibrated device that has the capability to view entire radio spectrums) is used, it will be necessary to understand how to interpret the data collected as well as how to operate the device.

It is important to note that IEEE 802.11 FHSS wireless LANs are legacy networks and unlikely to found with today's wireless installations. I mention this here because although it is unlikely, you could possibly encounter them. Also, FHSS is used with other wireless technologies such as Bluetooth devices.

> **Standards vs. Industry Definitions of Interference**
>
> The 802.11 standard loosely defines an adjacent channel as any channel with nonoverlapping frequencies for the DSSS and HR/DSSS PHYs. With ERP and OFDM, the standard loosely defines an adjacent channel as the first channel with a nonoverlapping frequency space.
>
> This contradicts how the term *adjacent channel interference* is typically used in the marketplace. Most Wi-Fi vendors use this term to loosely mean both interference resulting from overlapping cells and interference resulting from the use of overlapping frequency space. For example, vendors typically use this term in a case where AP-1 (channel 1) is located near AP-2 (channel 2).
>
> Adjacent channel interference is a performance condition that occurs when two or more access point radios are providing RF coverage to the same physical area using overlapping frequencies. Simultaneous RF transmissions by two or more of these access point radios in the same physical area can result in corrupted 802.11 frames because of the frequency overlap. Corrupted 802.11 frames cause retransmissions, which result in both throughput degradation and latency.
>
> Co-channel interference is a performance condition that occurs when two or more independently coordinated access point radios are providing RF coverage to the same physical area using the same 802.11 channel. Additional RF medium contention overhead occurs for all radios using this channel in this physical area, resulting in throughput degradation and latency.

Spectrum Analysis for Cellular Communications

Most newer-model wireless mobile devices have multifunction capabilities that allow them to be used for voice and data applications. The capabilities may include cellular voice communications, 3G, 4G, LTE, WiMAX, and Wi-Fi.

In the previous section you learned the importance of understanding the type of radio frequency coverage and interference sources that are contained within the physical area you intend to use for wireless networking. As wireless technology continues to evolve, the same concepts apply to cellular communications. One main difference between these technologies is which frequency bands are used in addition to what or who generates the RF signals. In a wireless local area networking deployment, the RF that is intended to be used for communications is generated from within the buildings from wireless infrastructure devices such as wireless access points, bridges, and the like. With cellular communications, the signal is generated from wireless cellular carriers and you are relying on the signal from outside to propagate throughout the building. The antennas or towers that provide the RF for cellular devices will encompass very large areas that may or may not provide adequate signal quality within the same physical space where the wireless local area network is installed. Therefore, having a good understanding of the RF spectrum as it pertains to wireless cellular technologies is equally important, and it is difficult to rely on cellular signals from outside to cover all areas inside of a building. A lower level or basement level of a building is one good example where you could experience weak cellular signal levels.

In Chapter 3 we looked at distributed antenna system (DAS) technology. You learned that RF signals such as cellular phone communications that are received from outside a structure or building are already weak or barely above the desired receive signal threshold. Bringing that signal in from the outside and allowing it to be at an acceptable level in a building can be accomplished using DAS. This antenna technology may use several antennas that create small cells as opposed to a single antenna to provide wireless coverage throughout an entire area or building.

A spectrum analyzer that is designed for these frequencies will allow you to test the entire building for the cellular coverage and determine if it is at an acceptable threshold level. Figure 8.12 shows a commercial off-the-shelf laptop-based USB spectrum analyzer that is used to capture radio frequency bands used with cellular communications.

FIGURE 8.12 The AirMagnet ES from Fluke Networks shows the USB spectrum analyzer for cellular networks in action.

You will notice in the figure the spectrum analyzer device connects to a USB port in a laptop computer or tablet device and uses an ultra-wideband antenna. The supplied antenna is ground plane independent with an SMA male connector and swivel mechanism that allows the antenna part to be rotated in various directions. This antenna is backward compatible with cellular applications such as GSM, UMTS, and LTE and covers all available cellular bands across the globe, signals between 698 MHz and 2690 MHz. In Exercise 8.2, you will install a demonstration version of the AirMagnet Spectrum ES from Fluke Networks. Keep in mind that this will allow you to view a recording of a capture. You will need to purchase the product from Fluke Networks to perform live captures.

EXERCISE 8.2

Using Spectrum Analysis Tools for Cellular Communications

In this exercise, you will install a demonstration version of Fluke Networks's AirMagnet Spectrum ES, a professional-grade RF spectrum analyzer that will speed in-building cellular deployments.

The AirMagnet Spectrum ES has the following system requirements:

- Microsoft Windows 7 Enterprise/Professional/Ultimate 32-bit and 64-bit or Microsoft Windows 8 Pro/Enterprise 32-bit and 64-bit
- Intel Core 2 Duo 2.00 GHz or higher
- 1 GB RAM required (2 GB recommended)
- 150 MB of available hard drive space
- Microsoft .NET Framework 4.0

Here are the steps:

1. Point your browser to the Fluke Networks website download page at

   www.flukenetworks.com/content/free-trial-airmagnet-spectrum-es

2. Complete the form and click the Submit button. You will receive an email at the email address you provided with instructions on how to download the file. At the time of this writing, the name of the file is SpectrumES_DemoSetup.exe.

3. Start the setup process by executing SpectrumES_DemoSetup.exe. Depending on the version of the Microsoft OS you are using, you may see a User Account Control dialog box. Click the Yes button to continue. The first screen of the installation wizard will appear.

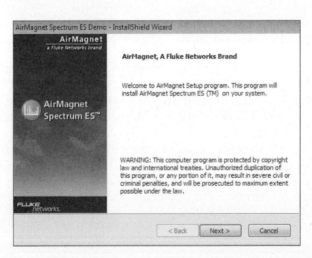

4. Click Next to start the installation process. The License Agreement dialog box will appear.

5. Read the contents of the license agreement dialog box and verify that the "I accept the terms of the license agreement" radio button is selected.

6. Click Next to continue the installation process. The Choose Destination Location screen will appear.

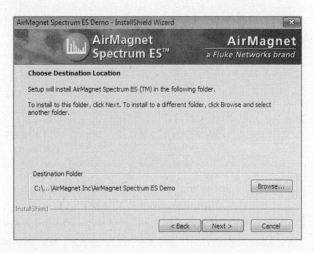

7. Click Next to select the default destination folder and continue the installation. The Ready To Install AirMagnet Spectrum ES screen will appear.

8. Click Install to continue the installation. The AirMagnet Spectrum ES Installation Completed screen will appear.

9. Click Finish to close the wizard.

10. Start the AirMagnet ES Spectrum ES program by choosing Windows Start ➢ All Programs ➢ AirMagnet Spectrum ES Demo ➢ AirMagnet Spectrum ES Demo. Depending on the version of the Microsoft OS you are using, you may see a User Account Control dialog box. Click the Yes button to continue. The AirMagnet Spectrum ES home screen will appear.

EXERCISE 8.2 *(continued)*

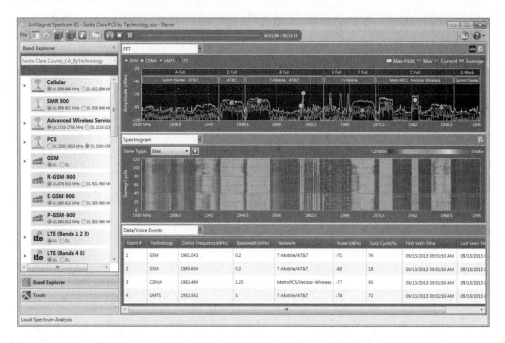

11. While observing the screen, let the sample recording run for a few minutes and notice the capture change as the program displays the recorded session. If desired, you can stop the capture by using the controls on the top-center portion of the program window.

12. Click the Tools button on the lower-left side of the screen. The tools window will appear.

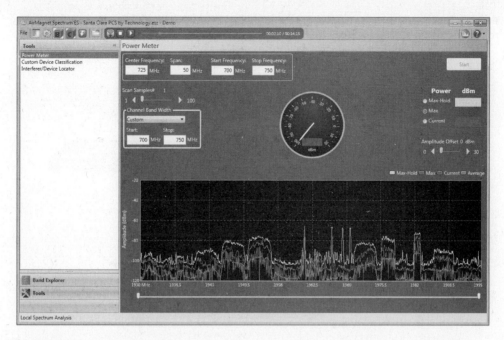

13. Explore the available tools by clicking the Custom Device Classification and Interferer/Device Locator options.

14. Click File and then Exit to close the AirMagnet Spectrum ES program. Uninstall the program using the Control Panel when you are finished using it. You can also purchase the program online.

Understanding radio frequency as it pertains to cellular technology is becoming more important in relation to indoor wireless networking. Many wireless infrastructure devices now contain cellular capabilities that will allow for 3G or 4G failover connectivity in the event of a WAN or Internet outage. This failover will allow a company to maintain WAN connectivity even though the wired connection to the company has failed for whatever reason. Several manufacturers of IEEE 802.11 wireless networking equipment provide branch office or remote office solutions that provide this type of failover functionally. Figure 8.13 shows a solution that will provide wireless cellular WAN connectivity in the event of a wired WAN (cable modem, DSL, T1, and so on) disruption or outage.

FIGURE 8.13 Different views of CradlePoint MBR1400 mission-critical broadband router for Verizon Wireless

Received Signal Strength

The main objective of a wireless network is to provide access to computer network resources and services for wireless devices that are connected to the network. This means it is essential for wireless client devices to have reliable connectivity to the wireless network infrastructure. To provide this connectivity, the wireless site survey process should include testing to verify that the received signal is adequate for the application of the device.

Devices that communicate wirelessly require two-way communications in order to operate correctly. This means the receiver must be able to receive enough of the signal to determine the data that was sent from the transmitter. Wireless client devices use what is known as the *received signal strength* to show the amount of power received from a transmission. Figure 8.14 shows a client utility that displays specifics regarding the received signal.

The amount of received signal strength required will be determined by the type or application of the wireless device as well as the amount of radio frequency noise in the area. RF noise consists of extraneous undesired radio signals in the area emitted by a variety of devices other than the transmitter. You should check with the device manufacturer to determine the minimum amount of received signal that is acceptable for a specific application. Some applications will require more received signal than others. For example, the manufacturer of a voice handset used with a Wi-Fi network may recommend a minimum of −65 dBm to −67 dBm, whereas a computer network card may require a minimum of −70 dBm.

FIGURE 8.14 Broadcom Wireless Utility shows signal, noise, and data rate.

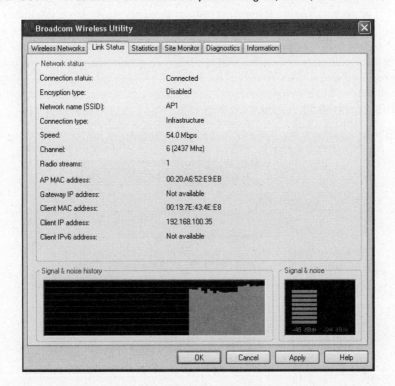

The *signal-to-noise ratio (SNR)* is the difference between the received signal and the noise floor. For example, if the received signal is –65 dBm and the noise floor is –95 dBm, then the signal-to-noise ratio will be 30 dB. This value is calculated by subtracting the received signal from the noise. In this case, –65 dBm – (–95 dBm) = 30 dB.

 The recommended signal-to-noise ratio (SNR) for most wireless networks is a minimum of 20 to 25 dB.

 In wireless network technology, the received signal strength indicator (RSSI) value is an arbitrary number assigned by the device manufacturer. There is no standard for this value, and it will not be comparable between devices from different manufacturers. When performing a site survey, whenever possible it is beneficial to survey with the same network adapter model that will be used in the deployment. This will provide more accurate results for the client devices using the network.

With respect to a cellular signal, an acceptable received signal value will be a little more liberal. The acceptable signal for most cellular devices is three bars; I'm kidding, but to some people that would not be funny. Most cellular phones have a screen that shows between one to five bars that represent signal strength. This graphical representation is derived from a formula in the firmware that is contained within the device. The formula will use the received signal value to represent that signal in a graphical fashion using bars, dots, or other means. Most smartphones include a diagnostic screen that will allow you to view what the actual value is, as shown in Figure 8.15.

FIGURE 8.15 iPhone 5 showing the received signal value in dBm

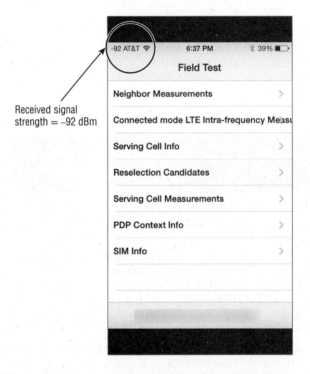

The Verizon Wireless website states that "the RSSI should be greater than −58 dBm (e.g., −32 dBm). A value of −96 dBm indicates no signal. If the signal is between −82 dBm and −96 dBm, move the device to an alternate location (an outdoor location is preferable)." You can see the difference between what one cellular provider states is an acceptable value and what you learned earlier in this section pertaining to IEEE 802.11 wireless networking.

Performing a Manual Radio Frequency Wireless Site Survey

Although some in the industry consider the physical manual wireless network RF site survey process "old school," it provides some of the most accurate results when it comes to obtaining certain information. A manual site survey requires a physical walk-through of the area, recording information to determine the performance of clients and devices that will be connected to the wireless network. This type of site survey can be very accurate because the surveyor is recording actual statistics, such as the following:

- Signal strength
- Signal-to-noise ratio (SNR)
- Data rate of connected devices
- Radio frequency interference

These measurements are recorded while the site surveyor is physically moving through the facility or location. Recommended received signal strength and signal-to-noise ratio (SNR) values were discussed earlier in this chapter. Small-scale manual site surveys can be performed inexpensively by using a wireless client adapter with site survey functionality, one of several freeware/shareware utilities, or commercial software designed specifically for this type of application. Figure 8.16 shows an example of a free site survey software program that can be used for very basic measurements.

FIGURE 8.16 Ekahau HeatMapper, a Wi-Fi coverage mapping site survey software utility, is a free download from Ekahau.

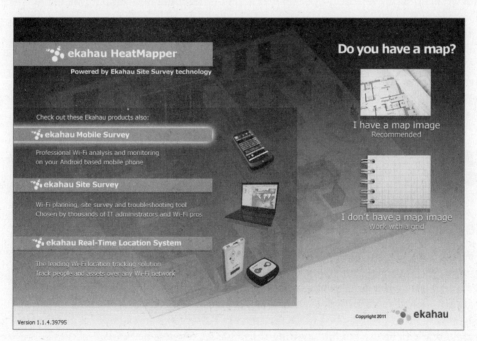

The manual site survey process can be used for recording measurements to determine the actual placement of hardware infrastructure devices and for spot checking, locating other IEEE 802.11 wireless networks, or verification after a predictive site survey has determined the placement of these devices. Not too long ago this manual process was the only way to perform a wireless site survey. The main advantage of the manual process is accuracy. One disadvantage is that it can be very time consuming, depending on the extent of the installation.

The process for a manual RF site survey will vary depending on the individual who is performing the survey. However, there are some common components and considerations that may be included:

- Obtaining a floor plan or blueprint
- Identifying existing wireless networks
- Testing access point placement
- Analyzing the results
- Considering the advantages and disadvantages of manual site surveys
- Performing a software-assisted manual site survey
- Using a manual site survey toolkit

Understanding each of these components will help provide a more accurate site survey and successful wireless network deployment.

Obtaining a Floor Plan or Blueprint

The first step in a manual site survey is to review a floor plan or blueprint of the location. Many in the industry refer to a floor plan as a map. This could be a simple sketch and is required to note access point placement as well as readings. The readings to note on the floor plan include signal strength and signal-to-noise ratio values from client devices. Figure 8.17 shows a basic floor plan of a small office building. The floor plan should be marked with approximate access point locations. In some cases, a floor plan or CAD drawing of the location may not be available. Depending on the size of the location, a fire evacuation plan could be photographed or scanned in and used as a starting point to create a drawing and may be adequate for the site survey.

The floor plan and approximate access point locations used in these examples are for illustration purposes only.

Identifying Existing Wireless Networks

As part of a manual site survey, you should identify existing wireless networks in the area, noting locations that include possible sources of radio frequency interference and may have an impact on the wireless network that will be installed. Using a device or software program designed specifically for wireless networks would be ideal. However, freeware or shareware programs may be satisfactory. Figure 8.18 shows an example of a software program that can be used to find existing IEEE 802.11 wireless networks.

FIGURE 8.17 Approximate access point locations

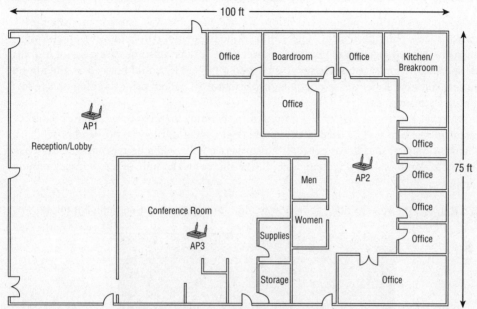

FIGURE 8.18 The AirMagnet Wi-Fi Analyzer can be used to view existing IEEE 802.11 wireless networks.

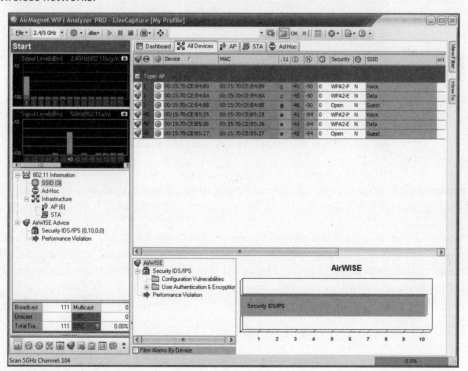

Testing Access Point Placement

Once you have identified potential locations for access points, you can start testing the proposed access point placement. This can be done by temporarily mounting an access point at the desired location to get the most accurate results. In some cases, a tool or fixture made specifically for wireless site surveys or even a tall ladder could be used as a temporary mounting solution. Refer to the documentation that was created prior to going on site for mounting locations.

It is always best to use test access points from the same manufacturer and, if possible, the same model that will be used in the actual deployment during the pre- and post-deployment manual site survey process. This practice will yield a better outcome than using generic access points during testing. Figure 8.19 shows an example of a temporary mounting solution.

FIGURE 8.19 Access point temporarily mounted using an expandable light pole

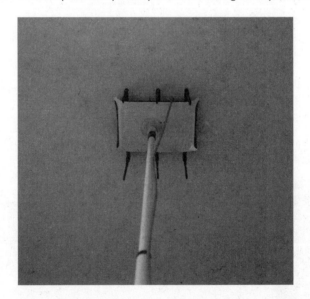

Using one or possibly two temporarily mounted access points, document the results of the testing while moving around the facility. One testing method is the active survey process and requires the client device used for the survey to perform an IEEE 802.11 authentication and association to the test access points. This will provide the active connection. The passive survey process (listening to all access points in the area) may also be used for this type of site survey. The area to be tested should include a reasonable proximity around the temporarily mounted access point. The extent should be the desired minimum received signal threshold; –67 dBm, for example. The information recorded and documented will depend on the surveyor but commonly includes signal strength and signal-to-noise ratio. It may be necessary to make adjustments to access point mounting during the testing process, based on the results of the onsite testing process. It would be beneficial to test an actual wireless client device and software application that will be used with the network to see if the performance meets expectations. Software-assisted surveys, as described later in this chapter, are commonly used for a site survey of any size in today's wireless deployments.

Analyzing the Results

Once all the testing is complete, it is necessary to perform an offline analysis to determine the final placement of access points or other infrastructure devices that may be used in the wireless deployment. This may include making adjustments from the original plan prior to testing or possibly adding and removing access points. The use of various antennas may also need to be analyzed and documented.

Advantages and Disadvantages of Manual Site Surveys

There are several advantages and disadvantages to the manual site survey process:

- Advantages
    - It is accurate because it is based on actual readings.
    - Physical characteristics of building are physically tested (attenuation values).
    - Allows verification of actual RF signal coverage and interference.
    - Allows marking exact installation locations of infrastructure hardware while onsite.
- Disadvantages
    - It can be time consuming.
    - Usually only one access point is used for testing, so readings will need to be merged.
    - Requires a physical walk-through of entire location.
    - Many areas require an escort and special clearances for access.

WLAN infrastructure equipment used in a manual site survey should be from the same manufacturer as the hardware that will be installed in the actual deployment. Although in some cases it may be difficult to accomplish, using a site survey client device to replicate what will be used in the actual environment is always recommended. This will minimize any potential issues from variations between manufacturers' devices.

Software-Assisted Manual Site Survey

The manual site survey technique discussed earlier may be adequate for smaller deployments or for organizations that have limited resources and budget. Manual site surveys can also be accomplished with the aid of commercial software programs designed specifically for this process. These programs vary in cost, complexity, and features and contain many advanced features:

- Capability to perform both passive and active surveys
- Capability to import floor plans with support for many graphic formats, including JPG, BMP, and CAD formats
- Capability to record critical data such as signal and signal-to-noise ratio
- Visual representation of RF signal propagation of surveyed areas
- Post-survey offline analysis of collected data

The Passive Wireless Site Survey

A *passive site survey* consists of monitoring the air and recording the radio frequency information from all wireless access points and wireless client stations or devices in the "hearing range" of the surveying station and includes radio frequency information from your own and neighboring devices. This type of site survey does not require an IEEE 802.11 authentication and association to an access point, and no traffic is passed between the survey station and the access point. A passive survey will provide an overall snapshot of the RF in use in or around the location, including RF noise and other IEEE 802.11 wireless networks in the area. It is used to get an overall picture of the wireless network access points that are transmitting within the area being surveyed. Figure 8.20 shows an example of a commercial site survey application in passive mode.

Since passive site surveys do not require an association to an access point, all radio frequencies will be detected, displayed, and recorded. Most commercial software applications with this functionality will have the capability to filter the coverage map on specific access points from the information that was recorded.

The Active Site Wireless Survey

An *active site survey* consists of a survey device such as a notebook computer associating to an access point prior to taking readings and collecting the data. Some claim this type of manual survey will provide more accurate results because of this direct IEEE 802.11 authentication and association to an access point. The association of the survey device to an access point will actively send and receive some basic RF information to and from the access point, allowing the information about signal strength and noise levels to be recorded. Figure 8.21 shows a commercial site survey performing an active site survey.

FIGURE 8.20 Ekahau Site Survey (ESS) showing a passive survey

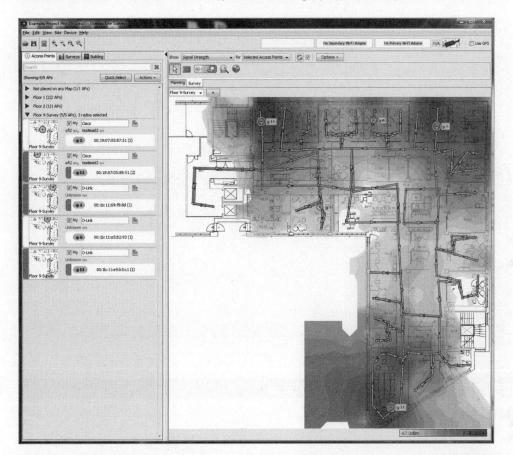

Performing either a passive or active site survey using software involves many of the same steps discussed in the section "Performing a Manual Radio Frequency Wireless Site Survey" earlier in this chapter. The main difference is the features and capabilities that are available in these commercial software packages. These elaborate features will eliminate the need to manually document and record all the information obtained while walking the area because the program handles this task for the surveyor.

In Exercise 8.3 you will explore the Ekahau HeatMapper site survey program. If you have a floor plan of your area, you can perform an actual manual wireless site survey.

FIGURE 8.21 AirMagnet Survey showing an active site survey

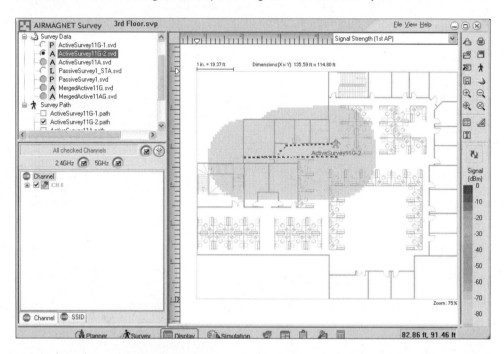

EXERCISE 8.3

Installing Ekahau HeatMapper

In this exercise, you will begin by installing the Ekahau HeatMapper, a free Wi-Fi site survey tool.

Ekahau HeatMapper has the following specifications:

- Wi-Fi coverage mapping tool
- Free of charge
- No official support provided

The requirements for Ekahau HeatMapper are as follows:

- Laptop computer running Windows Vista, Windows 7 or Windows 8 (64-bit or 32-bit)
- 1 GB of RAM
- 2 GB hard disk space
- 1 GHz or faster processor
- Wireless network adapter (internal or external)

The Ekahau HeatMapper program is available free of charge from Ekahau at www.ekahau.com. It allows you to see your wireless network: It shows the wireless coverage of any access point on the map. It also locates all the audible access points and shows their configurations and signal strength—in real time and on the map (floor plan).

1. Download the Ekahau HeatMapper program, which at the time of this writing has the filename Ekahau-Heatmapper-Setup.exe, from the Ekahau website and execute the program. Depending on the version of the Microsoft OS you are using, you may see a User Account Control dialog box. Click the Yes button to continue. The Ekahau Heat-Mapper Setup wizard will appear.

2. Click Next to start the installation process. The license agreement will appear.
3. Read the license agreement, and then select the I Agree button. The Choose Install Location screen will appear.

EXERCISE 8.3 *(continued)*

4. Click Next to accept the default installation folder. Depending on the version of the Microsoft OS you are using, you may see a Windows Security dialog box appear.

5. Click the Install button to continue if you choose to continue with the installation. The Ekahau HeatMapper program will now be installed on your computer. The Installation Complete screen will appear.

6. Click Finish to complete the installation process and exit the installer program. The Ekahau HeatMapper program is now installed on your computer, and the program will start. The Ekahau HeatMapper home screen (shown in Figure 8.16 earlier) will appear.
7. Click the "I have a map image" box to start using the program.
8. Browse to the location containing the floor plan image that you plan to use for the survey.
9. Select the image and click the Choose Image button. The Ekahau HeatMapper home screen will return, displaying the floor plan image you imported. If the Quick Help screen appears, you can click the small arrow button in the right center of the screen to minimize the help window.

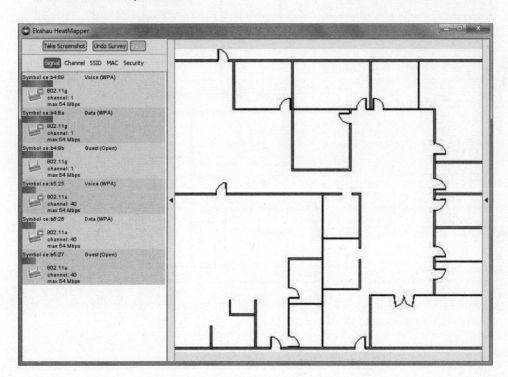

EXERCISE 8.3 *(continued)*

10. Click on the area of the floor plan where you are currently located. You will notice access points on the left side of the screen that the HeatMapper program detects.

11. Take your laptop and start walking around the facility slowly.

12. Left-click your current location on the map as you walk.

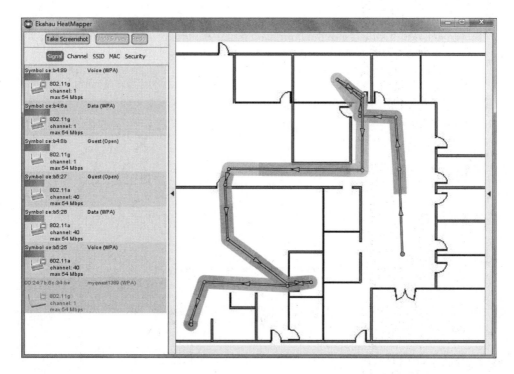

13. Right-click when you stop walking. After right-clicking, you will see the combined coverage heat maps of all access points.

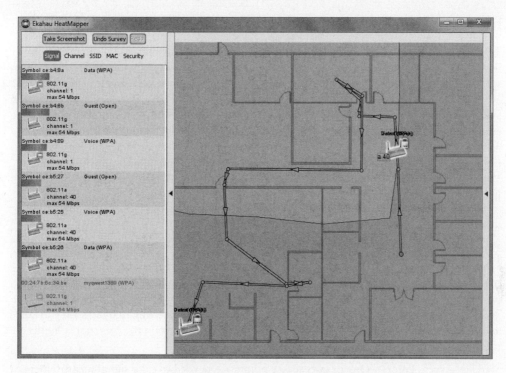

14. Move the mouse over an AP to see the coverage of a single access point.

15. Hide and show the AP list by clicking the small arrow button on the left side center of the screen.

16. Experiment with the HeatMapper program by clicking on the various access point options: Signal, Channel, SSID, MAC, and Security.

17. Click the Take Screenshot button to save an image of the current survey.

18. Close and exit the Ekahau HeatMapper program.

Manual Site Survey Toolkit

To perform a manual site survey, you need a toolkit containing essential components. A floor plan or blueprint is required to record data collected during a manual site survey. If a software program is used, both an electronic version of the floor plan and a paper printout are necessary. Site survey toolkits will vary in complexity and cost. Planning the equipment and tools you will need for a manual site survey is very important. I recommend you take everything you will need and not take items you will not need. Taking items you do not need creates an unnecessary burden. Many times you may be the only person performing the physical survey, and therefore you need to realize your physical and time limitations. You can create your own site survey kit based on what you will need, or you can purchase from a manufacturer a kit that is built specifically for this purpose.

A variety of items might be included in a wireless site survey kit. Now take a look at some of the most common items, and keep in mind that the kit will vary based on the site survey and the person performing the site survey.

Spectrum Analyzer (Optional but Highly Recommended)   A spectrum analyzer is an optional but recommended item that may be used in an RF wireless site survey. This device will sweep the area to look for devices that may cause interference with the deployment and locate all the radio frequency information in the area. Spectrum analyzers vary in cost and features, as discussed earlier in this chapter. Although "optional," this tool is highly recommended.

Wireless Access Points   One or two access points are needed to take signal measurements during the manual site survey process. It is important to use access points from the same manufacturer and if possible the same model that will be installed, to simulate what will be used in the actual installation. If this cannot be achieved, try to get an access point that is as close as possible to those that will be used in the deployment.

> I recommend that you set the test access points' output power level at about 50 percent or less during the site survey. This usually requires lowering the default power, which in many cases is set at maximum. This will allow for adjustments after installation to help compensate for potential differences from when the space was originally surveyed. The output power settings used will be determined by the individual performing the site survey. It is also important to keep in mind that if you are performing a site survey for both the 2.4 GHz and 5 GHz bands, the amount of RF transmit power as well as the gain of the antennas must be taken into consideration because of the difference in wavelength, which will in turn affect the range.

Wireless Client Device Such as a Notebook Computer, Wi-Fi-Enabled Smartphone, or Tablet   A device to take measurements from the client side is an important component of the site survey kit. This can be a notebook computer or tablet with appropriate software to take signal measurements. Size and weight should be considered as well. A Wi-Fi, VoIP handset may also be used for locations that are considering WLAN Voice over Internet Protocol (VoIP) capabilities. It is best to try to match the survey client devices with those that will actually be used in the wireless devices connecting to the wireless network. However, this is not required.

Portable Battery Packs or Extension Cords   Portable battery packs to temporarily power the wireless access points are highly recommended. However, if these are not available, extension cords are a good substitute. The disadvantage of extension cords is that they require accessible AC power outlets, which might not be available at all sites. Also, they may be difficult to use and can be cumbersome. An extra battery pack for the survey device is also recommended. Battery packs specifically designed for wireless site surveys are available from a variety of manufacturers. Most battery packs will last longer than the surveyor.

Various Antennas (If Required)   If antennas other than those mounted directly to an access point will be used in the deployment, a variety of them should be on hand for testing during the site survey process. If 802.11n technology will be tested during the site survey, appropriate 802.11n antennas will be required as part of the site survey kit. If you do not plan to use different antennas, there is no value in taking them along during the survey.

Temporary Mounting Hardware   Mounting hardware for temporarily mounting access points and antennas should also be considered, including expandable poles, brackets, tape, and nylon tie straps. You may have to be creative when it comes to temporary mounting. It is critical to make safety a number-one priority when performing this task.

Measuring Device, Either Tape or Wheel   A measuring tape or wheel is recommended for measuring distance to wiring closets and distance between access points and any other areas where distance is important. Laser measuring tools are also available, but many are capable of only short distances. If possible, it would be beneficial to have both a tape (or wheel) and a laser measuring device.

Digital Camera   As they say, a picture is worth a thousand words. Using a digital camera to take photographs of situations that may be difficult to explain in writing is a great help. This also adds quality to wireless site survey reports. Digital camera technology is now inexpensive, and most smartphones and tablets have this capability built in. I recommend taking many digital photographs when possible, especially in areas that may be difficult to describe in writing. It is better to have the photographs if the need arises than to wish you had them. Photographs of potential access point locations are valuable should physical limitations of the facility come into question. Also, photographs of existing devices are useful to determine if they are damaged or poorly deployed and need to be revisited as part of the new device deployment.

Pens, Pencils, Markers, and Paper   Documentation is a critical part of wireless site surveys as well as all other areas of the network. Keeping thorough and accurate notes will allow the surveyor to document areas of importance. Even with the advanced portable technology you have in your possession, it is always a good idea to have paper and writing utensils. Markers can also be used to identify the locations where access points will be installed.

Ladders or Lifts Ladders or lifts are required for mounting temporary access points and antennas. The height of the ladder or lift will depend on the mounting locations for the access points. Keep in mind that the surveyor may not be allowed to use a ladder or lift at certain locations or in some areas. You may have to work with the facility manager to get someone who will be able to perform this task.

> If heavy equipment is required to install an access point in the final location, several items should be considered. What is the cost of the equipment and does this cost fit into the budget? Will the equipment physically fit into the area where the access point needs to be installed? Will any special training be required to operate the equipment, or do you need to factor in the cost to hire an experienced operator? In some cases an alternate location to mount the access point may be a more cost-effective solution.

Movable Cart A movable cart or dolly to move all equipment used for the site survey will ease the burden of getting the equipment moved safely around the facility. Depending on the amount of site survey equipment and the number of access points, batteries, poles, ladders, and so on, it can be a cumbersome task to move all this around. This is especially true in larger facilities, and carts make this task much easier.

Performing a Predictive Modeling Site Survey

A predictive modeling site survey can be an accurate way to design a wireless network without having to spend time on site performing a physical walk-through and testing of the entire location. A *predictive modeling site survey* is a software-based site survey in which a floor plan (map) or drawing of the area to be surveyed is imported into a software program. The program then simulates the radio frequency propagation from access points by using the attenuation properties of various structural components and provides a coverage map based on the information that was input into the program. The surveyor can trace over walls, windows, doors, and other elements of the physical location in order to estimate the RF attenuation and show how the RF propagates based on the environment. This type of site survey can be performed in several different ways depending on the equipment used. You can use a stand-alone commercial software program designed specifically for this purpose, or in some cases, manufacturers build this site survey functionality directly into a wireless controller. Some manufacturers now have Internet-based wireless site survey programs, and the design is performed online. You will see more information on Internet-based survey tools later in this chapter.

A predictive modeling site survey requires the wireless engineer or surveyor to input a floor plan drawing of the facility, such as a CAD, JPEG, BMP (or other format) file, directly into the software program, in the controller, or to the online program. Then, details such as the attenuation values of the facility are added. The details can include the following information:

- Type of walls, such as drywall, brick, or poured concrete
- Thickness of walls
- Types of windows, including glass, thickness, and coating
- Type of doors, such as hollow core, solid core, fire doors, wood, steel
- Location of certain types of furnishings, such as cubicle offices
- Height of the ceiling

One thing to keep in mind about a predictive modeling site survey is that the accuracy and final results are only as good as the information that was input into the program. Basically it equates to "garbage in, garbage out." It is essential for the surveyor to use accurate information about the location, including the attenuation value of all the building materials. Figure 8.22 shows a predictive modeling wireless site survey tool.

FIGURE 8.22 Motorola LANPlanner showing a predictive modeling site survey

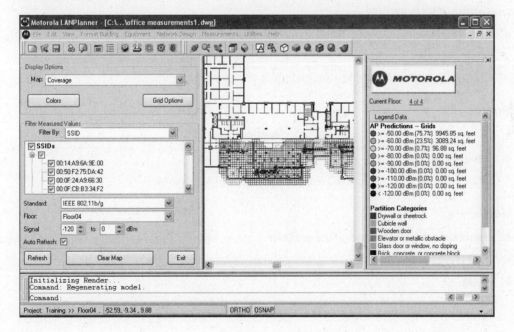

Here are some the advantages and disadvantages of using the predictive modeling site survey process:

- Advantages
  - Limited time on site.
  - Does not require a complete physical walk-through for testing, but a site visit is recommended.
  - Allows for easy adjustment of access point locations and settings.
  - Can model different scenarios.
- Disadvantages
  - Surveyor may be unfamiliar with the location's physical characteristics.
  - Accuracy limited to data input within the software program.
  - Requires extensive knowledge of physical properties of the installation area, including attenuation values.

In Exercise 8.4, you will explore a predictive modeling site survey program.

EXERCISE 8.4

Installing RF3D WiFiPlanner2

In this exercise, you will install the RF3D WiFiPlanner2, a cost-effective, full 3D design tool for planning or upgrading wireless networks.

RF3D WiFiPlanner2 has the following system requirements:

Software:

- Windows 7/Vista, Windows XP SP2, or Windows 2000 SP3
- Internet Explorer 5.01 or later (for online licensing)
- Microsoft .NET Framework 4.0 (free download from Microsoft)

Hardware:

- Processor: Intel Pentium 2.0 GHz or faster
- Display: 1024 × 768 or more
- RAM:
  - 1 GB for small and medium plans (<50 APs)
  - 2 GB recommended for larger plans

The RF3D WiFiPlanner2 demo version is available from Psiber (www.psiber.com).

1. Download the RF3D demo version from the Psiber website. As of this writing, the name of the compressed file is RF3D2.0.18DemoPackage.zip.

2. Extract the files to a folder on your computer.

3. Browse to the folder containing the extracted files and open the readme.txt file to read the contents.

4. Start the setup process by executing the RF3DWifiPlanner2_Installer_Demo.exe program. Depending on the version of the Microsoft OS you are using, you may see a User Account Control dialog box. Click the Yes button to continue. The Installing RF3D WiFiPlanner2 wizard screen appears.

5. Click Next to start the installation process. The Destination Folder screen will appear. Click Next to accept the default location and the program will be installed on your computer.

EXERCISE 8.4 *(continued)*

6. A screen that says "The RF3D WiFiPlanner2 has been successfully installed!" will appear.

7. Click Finish to complete the installation. The program will start and the RF3D WiFi-Planner2 home screen will appear. Read the EULA, click the I Agree radio button, and click OK. The program is now ready to use.

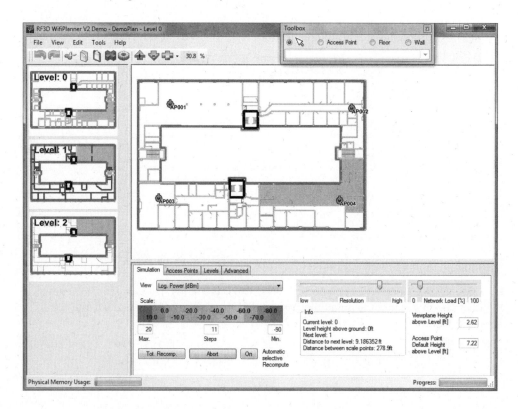

Performing a Predictive Modeling Site Survey

8. Click the Access Point radio button in the Toolbox dialog box.
9. Click within the floor plan image to place a few access points.
10. Click the Wall radio button in the Toolbox dialog box. Click on the drop-down box to view the options for the building material and their attenuation values.

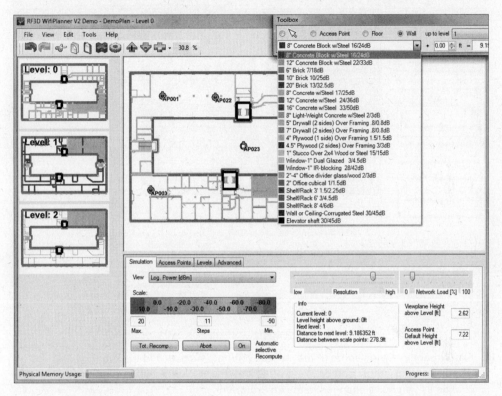

11. Click on the floor plan images on the left side of the screen to view the various RF propagation heatmaps.

EXERCISE 8.4 *(continued)*

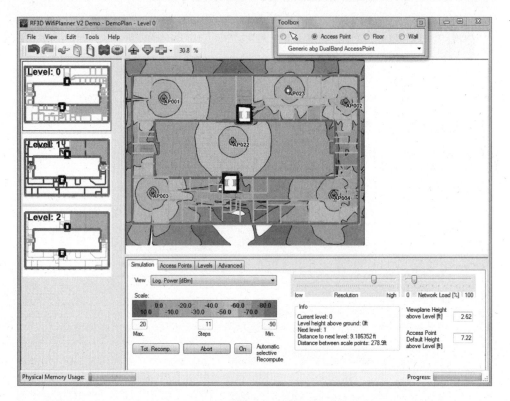

12. Experiment by trying some of the various options, such as adding, moving, and deleting access points and drawing walls and doors. Notice that the heatmaps change as you make changes to the environment.

13. To see their options, click the Access Points, Levels, and Advanced tabs near the bottom of the screen.

14. Click File and then Exit to close the program.

15. Uninstall the program using the Control Panel when you are finished using it. You can also purchase a license online.

There are many different site survey programs to choose from, both commercial and freeware or shareware. You should compare the programs and determine which would be best suited for your environment based on features, cost, and capabilities. Here are many common stand-alone programs for predictive modeling or manual site surveys that are available on the market today:

Site Survey Software Program	Website
AirMagnet Survey/Planner	www.flukenetworks.com
Berkeley Varitronics Systems (BV Systems) Swarm	www.bvsystems.com
Ekahau Site Survey	www.Ekahau.com
Fluke Networks InterpretAir	www.flukenetworks.com
Motorola LANPlanner	www.motorolasolutions.com
Motorola SiteScanner	www.motorolasolutions.com
Psiber RF3D WiFiPlanner	www.psiber.com
TamoGraph Site Survey	www.tamos.com
VisiWave – AZO Technologies	www.visiwave.com

Some enterprise wireless client adapter utilities can also be used in manual RF wireless site surveys to provide basic information such as signal strength and signal-to-noise ratio.

Internet-Based Predictive Modeling Site Survey

Some manufacturers of wireless network equipment have Internet or web-based predictive modeling wireless site survey programs available. This type of program is based on the assumption that you will be using their equipment in your wireless deployment. The programs are robust and free to access. Some also have free online training on their use. The process in using these tools is fairly straightforward. You need to request access or create an account on the manufacturer's website. You will then be able to log in and upload a floor plan map to the website program. Here is a list of such programs:

Site Survey Internet-Based	Website
Aerohive Wi-Fi Planner	www.aerohive.com/planner
Meraki WiFi Mapper	meraki.cisco.com
WLAN Coverage Estimator	www.airtightnetworks.com

>
> **Real World Scenario**
>
> **Hybrid Approach to Wireless Site Surveys**
>
> Another common approach a wireless network site surveyor will use is a hybrid method. This will combine both the physical manual site survey and the predictive model site survey. This method can give "the best of both worlds" and will allow you to check the physical installation location, make notes, place an access point to take signal readings in key areas, and get some spectrum analysis data. You will be able to apply the information collected to a predictive model to get more accurate results.

Performing a Post-Site Survey

Regardless of which survey method is used for the RF wireless site survey, you should conduct a final post-site survey. This task should be included in the initial scope of work. The post–site survey can be a walk-through where you spot check the coverage and signal strength. You should test known or suspected problem areas. This will give you a snapshot of the overall wireless installation. A better and more thorough method would be to do a complete walk-through in every area checking and recording the coverage and providing a coverage map. Using one of the software-based site survey programs previously discussed will allow you to do a passive, active, or predictive site survey and give you an accurate footprint of the overall wireless coverage. A post-site survey will allow you to test the roaming capabilities of different wireless clients and mobile devices that may be used in the actual wireless installation. This type of survey may be time consuming on a large facility, but it will give you not only a footprint of RF coverage but also a baseline for ongoing monitoring and troubleshooting. When performing a post-site survey, use the same devices used to perform the initial manual site survey. Include a walkabout on the exterior perimeter of the building in order to record the RF signal around the building. During the post-site survey, not only can you record the RSSI and SNR of the RF signal, you can detect and record any dead spots or areas of RF interference to validate the design. The goal of the post-site survey is to ensure that the signal level meets the design goal. Most industry experts recommend a threshold of –65 dBm to –67 dBm as a minimum for best performance for both data and voice.

Protocol Analysis

Wireless network protocol analyzers are becoming a common tool, and many network administrators will have one as part of their wireless network toolkit. A *protocol analyzer* allows a network administrator or engineer to view all wireless frames that are traversing across the air within the hearing range of the analysis device. At one time, performing a protocol analysis was a specialty task, and without extensive training few people had the skills to perform it. Along with the evolution of wireless networking technology in recent years, protocol analyzers are becoming more mainstream, more affordable, and easier to use. Many variations of analyzers are available in the market today. Here are some of the manufacturers and their products:

Manufacturer	Product	Website
AirMagnet	WiFi Analyzer	www.flukenetworks.com
BV Systems	Yellowjacket	www.bvsystems.com
Motorola	AirDefense Mobile	www.motorolasolutions.com
MetaGeek	Eye P.A.	www.metageek.net
NetScout	Sniffer	www.netscout.com
Network Instruments	Observer	www.networkinstruments.com
TamoSoft	CommView for WiFi	www.tamos.com
WildPackets	OmniPeek	www.wildpackets.com

Most wireless protocol analyzers require the use of certain network adapters and in many cases a special device driver. I recommend that you verify that you have access to an adapter supported by the protocol analyzer's manufacturer.

Protocol analyzers are available in software programs that can be installed on a notebook computer, and they're are also available in specialty dedicated handheld devices. Many analyzers are feature rich, with the capability to view security information, perform legislative compliance analysis and reporting, and generate a variety of reports in addition to performing protocol analysis or frame decoding. Figure 8.23 shows an example of a wireless network protocol analyzer. Some manufacturers of enterprise wireless equipment provide remote packet analysis capability that allows protocol analysis from any location. Most wireless intrusion prevention system (WIPS) sensors also have this capability.

FIGURE 8.23 OmniPeek by WildPackets identifies nearby wireless networks.

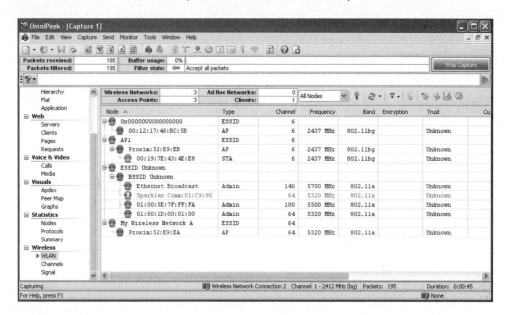

The main goals of a wireless network analyzer, or any protocol analyzer, are to troubleshoot network problems, gather information about security issues, and optimize the network's performance. When it comes to wireless site surveys, a protocol analyzer is a valuable tool for evaluating which wireless devices are currently in the same RF space where the proposed wireless network will be deployed. They can also be used to view the signal strength, security implementations, network name or SSID, and the channels on which access points and other devices are currently operating. An analyzer will show not only access points but any wireless device that may have an impact on the site survey and deployment. The following devices are among those an analyzer is able to locate and identify:

- Access points
- Ad hoc networks
- Wireless bridges
- Mesh networks
- Client devices

In Exercise 8.5, you will explore a protocol analyzer.

EXERCISE 8.5

Installing a Protocol Analyzer

In this exercise you will install the evaluation version of the CommView for WiFi protocol analyzer by TamoSoft.

Protocol Analysis

CommView for WiFi has the following system requirements:

- A compatible wireless adapter. For the up-to-date list, visit www.tamos.com/products/commwifi.
- Pentium 4 or higher.
- Windows XP, Vista, 7, 8, 8.1; Windows Server 2003, 2008, 2012 (both 32- and 64-bit versions).
- 10 MB of free disk space.

Microsoft Windows 7 Professional was used for this exercise. The steps may vary slightly depending on the version of the operating system used.

1. Download the CommView for WiFi setup program from the TamoSoft website at www.tamos.com or at the download page for this book, www.sybex.com/go/mobilityplus. As of this writing, the name of the compressed file is ca7.zip.

2. Extract the files to a folder on your computer.

3. Browse to the folder containing the extracted files and open the readme.txt file to read the contents.

4. Start the setup process by executing the setup.exe program. Depending on the version of the Microsoft OS you are using, you may see a User Account Control dialog box. Click the Yes button to continue. The CommView for WiFi Setup Wizard will appear on the screen.

5. Click Next to start the installation process.

6. The license agreement will appear. Read the license agreement, select the "I accept the terms in this license agreement" radio button, and click Next. The Select License Type screen will appear.

EXERCISE 8.5 *(continued)*

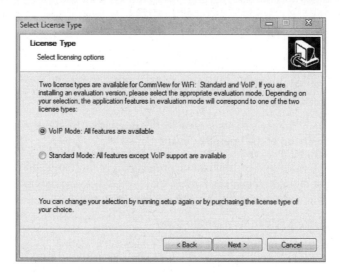

7. Accept the default and click Next to continue.

8. The Destination Location screen will appear. Click Next to accept the default location.

9. The Additional Settings screen will appear. Select the appropriate language and deselect the "Launch CommView for WiFi once the installation has been completed" check box.

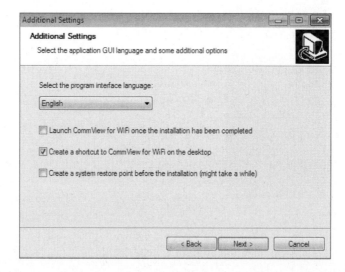

10. Click Next; the Ready To Install screen will appear. Click Next to start the installation.

11. After the installation is complete, the CommView For WiFi screen will appear. Click Finish to complete the installation.

Note: I recommend that you have a Wi-Fi adapter that is on the CommView compatibility list installed prior to completing the next step.

12. Click the CommView for WiFi icon on your Desktop to start the program. Depending on the version of the Microsoft OS you are using, you may see a User Account Control dialog box. Click the Yes button to continue. The program will start and the Driver Installation Guide dialog box will appear in the foreground for the driver installation. Verify that the "I want to install the driver for my compatible adapter" radio button is selected.

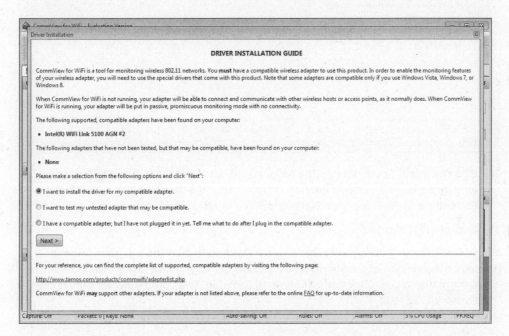

13. Click the Next button to continue if you choose to continue with the installation. The Driver Installation dialog box will appear. Verify that your Wi-Fi adapter is selected.

EXERCISE 8.5 *(continued)*

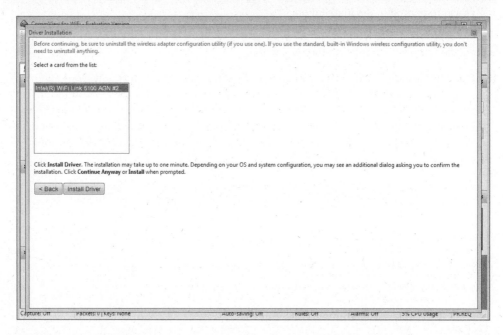

14. Click the Install Driver button. The program will attempt to install the device driver for your supported wireless network adapter. After the device driver is installed the CommView for WiFi program will instruct you to restart the application.

15. Click the Close button.

16. Click the CommView for WiFi icon on your Desktop to start the program. Depending on the version of the Microsoft OS you are using, you may see a User Account Control dialog box. Click the Yes button to continue. The CommView for WiFi home screen will appear.

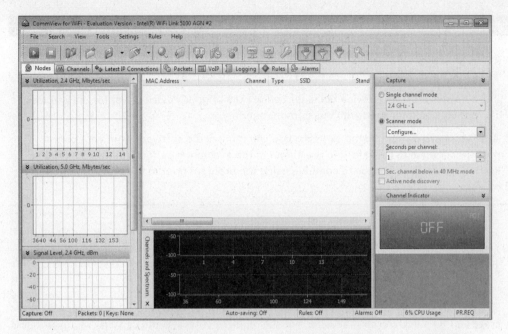

17. Click on the arrow in the upper-left corner of the program window or click the File Menu and select Start Capture to start a capture with the program. The Evaluation Version dialog box will appear.

18. Click the Continue button to use the evaluation version of the program. Select the channel you wish to view on the right-hand side of the screen. The screen will show all wireless network devices captured when the channels were scanned.

19. Click various buttons to see the features associated with this program. When you're finished with the demonstration, exit the program by clicking the File drop-down menu and choosing Exit.

> **EXERCISE 8.5** *(continued)*
>
> 20. A "Thank you for trying this evaluation version" dialog box will appear. Click OK to close the program.
> 21. If necessary, using Device Manager, restore the original device driver for the wireless network adapter used with this demonstration.
>
> After you have completed the exercise, you can remove the software from your computer or continue to use it for the remainder of the evaluation period. You can purchase a license online if you choose to continue using the program prior to the evaluation period ending.

RF Coverage Planning

As mentioned previously, one goal of a wireless site survey is to determine areas of RF coverage and RF interference. The details of the required coverage and capacity in any wireless network are part of the wireless network design, which involves determining the need, use, and business requirements of the network. This was discussed earlier in this chapter. These factors need to be taken into consideration during a physical site survey:

- Purpose of wireless network: data, voice, video and/or location services
- Size of the physical location
- Number of users and wireless devices, including bring your own device (BYOD) users
- Obstacles that may impact RF signal propagation
- Radio frequency range
- Bandwidth requirements of applications to be used
- Selected radio frequency of wireless network hardware

The planning process will provide some of the answers, whereas a site visit and on-site testing will determine others, such as obstacles, signal propagation, and RF range.

Infrastructure Hardware Selection and Placement

In addition to identifying areas of RF coverage and interference, a site survey will determine the best locations for wireless access points and other infrastructure devices. The correct placement of these devices is important in order to allow clients and devices to benefit fully from the deployment of the wireless network. The location of these devices is traditionally based on the following criteria:

- Desired RF coverage
- Required bandwidth
- Aesthetics requirements
- Applications, both hardware and software
- RF cell overlap
- Channel reuse patterns

Mapping out the infrastructure device placement is considered part of the planning and design process. Manual, software-assisted, or predictive modeling site survey processes will help identify the proper locations based on the items mentioned. In most cases, a preliminary visit to the location is highly recommended regardless of the size and the site survey method that will be used. Knowing the physical location will benefit the entire site survey process because it will help identify areas of concern (see Figure 8.24).

FIGURE 8.24 Motorola Planner

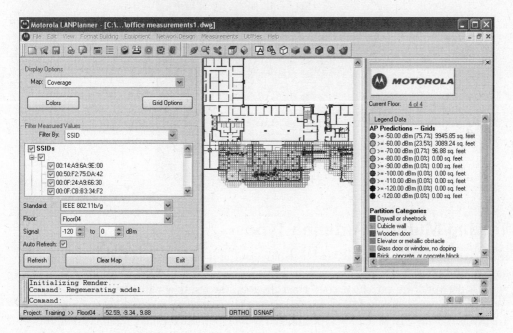

Testing Different Antennas

In Chapter 3, we looked at various types of antennas, including omnidirectional, semidirectional, and highly directional models. The characteristics and features of these antennas were discussed, as well as the best use for each based on a specific scenario. Here we'll consider how best to conduct antenna testing in a wireless site survey.

As part of a manual site survey, it may be necessary to test different antennas to determine the best RF coverage. This usually requires the surveyor to connect and temporarily mount various types of antennas to access points in order to determine the proper radiation pattern and to verify RF coverage within the desired area. Some predictive modeling site survey programs will allow you to change different antennas so you can see how that will change the RF propagation pattern. A temporary antenna mounting example is shown in Figure 8.25.

FIGURE 8.25 Expandable light pole used for temporary mounting and testing of yagi and patch antennas

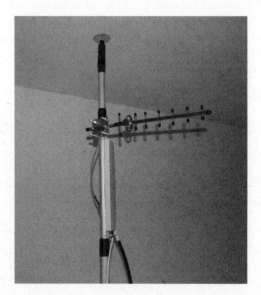

Testing Multiple Antenna Types

If different antennas will be used in the deployment, it is important for the surveyor to have several types of antennas as part of the site survey kit. Some organizations that deploy wireless networks are extremely concerned about aesthetics; they will allow only specific types of antennas to be used. For example, if access points with removable antennas are used, in many cases the only type allowed will be omnidirectional antennas that are attached directly to the access point with a gain of about 2 or 3 dBi. Other deployments may use access points in which antennas are permanently attached and cannot be removed or changed.

Keep in mind that polarization of antennas must be taken into consideration during a manual site survey. Polarization of antennas for infrastructure devices is critical, and tests should mimic as closely as possible what will actually be installed. With the variety of wireless client devices that may be used in a wireless network, it is a challenge to predict the polarization of all the devices that might be used. However, it is advisable for the surveyor to take this into consideration during the site survey.

Choosing the Correct Antennas

Choosing the correct antenna to be used in a specific deployment is part of the wireless design and site survey process. Many factors play a role in determining which antenna will be best for the application. Some locations have strict requirements about the type of antenna that may be used. Therefore, the surveyor may have to work with specific antennas to ensure proper RF coverage. Take the following factors into consideration when choosing antennas:

Manufacturer's Recommendations The manufacturer of an access point may recommend only a specific type of antenna. If this is the case, it is important to perform the site survey with the same type of antenna.

Customer Requirements A customer may require that only specific types of antennas be used. In this case, the survey should be performed with the type of antenna required by the customer. For example, a deployment consisting of walled offices may require the use of thin access points. Usually this type of access point uses an omnidirectional internal antenna.

Environmental Conditions The environment where the wireless network is installed may also determine the type of antenna to be used. If the location is a factory with harsh environmental conditions, that could have an impact on the type of antenna and may also call for an enclosure for the access point.

Aesthetics Many organizations are sensitive to the type of devices that are seen by customers and clients. Therefore, aesthetically pleasing antennas may be required by the customer in order to be a good fit for the location in which they will be used.

Required Coverage The required RF coverage will also affect the choice of antenna as well as the gain of the antenna. For example, a large office area may require the use of omnidirectional antennas that are physically attached to access points. A manufacturing facility may require semidirectional antennas to cover areas of the manufacturing floor. Keep in mind that in addition to being passive bidirectional amplifiers, antennas shape the coverage by either an increase or decrease in antenna gain.

Number of Wireless Access Points The number of access points to be installed in a location will also be a determining factor in the type of antenna to be used. An office building with a combination of walled offices and cubicles may have a dense deployment of access points with a limited number of users connecting to any particular access point. Omnidirectional antennas connected directly to the access point may be an adequate solution for this type of deployment. The purpose of the wireless network also affects the number of access points. For example, a wireless network that will support voice clients will typically require more access points than a network supporting only data. If the wireless network will support location services such as RFID tracking or wireless intrusion prevention mitigation, additional access points with semidirectional antennas may be required due to the design aspects of the technology.

Physical Geometry of Location   The attributes of the physical location will have an effect on the type of antenna to be installed and therefore should be tested during the site survey. This includes propagation of the signal and attenuation of obstacles. In the case where a building has long hallways or corridors, a yagi antenna would be a good candidate for a solution.

Regulatory Agency and Local Codes   When selecting antennas, both indoors and outdoors, it is important to become familiar with and completely understand any radio frequency limits or regulations that apply from the governing body or regulatory agency or any local codes that exist. In addition to the radio frequency codes, these may include, for example, height limitations and proximity to airports. Violations of any of these may result in large fines or other penalties.

Wireless Channel Architectures

There are two common types of channel architectures available today that pertain to IEEE 802.11 wireless networking. In wireless technology, channel architecture is the design, layout, or channel plan in use. In the 2.4 GHz band, for example, the use of nonoverlapping channels 1, 6, and 11 would be considered a channel architecture. The two wireless network architectures are commonly identified as multiple-channel architecture (MCA) and single-channel architecture (SCA).

Multiple-Channel Architecture

Most wireless network deployments today use *multiple-channel architecture (MCA)*. This type of installation will use access points set to different RF channels to avoid overlapping channel interference, as shown in Figure 8.26. A channel plan may be used with access points set to specific channels, or in many cases automatic channel selection allows the devices to choose the best channel in which to operate.

FIGURE 8.26 An example of multiple-channel architecture deployment

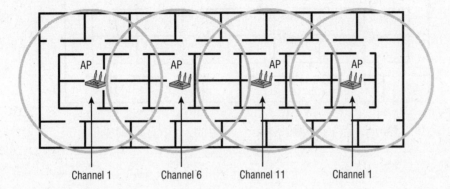

Single-Channel Architecture

Single-channel architecture (SCA) is a wireless networking technology available from only a couple of manufacturers. SCA allows all access points to communicate on the same RF channel. The controller that the access points are connected to manages these access points and helps to avoid co-channel interference. In single-channel deployments, not all access points are transmitting at the same time. The controller will determine which access points can transmit simultaneously based on the wireless devices that are in a specific area. A wireless site survey for SCA equates to providing radio frequency coverage based on the access point placements as the wireless controller manages all the radio frequency.

There are a few terms you should know with respect to single-channel architecture: *stacking*, *spanning*, and *blanketing*. They all refer to a means of managing coverage in a single-channel architecture.

For example, let's look at a three-story building. Each floor in the building may be assigned a channel to use; with SCA architecture, this is known as *stacking*. The first floor would be set to channel 1, the second floor would be set to channel 6, and the third floor would be set to channel 11. Since all access points on the same floor are set to the same RF channel, co-channel interference or overlapping channel interference is not a significant issue because of how the SCA technology works.

Using single-channel architecture may help save some time when it comes to the site survey. It is best to follow the manufacturer's recommendations for deployment for this type of system. Figure 8.27 shows an example of single-channel architecture.

FIGURE 8.27 An example of single-channel architecture deployment

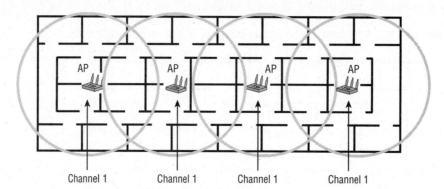

Wireless Device Installation Limitations

The installation limitations of any device that will be installed and used on the wireless network need to be evaluated. Sometimes things look great on paper, but when it comes time to actually perform the installation, it may be a little different. Various wireless devices may have the following installation limitations:

Wireless Access Points   An installer may run into limitations when it comes to mounting access points or other infrastructure devices. For this reason, the surveyor must pay close attention to the location where these devices will be mounted. The type of ceiling and mounting hardware needs to be considered. A wireless site survey should include a physical walk-through evaluation of the proposed mounting locations to observe any issues that may affect where access points, bridges, or repeaters can be mounted.

Antennas   If antennas other than those that are designed to be connected directly to an access point are used, special circumstances may need to be taken into account:

- Mounting issues
- Cabling issues
- Aesthetics
- Height restrictions

Ethernet/PoE   As mentioned earlier in this chapter, the maximum distance an Ethernet unshielded twisted-pair cable can run is 328 feet, or 100 meters, per the IEEE 802.3 standard. Part of the wireless site survey process includes verification of the distance infrastructure devices will be mounted from the wiring closet to be certain they do not exceed the maximum. The capabilities of the Ethernet system must also be evaluated to verify the capacity of the wiring system. This will ensure that there is adequate connectivity for the new wireless infrastructure devices.

If PoE will be used, it is important to verify that the infrastructure will be able to support the number of devices that require DC power from the PoE infrastructure devices. If the current wired infrastructure is not PoE compliant, the infrastructure may need to be upgraded prior to wireless deployment. If an upgrade is not feasible, an alternate solution such as single port power injectors or patch panels may be required.

Local Code Restrictions In some cases, places like elevator shafts and stairwells have become areas where local codes or regulations will not allow devices to be installed. This must be identified up front for expectation purposes, especially where voice over wireless LAN is used so that users are aware of the limitations that may be beyond the survey's or installer's control. Also, it is important to be aware of the fact that some fire codes may require plenum-rated (fire-retardant) cable and devices in certain places where wireless networks will be installed.

Site Survey Report

The survey report should be a complete document itemizing all components of the wireless site survey. This includes the business aspects of the wireless site survey as well as the physical aspects. This report should include but not be limited to notes, charts, graphs, photos, test results, and any other pertinent data that will have an effect on the wireless network deployment. Most reports will be customized based on the individual needs and requirements of the customer. Most commercial site survey application programs have built-in reporting features that can be included in the site survey report. Figure 8.28 shows a sample page from a commercial site survey application with report-generation features.

The main content of the site survey report should include the following items:

- Customer requirement analysis
- Radio frequency interference source analysis
- Radio frequency coverage analysis
- Device capacity and application analysis
- Infrastructure device placement and configuration information

The RF dynamics of many environments are constantly changing as more devices using radio frequency for communications become a part of the wireless network or share the same RF space. It is a wireless network engineer's responsibility to understand that site requirements may also be constantly changing. Therefore, the RF site survey is becoming an ongoing process, and the site survey report can be seen as a living document.

You can download a sample site survey report from Ekahau at this book's companion website:

www.sybex.com/go/mobilityplus

FIGURE 8.28 Ekahau site survey built-in reporting features

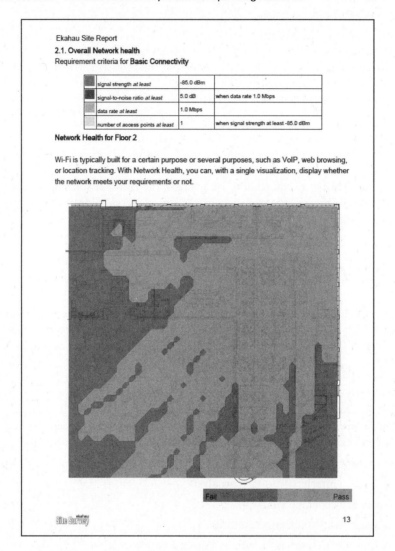

Summary

In this chapter, we explored the business aspects of wireless network site surveys. The objective of a site survey is to find areas of RF coverage and interference, understand capacity planning, and determine hardware installation locations. We explored some of the factors that determine the complexity of a site survey.

You saw the importance of gathering information as well as the types of information required to successfully perform a wireless site survey and design a wireless network. Anyone designing a wireless network needs to understand completely the expectations of how the network will perform in the environment before deploying the network. These expectations will be met by asking the right questions of the right people—managers and users of the wireless network—because they are the ones who will be using it. Having the appropriate documentation is a key element to a successful deployment. Accurate documentation will also help streamline some of the site survey process and the deployment of wireless networking hardware. Accurate documentation will help with the installation of cabling and access points and minimize questions about the installation.

It is essential to know the location and type of existing networks, both wired and wireless. If existing wireless networks are in place, you must determine what role, if any, they will play in the new deployment or upgrade.

In this chapter you learned about RF coverage and capacity requirements and the factors to be taken into consideration to ensure proper coverage throughout the location where the wireless network is installed. Client connectivity requirements and other considerations were discussed. The types of wireless client devices that will be used, such as notebook computer, tablet, or bar-code scanner, must be determined. Antenna orientation to ensure correct polarization will need to be considered during the site survey process. We also looked at the type of antennas commonly used in a particular deployment scenario.

The wireless site survey process is one of the most important components of a successful wireless deployment. Many areas need to be considered, including an RF spectrum analysis to determine RF interference sources that may affect the wireless deployment. Although a spectrum analysis can be considered optional, it is advisable to complete one whenever possible. Spectrum analyzers are available in various types, such as instrumentation devices and USB adapters, and now may be part of the wireless access point.

In this chapter we compared Wi-Fi and non-Wi-Fi sources of interference, both of which can have an effect on the performance of a wireless network installation.

We looked at the various types of wireless site surveys that can be performed, including manual and predictive modeling. We also explored the two types of manual site surveys, passive and active. You saw some of the items that may be included in a survey toolkit as well as temporary mounting examples for access points and antennas. We looked at the use of a protocol analyzer in a wireless site survey and how one could be used to identify existing wireless networks in the area of concern.

Finally, you learned about both multiple-channel architecture (MCA) and single-channel architecture (SCA) and the differences between them. Best practices for hardware placement were also reviewed, including access points and antennas, as well as some of the limitations that may be encountered during the wireless network site survey process.

Chapter Essentials

Understand the business requirements of a wireless network. Be familiar with the business information required for a successful wireless network site survey. These business requirements include bandwidth needs, expected coverage area, and hardware and software applications used (for the devices and the different technologies). Understand that the site survey process will vary based on the business model in which a wireless network is deployed. These will include enterprise, manufacturing, healthcare, and public access, to name a few.

Understand the interview process. Know who you should interview and the type of questions to ask during the site survey and design process. This will help ensure a more successful wireless network deployment.

Identify the importance of site-specific documentation. Know the various types of documentation required based on the business model for the wireless network site survey. This includes blueprints and floor plans as well as other important documentation.

Know the importance of identifying existing networks. Understand the details of existing wired and wireless networks and be able to define the characteristics of both, such as wiring closet location and power requirements.

Understand the application requirements for the wireless network. Know that different applications, both software and hardware applications, will have an effect on the wireless network design and the site survey must take this factor into consideration.

Understand RF coverage and capacity requirements. Know the factors involved with providing adequate RF coverage within a wireless network deployment. Understand that with a wireless site survey, capacity requirements must be met to ensure acceptable performance.

Understand the need for an RF spectrum analysis and how to locate sources of interference. An RF spectrum analysis will allow you to "see" RF in an area proposed for a wireless network. Identify different types of RF interference that can have an effect on a wireless network. Cellular communications should be validated in areas that rely on cellular connectivity.

Know the differences between manual and predictive modeling site surveys. A manual site survey typically requires a complete walk-through and testing throughout the proposed area where a wireless network will be deployed. A predictive modeling site survey may require minimal time on site and is a software-based analysis solution.

Identify two different types of manual site surveys. Know that manual site surveys can be passive or active and understand the differences between each.

Know how a protocol analyzer can be used as part of wireless network site survey. Understand how a wireless protocol analyzer can be used to help identify existing wireless networks during a site survey and how the existing wireless networks may have an impact on the wireless network after deployment.

Be familiar with the limitations of placement regarding wireless infrastructure devices. Explain some of the limitations regarding placement of wireless network devices, including access points, bridges, and antennas.

Understand the factors regarding proper antenna use. Identify the different uses of antennas based on the customer requirements and characteristics of the environment.

Understand different wireless architectures Two common wireless network architectures exist, multiple-channel architecture (MCA) and single-channel architecture (SCA). Know the differences between them.

Chapter 9

Understanding Network Traffic Flow and Control

TOPICS COVERED IN THIS CHAPTER:

✓ Local Area Network and Wide Area Network Traffic Flow

✓ Network Subnets

✓ IP Subnetting

✓ Network Traffic Shaping

This chapter explores network traffic flow in both local area and wide area network installations. You will learn more about Network layer (Layer 3) logical addressing, including IP addresses, subnet masks, and how to subnet a network. Understanding subnetting is an important topic for any network infrastructure engineer or administrator. Although there are plenty of tools available to help design an IP network, a basic understanding of the manual process of IP subnetting is good to have. As the use of mobile devices continues to grow quickly, the need for more network capacity and bandwidth control is becoming a big factor. Different types of technology and controls are available for network administrators and service providers to manage the available bandwidth and allow technology such as voice and video services to effectively provide a quality user experience. We will look at some of the traffic-shaping techniques, including bandwidth restrictions and quality of service.

Local Area Network and Wide Area Network Traffic Flow

To learn about how network traffic is handled, you will need to have an understanding of some basic networking concepts. Even if you are already familiar with some of this information, this will be a great review. In Chapter 1, "Computer Network Types, Topologies, and the OSI Model," you learned about various types of networks, including the local area network (LAN), the wide area network (WAN), the metropolitan area network (MAN), and others. Recall that LANs are typically contained within the same physical area and usually are bounded by the perimeter of a room or building. However, in some cases a LAN may span a group of buildings in close proximity that share a common physical connection. The WAN will connect networks over a much larger geographic area that consists of point-to-point or point-to-multipoint connections between two or more LANs. The WAN links that connect LANs together may consist of leased lines, fiber-optic connections, and even wireless technology. Because the physical infrastructure is scaled to a much larger area within the WAN, routable protocols such as TCP/IP and more "intelligent" infrastructure devices such as Layer 3 routers are required for WAN traffic flow.

Local Area Network Traffic Flow

The networking technology in use, IEEE 802.3 Ethernet, for example, specifies a maximum cable length of 100 meters for 10Base-T networks and is the reason this physical limitation exists for many LANs. Early Ethernet LANs used coax cable for the physical connections that made up the bus topology you learned about in Chapter 1. This physical connection was known as 10Base2 which is 10 Mbps, baseband transmission, and 200 meters. In reality, it is 185 meters, or 607 feet, long. The baseband transmission allowed a signal (computer data) to travel across a wire without requiring any type of modulation. These LANs also used the Carrier Sense Multiple Access with Collision Detection (CSMA/CD) access method. With this access method, the devices that were connected to the wired medium (usually PCs) would first sense the medium to see if it was clear, basically to check that there were no other transmissions occurring. At that time, the device would send the data. If by chance two devices performed this process at the exact same time, a collision would occur. The transmitting devices would know the collision happened (collision detection), a "jam" signal would be sent, and they would back off for a random period of time and try again. Eventually, the information would get to the destination. This technology is considered a single collision domain, which means all nodes can hear all traffic on the network. Figure 9.1 illustrates the CSMA/CD process.

FIGURE 9.1   Ethernet LAN and CSMA/CD

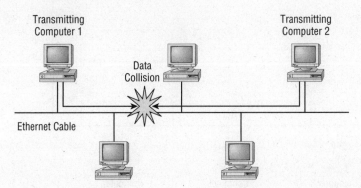

Because of this collision-detection method, a LAN was limited to the number of devices (also known as nodes) that could be connected to it. Common sense will tell you that the more devices connected, the more collisions you have and the poorer the performance will be. One way to extend this type of LAN would be to utilize a device known as a repeater. This infrastructure device physically connects LANs together, extending them beyond the IEEE 802.3 standard distance limit, and has Physical layer (Layer 1) intelligence. The repeater lacks any kind of intelligence to manage network traffic and will forward all of the Ethernet frames it encounters. So basically, the repeater is just a way to extend the length of a physical network cable. Just as in a single LAN scenario, collisions will occur and degrade the network's performance. However, in this situation, with even more nodes connected, the performance will be worse yet because of the increased number of nodes on the larger LAN. Figure 9.2 shows a repeater connecting two LANs together.

FIGURE 9.2 Two LANs connected with a Physical layer repeater

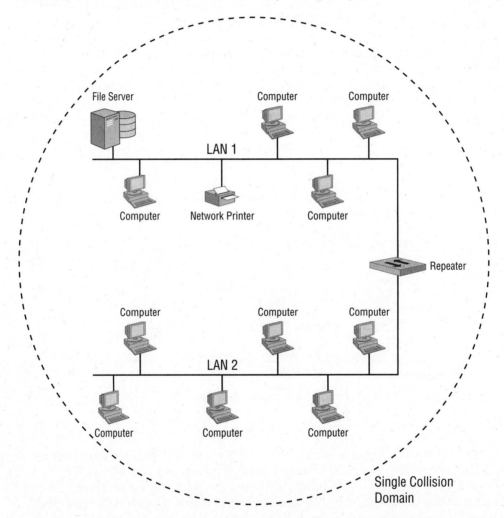

In Chapter 1, you also learned about the star topology. It was mentioned that the star topology is the most commonly used method of connecting devices together on a LAN today. Early LAN installations using the star topology used an infrastructure device known as a hub. Recall that with the high-speed linear bus topology, a break in any part of the cable or bad connection at any point would cause the entire network to cease to function. This is not the case with the star topology. If a cable or connection breaks, only the single connected device would be affected. This type of Ethernet technology is known as 10Base-T, which is 10 Mbps, baseband transmissions using unshielded twisted-pair wire. According to the IEEE 802.3 standard, the physical imitation with this medium is 100 meters, or 328 feet. This is the distance for each connection to the central point of all connections, or the hub. If by chance the hub were to fail, then again, the entire network would cease to function. Like the bus topology, the star is a single collision domain. This means that if a connected device (computer, for example) were to send a transmission to the hub, it would be heard by every port and every device that was connected to the network. Basically, this means that all connected devices are sharing the bandwidth. Whether it be a 10 Mbps hub or a 100 Mbps hub, the entire bandwidth is shared by all devices. Hubs are rarely if at all used with today's LANs for network infrastructure connectivity. Figure 9.3 shows a star topology using a hub and single collision domain. One thing to keep in mind is that small hubs are great to have on hand as a tool for troubleshooting purposes.

FIGURE 9.3 Star topology using a hub and single collision domain

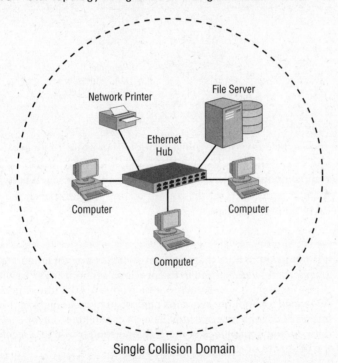

With the evolution of networking technologies, the hub was eventually replaced with the Ethernet switch. Unlike the repeater or hub, an Ethernet switch is a device that functions at the Data-Link layer (Layer 2) of the Open Systems Interconnection (OSI) model. However, the Ethernet switch is Media Access Control (MAC) based and does have the ability to control network traffic flow by processing and forwarding network data. This is because the switch with Layer 2 intelligence has the capability to learn the MAC addresses of the connected devices. Let's look at an example, illustrated in Figure 9.4. The figure shows a 16-port Ethernet switch with two connected devices, computer 1 and computer 8. Computer 1 is connected to port 1 and computer 8 is connected to port 8. At this point, the switch has learned the MAC addresses of the two connected devices. Therefore, when computer 1 sends data to computer 8, it will send the data only to the switch port that computer 8 is connected to, port 8.

FIGURE 9.4 Star topology with an Ethernet switch sending a frame based on learning MAC addresses

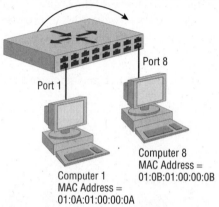

This is a much more efficient way of communication than using a hub because each connected device will essentially have its own dedicated link to the infrastructure, providing the maximum amount of bandwidth. This will allow for much better performance and a better experience for the connected users. At this level, the traffic that passes across the LAN is organized in frames.

 It's important to note that Ethernet switches are not limited to only Layer 2 devices. As you learned in the previous section, Layer 2 switches are Data-Link layer devices that operate from a MAC address perspective. Advanced technology switches operate at higher layers of the OSI model and have Layer 3 and possibly Layer 4 capabilities. Layer 3 switches have data routing capabilities and are able to send traffic to a receiver based on the best path or route.

Wide Area Network Traffic Flow

In the previous section, you saw that network traffic in a LAN is MAC address based and broadcast based; therefore, without the use of additional technology, it will commonly be restricted to devices that are connected within the physical limitations of the connected medium. In this section, you will see how traffic flows in a WAN using higher-layer protocols and technology.

Recall from Chapter 1 that as networking technology evolved, WANs were needed to expand the LAN beyond the physical limits of a single room or building. These LANs began to expand into larger geographical areas and were connected by point-to-point or point-to-multipoint connections. A WAN can be made up of private connections between locations or use a public network infrastructure such as the Internet. Either way, the traffic must extend beyond the physical limitations of the LAN. In addition to using the physical MAC address of a device, we must also use a logical address or IP address of a device in a routable network. We use routers to connect LANs together, which will form the WAN. As shown in Figure 9.5, routers have more intelligence and operate at the Network layer (Layer 3) of the OSI model. A router differs from a hub or a switch in that it will forward traffic from one LAN to another LAN using upper-layer protocols. At this point the traffic is organized into packets. Recall at Layer 2 the traffic is organized in frames.

FIGURE 9.5 Routed WAN with Network layer logical addresses

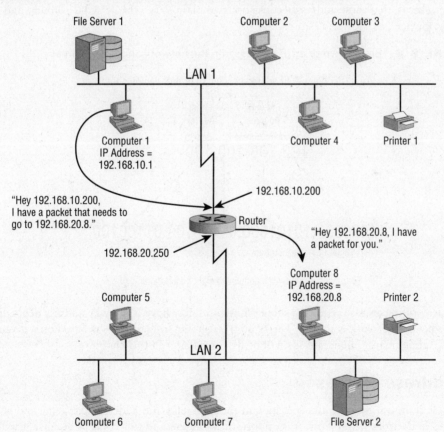

Network Subnets

In Chapter 1, you learned that the Network layer (Layer 3) of the OSI model is where the Internet Protocol (IP) resides, which is responsible for addressing and routing data by determining the best route to take based on what it has learned or been assigned. An IP address is defined as a numerical identifier, or logical address, assigned to a network device that is configured to use Transmission Control Protocol/Internet Protocol (TCP/IP) on a LAN. Network devices that use TCP/IP and are contained within the same LAN are known to be on a subnetwork, which basically is a smaller part of an IP-based internetwork. You learned in Chapter 1 that an IPv4 TCP/IP address is a logical address (IP address) that is a binary 32-bit dotted-decimal address usually written in the form of www.xxx.yyy.zzz. An example of an IP address is 192.168.100.100. This is what is known as a Class C IP address. Recall that each of the four parts is a byte, also known as an octet, and is 8 digital bits in length. There are two main IP address types: public addresses and private addresses. Private addresses are unique to an internal network, and public addresses are unique to the Internet. Recall that these addresses consist of two main parts: the network (subnet) and the host (device), as shown in Figure 9.6.

FIGURE 9.6 Breakdown of an IP address into the network and host address

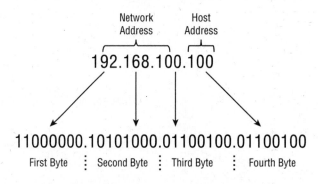

Logical addresses also require a subnet mask and may have a gateway address depending on whether the network is routed. The IP address shown in Figure 9.6 is based on a default Class C subnet mask. Subnet masks will be discussed later in this chapter.

IP Address Classes

In this section, we will take a deeper look at the IPv4 (classful) addressing scheme and how it relates to network traffic flow. It is important to understand this concept because it will help with the more complex concept of traffic routing. TCP/IP became a more popular network protocol with the expansion from the LAN to the WAN. Because TCP/IP is the native protocol of the Internet, the demand was even greater with the development of the World Wide Web and Hypertext Transfer Protocol (HTTP), a Layer 7 Application layer protocol. IPv4 addresses are divided into what are known as address classes. The address classes are identified as Class A through Class E addresses. Table 9.1 shows how the IPv4 address classes are identified with the number of network addresses and the number of host addresses.

TABLE 9.1 IPv4 address classes

Address Class	Network Address	Number of Networks	Number of Hosts
Class A	0–126	128	16,777,216
Class B	128–191	16,384	65,536
Class C	192–223	2,097,152	256
Class D	224–239	Used for multicast	
Class E	240–255	Reserved for future use	

The Class A address of 127 is reserved for testing and diagnostics and therefore is not included in Table 9.1. You may be familiar with the 127.0.0.1 address, which is used as a loopback address to test the IP stack on the local device.

You can see in the table that the Class A addresses offer the highest number of hosts and the Class C addresses offer the lowest number of hosts. The Class D addresses are used for multicast, which is reserved for data that will be sent to groups of devices that are part of a multicast group, and the Class E addresses are not used and are reserved. The Internet Assigned Numbers Authority (IANA) is the organization that was given the authority for IP address allocation. Class A addresses were reserved early on and are held by large company and government organizations, among them the U.S. Department of Defense, International Business Machines (IBM), Hewlett Packard, and Ford Motor Company. The Class B address range was the next to get depleted. As Internet technology advanced, and with the continued need for IPv4 public IP address ranges, it was not easy for individuals or companies to be assigned entire IP address ranges. Instead, IP address allocations for the most part are handled by Internet service providers (ISPs) because with technologies such as NAT, typically a single IP address is all that is required for Internet connectivity.

Special Use IP Addresses

There are several ranges of IP addresses that are considered special use and are not available as publicly assigned IP addresses. These are commonly used within an organization and are referred to as private IP address ranges. Table 9.2 shows the ranges reserved for this purpose.

TABLE 9.2 Private IP address ranges

Address Class	Starting Address	Ending Address
Class A	10.0.0.0	10.255.255.255
Class B	172.16.0.0	172.31.255.255
Class C	192.168.0.0	192.168.255.255

The Request for Comments (RFC) that provides the details of these private address ranges is RFC 1918. Figure 9.7 shows how these address ranges can be used.

FIGURE 9.7 Company network using private IP address range for internal usage

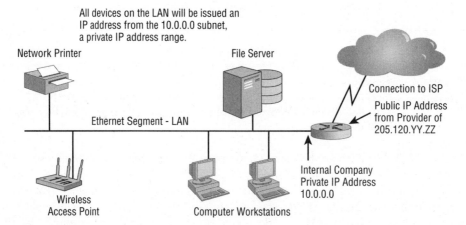

IP Subnetting

As mentioned earlier, a subnetwork is a smaller part, or section, of an internetwork. A subnetwork is typically defined by the LAN boundaries or the LAN segments. There are several reasons you would want to create subnetworks:

- Traffic types (broadcast traffic, for example)
- Performance

- Logical separation of the organization
- Troubleshooting
- Security

One reason to subnet a network is to keep certain types of network traffic contained within the LAN boundaries, such as network broadcast traffic. This type of traffic is heard by every device on the network segment (LAN) and can cause performance issues depending on the number of nodes or devices that are connected to the LAN. Typically, broadcast traffic does not cross a LAN boundary unless the infrastructure devices are configured to do so. Subnets could be defined by floors in a building, separate buildings, or different physical locations. Subnetting will also aid with troubleshooting network problems because you will know in which LAN segment a troubling device is physically located. Security is another reason to subnet your networks. The division of networks will allow for infrastructure devices to be installed, preventing access between subnets.

With newer technologies like IP subnet calculators, dividing a network into subnets is not as complicated as it once was. Although these online and downloadable calculators make the job easier, it is still good to understand the basic concepts of IP subnetting. A basic understanding of the binary, or base-2, numbering system is really all that is required. The binary system allows us to represent numbers using two different symbols, a logical 0 (zero) and a logical 1 (one).

Subnet Mask

Before we explore the process of IP subnetting further, it is important to understand the function of the subnet mask. In an IPv4 address the subnet mask is a 4-byte, 32-bit dotted-decimal address that will identify the IP network that the host is part of. Earlier in this chapter you saw the different classes of IPv4 addresses that can be used within a TCP/IP-based network. In addition to defining an IP address, you need to define a subnet mask. Table 9.3 shows the default subnet masks for the IPv4 system.

TABLE 9.3 Default subnet masks for IPv4

Address Class	Default Subnet Mask
Class A	255.0.0.0
Class B	255.255.0.0
Class C	255.255.255.0

The subnet mask will be used in conjunction with the IP address to determine which network the device belongs to. Other TCP/IP parameters, such as the default gateway and the DNS server, may be required in order for the device to effectively communicate on the internetwork. The main goal of this section is to learn about IP addressing and subnet masking, so the additional parameters will not be discussed at this time. The IP address can be logically combined with the subnet mask, and by performing a Boolean algebra function known as a logical AND, you can determine which IP network the node belongs to. To do this, you will first need to understand the method of logical *ANDing*. In this example, we will combine two digital bits and get a result:

Input A	Input B	Output
0	0	0
0	1	0
1	0	0
1	1	1

As you can see in the table, combining the two inputs (Input A and Input B) will result in an output. To get an output of a logical 1, both inputs *must* also be a 1. Any other combination will result in an output of 0. Now let's look at this in the form of an IP address and subnet mask. We will use the IP address example of 192.168.100.100 and a default Class C subnet mask of 255.255.255.0. Remember, an IP address is actually a 32-bit binary number that is represented in decimal form because that is much easier for humans to comprehend and use. This 32-bit address is made up of four 8-bit sections known as octets, which are each 8 bits in length. Each bit in the base-2 numbering system is identified in decimal form by using 2^n power, where n is the multiplier. For example, 2^2 will be 2×2, which equals 4. Another example is 2^3, which is $2 \times 2 \times 2 = 8$. From this logic, we can build the following table:

2^7	2^6	2^5	2^4	2^3	2^2	2^1	2^0
128	64	32	16	8	4	2	1

Rather than raising 2^n power, an easier way would be to start with the number 1, double it each time, and you will yield the same result:

- 1
- 1 doubled = 2
- 2 doubled = 4
- 4 doubled = 8

And so on. From this method you can build the same table.

Now let's apply this to the IP address in our example of 192.168.100.100 for each octet. Starting with the left, or most significant, octet and using the base-2 numbering system, 192 would be the two left bits set to a 1 and all others set to a 0.

2^7	2^6	2^5	2^4	2^3	2^2	2^1	2^0
128	64	32	16	8	4	2	1
1	1	0	0	0	0	0	0

Now add the decimal numbers together based on if it is a logical 1 or a logical 0. If there is a 1 in the column, you will add that number; if there is a 0 in that column, you will not add that number. Therefore, based on the preceding table, the answer is 128 + 64 + 0 + 0 + 0 + 0 + 0 + 0 = 192. We can now do this conversion for the remaining three octets.

The second octet is as follows:

2^7	2^6	2^5	2^4	2^3	2^2	2^1	2^0
128	64	32	16	8	4	2	1
1	0	1	0	1	0	0	0

128 + 0 + 32 + 0 + 8 + 0 + 0 + 0 = 168

The third octet:

2^7	2^6	2^5	2^4	2^3	2^2	2^1	2^0
128	64	32	16	8	4	2	1
0	1	1	0	0	1	0	0

0 + 64 + 32 + 0 + 0 + 4 + 0 + 0 = 100

Finally, the fourth octet:

2^7	2^6	2^5	2^4	2^3	2^2	2^1	2^0
128	64	32	16	8	4	2	1
0	1	1	0	0	1	0	0

It happens to be the same as the third octet, so that will be easy in this case, 0 + 64 + 32 + 0 + 0 + 4 + 0 + 0 = 100. Now, placing these octets together from left to right will yield the following binary number for the dotted-decimal address IP address:

192.168.100.100 = 11000000.10101000.01100100.01100100

You can see why the decimal equivalent to the binary numbers is much easier to work with.

 Real World Scenario

Using the Microsoft Windows Calculator to Perform Decimal-to-Binary Number Conversions

You just learned the manual method of performing a decimal-to-binary conversion. The conversion can be accomplished much easier by using the calculator with a conversion feature. One such calculator is built into the Microsoft Windows operating system. Let's use the same example we used for the manual method earlier, 192.168.100.100. First you will need to launch the Microsoft Windows calculator. Unless changes are made to your default settings, it should show up as a standard calculator view.

To perform the conversions, you will need to set the calculator to the programmer view. You can do this by clicking the View drop-down and selecting Programmer. The Alt+3 key combination is a shortcut that will provide the same view.

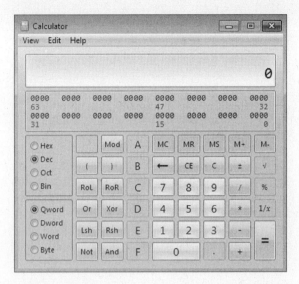

First, select the radio button for the number type you plan to enter. In this case, you want to enter the decimal system number 192, which represents the first octet of the IP address. Select the Dec radio button and enter **192** on the keyboard.

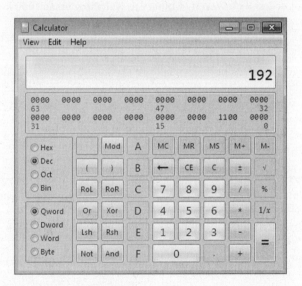

Next, select the radio button for the number type you want to display, in this case Bin for binary. With the click of a button, you can see that the 192 decimal system number has been converted to 11000000 in binary.

(continued)

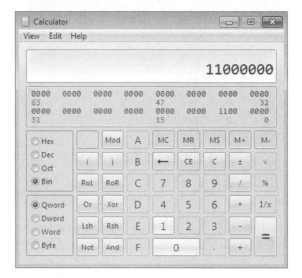

One important note regarding the Microsoft Windows calculator is that it will not show any 0s in the 8-bit placeholders that are preceding any 1s. For example, if you enter the decimal system number 8 and convert it to binary, it will show up as 1000 instead of 00001000. You always need to think in bytes, which are 8 bits.

Now that you have an understanding of subnet masking, you can use the logical ANDing process to combine an IP address and the subnet mask to determine the network address of a device. We will use an IP address from the 192.168.100.0 network and the default Class C subnet mask of 255.255.255.0 from the earlier example. Remember that a logical AND requires both bits to be a logical 1 in order to get a result of a logical 1. Any other combinations of 1s or 0s will result in a logical 0.

	Dotted Decimal	First Octet	Second Octet	Third Octet	Fourth Octet
Host IP Address	192.168.100.70	11000000	10101000	01100100	01000110
Subnet Mask	255.255.255.0	11111111	11111111	11111111	00000000
ANDing Result	192.168.100.0	11000000	10101000	01100100	00000000

You can see from the logical ANDing result that the host with an IP address of 192.168.100.70 is on the 192.168.100.0 subnet. Now, using the same IP address with a different subnet mask, 255.255.255.224, will yield a different result for the network address.

IP Subnetting

	Dotted Decimal	First Octet	Second Octet	Third Octet	Fourth Octet
Host IP Address	192.168.100.70	11000000	10101000	01100100	01000110
Subnet Mask	255.255.255.224	11111111	11111111	11111111	11100000
ANDing Result	192.168.100.64	11000000	10101000	01100100	01000000

You can see from the logical ANDing in this example that the host with the IP address of 192.168.100.70 and a subnet mask of 255.255.255.224 is on the 192.168.100.64 network.

Creating Subnets

Now that you have learned more about IP addresses and subnet masks, I will now explain how to create IP subnets using a single IP address. Recall that subnetting is dividing a network into smaller parts and is done for various reasons, such as traffic types (including broadcast traffic), network performance, and logical separation of an organization. Based on the size of the network, the network infrastructure and the logical organization will determine how the subnetting will be put in place. Remember, the class of the IP address you use will determine the number of available networks and hosts. Let's look at one possible network solution in the following case study.

Real World Scenario

Creating IP Subnets

You are the network administrator for a company named ABC Insurance Company. This company consists of four main departments: sales, marketing, customer service, and human resources. The company has a total of 65 employees who work at the single location, and the building comprises four floors. The organization is laid out as follows:

- Sales: 15 employees
- Marketing: 6 employees
- Customer service: 20 employees
- Human resources: 3 employees
- Other: 21

(continued)

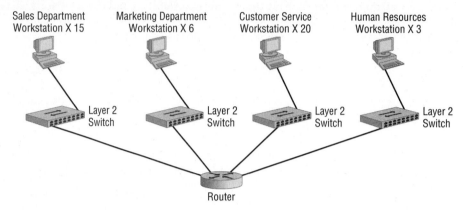

You need to design an adequate subnet strategy for ABC Company's network. The employees are distributed throughout the building and the counts are as follows:

- 4th floor: 15 sales employees, 5 others
- 3rd floor: 6 marketing employees, 3 others
- 2nd floor: 20 customer service employees, 7 others
- 1st floor: 3 human resources employees, 6 others

Based on the network traffic metrics you complied over a several-week period, it is determined that the network traffic appears to be evenly distributed between the sales, marketing, and customer service departments. These three departments use computers and Voice over IP (VoIP) telephones for their jobs. Company policy states that BYOD devices are not allowed. This prevents personal cell phones, tablets, and other personal mobile devices from being used on the corporate network. You have also determined there are eight TCP/IP-based networked printers throughout the building.

Based on these numbers, you will need a minimum of 138 IP addresses, 2 for each user plus 8 for the network printers. This does not count addresses for files servers, Layer 2 network switches, routers, or other possible IP-enabled infrastructure devices. You know that a single Class C address will allow for 256 hosts and therefore would be adequate for your organizational needs. You have decided on a private Class C address of 192.168.100.0 for internal use and as you know a Class C address will allow for a total of 256 hosts address if no subnets are in use. Using this address, you can now design an IP addressing solution that will work for your company.

Based on your analysis, it was determined that you will need at least four subnets, one for each department. As a precaution, you decide to allow for extra subnets for potential growth and advances in technology and decided on a minimum of six subnets. Remember, when you subnet an IP address range, you will be "borrowing" bits from the host address range in order to obtain the needed network addresses. The result will be more networks but fewer hosts. In this case, you will need to borrow from the fourth octet. Recall that you use the binary base-2 numbering system for IPv4 addressing. You can determine how many bits you need to borrow by using 2^n:

$2^1 = 1$

$2^2 = 4$

$2^3 = 8$

It looks like you will need to borrow 3 bits, which will give you eight networks, more than what is required based on the current configuration. Now you can create the subnet mask in the fourth octet, which will be 11100000. Remember that a Class C default subnet mask is 255.255.255.0 and the fourth octet will be 11100000 because you borrowed 3 bits for the host address range. As you learned earlier, this will create a third octet in your subnet mask of 128 + 64 + 32 + 0 + 0 + 0 + 0 + 0 = 224, and the final subnet mask is 255.255.225.224. This will leave 5 bits for host addresses; raise 2 to the 5th power to get the number of hosts per subnet.

$2^5 = 32$ hosts per subnet

Now it is time to determine the network numbers and the host addresses for each network.

Subnet #	Subnet	Subnet Mask	Host Range
1	192.168.100.0	255.255.255.224	192.168.100.1–192.168.100.30
2	192.168.100.32	255.255.255.224	192.168.100.33–192.168.100.62
3	192.168.100.64	255.255.255.224	192.168.100.65–192.168.100.94
4	192.168.100.96	255.255.255.224	192.168.100.97–192.168.100.128
5	192.168.100.128	255.255.255.224	192.168.100.129–192.168.100.158
6	192.168.100.160	255.255.255.224	192.168.100.161–192.168.100.190
7	192.168.100.192	255.255.255.224	192.168.100.193–192.168.100.222
8	192.168.100.224	255.255.255.224	192.168.100.225–192.168.100.254

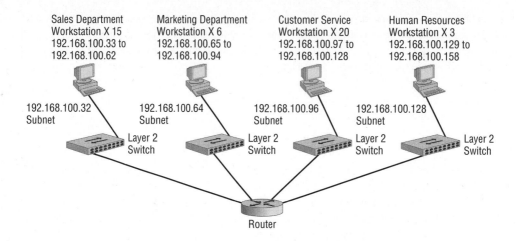

 I want to point out that there two special subnets known as the all ones subnet and subnet zero. As the names imply, the all ones subnet will consist of a 1 in every bit location and subnet zero will include a 0 in every bit location. RFC 950 recommended not using all 1s or all 0s. Therefore, in many cases you will see 2 subtracted from the number of networks or the number of hosts. For example, instead of providing eight subnets by borrowing 3 bits, $2^3 = 8$, you would subtract 2 from the number of networks, resulting in $2^3 - 2 = 6$. This would yield six usable subnets instead of eight. Depending on the technology used, some will use these special subnets.

You just learned the manual process for creating different subnets from a single IP address and creating a subnet mask for the different IP networks. There are tools available (many of them free) that will allow you to perform the same task with very little effort. Even though these tools do exist, it is still beneficial to understand how to perform IP subnetting using the manual method you learned earlier. In Exercise 9.1, you will install a free subnet calculator from SolarWinds and create subnets using the same scenario from the previous case study.

EXERCISE 9.1

Using an IP Subnet Calculator

In this exercise you will use a free subnet calculator software program from SolarWinds to perform IP subnetting. This program will allow you to easily create several IP subnets from a single IP address. To perform this exercise, you can download the program from the SolarWinds website at www.solarwinds.com. In this exercise, the program was installed on a computer running the Microsoft Windows 7 Professional operating system.

1. Point your Internet browser to www.solarwinds.com.
2. Navigate to and click the Free Downloads tab.
3. Scroll down to the Free Tool Downloads - Free Network Management Tools and click on the Browse All System Management Free Tools hyperlink area on the tools web page.
4. Click on the Download Free Tool button in the Subnet Calculator field. You will be presented with a window to enter your name and email address. Enter the required information and click the Proceed To Download button. The program you will download is named SolarWinds-Subnet-Calculator.exe.
5. Navigate to your download folder and execute the program you downloaded to start the installation process.
6. Depending on the version of the Microsoft OS you are using, you may be presented with a User Account Control dialog box window. Click Yes to continue and the setup wizard dialog box will appear on your screen.
7. Click Next at the Welcome To InstallShield Wizard For SolarWinds Advanced Subnet Calculator window to continue with the installation. The License Agreement window will appear on your screen. Read the agreement and click the "I accept the terms of the license agreement" radio button.

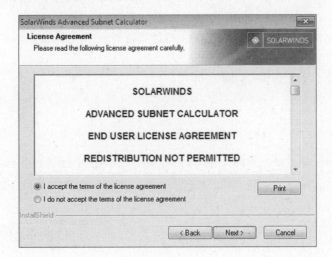

EXERCISE 9.1 *(continued)*

8. Click Next to continue with the installation. The Customer Information dialog box will appear on your screen.

9. Click next to accept the default settings and continue with the installation. The Choose Destination Location dialog box will appear on your screen.

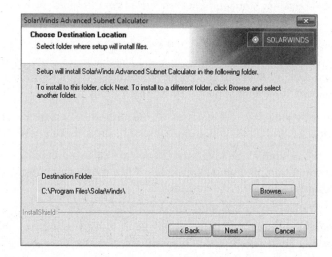

10. Click Next to accept the default location. The Ready To Install The Program dialog box will appear on your screen.

11. Click Install to install the program. The InstallShield Wizard Complete dialog box will appear on your screen.

12. Click Finish to complete the installation.

13. Start the Subnet Calculator program by choosing Windows Start > All Programs > Advanced Subnet Calculator > Advanced Subnet Calculator. The Advanced Subnet Calculator program will appear on your screen.

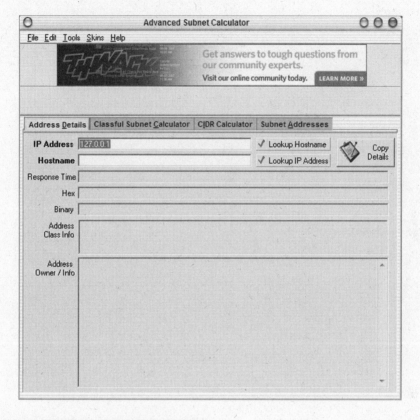

14. Click the Classful Sublet Calculator tab.
15. Enter **192.168.100.0** in the IP Address field.

EXERCISE 9.1 *(continued)*

16. Click the Number Of Subnets drop-down box and select 8 as shown in the following screen shot.

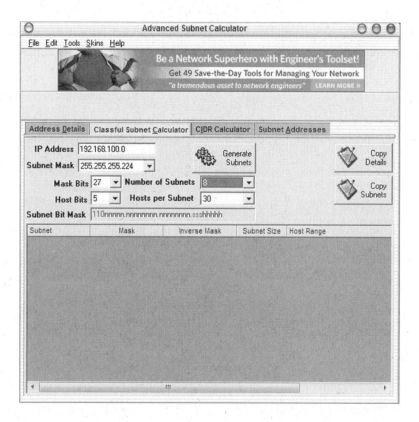

17. Click the Generate Subnets button. The available calculated subnets with the host range will appear in the window.

In addition to the subnets that were created, notice that the Subnet Mask of 255.255.255.224 was calculated. Also notice that the value for Mask Bits is 27, the value for Host Bits is 5, and the value for Hosts Per Subnet is 30. This is a free program you can keep installed on your computer for future use if desired.

Routing Network Traffic

Moving traffic across a WAN or an internetwork is known as routing. A network router is a device that operates at the Network layer (Layer 3) of the OSI model and has more intelligence than Data-Link layer devices or Physical layer devices, including network bridges, Layer 2 Ethernet switches, and legacy Layer 1 hubs. Where Layer 2 bridges or switches will use MAC addresses to handle the traffic flow, a Layer 3 router uses IP addresses to control the traffic flow. Routers will use various protocols to manage the traffic, including these common routing protocols:

- Border Gateway Protocol (BGP)
- Enhanced Interior Gateway Routing Protocol (EIGRP), which is Cisco-proprietary

- Interior Gateway Routing Protocol (IGRP)
- Intermediate System to Intermediate System (IS-IS)
- Open Shortest Path First (OSPF)
- Routing Information Protocol (RIP)

Traffic can be routed through either static routes or dynamic routes. Both methods have their advantages and disadvantages:

Static Routing

This type of routing will route IP traffic based on each device's individual routing table. With router hardware, a network administrator will define the routes manually. One advantage to this type of routing is that it is always known how internetwork traffic will flow.

Dynamic Routing

This type of routing uses protocols that will determine the best route for traffic to take across an internetwork. This differs from static routing in that the router will learn the best paths, or routes, based on the protocols that are used. Figure 9.8 illustrates the dynamic routing process.

FIGURE 9.8 Network routers can learn routes dynamically.

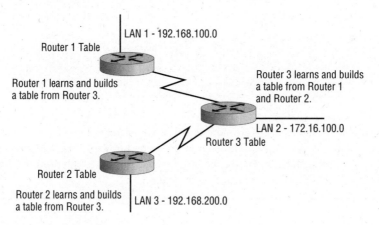

Network Traffic Shaping

You learned earlier in this chapter about network traffic flow for both LAN and WAN technology. Getting data or other information from one device to another can be accomplished in different ways, depending on whether the devices are located on the same local network/subnet or are across a routed internetwork. Although network infrastructures are getting better by becoming faster and more reliable, in many cases we still have a limited amount of capacity, or bandwidth, available. The increasing demand for mobile access and networking services only adds to the concern. Network traffic shaping, also known as packet shaping, allows for infrastructure managers and network administrators to manage the way traffic flows across a network, with the hope of providing a better overall experience for all users and their connected devices. The following traffic management techniques are commonly used in LANs and WANs:

- Backhauling network traffic
- Bandwidth and user restrictions
- Quality of service (QoS)

The following sections provide an overview of these methods of traffic management.

Backhauling Network Traffic

The term *backhaul* has different meanings, depending on whether it is being used in a wired networking, satellite, cellular, or wireless LAN communication context. In a wired network, a backhaul is defined as the network backbone or infrastructure that connects servers and other core devices together. With satellite communications, the term backhaul refers to a broadcast is center receiving raw, unedited TV signals from a satellite transmission and rebroadcasting over a cable network infrastructure. In wireless cellular communications, backhaul the communications link connecting the cellular base station tower to the central office location or a remote site, or even tower-to-tower communications. Therefore, the technology used will determine the backhaul definitions and methods.

As the use of mobile devices continues to increase, so does the need for more network backhaul capacity. According to the *Cisco Visual Networking Index* white paper, annual global IP traffic will exceed the zettabyte threshold (1.4 zettabytes) by the end of 2017. The report also mentions some other key highlights:

- Global IP traffic has increased more than fourfold in the past five years, and will increase threefold over the next five years.
- Busy-hour Internet traffic is growing more rapidly than average Internet traffic.
- Metro traffic will surpass long-haul traffic in 2014, and will account for 58 percent of total IP traffic by 2017.
- Content delivery networks (CDNs) will carry over half of Internet traffic in 2017.
- Nearly half of all IP traffic will originate with non-PC devices by 2017.
- Traffic from wireless and mobile devices will exceed traffic from wired devices by 2016.
- In 2017, the gigabyte equivalent of all movies ever made will cross global IP networks every 3 minutes.
- The number of devices connected to IP networks will be nearly three times as high as the global population in 2017.

- It would take an individual over 5 million years to watch the amount of video that will cross global IP networks each month in 2017.
- Internet video to TV doubled in 2012.

 You can get and read the *Cisco Visual Networking Index* white paper in its entirety at this book's companion website at www.sybex.com/go/mobilecomputing.

This list gives you an idea of the amount of network traffic that will be backhauled over the next several years, and how the infrastructures will need to be built or retrofitted to handle the large amounts. Figure 9.9 shows one example of a backhaul using cellular technology.

FIGURE 9.9 Cellular backhaul diagram

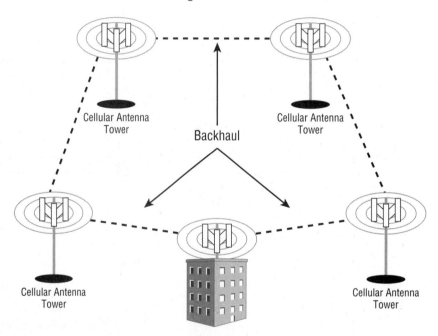

Bandwidth and User Restrictions

As mobile technology continues to advance, we are using it more, and the need for limiting and controlling the bandwidth available to users is becoming more important. Within a LAN, gigabit speeds are becoming more commonplace, and gigabit wireless networks are following. IEEE 802.11ac technology is beginning to work its way into enterprise networks, providing much higher speeds to wireless devices operating in the 5 GHz frequency band. Local network infrastructures are starting to be upgraded or are being built to handle the newest high-speed technology. Therefore, in many cases the local network can easily provide higher data rates and handle large amounts of data.

The main problem lies with the wide area connections outside of this LAN. Because of limitations either from the service provider or the technology in general, such as Digital Subscriber Link (DSL), these wide area connections often become bottlenecks for users connecting outside of their local network.

To provide efficient use of the available bandwidth, limits and restrictions may need to be put in place. The industry uses a variety of terms, such as *bandwidth restrictions*, *bandwidth throttling*, and *bandwidth caps*, but they all essentially mean the same thing. Regardless of what you call it, the objective is to restrict how much of the available bandwidth a user is allowed, either at a point in time or over a period of time. For an enterprise network, this will be put in place by the IT department to provide a better overall experience for all of the users of that network. From a service provider perspective, these restrictions are usually in place based on the plan you have purchased from the service provider. The more you pay monthly for the service, the more bandwidth you will be allowed to use. Bandwidth restriction can be accomplished through the use of either hardware or software solutions. Some of the restrictions are put in place by using policies, role-based access control, or dedicated software programs, while others are put in place through settings in the infrastructure hardware that is in use. Figure 9.10 shows how enterprise-grade wireless access point management software can limit the bandwidth for the connected users. In this example, a wireless user connected to the DATA SSID will be able to upload at 1 Mbps and download at 3 Mbps.

FIGURE 9.10 AirTight Network's cloud-based Management Console interface allows for limiting bandwidth by restricting upload and download traffic for user on a per-SSID basis.

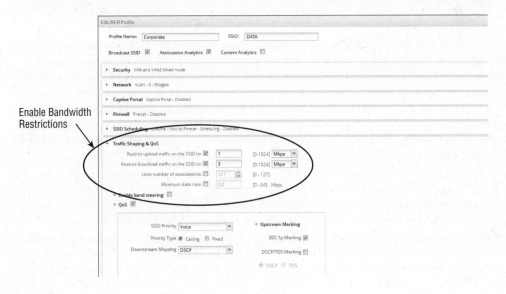

In addition to the amount of bandwidth that can be used at any point in time, network providers, including cellular service providers, have the capability to limit how much bandwidth you are allowed to use per month, or for another specified period of time. Many cellular wireless service providers offer all-in-one plans that will allow for a specific amount of voice, text, and data that can be used in a time period, and is usually measured in megabits or gigabits per month. Once the allowed limit is reached, the user will be charged an additional amount based on the usage. Internet service providers may also allow for a maximum amount of bandwidth per month, with additional penalties for those that exceed their own limit. Penalties may include warnings, suspension of service, and additional fees.

Open Internet

The Federal Communications Commission (FCC) in the United States issued a change order in 2010 for Open Internet, also known as *net neutrality*. The objective of this order is to require Internet service providers (ISPs) to treat all Internet traffic equally. This would restrict ISPs from specifically controlling the type of traffic you send or receive. For example, if you were to watch a streaming movie from an online content provider, the ISP could not slow it down to allow priority to other traffic types of traffic. This order would also prevent ISPs from blocking access to specific Internet content and other features. In 2014, a United States federal appeals court struck down the FCC's Open Internet (net neutrality) rules. This will give ISPs more control on how they regulate Internet traffic and access to content.

Quality of Service

Certain types of network traffic require special attention based on how they are used. Other types of traffic do not.

Let's look at network printing, for example. When a user sends a file to a network printer, speed is not a major concern. In many cases, there may be one printer for many users of a LAN. The printers are usually in a central location and easily accessible to all the users. After a user sends a document to the printer, they will need to get up from their desk and walk over to the printer to retrieve the print job. In a busy environment, there may be several print jobs queued up that were sent by different users. Depending on the urgency, the user may not retrieve the print job for several minutes or even hours. The point is that this kind of traffic usually has a low priority.

Conversely, other types of network traffic may have a high priority, such as real-time applications like voice, video, and some streaming content. Using VoIP is common with almost all enterprise wired computer networks and cellular networks, and it's becoming common with most enterprise wireless networks. The same is true for certain types of video communications. This type of traffic requires special attention and priority in order to provide the user with a good networking experience. As the name implies, quality of service (QoS) will provide a better quality of network communications for traffic such as voice and video. Unlike sending a file to a printer or saving a document to a network file server, using voice and video technology involves human senses such as hearing and sight. If you are having a telephone conversation with another person, you want good, reliable communication. The same goes for attending a live, real-time videoconference. The best way to accomplish this is to incorporate quality of service with the network communications.

One example of technology that can benefit from QoS is using voice on a network. Voice communications are not bandwidth intensive, but they are subject to issues such as latency and jitter.

Latency

Voice and video communication are two types of networking applications that are subject to latency. Latency is basically a delay in the communications from the source device to a destination device. The latency is measured as the time value for a frame to get to a destination device. Most network engineers try to achieve a total one-way, end-to-end delay for voice communications of less than 150 milliseconds (ms). High latency can result in poor voice or real-time video communications.

Jitter

Jitter is the variation in the delay or latency of frames received by a destination device. Jitter can be controlled by using something known as a jitter buffer. Buffering of information will allow for the data frames to be received as a steady stream of information by the receiving device. Jitter buffers may add 30 to 50 ms to the latency to reduce the effects of jitter. Excessive jitter can cause poor quality VoIP calls and video delays.

Providing end-to-end QoS on a network is not necessarily a simple task, and depending on the device types, number of devices, and network infrastructure, it may be quite complex. Quality of service is a complex topic and the technical details are beyond the scope of this book.

Summary

In this chapter you learned about the flow of network traffic, as well as the different types in both LAN and WAN installations. You saw that local traffic stays within the LAN boundary and wide area traffic will cross network boundaries in an internetwork. We explored Network layer logical addressing, including IP addresses and subnet masks, and how to subnet a network. You also learned about IPv4 address classes and the number of networks and hosts within the different classes. You learned about IP subnetting, and how to subnet an IP address to support different physical networks. You had a chance to see the manual process of creating IP subnets, and used a free tool to perform the same task in a much easier way. We explored the different types of technology and controls that are available for network administrators and service providers to manage the available bandwidth and allow technology such as voice and video services to operate effectively to provide a quality user experience.

Chapter Essentials

Understand how traffic flows in local area networks and wide area networks. Local area network traffic is confined within the boundaries of the LAN. Modern LANs use TCP/IP for communications on the local network, which may also be used for WAN communications. Wide area network traffic flow will include routing traffic from one LAN to another LAN, and will require the use of more extensive network infrastructure devices.

Understand network subnets. Know that a subnetwork is a smaller part of an IP-based internetwork. A subnet can be created by borrowing bits from the host addresses to provide more small IP-based networks.

Know the different IP address components and classes. Know IP address classes used with IPv4 technology. There are a minimal number of Class A IP network addresses and a large number of host addresses. A Class C network has a large number of network addresses with a smaller number of available host addresses.

Understand the subnetting process. Know that all devices using the TCP/IP protocol suite have, at minimum, an IP address and subnet mask. The IP address is the logical identifier, and the subnet mask is used to identify which IP network the device is part of.

Understand the basics of routing network traffic. Moving traffic across a wide area network or an internetwork is known as routing, and network routers are Layer 3 devices. Know the two different methods of routing: static routing and dynamic routing. Be familiar with the common routing protocols.

Be familiar with different traffic shaping techniques. Network traffic shaping, or packing shaping, allows for infrastructure managers and network administrators to manage the way traffic flows across a network, with the hopes of providing a better overall user experience.

Understand the various network traffic backhauling types. Backhauling traffic is getting data from one location to another. This could include large files such as device firmware, network device configuration files, and other administrative data. Understand the different meanings of the term *backhauling* based on the technology that is used.

Know how providers and network administrators can add bandwidth and user restrictions. These restrictions will allow better control of network traffic flow and prevent users from monopolizing all of the available bandwidth. Also know that network service providers can limit the amount of bandwidth users are allowed over a period of time, such as a month.

Understand the benefits of quality of service. Know that quality of service technology is beneficial to time-sensitive data communications such as voice, real-time video, and certain streaming content.

Chapter 10

Introduction to Mobile Device Management

TOPICS COVERED IN THIS CHAPTER:

- ✓ Mobile Device Management Solutions
- ✓ Common Mobile Device Operating System Platforms
- ✓ The Mobile Application Store
- ✓ Pushing Content to Mobile Devices
- ✓ MDM Administrative Permissions
- ✓ Understanding MDM High Availability and Redundancy
- ✓ MDM Device Groups
- ✓ Location Based Services
- ✓ Mobile Device Telecommunications Expense Management
- ✓ Captive and Self-service Portals

Over the past several years we have seen the influx of mobile device technology and the concept of bring your own device (BYOD) within the corporate world. This has created something of an administrative challenge for the information technology professional. While some organizations embrace this concept, others are skeptical of the adoption for various reasons, including technical support and security concerns.

In this chapter we will explore the basics of mobile device management options, including both on-premise and cloud-based Software as a Service (SaaS) solutions. Mobile device management (MDM) solutions will provide the capability to manage mobile devices of various platform types that connect to the corporate network. This includes access to digital applications (apps) via online stores. You will see that with MDM solutions, you will require special permissions to perform specific tasks on mobile devices. We will explore how mobile apps and other device information can be pushed to devices. You will also see that devices can be grouped based on management requirements and why high availability is an important feature of MDM.

Geo-fencing (building a virtual electronic boundary) and geo-location (the ability to physically locate a device) are two components of the mobile location services you will learn about. A captive portal will allow network administrators to restrict access to a network until authentication has occurred or terms and conditions have been accepted. This prevents users from accessing resources or services unless they have been properly authorized to do so. Finally, self-service portal solutions give users the ability to register their devices to the network, enabling quick access to the network without the need to open a service ticket or contact the help desk for technical support. We will take a look at both the captive portal and self-service portal.

Mobile Device Management Solutions

The extensive increase in the number of multifunction wireless mobile devices gives users the opportunity to access information of all types from anyplace that provides infrastructure network access via technology such as cellular connections and wireless LANs (Wi-Fi). This includes the home, public venues and the workplace. Although very convenient and popular, this emerging mobile technology may create something of a challenge for the enterprise and corporate IT services professional and help desk support personnel. This is where mobile device management (MDM) solutions stand out. MDM allows network managers, administrators, engineers, and other information technology personnel to manage, control, and secure multiplatform device environments that may be used on the network.

Choosing the correct MDM solution for your networking requirements involves some contemplation. The correct solution must meet the needs of the organization and will require careful consideration to validate it does so. When it comes to MDM, there are two main options that exist. In-house (on-premise) solutions or cloud-based solutions, which are provided as Software as a Service (SaaS) technology, are widely available. MDM solutions include, but are not limited to, the following features:

- Application distribution (push technology)
- Compliance reporting
- Content management
- Device registration (self-service portal)
- Location-based services
- Multiplatform management and support
- Password control
- Policy enforcement
- Remote lock
- Remote wipe
- Remote unlock
- Secure communications (virtual private network)
- Telecommunications expense management

You will learn more about these and other MDM features later in this chapter and in Chapter 13, "Mobile Device Operation and Management Concepts." There are many MDM solutions available on the market today. If your organization allows employees, contractors, vendors, and others to use their own personal devices or company-owned devices to access your network, you will need to ensure that the platforms are supported in order to lessen support calls and other associated costs.

Most, if not all, mobile device management providers offer a free trial period for their software solutions and packages. I highly recommend you try these products prior to purchasing a specific provider's solution. It is best to know that the solution you choose will meet your specific needs. There is no "one size fits all" when it comes to mobile device management solutions. Choosing the correct solution will depend on many factors, including business model, number of employees and devices, and company policy requirements.

You will now see an overview of both cloud-based and on-premise solutions and how each has its own distinct advantages and disadvantages.

The Software as a Service (SaaS) Solution

When it comes to information technology services, the term *cloud* seems to be the newest buzzword. Cloud-based solutions are commonly known as *Software as a Service (SaaS)* solutions and are not limited to mobile device management (MDM). In many cases, the SaaS MDM solution is a virtual on-premise solution that has many if not all of the same features but without the extra administration overhead and costs that go with the on-premise model. (On-premise MDM will be discussed in the next section.) Other SaaS solutions include products that are truly managed services provided from certain software manufacturers.

Some of the advantages and disadvantages of a cloud-based solution are as follows:

- Advantages
  - Hardware investment is not required.
  - Installation and setup time is minimal compared to an on-premise solution.
  - Less administration overhead and potentially better availability compared to an on-premise solution.
  - Can be administered from any location where an Internet connection is available.
- Disadvantages
  - Recurring monthly subscription fees.
  - Internet connection is required for all management and control.
  - Concerns for unauthorized access of data stored on cloud provider's server if not properly secured.

Depending on the environment where it is deployed, the SaaS MDM solution will be a great fit in many cases. I recommend additional research and reviewing the advantages and disadvantages to determine the best solution.

The On-Premise Solution

Many network managers and administrators have a certain comfort level with the idea of an *on-premise* mobile device management solution as with other on-premise technology. This is because these solutions provide direct access and more control with the installed products. Some of the advantages and disadvantages of an on-premise solution are as follows:

- Advantages
  - Can be maintained within the organization's data center.
  - No recurring monthly subscription fees.
  - Internet connection is not required for management and control.
  - Ownership of the investment including associated hardware and software.

- Disadvantages
    - Initial investment with required hardware and software.
    - Installation requires setup time and may require additional resources.
    - Administration overhead, backup solution, and technical support.
    - May require additional setup resources and administration from outside the organization.

Choosing either a SaaS (cloud-based) solution or an on-premise solution requires considering many factors, such as corporate policy, security concerns, and cost. It is best to compare all features, advantages, and disadvantages prior to purchasing any one specific solution. Following are some MDM solution providers that offer on-premise solutions, cloud-based solutions, or both.

MDM Solution	Website
AirWatch	www.air-watch.com
Amtel	www.amtelnet.com
Apple Profile Manager	www.apple.com/support/osxserver/profilemanager
BoxTone	www.boxtone.com
Centrify	www.centrify.com
Citrix – Zenprise	www.citrix.com/products/xenmobile
Cloudpath Networks	cloudpath.net
Good Technology	www1.good.com
IBM	www.ibm.com/us/en
LANDesk	www.landesk.com
Microsoft Intune	www.microsoft.com/intune
MobileIron	www.mobileiron.com
Notify Technology	www.notifycorp.com/products/notifymdm
SAP/Sybase	www.sap.com
Sencha	www.sencha.com

(continued)

MDM Solution	Website
SOTI	www.soti.net
Symantec	www.symantec.com
Wavelink	www.wavelink.com

Common Mobile Device Operating System Platforms

There have been many mobile device operating systems over the past decade. Some have come and gone and others have survived. For the corporate or enterprise mobile device user, there are four main players in the current market: Android, BlackBerry, iOS, and Windows Phone.

Android   Android is an operating system based on Linux and designed for smartphones and tablet devices that are mostly touchscreen capable. This O/S was originally developed by Android Inc. in the 2007 time frame and has undergone many revisions since the first version was released in 2008. The Open Handset Alliance (OHA) continues to update the operating system. The OHA is a consortium of approximately 84 firms that was created to develop open standards for mobile devices. According to a report published in early 2014 by Good Technology, enterprise Android activations were at about 26 percent compared to 72 percent for Apple iOS during the fourth quarter of 2013.

BlackBerry   The first BlackBerry operating system was released in 1999 for the Pager 580. This operating system was designed by BlackBerry Ltd. and is proprietary to BlackBerry mobile devices. Synchronization of various corporate email services is where the BlackBerry operating system rose in popularity. The Good Technology report mentioned earlier did not include the BlackBerry operating system because of the lack of reporting by the company. BlackBerry 10 is the most recent release of the operating system. Although it has had its share of ups and downs, BlackBerry still continues to be a popular operating system in the corporate world.

iOS   In 2007 Apple Inc. introduced the iOS mobile operating system for the iPhone. This OS is now incorporated into other Apple devices, including iPod Touch, iPad, iPad Mini, and Apple TV devices. According to a report from Good Technology, 73 percent of all enterprise activations in the fourth quarter of 2013 were Apple iOS. In the third quarter of 2013 Apple iOS accounted for 72 percent of enterprise activations and 69 percent in the second quarter of 2013. These numbers show an increase in activations over the 2013 time period.

Windows Phone   Microsoft released its proprietary mobile operating system Windows Phone, also known as WP, in the fourth quarter of 2010. Unlike the Apple iOS and Android operating systems, Windows Phone is targeted for consumer use. Windows Phone replaced Windows Mobile (last updated in February 2010); however, it is not backward compatible with Windows Mobile. According to the same report by Good Technology, enterprise activations for the Windows Phone OS were about 2 percent over the fourth quarter of 2013, a very small number compared to Android and Apple iOS.

Mobile device multiplatform support must be considered prior to purchasing and implementing a mobile device management solution. The type of business in which the solution will be deployed and whether employee owned devices are allowed are factors that must be considered in the selection.

The Mobile Application Store

The *mobile application store*, also known as an "app" store, is a digital distribution platform consisting of different websites that allow users to obtain applications (*apps*) for their mobile devices. These applications may be available at no charge or may be fee based. Application stores fall under two different categories: enterprise app stores and consumer app stores.

The Enterprise Mobile Application Store   The enterprise app store is a site that allows employees of an organization to visit for company-standard mobile applications. The enterprise application store helps an organization control and regulate the apps that are used within its environment. The development of an on-premise app store can be involved and expensive, and therefore organizations that may not have the resources or funds for an on-premise solution can create their own store by using an online provider. The Gartner Group has reported that by the year 2017, 25 percent of enterprises will have an enterprise app store.

The Consumer Mobile Application Store   The consumer mobile application store is intended for the common user to download or purchase commercial, off-the-shelf mobile applications. The following list shows some of the available categories within a consumer app store and is only a small sample of what is available:

- Business
- Education
- Games
- Health and fitness
- Medical
- Music
- Productivity
- Sports
- Weather

Although there are stores available for a variety of device types, the most popular supported platforms with stores are the Android, BlackBerry, Apple iOS, and Windows Phone. The following list shows popular application stores. These stores may host apps for both the enterprise user and the consumer user.

Application Store	Website
Apple App Store	store.apple.com
BlackBerry World Store	appworld.blackberry.com
Google Play Store	play.google.com
Nokia Store	store.ovi.com
Samsung Store	apps.samsung.com
Windows Phone Store	www.windowsphone.com

The use of mobile apps continues to grow at a very fast pace. As people become more dependent on mobile device technology, the need and variety of apps is constantly expanding. Technology research groups estimate that in 2013, well over 100 billion apps were downloaded, with revenues from these apps exceeding more than $25 billion.

As of the third quarter of 2013, the Android Google Play Store had officially reached over one million apps and had beaten the Apple App Store, which had 900,000 available applications.

An emulator can be a valuable tool for testing prior to deploying MDM policies, applications, or other content. Exercise 10.1 shows you how to install a free emulator that you can use for a trial period.

EXERCISE 10.1

Installing Android Emulator Software in Microsoft Windows

In this exercise you will install a free trial of an easy-to-install-and-use Android emulation program on a computer using the Microsoft Widows operating system. To perform this exercise, you will download a Microsoft Windows–based Android emulator program from YouWave at youwave.com. For this exercise, the emulator program was installed on a computer running the Microsoft Windows 7 Professional operating system.

1. Point your Internet browser to youwave.com.

2. Navigate to and click the Download button. The Download YouWave For Android landing page will appear. This will allow you to download a 10-day free trial.

3. Select the desired version and click the Download button. In this exercise, the Home 3.15 version was selected.

4. Save the file that contains the program to your download directory. The file is named YouWave-Android-Home-3-15.exe.

5. Navigate to the folder on your computer that contains the program you downloaded in step 3 and execute the file. Depending on the version of the operating system you are using, a User Account Control dialog box may appear.

6. Click Yes to continue. The YouWave Android Setup dialog box will appear.

7. Click Next to continue. The license dialog box will appear.

8. Read the agreement and click the I Agree button to continue the installation. The Choose Install Location dialog box will appear.

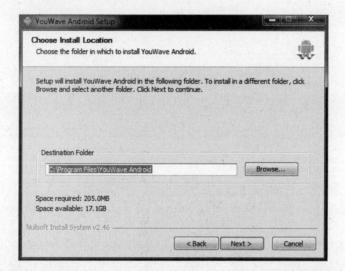

9. Click Next to accept the default installation location. The Choose Start Menu Folder dialog box will appear.

10. Click the Install button to accept the default folder and continue the installation. The Completing The YouWave Android Setup Wizard dialog box will appear.

11. Click the Finish button to end the installation. A YouWave Android Setup dialog box will appear.

12. Click OK to end the installation.

13. Start the YouWave Android emulator program by choosing Windows Start ➢ All Programs ➢ YouWave Android ➢ YouWave Android. The trial period dialog box will appear on your screen.

450 Chapter 10 ▪ Introduction to Mobile Device Management

14. Click OK to continue. The YouWave Android emulator will appear on your screen.

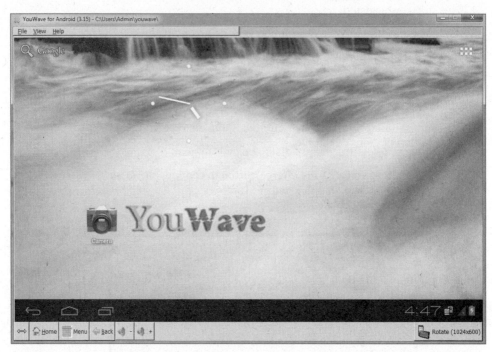

15. Explore some of the program features by clicking the various options.

> You can continue to use the program for the remainder of the 10-day evaluation period. When this trial period expires, you can uninstall the program from your computer or purchase an activation key from the YouWave website.

Although using actual devices is a great technique for development and testing, emulators of any type can be a valuable tool for many aspects of technology. Computer-based Android and other mobile device emulators aid in software development, debugging, pilot test beds, and troubleshooting.

Pushing Content to Mobile Devices

Pushing content to mobile devices is a common requirement in enterprise device deployments. MDM solutions allow for corporate and other applications to be pushed out to all specified mobile devices that connect to the corporate network, regardless of whether the device is a company-owned asset or an employee-owned personal device. This lessens the burden of requiring employees to visit the enterprise application store and download a specific app. In addition to apps, the following information can be pushed to mobile devices (among other types of information):

- Operating systems and updates
- Configuration information such as Wi-Fi profiles and settings
- Security information and updates
- Videos
- Email configurations

Some mobile device operating systems include platform-dependent native push services such as the Apple Push Notification Service (APNS), which is required for Apple device MDM enrollments. You learned about these in Chapter 2, "Common Network Protocols and Ports."

MDM Administrative Permissions

You will need to have the proper privileges to manage mobile devices from a corporate network, depending on the action to be performed. This is really no different than common day-to-day network administration tasks that are usually defined with the corporate network policy.

Mobile devices used to access corporate network resources may not be a corporate-owned assets and may be personal property of the employees. You may need special privileges, or *administrative permissions*, to use MDM solutions to perform the following actions:

- Creating and maintaining mobile device policies
- Grouping devices with the correct policy
- Allowing for remote wipe, remote lock, remote unlock, and password control

The privileges will be defined as part of the network policy, and that task may be delegated to the organization responsible for computer network administration. Assigning administrative permissions can be accomplished in different ways, depending on the MDM solution you choose. In the case of bring your own device (BYOD) solutions, device owners will need to be familiar with the corporate policies regarding the use and administration of employee-owned devices. Understanding the End-User License Agreement (EULA) process and how it is accepted will need to be defined for several reasons, including for legal purposes. The EULA and additional pertinent information will be discussed in more detail in Chapter 11, "Mobile Device Policy, Profiles, and Configuration."

Understanding MDM High Availability and Redundancy

One important component of MDM services is to ensure that access will always be readily available. Consistent or *high availability* for any type of networking service is commonly accomplished by redundancy of some type and is not a new concept. Computer networking services redundancy dates back to the early days of local area networking. At that time, it was important to ensure that various types of data were always readily available and not just accessible from a data backup solution. This was handled in a variety of ways, including hard drive storage failure protection by disk mirroring or using RAID technology. RAID is an acronym that is spelled out a couple of different ways, including redundant array of inexpensive disks. In the event of a hard drive or disk storage failure, the file server would be able to continue to operate and provide access to any stored data. Without this type of fault tolerance, the hardware would need to be repaired or replaced and the data restored from a backup resulting in potentially significant network downtime.

File server redundancy, including grouping, or "clustering," servers together for fault tolerance purposes is another method that would allow for high availability. The cost for this type of redundancy is higher because you would need to have completely redundant servers in the event of a catastrophic hardware failure. The benefit of this type of solution is 100 percent uptime for the network services; hence a higher cost.

The requirement for information to be always available with no downtime is becoming more common and will be part of a company's best practice policies. MDM solutions are no different than other networking services such as email and the reliable storage of information. Therefore, redundancy will need to be considered to ensure the availability of appropriate services. This is something that should be considered when evaluating an

MDM solution for your organization and is more of a concern with on-premise MDM solutions. With Software as a Service (SaaS), or cloud-based MDM solutions, the service provider will be responsible for the redundancy by providing redundant data centers, an uninterruptible power source (UPS), and redundant Internet connectivity in addition to other common redundancy best practices.

MDM Device Groups

Creating *device groups* within your mobile device management solution will help to streamline administration. Creating groups will allow the MDM administrator to identify profiles, applications, and other content that will need to be made available to the device user. This concept is similar to computer network administration. Organizing device groups will simplify management by providing common access to devices based on the requirements of the organization. Different options are available for an administrator to create device groups, which will depend on the specific network operating system and the *directory service* in use. The following directory service implementations are common:

- Microsoft Windows Active Directory
- Red Hat Directory Server
- Novell eDirectory
- Apple's Open Directory

These directory services can be queried using *Lightweight Directory Access Protocol (LDAP)* and will aid in the simplification of the administration process.

Group management with MDM is an important component of the entire practice of mobile device management. Improper group development and assignments will potentially lead to many technical support incidents and troubleshooting issues. You can also create pilot and test groups that can be used prior to a large-scale deployment for policies, mobile applications, and other configuration information. Troubleshooting group-related issues will be discussed further in Chapter 19, "Mobile Device Problem Analysis and Troubleshooting."

Location-Based Services

Location-based services (LBS) is a technology that can be used many different ways. One use of LBS allows for various types of assets and devices and even people to be geographically located by different methods of radio frequency technology. These methods include radio frequency identification (RFID), Global Positioning System (GPS), and Wireless LAN (Wi-Fi) technology. When it comes to mobility and mobile device technology, LBS is widely used with social media and mapping services to help identify the location of a mobile device for various purposes, including personal, work, and entertainment. The following sections will provide a brief overview of LBS technology that is used with mobile device technology and a focus on geo-fencing and geo-location. LBS solutions can be software based, hardware based, or a combination of both.

Geo-fencing

A fence in any sense of the word is something that will provide an enclosure for the perimeter of an area. This enclosure is designed to keep items from leaving and entering an area and to know what is contained within an area. Geo-fencing from a mobile computing perspective is taking this concept to an entire new level. This virtual electronic boundary will allow network administrators to identify when specific mobile devices enter or leave an area. With the explosion of the number of mobile devices that are now used, keeping private company information within the boundary of the organization is critical. This is just one example of how geo-fencing is used. Figure 10.1 shows how *geo-fencing* is a boundary surrounding a specific geographic area.

FIGURE 10.1 Geo-fencing example

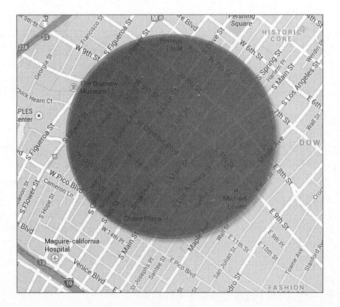

Commonly used with Global Positioning System (GPS) or wireless LAN technology, geo-fencing-enabled devices can be managed from within the MDM solution. Tasks include managing device features, device activation and deactivation, restricting access to specific apps such as games and social media, and protecting private company information. Disabling the use of other device features such as built-in cameras can also be accomplished with MDM.

From the consumer side, geo-fencing is also popular with social media applications. These applications will allow a user to notify their friends or colleagues when they leave a specific location or when they arrive at a new location. The opportunities are endless as to how this can be used, including intentionally informing someone of a location, iOS reminders (which are intended to provide an alert when arriving at or leaving a specific location), and even shopping.

Geo-location

Geo-location technology allows for many different device types to be geographically located. The following devices are among those that can use this technology:

- Smartphones
- Tablets
- Computers

With geo-location, different methods can be used to locate a device. One example is Time Difference of Arrival (TDoA), which is a multilateration technology technique. TDoA will measure the difference in distance between two stations from known locations. This provides an accurate method for locating a device of some sort. Another method of identifying the physical location is by using the device's assigned IP address to provide specific information about the physical location of the connected device. Figure 10.2 shows how geo-location can pinpoint a device at a precise location.

FIGURE 10.2 Geo-location can identify the physical location of a device

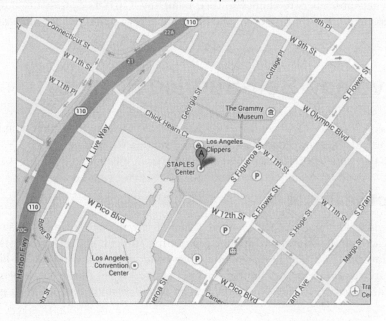

The use of geo-location technology covers a wide range of applications. Although with consumer mobile devices, the user has the capability to disable geo-location technology, many opt not to do so. This is because they are either not aware that the technology exists and is enabled or they plan to use it for social media purposes. Geo-location technology can be used for marketing, social media, and enterprise device security.

Mobile Device Telecommunications Expense Management

Managing expenses for mobile device technology can be a daunting task. Many mobile device management solutions, either on-premise or cloud-based SaaS, offer *telecommunications expense management* features. One common concern with mobile device technology is that users exceed the amount of data communications available to their device or their data plan. Unlimited data plans are becoming a thing of the past, and data use will need to be strictly managed. This is where telecommunications expense management plays a role. Enabled devices allow MDM systems to handle data plan management and other tasks including invoicing, identifying excessive roaming, and disabling data capability for lost or underutilized devices. While many carriers are doing away with unlimited data plans, some are increasing the data cap allowed for customers while keeping the price the same. Therefore, the battle to retain or get new customers continues to build amongst the wireless carriers. Some will even pay an early termination fee to try to get subscribers to change to their company. Carriers continue to build out their wireless network infrastructure, providing the ability to handle the increased data usage.

Captive and Self-Service Portals

In the following sections, I will compare two different available methods that mobile devices can use to access resources on a network, including wireless networks. These methods are the captive portal and self-service portals.

The Captive Portal

A *captive portal*, sometimes called a walled garden, is a web page redirection that a wireless device user is presented with after performing an IEEE 802.11 open system authentication and IEEE 802.11 association when connecting to a wireless (Wi-Fi) network. Captive portals are not limited to Wi-Fi networks, but this is one network type where they are often used. In order for the user to access permitted resources or gain wireless network access, a web page will require them to authenticate in some way, which may include the following:

- Entering user credentials (username and password)
- Inputting payment information
- Agreeing to terms and conditions

When one or more of these methods is used, the wireless device will be able to access the network and use whatever resources they have permissions to use. Most if not all public access wireless networks have some type of captive portal enabled. This includes networks at public venues such as hotels, coffeehouses, restaurants, and airports. This will help to protect both the provider (host) and the user of the wireless network. Private corporate networks also use captive portals when a user connects to the guest service set identifier (SSID). Using captive portal authentication with enterprise guest networks will ensure that a user is connecting to an authorized company access point and not a potential rogue access point. Most enterprise-grade wireless access points, including cloud-based access points and wireless LAN controllers, have built-in captive portal capabilities that are fairly straightforward and easy to implement.

Keep in mind that a captive portal does not offer security of any data and only provides a way for network infrastructure devices to restrict device and user access to a network until some type of successful authentication has been provided. This "authentication" could be as simple as agreeing to a "Terms and Conditions" web page. After a user connects to a wireless network through a captive portal, additional security measures should be put in place. For public networks, this includes options such as virtual private network (VPN) connections or using a secure protocol like HTTPS, at a minimum. With a corporate network, additional authentication may include WPA2 passphrase authentication or IEEE 802.1X/EAP user/certificate-based authentication. These and other security concepts will be discussed further in Chapter 16, "Device Authentication and Data Encryption."

The Self-Service Portal

Mobile device management platforms often include self-service portal functionality to allow mobile device users to register their device and connect to a network. With the concept of bring your own device (BYOD) continuing to grow and more personal devices being allowed to connect to a corporate network, a self-service portal will help by requiring less administration overhead and therefore providing a cost savings to the organization. The MDM platform that is used will determine which management features the user will have access to.

A *self-service portal* is a method of onboarding a device and allowing it to comply with the appropriate company/device policy. The way these portals operate will vary based on how each MDM platform incorporates the technology. If a user already has a user logon account such as a Microsoft Windows Active Directory account, it may be as simple as entering their credentials to start the onboarding process. If they are a guest or new user, they will need to have an account created.

Here is one example of the self-service portal process.

1. Create an account for the device user. This is usually done from an administrative console page.
2. A username, first and last name, email address, and phone number is usually required.
3. An email will be sent to the user with information regarding the device registration.
4. The user will then access the self-service portal web page logon screen, which requires them to accept any terms and conditions or specific device policies.
5. The user will then select their device type (iOS, Android, or other).
6. Depending on the use of device, a guest network or company network will determine the capabilities and policies.
7. A text message will be sent to the device with a password or PIN, which will then allow the user to access to the network.

The preceding steps represent a generic process, and it is important to understand that the steps and processes will vary depending on the MDM platform you are using.

Users may also be able to use their company credentials to log on to a guest network portal. If the company has an agreement with the provider or wireless carrier, the credentials may be transferable. For example, when the user enters an area that charges for Wi-Fi service such as an airplane or a stadium with which their company has an agreement, instead of paying for access, the user can log in using their company credentials. Many users may not even be aware of agreements such as this that exist.

There are many benefits that are available to the user of the device that is now being managed (including but not limited to):

- Locating a device
- Locking a device
- Requesting information from a device
- Remotely wiping a device

One example where these features will prove helpful is that when a device is lost or stolen, they can help protect the integrity of the information contained within the device.

Summary

In this chapter you learned about how the influx of mobile device technology and the concept of bring your own device (BYOD) create administrative challenges and how these challenges can be addressed by implementing a mobile device management (MDM) solution.

We explored the basics of mobile device management solutions including both cloud-based (SaaS) and on-premise solutions. You saw that various platforms can be managed using an MDM solution, and the management options included access to applications (apps) via online stores. As with computer network administration, MDM solutions require special permissions to perform specific tasks on a mobile device.

We also explored the pushing of content to mobile devices. Devices can be grouped based on management requirements. By grouping servers, you can provide a high availability solution to help ensure maximum uptime for the services provided. We looked at how location-based services, including geo-fencing and geo-location, allow a virtual electronic boundary to be built or a device's location to be pinpointed within a specific area. Finally, we explored the concepts of both the captive portal and the self-service portal.

Chapter Essentials

Be familiar with different mobile device management (MDM) solutions. Know the main differences between an on-premise and Software as a Service (SaaS) cloud-based mobile device management solutions. Understand the advantages and disadvantages of each type.

Understand the concept of pushing content. Know that mobile applications and device configuration information can be pushed to mobile devices. This will aid in management, troubleshooting, and changes to device configuration.

Know that administrative permissions allow for mobile devices to be managed from a corporate network. MDM solutions may need special privileges in order to perform the specific actions on the device.

Be familiar with mobile device grouping. Device groups will help to streamline administration. They will help to identify profiles, applications, and other content that will need to be made available to the device user.

Understand the concept of location based services. Know the difference between geo-fencing and geo-location and how they are used with mobile device technology. These concepts allow a user to build a virtual electronic boundary and will allow networks to be able to identify when mobile devices enter or leave an area. They are also useful for gathering specific information about the location of connected devices.

Know the differences between a captive portal and self-service portal. Understand that a captive portal is a way of restricting access to network resources unless proper authorization has occurred and that a self-service portal allows users to register their own personal devices. This will give them the capability to use the network resources they have been assigned.

Chapter 11

Mobile Device Policy, Profiles, and Configuration

TOPICS COVERED IN THIS CHAPTER:

- ✓ General Technology Network and Security Policy
- ✓ Information Technology and Security Policy Implementation and Adherence
- ✓ Backup, Restore, and Recovery Policies
- ✓ Operating System Modifications and Customization
- ✓ Technology Profiles
- ✓ Understanding Group Profiles
- ✓ Policy and Profile Pilot Testing

With respect to computer networking, a policy is documentation that defines rules, restrictions, acceptable use, liability, and more. In this chapter we will look at some of the basic components of a network security policy. We will also explore some of the most common industry regulatory legislation and the importance of compliance. The technology user must have a clear understanding of the policy and how to adhere to it in order to effectively utilize the network and services it provides. The contents of a policy will vary based on the business model or the organization type. There is really no one-size-fits-all policy.

This chapter will provide a basic outline of some of the more common policy components that are part of the framework for most organizations. If a policy is too strict, it may hamper the ability for a user to perform the needed tasks while using the network; therefore, the appropriate balance between security and usability is important. Policies exist for many areas of mobile technology including passwords/passcodes, remote access, data backup, restore, and more. You will see an overview of and the importance of a backup and restore policy. Knowing how the mobile device's operating system can affect the adherence to policy is another topic we will explore. Changes to the mobile operating system may also be limited based on the network security policy. We will explore profiles, which allow a device to connect to a network and follow specific configurations, requirements, or settings for the connected session and how applying grouping concepts will help streamline mobile device administration. Finally, we will look at testing and pilot programs as they pertain to validating technology and changes within an existing network environment.

General Technology Network and Security Policy

The definition of the term *policy* will vary based on its specific use. For example, an automobile insurance policy is a contract between an insurance provider and an individual known as the policyholder. The insurance policy will identify details about specific agreements concerning what the policy covers and the responsibility of the provider if a claim is filed by the policy holder. With regards to technology, a security policy defines an agreement between a corporation or the technology provider and the technology user. This agreement defines the specific use as well as the terms and conditions with which the user will comply while using the provider's network and services.

The specific use of the network or services will determine the type of policy and the specifics that are contained within. For example, companies that work within the medical field will need to understand compliance with the Health Insurance Portability and Accountability Act (HIPAA) in addition to the general policy framework. Another common example is that organizations that process, store, or transmit credit card information must comply with the Payment Card Industry (PCI) standards.

Industry Regulatory Compliance

It is very important for companies, organizations, and businesses that collect private or personal information from individuals to appropriately secure all of the information. In recent years, there have been several legislative regulations that businesses are required to conform to. This compliance, of course, will depend on the type of business or the business model. The legislative regulations determine how data containing personal information is handled for health care, retail, financial, and other types of businesses. Plans for complying with them are known as legislative compliances.

When using mobile devices of any type that access wireless network infrastructures, it is imperative for companies and organizations to verify any additional requirements from a security perspective that may be needed when dealing with regulatory compliance.

Two popular legislative compliance requirements are PCI and HIPAA.

Payment Card Industry (PCI) Compliance

Payment Card Industry (PCI) regulations require companies to adhere to security standards created to protect card information pertaining to financial transactions. According to the PCI Standards Council, in order to be PCI-compliant, a company must meet the following six requirements:

- Build and maintain a secure network.
- Protect cardholder data.
- Maintain a vulnerability management program.
- Implement strong access control measures.
- Regularly monitor and test networks.
- Maintain an information security policy.

Businesses that perform payment card financial transactions must follow all regulations and maintain compliance in order to ensure the cardholder will be adequately protected.

For more information on PCI, visit the PCI Security Standards Council website at www.pcisecuritystandards.org.

Health Insurance Portability and Accountability Act (HIPAA) Compliance

HIPAA stands for the Health Insurance Portability and Accountability Act of 1996. The goal of HIPAA is to provide standardized mechanisms for electronic data exchange, security, and confidentiality of all healthcare-related computer information and data. HIPAA consists of two parts:

- HIPAA, Title I
- HIPAA, Title II

If someone loses or changes their job, Title I of HIPAA protects their health insurance coverage. This will allow them to maintain coverage for a limited time.

In the information technology industries, Title II is what most people mean when they refer to HIPAA. It establishes mandatory regulations to govern how healthcare providers conduct business by securing computer data and ensuring that confidential personal information stays that way.

For more information on HIPAA, visit the U.S. Department of Health and Human Services website at www.hhs.gov/ocr/hipaa.

PCI and HIPAA are just two examples of legislative compliances that organizations must follow. Depending on the type of business or field they are in, organizations might also need to comply with the following laws and regulations:

- Gramm-Leach Bliley Act (GLBA) — Banking/Financial Services
- International Organization for Standardization (ISO)/International Electrotechnical Commission (IEC) ISO/IEC 27002 — Information Security Standard
- Family Educational Rights and Privacy Act (FERPA) — Education
- Federal Information Processing Standards (FIPS) — U.S. Government
- Sarbanes-Oxley Act (SOX) — Public Accounting

Security Policy Framework

Keep in mind that a network security policy's contents will vary based on the specific type of business or application in which it is applied. For example, the policy for an organization that handles financial information will be different from the policy for a manufacturing company. However, most policies are based on a general framework. Using this framework will enable you to build a policy that will meet the needs of the organization.

Typically, computer network security policy consists of two main parts, general policy and functional or technical components. The general policy components consist of the policy purpose and business justification, management approval, policy definition and documentation processes, training policies, revision processes, violations and response, legal concerns, and potential threat assessment. The functional or technical side commonly consists of administrator and user training, monitoring, auditing, response and forensics, defining acceptable use, and security requirements including authentication, encryption, passcode and password policies. Keep in mind this is a general framework that can be used as a guide to assist in developing an actual security policy.

The best way to learn the components and features of mobile device management (MDM) solutions including defining profiles to help enforce security policy is to actually evaluate one and use it. Chapter 10, "Introduction to Mobile Device Management" lists several manufacturers of MDM solutions. Most if not all of these companies provide evaluation or trial periods of their products. In this chapter we will be using examples from Windows Intune, a cloud-based management solution from Microsoft. You can register for a free trial to evaluate this product at www.microsoft.com/intune.

Information Technology and Security Policy Implementation and Adherence

Every company or organization that provides technology resources of any type should already have an information technology and security policy in place for their computer network. If this is not the case then it should be a very high priority to get one in place as quickly as possible. Given that various wireless technologies are becoming a major part of all networks, it is critical for wireless and mobility technology to be part of this security policy.

If a security policy already exists, adding the wireless technologies that are used will be a fairly straightforward process. Getting the right people involved is essential for the implementation and success of a security policy. This includes the Information Technology group as well as the management team.

A well-designed policy is critical in order for network security to be successful, balanced, and maintained. Because wireless networks use the radio frequency (RF) spectrum for communications, the legal use of unlicensed and licensed RF bands will also need to be understood to ensure compliance with the local governing regulating bodies and the appropriate authorities. The wireless communications/network portion of the security policy should include, at a minimum, sections defining the purpose, scope, policy details, enforcement information, and revisions/maintenance.

Purpose   This section will provide an overview of why the policy has been put in place and define the importance of compliance to ensure that everyone understands that misuse could result in severe interference with the intended use of the network and may put the organization and users at serious security risks.

Scope   The scope will provide an overview of who the policy will apply to, the use of the wireless spectrum, the acceptable use policy, and the importance of compliance.

Policy Details   This section should contain a basic overview of the wireless technology that will be used, including the allowed RF spectrum and the wireless technology and devices that can or cannot be used. This section should also identify some of the more technical aspects such as user authentication and data encryption, the tolerance for specific types of network traffic and RF interference, and acceptable applications.

Policy Enforcement   This section should define any actions, including potential litigation, that will be taken against those who are not in compliance.

Revisions/Maintenance   With the frequent changes in technology, a security policy can be somewhat of a moving target. Revisions and maintenance should be documented including a change record or revision history page. Figure 11.1 shows an example of a cloud-based mobile device management (MDM) program that can be used to create device profiles based on policy and provide policy enforcement.

FIGURE 11.1   Windows Intune mobile device management (MDM) dashboard

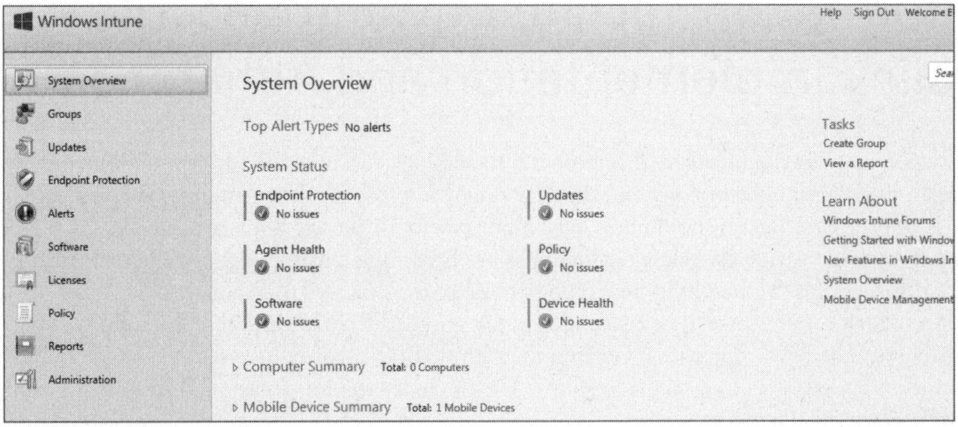

Acceptable Use Policy

One of the most basic, common, and important policies for computer networking is an acceptable use policy. This is typically a policy that the user must comply with in order to access and use any computer network resources or services. At a minimum, the acceptable use policy will outline the rules regarding what a user can and cannot do. In most cases this is an abbreviated version of the complete policy that the user will be referred to and have access to if further detail or clarification is needed. Most organizations require all users to sign and agree to this policy. This could be in the form of a physical document or in an electronic format. For the mobile wireless guest user, the acceptable use policy may come in the form of a web-based captive portal and usually includes restrictions and limitations. You learned about captive and self-service portals in Chapter 10.

Balancing Security and Usability

When it comes to policy, the ability to maintain a good balance between security and usability is a key factor. Some people believe that you can have too much of a good thing. If a policy is too tight, it may limit productivity or a user's ability to utilize their devices correctly and to get their job done. On the other hand if it is too loose, it could potentially compromise the security and integrity of the corporate network. Therefore, providing an accurate balance between the two must be closely evaluated. A poorly balanced security policy is directly proportional to a policy failure. One example is a passcode or password policy. Most industry best practices agree that the longer and more complex a passcode is the more secure it will be. However, if it is too long or complex, a user may have difficulty remembering it, or depending on the device, it may be a challenge to type it in. Another example is what exactly the user of the device is allowed to do. Can they install and remove apps, and can they modify the system's or device's settings? All options must be carefully considered in order to maintain a good balance. Much of this will also depend on the type of organization where the device is used. Finding the correct balance will be a challenge and is contingent on understanding what you are trying to protect and what you are trying to prevent.

Backup, Restore, and Recovery Policies

Regardless of whether a device is a company asset or is employee-owned, the importance of backup, restore, and recovery cannot be overestimated. A well written backup policy that can be easily understood and implemented will make what is usually considered an unpleasant or often ignored task much easier to comply with. Mobile devices that are used to store corporate data and intellectual property information such as trade secrets must correctly implement and adhere to the specified policy. Automating this task as much as possible will lessen the reliance on the device's user interaction and allow for a better chance of policy compliance.

Many may think of device backup as only something that is used for recovering data that may be lost because of device failure or unintentionally deleted. With mobile devices you must also take into consideration that a device can be stolen or lost, which will then put all information that is stored on that device at risk if the device falls into the wrong hands. In such a situation, a network administrator can use an MDM solution to easily locate the device over the Internet using location services and remotely backup the data stored on the device to a corporate file server or cloud-based storage network. After a backup is complete, the device can be securely locked or even remotely wiped to provide additional security.

Backup concepts and processes are discussed further in Chapter 18, "Data Backup, Restore and Disaster Recovery."

Operating System Modifications and Customization

Mobile device operating systems are designed to offer specific features and capabilities. This allows for consistent functionality and operation of the devices that run the operating systems. However, many hardware device manufacturers customize the device to operate with some of their specific features and to include their company branding. In addition, the owner of the device may be able to add their own customizations which may include:

- Fonts
- Home screens
- Ring tones
- Wallpaper
- Widgets

The fact that some operating systems such as Android are open source provides many opportunities for the device manufacturer or the owner to add their own unique flair.

Manufacturers of mobile device technology usually have certain features "locked down" which means they cannot be changed. However, some devices may be "rooted" or "jailbroken" to remove manufacturer's restrictions. Android rooting and iOS jailbreaking can void the device manufacturer's warranty and cause other issues. These are both discussed further in Chapter 15, "Mobile Device Security Threats and Risks."

Operating System Vendors

In Chapter 10, we explored some of the popular mobile device operating systems. These include Android, BlackBerry, Apple iOS, and Windows phone. Chances are very high that organizations that allow employee-owned devices to access the company network will have a mix of these and other various devices with different operating system vendor platforms. The ability to integrate the various devices and operating systems into the environment may come with certain challenges that limit capabilities among the different vendors. These capabilities include the ability for remote management of the devices, general support for different operating systems features, and specific technology capabilities such as passwords and resets.

Because of the diverse number of devices and available operating systems installed, the MDM administrator should be proactive with keeping up to date with any technology changes. Some of the methods an administrator can use to stay current with mobile technology changes include:

- Social media
- Subscribing to available services from the manufacturers
- Rich Site Summary (RSS) services or feeds
- Online webinars and available training sessions

The proactive approach will allow the administrator to prevent potential issues or problems from occurring by becoming familiar with various limitations between the mobile device operating systems in use.

Original Equipment Manufacturer (OEM)

The definition of original equipment manufacturer (OEM) may vary depending on the context in which it is used. In the automobile industry the term "OEM part" is a replacement part that is manufactured by the same company that made the original part. This gives the buyer confidence that the part will be an exact replacement and will work as designed.

With regards to mobile technology, one definition is when a company purchases the rights to resell a product under its own brand name. One example is the ability for different manufacturers to sell smart phones that run the Android operating system. Integrating directory services into an MDM solution may allow an administrator to utilize an existing directory database and help streamline user/device authentication and enrollments and can help to gather information from the managed device and the connected user.

Vendor Default Device Applications

Mobile device operating systems often have default applications that are installed by the manufacturer. These applications may include Internet browsers, email applications, video, audio/music, documents viewers, and others. The type of file that is accessed will determine which application is launched to manage or open that file. On devices with the Android operating system installed, you will be able to select which application you want to use to complete the task. If you would like to, you can use the selected program to "always" open future files that are of that same type or you can specify to open the file "just once." If you choose the latter option, you will be presented with the option again when a file of the same type is opened. If you change your mind and later want to use a different application, this can easily be done by resetting the default application. In Exercise 11.1 you will reset the default application for a device that is running the Android operating system.

> **EXERCISE 11.1**
>
> **Resetting the Default Application—Android**
>
> This is a short exercise to demonstrate one way to reset the default applications on a device using the Android operating system. Keep in mind these steps may vary based on the operating system and the version that is in use.
>
> 1. Select the Android settings icon.
> 2. Select applications.
> 3. Select the application that is currently set to open the file type you would like to change.
> 4. Scroll down to launch by default and tap clear defaults.
> 5. You will be prompted to select an application type the next time that file type is launched.

Technology Profiles

When working with various types of technology such as wireless networking, profiles are used to provide configuration or settings information and to help with security policy compliance. Saving this information within a profile will help to lessen the overhead of entering specific information every time you wish to connect to a network or access a network resource. Profiles can be used in a variety of scenarios, for example with IEEE 802.11 wireless networking client devices or connecting to a virtual private network (VPN) from a client to a VPN concentrator. Mobile device profiles using MDM will simplify and streamline access to an infrastructure and device management.

Mobile Device Profiles

A profile is a set of parameters or instructions that will provide a device the ability to connect to a network and follow specific requirements or settings for the connected session. These profiles will define specific settings and configuration information in addition to restrictions based on the device itself and who is using the device. The profiles can be assigned for operating system, individuals, and groups. Profiles may contain configuration information for:

- Email access
- Internet access
- Network configuration information
- Wi-Fi access
- VPN configuration information

In addition to configuration and settings, a profile will help maintain security policy compliance. Figure 11.2 is an example of a cloud-based MDM software application that can be used to set password policy enforcement.

FIGURE 11.2 Windows Intune password policy configuration screen

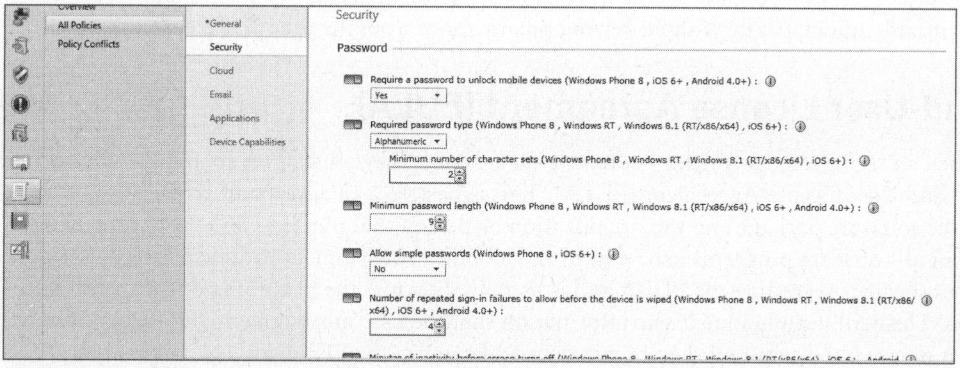

Directory Services Integration

From a computer networking perspective, a directory service is similar to that of a database where information about specific objects can be created, stored, and retrieved. An analogy is a simple address book contained within a smart phone. In this address book you could enter specific information about contacts such as their name, various phone numbers, an email address, and much more. Searching a person by name will provide any additional information that is documented about that contact.

Microsoft Windows Active Directory and Novell eDirectory are examples of computer network directory services. These are X.500-compliant directory services and are used to centrally manage users, groups, devices, and other resources. A network administrator can create user and group objects and allow them access to various network resources such as a shared folder on a file server or a network printer. Integrating directory services will allow a network administrator to assign mobile device applications, user and device profiles, and other data that can be associated with their directory services group membership. Depending on the MDM solution that is used, you can also synchronize any changes that are made to the directory database to the mobile device policy that it is connected to. This will include any changes that are made to the user, device, or group object that the device is associated with. Deleting a user object in the directory services database will allow the MDM software to take the appropriate designated action.

Issuing Digital Certificates

A digital certificate can be used as an access control method for various mobile devices, computers, and other devices and will ensure that only valid authorized users will gain access to the corporate network infrastructure. Certificates are one method of access control used to confirm that a person is who they say they are. This will help to restrict access to a system or resource only to those who are entitled to the access. Digital certificates can use a security concept known as a Private Key Infrastructure (PKI). Access control methods, certificates, and PKI are discussed further in Chapter 16, "Device Authentication and Data Encryption."

In a full PKI deployment, digital certificates are required on the RADIUS server and the client device. Installing digital certificates can be streamlined using MDM software to issue the certificates automatically without relying on any input from the user of the device.

End-User License Agreement (EULA)

When a user installs proprietary software on a device, they must agree to and comply with the End-User License Agreement (EULA). This is basically an agreement between the creator of the software package and the organization or person that purchased the software. Most if not all software programs used within the mobility environment include a screen that provides specifics regarding the EULA and is usually shown at the beginning of the install process. This will include some basic information that the user must agree to before continuing.

It is very important that the organization understands the EULA for legal and liability management regarding mobile device technology and employee-owned devices. In most cases these are customized by the legal department to ensure the EULA complies with the company objectives and covers the specific use of the software. Keep in mind that in many cases this is a scaled-down version of a more complete document that defines all the specifics in more detail.

Figure 11.3 shows an example of a cloud-based MDM administration console used to create a license agreement.

FIGURE 11.3 Windows Intune management console used to create a license agreement

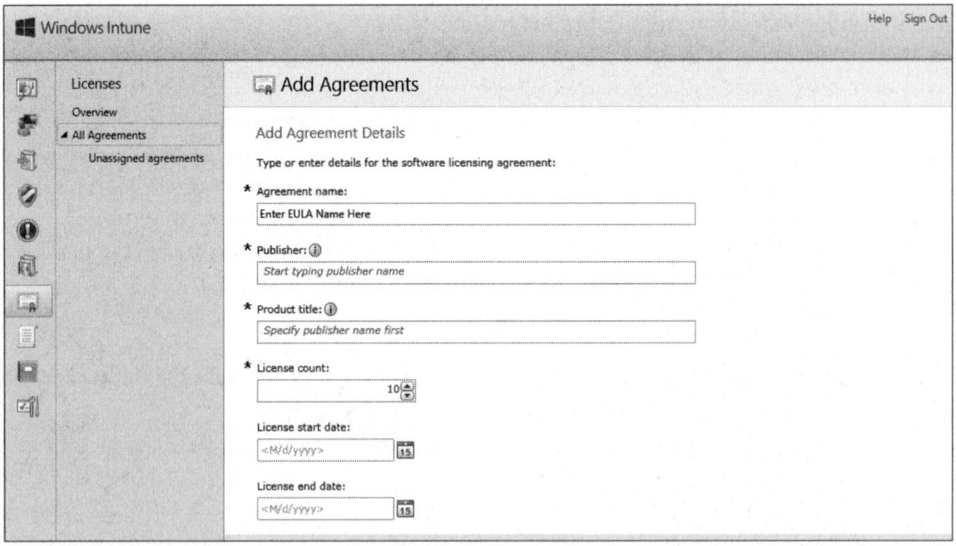

Notice in Figure 11.3 that this program allows an administrator to enter the license name, software publisher, product title, license count, and start and end dates. When a user connects to the portal, they will be able to agree to and install the program.

Understanding Group Profiles

For anyone who has performed basic computer network administration tasks, understanding group profiles should be fairly straightforward. Managing a network and access to its resources is much easier and streamlined when using a grouping methodology. See the case study to examine one specific scenario.

Assigning Access to Network Resources Using the Group Technique

You are a network administrator and have 3 departments in the company that will each require access to specific resources that are located within the computer network. The departments are sales (10 members), engineering (3 members), and consultants (2 members). The sales department will require access to a marketing database, the engineering department must be able to access product technical specifications, and the consultants department will require access to specific company briefings. There are two ways to approach this. One would be to explicitly assign permissions to the resources to the members of each department. The second way would be to create a group object for each department, add the members to each group, and assign the groups access to the resources. If a user will no longer require access to the network service or resource in which they have permissions via the group membership, the administrator can remove them from the group, thereby removing their access. Using groups helps to streamline the administration process of mobile devices by reducing repetitive administrator work in maintaining policy compliance and allowing access to network resources and services.

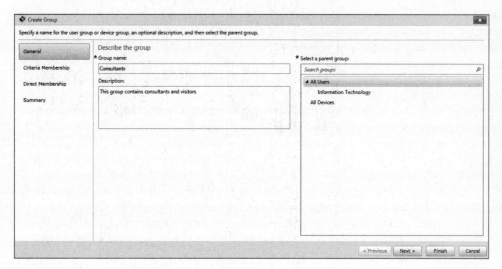

continues

> *continued*
>
> The first option is an inefficient method of administration because any changes are made on an individual per-user basis and would require potentially extra administration and overhead. The second option is more efficient because any changes are on a per-group basis.

The case study described here is very similar to the concept of creating profiles for devices within an MDM environment. The needs of the individual, whether the device is company-owned, and departments the user of the device belongs to will be factors in determining the capabilities of the device and the user. Some common profiles for specific devices and users with MDM techniques include:

- Corporate-owned devices
- Employee-owned devices (BYOD)
- Outside consultants

Corporate-Owned Mobile Device Profiles

Issuing corporate-owned mobile devices to employees provides a network administrator with much-needed control over how the device is managed. Unlike with employee-owned, BYODs (discussed next), there is less concern about the type of applications or data that is used or stored on the device since the network administrator can have complete control over ensuring that applications and profiles that are installed on the device comply with corporate security policy. Something to consider is whether personal applications and data will be allowed on a corporate-owned device. The corporate security policy will need to define exactly how this is handled.

Employee-Owned Mobile Device Profiles

Devices that are employee-owned will require special consideration as to what exactly a user can or cannot do with the device once it is connected to the corporate wireless network. The continued growth of the bring your own device (BYOD) concept has raised concerns regarding security and network capacity. However, the benefits of BYOD must also be considered. A company, organization, or school can leverage employee-owned or personal devices to provide a cost savings by allowing these devices to be used on the company network.

In cases such as this, the correct provisioning and policy compliance will need to be closely evaluated. If employee-owned devices are allowed to be used on the corporate network, containerization of applications and data should be considered. This will allow corporate applications and data to be stored, maintained, and managed separate from the employee's own personal data. In the event a device is lost, stolen or unaccounted for, the MDM software program can take the appropriate action that will mitigate any potential security concerns by performing a device lockdown or remote wipe of the corporate data that is installed on the device. The owner of the device that is used on the organization's network must be made aware that with certain security threats the appropriate action will be taken in order to secure the device and protect any confidential company data.

MDM providers offer different mechanisms or processes using containerization technology. It is recommended as part of the software evaluation and trial period that you take a close look at how the manufacturer deals with this concept and try to determine whether their way would be a good fit for your organization.

The practice of companies allowing employee-owned devices on the corporate network will continue to grow. The Gartner group estimates that by the year 2017 more than 50 percent of companies will require that employees supply their own devices for use on the corporate network, helping to justify the use of MDM solutions and technology.

Outside Consultants' and Visitors' Mobile Device Profiles

Depending on the business model or organization type, providing network access to outside consultants and visitors can be considered somewhat of a challenge. In many cases, connecting to the organization's network, wired or wireless, is not allowed under any circumstances. In situations such as this, it is not uncommon for the organization to provide a third-party service such as a cable modem or digital subscriber line (DSL) connection from an Internet Service Provider (ISP) to allow outside Internet access for the guest user. A benefit of this scenario is that the outside consultants will in no way be connected to the organization's network infrastructure, which may be a requirement in the network security policy. If this type of service is not provided or allowed, then the only option a consultant will have is to use a cellular data connection such as 3G, 4G, or LTE to get outside access.

For various reasons including security, many organizations allow physical wired access to a network only to its employees but will allow wireless network access via a connection to the guest wireless network. This guest network is typically on a completely separate virtual local area network (VLAN) to ensure it meets the requirements of the network security policy. In this situation it is important to define how the outside devices will be able to connect and what exactly they will be able to do once they are connected. Defining the requirements such as content filtering, the amount of bandwidth, and time restrictions are some examples of the limitations guest networks may be subject to. Most if not all of these restrictions can be achieved through the use of an MDM profile.

Policy and Profile Pilot Testing

Before any new technology, software, or changes are deployed, it is beneficial to perform pilot testing within key areas of the organization in order to validate that the technology will work as specified and designed. This is a concept that is true with various types of technology, including, but not limited, to Voice over Wi-Fi (VoWi-Fi) handsets, client device deployments, firmware updates, software patches, fixes and upgrades, and of course mobile device policy and profiles. Although some consider this an option rather than a requirement, this process must be carefully considered as it is best to find any issues that may arise on a smaller scale rather than after a full deployment or upgrade. The pilot program process will vary depending on how and where the technology is used, the scale of the deployment, the type of technology, and who is performing the testing. Some common steps that may be used as part of pilot program testing include:

- Selecting the individuals or groups to be part of the test
- Soliciting feedback and monitoring the testing phase
- Making adjustments or changes as needed based on the results
- Getting proper approval from change management
- Implementing full deployment throughout the organization

It is also important to monitor the devices and users after the implementation has occurred and to solicit feedback in order to identify any issues that arise that were not identified during the pilot program testing phase.

Summary

In this chapter we explored the importance of understanding computer networking policies, especially security policy. We briefly looked at a few industry regulations including HIPAA for organizations that work with healthcare and PCI for those who deal with financial credit card transactions. You saw that how policies are defined will depend on the organization in which they are used. You learned about the basic components of a network security policy, which include the purpose, scope, details, enforcement, and maintaining changes. I explained the importance of an acceptable use policy and that most if not all organizations require employees and guests to agree to the terms and conditions of this type of policy. You saw the importance of balancing security and usability and looked at some common networking policies such as device backup, restore, and recovery. In addition to policies, you learned about device technology profiles, which contain configuration or settings information for the device based on the user and that can also incorporate specific profiles. Finally, you saw the importance of testing technology, mobile device profiles, and policies by using pilot programs and validating any potential implementation issues prior to a full scale deployment.

Chapter Essentials

Understand basic technology policy. Know that a technology security policy defines an agreement between a corporation or technology provider and the technology user with instructions and rules that must be complied with.

Be familiar with policy regulatory compliance.   Understand that policy compliance is an important topic, regardless of whether it is for specific industries such as medical, financial, or education, or even for specific requirements within an organization such as mobile device data backup and restore.

Know the common components of a basic Information Technology policy.   Although policies will vary based on the type of organization, understand the basic components that will be part of the policy. These include the purpose, a scope, the policy details, policy enforcement, and revision/maintenance.

Understand the basic acceptable use policy.   The acceptable use policy is typically a requirement within most businesses or organizations. Users will agree to this policy in writing or electronically before any access is granted. One common practice is to require wireless guest users to agree via a web-based captive portal.

Define a balance between security and usability.   Understand that a security policy that is too strict may limit the ability for a user to utilize their devices correctly and to be able to get their job done, and a policy that it too liberal may potentially compromise the security and integrity of the corporate network.

Understand common security policies.   Policies such as data backup, restore, and recovery procedures should be seriously considered, implemented, and enforced. This will ensure the security and availability of corporate data.

Understand the concept of using profiles.   Know that device and user profiles can be used to provide a user with the ability to connect to a network and follow specific requirements or settings for the connected session. Also understand this will allow compliance with specific defined company policies.

Be familiar with the grouping concept.   Understand that managing a network and access to resources is much easier and streamlined when using a grouping methodology. Groups can be created based on the company's organization chart and will ease mobile device administration. Providing access to or removing access from various resources and services is easily accomplished using the group methodology.

Understand the concept of pilot testing.   Understand that testing new technology and profiles within key groups of the organization will help lessen full deployment issues. This allows the administrator to become familiar with any problems that may arise from using the new technology or how profiles and policies are implemented. This process usually includes selecting which groups will be tested, soliciting feedback from the users, making any needed changes, getting proper approvals, and finally implementing changes throughout the organization.

Chapter 12

Implementation of Mobile Device Technology

TOPICS COVERED IN THIS CHAPTER:

- ✓ System Development Life Cycle (SDLC)
- ✓ Pilot Program Initiation, Testing, Evaluation, and Approval
- ✓ Training and Technology Launch
- ✓ Documentation Creation and Updates
- ✓ Mobile Device Configuration and Activation
- ✓ Wireless LAN, Cellular Data, and Secure Digital Adapters
- ✓ Mobile Device Management Onboarding and Provisioning
- ✓ Mobile Device Management Offboarding and Deprovisioning

Implementing any type of new technology comes with its own unique set of challenges. Both wireless LAN technology and cellular technology have grown at a tremendous pace over the last decade. Mobile device technology allows a user to access information in ways that seemed impossible not too long ago. In this chapter, we will explore some of the foundational components of this technology, including the processes and methodologies known as the System Development Life Cycle. The implementation process includes pilot testing to ensure the technology meets the company requirements, getting approval or management buy-in, training for all staff and employees, and device activation.

It is important to be familiar with and understand the different devices that are used with over-the-air technology, including both infrastructure and wireless client devices for wireless LAN and cellular data. We will explore some of the different options that are available. I will explain the method used for bringing these client devices onto a network for the first time, a process known as onboarding, and taking devices offline or off the company network, a process known as offboarding. There are several ways that devices can be brought on to a network, both manual and automatically. The objective is to make this process as transparent to the end user as possible. With offboarding, it is important to make sure procedures are followed to secure company data when an employee leaves the organization.

Finally, I will explain the importance of preparing mobile technology devices for recycling or disposal after they have been decommissioned. Deleting data from a device may not really prevent it from being seen or accessed again, and certain technology factors must be considered.

System Development Life Cycle (SDLC)

The definition of *System Development Life Cycle (SDLC)* will vary based on the context in which it is used. With regard to information technology (IT), the SDLC consists of several different phases and processes. One example used with IT allows for network designers and engineers to plan, design, test, build, and deploy the information system technology. Another term used to describe this methodology is *application development life cycle*. The SDLC concept in this context will help ensure the success of an infrastructure and mobile device deployment. The following list describes some of these common components that are included in the SDLC. Keep in mind that this information may vary based on how a system is implemented.

Plan the system. The planning component is required to ensure that all parts of the organization are involved and adequately covered. This includes identifying concerns and addressing specific needs of various departments contained within the organization. The expectations should also be addressed to verify that all of the parties are on the same page, so to speak. Understanding the business requirements of the organization will help to provide a satisfactory solution.

Design the system. Design is based on the information that is gathered from the planning phase to ensure that all of the intended technology requirements are met. This phase will ensure that organization-, company-, and department-specific conditions are met.

Test the system. This phase is to verify that everything will work as planned and designed. The testing phase may include a pilot program, as explained later in this chapter. Adequate testing of a system will help provide positive results once the system is built and deployed. Including key departments and individuals will help ensure the system's success.

Build the system. With respect to information technology, the build phase will include procurement of all components per design and bill of materials. This is based on the results of the pilot testing. Once the design or proof of concept is complete, preparation for full deployment will begin.

Deploy the system. The next step is to deploy the system per the design, and make any changes due to feedback from the testing phase. In some SDLCs, the build and deployment phases are combined, while in others they are separated as two distinct components.

Maintain the system. Maintaining the system to provide acceptable performance, and to make certain the system will operate as planned and designed, is the last step. There are many solutions that can be put in place to help streamline the maintenance and support of information systems. This includes help desk software solutions, documentation, and system modifications based on environmental (physical) changes to the organization.

Figure 12.1 illustrates the System Development Life Cycle (SDLC) process as described in this book.

The abbreviation *SDLC* is used for various terms. In the context of this chapter, it refers to System Development Life Cycle. In the technology world, it also means Synchronous Data Link Control, which is a protocol that was developed by IBM in the early 1970s.

FIGURE 12.1 System Development Life Cycle (SDLC)

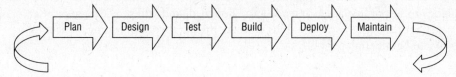

Pilot Program Initiation, Testing, Evaluation, and Approval

In Chapter 11, "Mobile Device Policy, Profiles, and Configuration," I explained that pilot testing of mobile device policies and profiles can be a beneficial way to ensure that a profile will work as designed, and meet the required goals. *Pilot programs* can also be used to ensure that the complete technology solution hardware and software can operate as designed or intended. This includes testing installed applications, verifying coverage and capacity requirements, and testing overall performance of the network. I will now expand on this concept and explain its steps in more detail.

Select the individuals or groups to be part of the test.   Spend some time researching individuals or groups to be part of the pilot program to choose those best suited. It may be best to use people from different departments, because the types of applications and work habits will differ based on what their job responsibilities are. This could range from light users who only use their device for email and other simple tasks, to heavy users who multitask with several bandwidth-intensive applications that can stress the system. You should specify a timeline and provide documentation of the expectations to the people who are involved in the pilot testing process.

Solicit feedback and monitor the testing phase.   It is important to have a plan in place that will allow you to receive feedback from users who are part of the pilot testing. This could be as simple as an anonymous feedback form on an intranet web page, telephone calls, or even personal visits if feasible. Make the users feel comfortable in knowing that their feedback is important and that the information they provide will help to make a deployment a better user experience for everyone in the organization.

Make adjustments or changes as needed based on the results.   It would be very nice if everything always worked as planned or designed. Unfortunately, this is usually not the case. This is where feedback from the pilot program will help. You can make adjustments or changes to address feedback you have received from users participating in the program. A network that functions well and addresses the needs of the user population will ensure user satisfaction, and help to improve the job performance of the user base. This will in turn make things easier for the network administrator and other support personnel.

Get proper approvals from change management.   Getting the proper approvals from the change management group will ensure that your research and pilot testing has met the desired goals, and the deployment can move forward as planned.

Implement full deployment throughout the organization.   This final step will allow the technology to be put in place as intended and designed, and will provide a great user experience for the entire user base. This is where the hard work of testing, collecting feedback, and evaluation pays off.

Training and Technology Launch

Training is an important component of all types of technology deployment. The training process for both the technical staff and the end user base must be taken into consideration. Network administrators will need to learn about the new mobile devices and associated hardware. This includes installation, configuration, setup, and ongoing support. The end user will need to understand the technology and learn how to use the devices. This is especially true when it comes to understanding the security solutions that have been put in place, and adhering to corporate security policy.

Training for Network Administrators

This type of training will include both vendor-neutral training to learn about the technology in general, and manufacturer-specific training to learn about the products. Many options are available, including online videos, online instructor-led training, and in-person instructor-led training. The online videos and online instructor-led training solutions may be available for free, or they could be fee based. In most cases, in-person instructor-led training is fee based. Hardware and software manufacturers often have their own training programs that can be alternatives to vendor-neutral training solutions. These are beneficial because they provide content specific to their products. Most manufacturers are appreciative if the individuals who attend the training sessions already have a basic understanding of the general technology. This enables the instructors to skip the basics and instead focus on the functions, features, and bells and whistles of their specific solutions.

Training for End Users

It is important not to forget about the end user of the deployed technology solution. They need to be adequately trained, and understand the technology that has been implemented. Assuming that the end user will be fine without proper training is a mistake that can lead to problems down the road. In addition, no matter how much money you spend on the technology, how fancy the technology is, and how proud you are of the implementation, if it does not improve the employee's productivity, it is a waste of money and could result in overall system failure. Provide adequate training to end users to ensure proper operation of the equipment and that the security requirements are met and maintained.

Technology Launch

After all the requirements are met, including those described in the SDLC and during design, and after pilot testing, adjustments, management buy-in, and training, it is time to deploy the technology solution. It is important to closely monitor the performance after the deployment has occurred. This will allow you to identify any problems or issues that may not have been identified during the pilot testing phase. A baseline analysis of the performance will provide statistics and metrics that can help you ensure the overall health of the system.

Documentation Creation and Updates

Quality documentation within the information technology arena will usually lead to a successful deployment. The task of documentation is often neglected or even forgotten, but its importance cannot be underestimated. From the initial gathering of information and business justification phases to the final deployment and buy-in phases, a well-documented process will help to provide technology that will work as designed and can be supported and maintained. Documentation is not just a one-time task. It is an ongoing process that must be maintained and updated. This is true with all areas of information technology. Quality documentation will help to ensure that a long-term deployment can successfully operate and be supported as intended.

Mobile Device Configuration and Activation

Many wireless mobile devices are multifunctional and can be used with Wi-Fi, Bluetooth, WIMAX, and cellular networks. For a device to be used on any of these networks, a configuration or activation of some sort is required. For Wi-Fi devices, configuration will consist of connecting to the appropriate wireless access point and providing the correct credentials if required. Once this is completed, the device will have access to any of the network resources to which the user of the device has been granted access. Bluetooth technology is commonly used to connect devices such as smartphones, tablets, and computers together and to connect other devices, such as headsets, cameras, keyboards, and computer mice, within a personal area network (PAN). These devices use what is known as the discovery phase to scan the air and find other Bluetooth-capable devices. Once other Bluetooth devices are found, the user can enter the correct personal identification number to allow the devices to be "paired." This pairing of devices will create a *piconet*, which is an ad hoc network for Bluetooth devices that are part of the same PAN. Both the Wi-Fi network and the PAN allow users to connect by having the correct credentials, and activation is not generally required. Figure 12.2 shows an example of a piconet created from connecting devices in an ad hoc Bluetooth network.

On the other hand, devices that contain cellular capabilities require some type of *activation* process to be used on the wireless carrier network. This activation will require the user to subscribe to a service that allows the device to function on the cellular network. Unlike Wi-Fi and Bluetooth networks, these services are fee based and are available with a

FIGURE 12.2 Bluetooth piconet

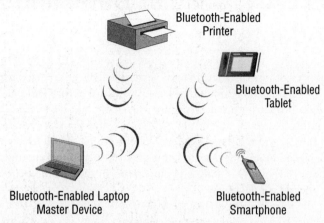

Bluetooth Piconet

yearly contract agreement or monthly billing. The activation process on a cellular network is fairly straightforward. Calling a specific telephone number or using an activation web page are both common methods used today. Figure 12.3 shows an example of a website used to activate cellular phones and other devices.

FIGURE 12.3 Verizon Wireless cellular phone activation web page

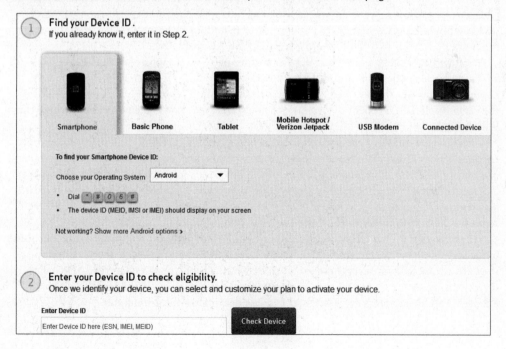

Wireless LAN, Cellular Data, and Secure Digital Adapters

A combination of network infrastructure and client-side devices are used for *over-the-air communication*. In Chapter 6, "Computer Network Infrastructure Devices," we explored some of the common infrastructure devices that are used to allow wireless devices to access a network using radio frequency for communications. These devices include wireless access points, wireless bridges, and wireless LAN controllers, which are all used to allow wireless LAN–capable client and mobile devices to connect to an infrastructure and access the available network resources. We will now explore some of the common client devices that are used for over-the-air connectivity.

When we hear the term *client devices*, we often think of computers—either desktop or notebook—connected to a computer network. However, there are many other devices, both wired and wireless, that can connect to a network. Wireless LAN client devices include various types of computers, tablets, smartphones, scanners, print servers, cameras, and other devices that are used to send data across the network.

Devices that connect to wireless networks use various types of adapters. Which adapter is used depends on the client device that it connects to. You can connect wireless adapters to such devices as a notebook computer, tablet, desktop computer, or bar-code scanner. Wireless LAN adapters are available in various types, both external and internal to the device. External adapters will connect to an available interface in the device, such as a USB port or card slot. Some devices use internal adapters that may require some level of disassembly or removal of a cover panel prior to installation. Examples of internal adapter types are PCI, Mini PCI, Full Mini PCIe, and Half Mini -PCIe.

Wireless LAN client adapters differ from other network adapters (such as Ethernet adapters) because they contain radio hardware and use radio frequency (RF) to send the computer data over the air. A wireless LAN design should be partly based on the needs of the client applications, client device types to be supported, and the environment where they will be used.

Cellular data adapters are used to connect certain device types to a wireless cellular carrier service. These devices include laptop computers, branch office wireless LAN routers, and others. Cellular data adapters are available in various form factors, and commonly use a direct Universal Serial Bus (USB) connection. They may also be built in to devices such as a laptop computer via a Mini PCIe adapter. Some external adapters are available that require a cable to connect to the device.

Secure Digital (SD) adapters were originally designed as memory storage cards, but some cards had wireless LAN technology added to them. Although these are not as common, they are available from several manufacturers. Many mobile devices can read SD cards directly using a port on the device or via an adapter of some sort. Both cellular data adapters and SD adapters are discussed in more detail later in this chapter.

USB Wireless LAN Adapters

Introduced in 1995, the *Universal Serial Bus (USB 1.0)* standard was designed as a replacement for legacy serial and parallel connections. *Serial communication* is the process of transmitting one data bit at a time. *Parallel communication* has the capability of transmitting several data bits at a time. Imagine a single-lane road compared to a four-lane highway. On a single-lane road, only one car at a time can travel, whereas on a four-lane highway, many cars can traverse the same path at the same time.

USB allows connectivity for keyboards, digital cameras, printers, computer networking adapters, and other devices that once used serial and parallel data connection ports. USB standards are implemented by the USB Implementers Forum (USB-IF). This organization consists of companies from the computer and electronics industries, including Intel, Microsoft, NEC, and HP. Over time, they have evolved the USB specification through several iterations.

USB 1.0 specified data rates from 1.5 to 12 Mbps and was replaced by USB 1.1 in 1998. Devices using version 1.1 of the standard were common in the market.

The USB 2.0 specification was released in April 2000. The first revision appeared in December 2000, and the standard has been revised several times since. USB 2.0 incorporates several changes, including connector types. Data rates now allow for a maximum speed of up to 480 Mbps. Keep in mind that the signaling rate is actually 480 Mbps. However, the effective throughput may only be as high as 280 Mbps, and is sometimes much less due to the bus access constraints of the devices. Figure 12.4 shows an example of a USB 2.0 port.

USB 3.0 technology was introduced in early 2010 and is commonly available to the point of being included on newer system boards. USB 3.0 devices have faster transfer rates and use a wider bandwidth, and they allow multiple logical streams and improved bus use with asynchronous readiness notification without polling. USB 3.0 greatly increases the transmission speed from the 480 Mbps of USB 2.0, up to 4.8 Gbps. This is more than 10 times that of the earlier standard, and as a result USB 3.0 is known as SuperSpeed. In addition to speed, the new USB 3.0 specification addresses improvements to the technology, including bandwidth, by using bidirectional data paths, power management, and improved bus utilization. USB 3.0 device ports are typically blue in color and are backward compatible to USB 1 and 2 ports.

 USB 3.1 will have rated speeds (signaling rate) of up to 10 Gbps. At the time of this writing, no devices have been released that support USB 3.1 technology.

FIGURE 12.4 USB 2.0 port on a notebook computer panel

USB 2.0 Port

Features of USB

USB uses a standard connector that replaces 9-pin serial, 25-pin parallel, and various other connector types. External configuration allows the user to plug in the USB device and power it with a single USB port. The computer operating system will guide the user through the device driver installation process. External installation minimizes the need to open up a computer case and make adjustments within the computer, such as switch or jumper settings. USB also supports hot-swapping of devices, allowing connection and disconnection without the need to power down the device or the computer. In some cases, USB allows for power to be delivered to the peripheral device, eliminating the need for an external power supply. Fewer and fewer devices are being built with PC Card or ExpressCard interfaces, and most use only USB ports for adding peripheral devices.

 For additional information and specifications regarding the USB standards, visit the USB Implementers Forum (USB-IF) at www.usb.org.

Installation and Configuration of USB Devices

Exercise 12.1 walks you through the steps to install the D-Link Wireless N USB 2.0 adapter. Many USB wireless LAN adapters use installation procedures similar to this one. Installation steps are specific to the manufacturer, and I recommend that you follow the manufacturer's installation instructions. Always read the manufacturer's manual regarding installation and safety before attempting installation.

EXERCISE 12.1

Installing a USB 2.0 Wireless LAN Adapter

To install the D-Link Wireless N USB 2.0 adapter on a computer running Microsoft Windows, follow these steps:

1. Insert the setup CD into the CD-ROM drive. The program should start automatically, and an Autorun screen will appear. Click to start the installation, and the Installation Wizard window will appear.

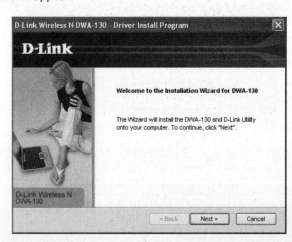

Wireless LAN, Cellular Data, and Secure Digital Adapters

2. Accept the default location to install the files, or browse for an alternate file location.
3. When prompted, insert the USB adapter into an available USB port on your computer.
4. When prompted, enter the network name (SSID) manually. If you don't know the SSID, click Scan to see the site survey page. The site survey page will also appear if the SSID is entered incorrectly. Click the network name (SSID) and click Next.
5. Click Finish to continue. If prompted to restart the computer, select Yes, Restart The Computer Now.

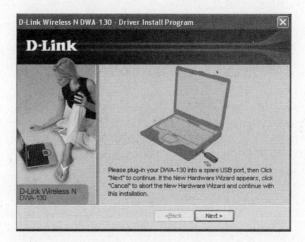

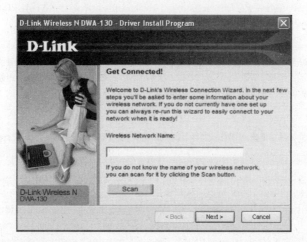

Mini PCI Wireless LAN Adapters

Mini PCI is a variation of the PCI standard, designed for laptops and other small-footprint computer systems. One common example of a Mini PCI card is the IEEE 802.11 Mini PCI adapter shown in Figure 12.5.

FIGURE 12.5 IEEE 802.11 Mini PCI adapter

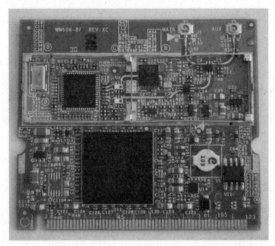

Mini PCI cards are common in many devices, such as Fast Ethernet networks, Bluetooth, modems, hard drive controllers, and wireless LANs. In the wireless world, Mini PCI cards are used in access points and client devices such as laptop or notebook computers.

Mini PCI Express (*Mini PCIe*) cards are a replacement for the Mini PCI cards and are based on PCI Express.

 Many notebook and portable computers with built-in wireless LANs use Mini PCI, Mini PCIe, or now Half Mini PCIe cards for wireless IEEE 802.11 wireless LAN connectivity.

Features of Mini PCI, Mini PCIe, and Half Mini PCIe Cards

Mini PCI cards are available in three types: Type I, Type II, and Type III. Types I and II use a 100-pin stacking connector. Type III cards use a 124-pin edge connector. Type II cards have RJ11 and RJ45 connectors for telephone and Ethernet network connections. These cards are commonly located at the edge of the computer or docking station so that the connectors can be mounted for external access, such as to a modem or computer network.

Mini PCIe cards are 30 × 56 mm and have a 52-pin edge connector, consisting of two staggered rows on a 0.8 mm pitch. These cards are 1.0 mm thick, excluding components. Table 12.1 summarizes the features of Mini PCI and Mini PCIe cards.

The *Half Mini PCIe* cards are 30 × 31.90 mm. The main difference between this card and the Mini PCIe card mentioned earlier is the length. The length of this new form factor is about half the length of the Mini PCIe card.

With the introduction of the Half Mini PCIe card form factor, the Mini PCIe card is now called the *Full Mini PCIe* card.

TABLE 12.1 Features of Mini PCI, Full Mini PCIe, and Half Mini PCIe cards

Card Type	Connectors	Size
Mini PCI Type IA	100-pin stacking	7.5 × 70 × 45 mm
Mini PCI Type IB	100-pin stacking	5.5 × 70 × 45 mm
Mini PCI Type IIA	100-pin stacking, RJ11, RJ45	17.44 × 70 × 45 mm
Mini PCI Type IIB	100-pin stacking, RJ11, RJ45	5.5 × 78 × 45 mm
Mini PCI Type IIIA	124-pin edge	5 × 59.75 × 50.95 mm
Mini PCI Type IIIB	124-pin edge	5 × 59.75 × 44.6 mm
Full Mini PCIe	52-pin edge, two staggered rows on 0.8 mm pitch	30 × 31.90 × 1 mm (excluding components)
Half Mini PCIe	52-pin edge, two staggered rows on 0.8 mm pitch	30 × 56 × 1 mm (excluding components)

Figure 12.6 shows a Mini PCIe adapter.

FIGURE 12.6 Intel 3945 IEEE 802.11a/b/g Full Mini PCIe adapter

Installation and Configuration of Mini PCI, Full Mini PCIe, and Half Mini PCIe Cards

As with the PCI card installation process, Mini PCI and Mini PCIe installation may require the user to physically install hardware in the computer. Location of the Mini PCI or Mini PCIe interface varies depending on the computer manufacturer. For example, you just have to remove a cover panel on the bottom of some notebook computers. And on some computers, you need to disassemble the computer case. Exercise 12.2 describes the typical installation steps.

Even if your computer is unplugged and the battery is removed, many components can still retain a dangerous electrical charge. Also, static electricity from carpet or other materials can potentially damage the computer. Always be sure to read the instructions carefully, follow proper safety procedures, and avoid standing on carpet when taking the panel off your computer and working inside its box, even for simple procedures.

EXERCISE 12.2

Installing Mini PCI and Mini PCIe Cards

The following steps are typical for installation of a Mini PCI and Mini PCIe wireless LAN card on a notebook computer. Exact installation steps are specific to the manufacturer, and I recommend that you follow the manufacturer's setup instructions.

1. Shut down the computer. Verify that the computer is not in Hibernation mode. If it is, turn on the computer and perform a complete shutdown.

2. Disconnect the AC power cord from the wall jack.

3. Disconnect all connected peripherals and remove the battery pack.

4. Remove the panel covering the Mini PCI or Mini PCIe compartment (details of this step will depend on the computer model).

5. Insert the Mini PCI or Mini PCIe card into the correct slot. Note the correct pin orientation.

6. Connect the wireless antenna cables to the Mini-PCI or Mini PCIe card.

7. Replace the panel for the Mini PCI or Mini PCIe compartment.

8. Replace all peripheral devices and the battery pack. Plug the AC power cord into the wall jack.

9. Power on the computer and insert the setup CD-ROM into the CD-ROM drive. The program should start automatically, and a welcome or Autorun screen may appear. When the screen appears, click Install Drivers and follow the onscreen instructions to install and configure the wireless MiniPCI or Mini PCIe card.

Always read the manufacturer's manual regarding installation and safety before attempting installation.

Upgrading an IEEE 802.11g wireless Full Mini PCIe card to a newer IEEE 802.11n or 802.11ac Half Mini PCIe card may require the use of a special adapter or bracket. Keep in mind that the Half Mini PCIe card is about half the length of the full card, and you may not be able to install the card securely. Therefore, you may need to purchase a special bracket to securely mount the new Half Mini PCIe card.

Cellular Data Adapters

One way to provide *broadband* Internet access to a mobile device without cellular capabilities, such as a notebook computer, is to add a cellular data adapter. The most common type of adapter uses Universal Serial Bus (USB) technology. Just like a cellular phone, a cellular USB data adapter requires fee-based, month-to-month or long-term service. One advantage to using this type of adapter is the ability to get Internet connectivity using 3G/4G LTE anywhere a cellular signal is available. Conversely, in order to get connectivity using wireless LAN technology, a signal from a wireless access point must be available. A few other options for cellular data access include wireless broadband hotspot devices and smart phone tethering.

Wireless Broadband Hotspot Device   These devices are becoming very popular because of their convenience factor. A wireless *hotspot* device operates in the same manner as a wireless access point, but instead of being connected to a wired infrastructure, it uses a 3G/4G LTE cellular connection for Internet access. One benefit is that this device will allow Internet access anywhere a cellular signal is available. The user will configure the device to provide a secure service set identifier (SSID), which will allow anyone with the correct credentials to connect using any Wi-Fi–capable device. These broadband hotspot devices will require activation and a fee-based monthly subscription, or a contract with a wireless carrier to provide a data plan.

Smartphone Tethering   For an additional fee, some cellular wireless carriers will allow you to *tether* a smartphone using a USB cable or Bluetooth technology to devices that do not have wireless broadband capabilities. For people who have an unlimited data plan with their wireless carrier, this is a nice option. It is important to understand that when you use this method, you will be charged depending on the amount of data your plan allows or the amount you use. If you frequently use the tethered device for cellular data, it can get expensive. One disadvantage to this option is that it allows only one device to connect to the Internet at any one time.

 Most broadband wireless hotspot devices allow only a limited number of devices (typically five) to connect to the Wi-Fi connection available from the hotspot device. This will help with the overall performance and provide a better experience for the user who is connected to the Wi-Fi link. Using this type of device for a large number of users with bandwidth-intensive applications is not usually recommended.

Secure Digital (SD) Adapters

Secure Digital (SD) was designed as a flash memory storage device, and now is available with storage capacities from 8 MB to 32 GB. The SD memory card was a joint venture among SanDisk, Toshiba, and Panasonic in 1999. SD adapters are available in different form factors, including what is known as MicroSD.

Even though the SD card was designed to provide flash memory, the slot will allow for connection of other devices such as cameras, Global Positioning System (GPS) units, FM radios, TV tuners, Ethernet networks, and of course wireless LANs. In this format the SD card is known as *Secure Digital Input Output (SDIO)*. This card is designed to provide high-speed data I/O with low power consumption for mobile electronic devices.

Features of SDIO Cards

SDIO cards are commonly available in three sizes:

- The full-size SDIO card is 24 × 32 × 2.1 mm—approximately the size of a postage stamp. This SDIO card is intended for portable and stationary applications.
- The mini SDIO is 27 × 20 × 1.4 mm in size and is used with wireless LAN and Bluetooth adapters.
- The micro SD is 15 × 11 × 1.0 mm and is used in smartphones, tablets, and mobile devices.

For additional information regarding SD and SDIO technology, visit the SD Association at www.sdcard.org.

Installation and Configuration of SDIO Cards

Physically installing an SDIO 802.11 wireless LAN card is similar to installing other adapters used in laptop computers and mobile devices. However, there are some differences. The differences may include connecting a mobile device to another host PC running ActiveSync in order to complete the installation process. I recommend that you follow the manufacturer's setup instructions for installing a specific card. Here are the typical steps for installing an SDIO wireless LAN card:

1. Connect the mobile device to the host PC running ActiveSync.
2. Install the software using the host PC.
3. Insert the SDIO wireless LAN card.
4. Start the program on the mobile device.
5. Find a wireless LAN to connect to, and create a profile.
6. Connect to the wireless LAN.

Always read the manufacturer's manual regarding installation and safety before attempting installation.

Mobile Device Management Onboarding and Provisioning

Mobile device *onboarding* is the process of connecting a device to a wireless network for the first time, and allowing a user to register it to gain access to company and other network resources. This may also include providing guest access to users such as visitors and contractors, so they can access specific company resources or the public Internet. The objective here is to make this process simple and transparent to the end user of the device, regardless of whether the device is company owned or employee owned (bring your own device). This process should require minimal, if any assistance, from the company's technical support department. The onboarding process will ensure that the device is in compliance with corporate security policy and will provide access to only the network resources that the user of the device has permission to access. Onboarding can be accomplished in several ways:

- Manual methods
- Self-service methods
- Certificate enrollment methods
- Role-based enrollment methods
- Other enrollment and identification methods

The method used will depend on several factors, including the size of the organization, number of devices, technical support constraints, whether the device is employee owned or company owned, and more. Figure 12.7 shows the basic flow for mobile device onboarding.

FIGURE 12.7 Mobile device onboarding flowchart

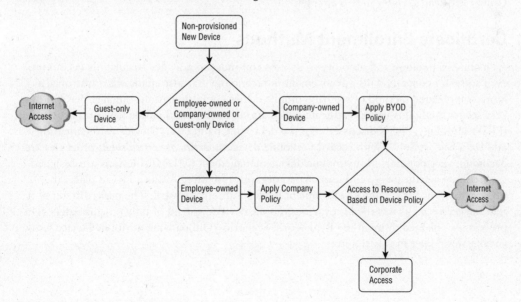

Manual Method

Depending on the device and its intended use, onboarding using a manual method can be a cumbersome and inefficient process for providing access to a network. However, this really depends on whether you are connecting a company-owned notebook computer to wireless LAN or a mobile device such as an Android tablet or an iPhone. Connecting a company-owned computer to a wireless network could be as simple as providing the service set identifier (SSID) and the proper credentials to the user. In this situation, chances are the user has been supplied with a computer that has the necessary software installed with the needed applications to effectively perform their tasks. In many cases, the organization's information technology (IT) team will use special utility software to create an exact replica of a computer hard disk that can be stored on a file server and reproduced onto many other computers. On the other hand, for more mobile devices such as smartphones and tablets, this might be not as easy a process because of several factors, including whether the device is company owned or employee owned.

Self-Service Methods

The self-service method of onboarding is just as the name implies. This is an automatic method that is simple and transparent to the user. A typical configuration is at least two different wireless service set identifiers (SSIDs) that are on two different virtual local area networks (VLANs). One will be an open or guest network, and the other will be a secured corporate network. In most cases, the user will connect their wireless device to the open SSID. They then launch their Internet browser and access the enrollment web page. At this point, the user will enter their corporate credentials and follow the onscreen instructions. This process is explained more in the Real World Scenario "Onboarding a Mobile Device Using Certificates," later in this chapter.

Certificate Enrollment Methods

As mobile technology and mobility of devices continues to advance, security is becoming even a greater concern. This advancement in technology has also made what was once a more-complicated method of securing mobile device communications much easier. One very secure method is using certificate-based authentication and public key infrastructure (PKI) technology. PKI is discussed in more detail in Chapter 16, "Device Authentication and Data Encryption." With regard to mobile devices, *certificate enrollment* processes are becoming very popular. Many mobile device management (MDM) solutions are designed to distribute certificates to end users. If an organization already has a PKI in place, the process to allow for certificate enrollment is fairly straightforward. The main difference is that the certificate is installed on various mobile devices instead of only on computers. This process is explained more in the Real World Scenario "Onboarding a Mobile Device Using Certificates," later in this chapter.

Role-Based Enrollment

Role-based onboarding uses the same concept as role-based access control. *Role-based access control (RBAC)* is a way of restricting access to only authorized users. This access is from authentication based on specific roles, rather than user identities. It was designed to ease the task of security administration on large networks. RBAC has characteristics like those of a common network administration practice—the creation of users and groups. RBAC may also fit well under the authorization part of authentication, authorization, and accounting (AAA) services because, again, it has similar characteristics.

To give a user on a computer network access to a network resource, best practices recommend creating a group object, assigning the group permissions to the resource, and then adding the user object to the group. This method allows any user who is a member of the group to be granted access to the resource. Role-based access control can be applied to various activities users may perform while connected to a wireless LAN, including limiting the amount of throughput, enforcing time restrictions, or controlling access to specific resources such as the Internet.

Let's look at an example. Your organization consists of several departments that use resources available on the wireless network. The departments are sales, engineering, and accounting. Each department has specific requirements for what they need from the wireless LAN, and when they need to access it. This is where RBAC would be a great fit. If the network administrator wanted to restrict access to the wireless LAN for the sales department from 8:00 a.m. to 5:00 p.m., she could create a role using this feature. If the engineering department was using too much bandwidth, a role could be created to restrict throughput for that department. These are a couple of examples where RBAC can work with a wireless LAN.

Other Enrollment and Identification Methods

There are other enrollment and identification methods that may be used for mobile device onboarding. These include using the International Mobile Station Equipment Identity (IMEI) number, the Integrated Circuit Card Identifier (ICCID), and Simple Certificate Enrollment Protocol (SCEP).

International Mobile Station Equipment Identity (IMEI)   The IMEI is a number that uniquely identifies cellular mobile phones. It can be used as part of the enrollment process to identify the device.

Integrated Circuit Card Identifier (ICCID)   The ICCD is a number that is assigned to a Subscriber Identity Module (SIM) card that will identify the SIM card internationally.

Simple Certificate Enrollment Protocol (SCEP)   *Simple Certificate Enrollment Protocol (SCEP)* is a protocol, which as you know is a set of rules. It was originally designed for installing certificates on infrastructure devices such as routers and Ethernet switches. SCEP is available as an add-on for Microsoft Windows Server Certificate Services, starting with Windows Server 2003. For mobile device enrollments, SCEP has been expanded to allow mobile device management (MDM) solutions to install certificates on various mobile device types. SCEP is natively supported by Apple iOS devices such as iPhones and iPads, starting with iOS 4, and is the preferred way for certificates to be installed on these device types.

Real World Scenario

Onboarding a Mobile Device Using Certificates

In this Real Word Scenario, I will be demonstrating some of the onboarding topics using a product from Cloudpath Networks called XpressConnect. Cloudpath Networks, founded in 2006, is the market leader for automated device enablement (ADE) and the inventor of the Wi-Fi onboarding paradigm. You can read more about Cloudpath Networks and see the product line at its website (cloudpath.net). The Microsoft Windows 7 operating system was used to perform these steps, and they may vary slightly depending on the device and operating system in use.

Access Point Setup

A common way to perform this type of onboarding is to have two service set identifiers (SSIDs) available. One SSID will be configured as an open wireless network. This would commonly be the "guest" SSID on many corporate wireless LANs. The second SSID is configured as a secure SSID using WPA2-Enterprise with CCMP/AES, and is configured to connect to RADIUS services and to query a directory service such as Microsoft Windows Active Directory. This SSID will provide the secure communications that allow the device to be properly onboarded.

The User Experience

The end user of the device will first connect to the open SSID, allowing them to get a wireless connection to the network. Then they will enter a URL that will allow them to connect to the XpressConnect enrollment portal used to activate devices. Once the URL is entered into a browser, the welcome page will appear.

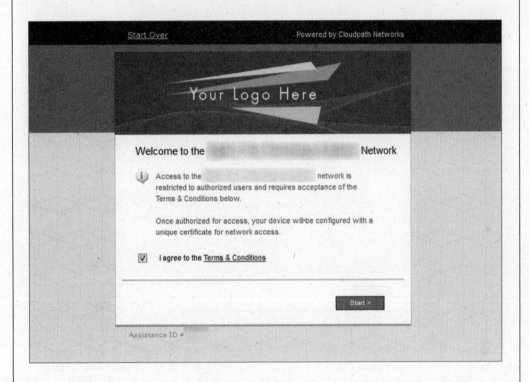

(continued)

The user will read and agree to the terms and conditions, and click the Start button to continue. The user will then be presented with a page allowing them to select either visitor access or employee access.

If the Visitors option is selected, the user will be prompted for a voucher code that can be obtained from an employee of the company. The employee that provides the voucher may be a single individual assigned to the task or the visitor's company contact.

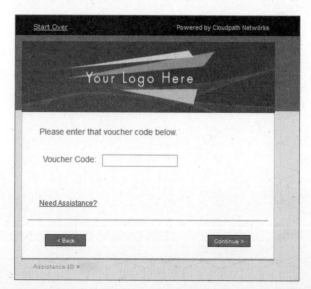

Once the voucher code is entered and the user clicks the Continue button, the XpressConnect software will load. Depending on the operating system in use (Microsoft Windows, Apple devices including the iPhone and iPad, Android devices, or others), the user may be prompted to download and install the XpressConnect software.

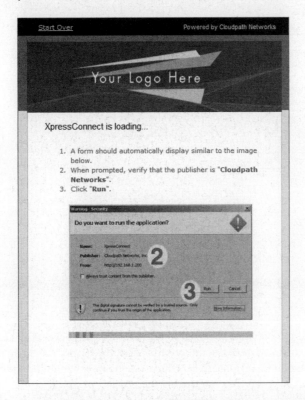

After the software has loaded, the wizard program will launch and the enrollment process will begin.

(continued)

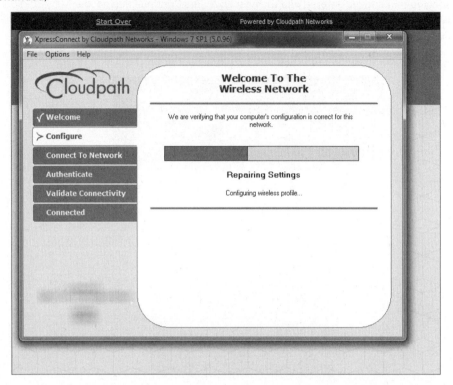

If the Employees option is selected, the user will be prompted to enter their credentials. These credentials will be used to query the organization's directory database and will allow for a Transport Layer Security (TLS) certificate to be installed on the mobile device.

After the correct credentials are entered, the user will click the Continue button and the XpressConnect software will load. After the software has loaded, the wizard program will launch and the enrollment process will begin. At this point, the user will be prompted that a certificate will be installed on the device. Once the user agrees to install the certificate, the wizard will continue.

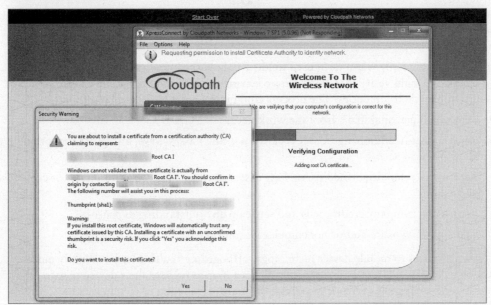

Once the wizard is finished, the user will have secure access to the network using an EAP-TLS certificate.

You can view a short video about the XpressConnect enrollment system by visiting the Cloudpath web site at `cloudpath.net` and clicking the Watch A Video box.

Voucher Methods

Organizations can also use a voucher distribution method for network access. Cisco Systems recently acquired Broadhop, which has a policy control product, Quantum Policy Suite (QPS). An organization can use this product to generate vouchers that employees or guests can use to access the wireless LAN or other network services. Another feature of the Cisco QPS product is something called One-Click. Users are granted access to the network after reading an acceptance policy and clicking a button, thereby agreeing to the policy. In addition to software applications, some organizations use hardware solutions to control access. For example, Cisco Systems Integrated Services Gateway is basically a router that controls access to the network. It can be used with various access methods, including wireless networking. A feature known as Media Access Control Transparent Auto Login (MAC TAL) allows devices to be automatically logged on to the network based on the device's Media Access Control (MAC) hardware address. This is typically used with web-based authentication on publicly accessible wireless networks.

Mobile Device Management Offboarding and Deprovisioning

Offboarding originally was a term used to identify how the human recourses department of a company would verify that an employee leaving the company would be properly dismissed from their job duties. The process could include the following steps:

- Remove all personal items from the workplace.
- Return identification badges, access cards, parking passes, keys, and other items that allow physical access.
- Return any company-owned technology assets, such as a notebook computer or cell phone.
- Return any company credit cards and settle up any outstanding expenses.
- Sign a nondisclosure and/or noncompete document.

When it comes to mobile device technology, offboarding is a similar but slightly more technological process. Just as onboarding allows a user to register their device to gain access to the company's network resources, offboarding is required to remove any access when a user leaves the company.

Before the popularity of current mobile device technology with employee-owned devices, offboarding was a more straightforward process. In most cases the company issued a notebook computer and possibly a cell phone to an employee when they started with the company, and those items were returned to the company upon the employee's departure. Now with the advancements in mobile technology and the "bring your own device" philosophy, offboarding is a little more challenging. If proper offboarding policies and procedures are not specified or followed, it could result in potential security issues for the organization. Depending on the business model or organization type, industry regulatory compliance with, for example, Payment Card Industry (PCI) standards or the Health Insurance Portability and Accountability Act (HIPAA) may also play a role in the offboarding procedure. PCI and HIPAA were discussed in Chapter 11.

There are different methods used with the technology offboarding process. At a minimum, a checklist of some type would be beneficial to ensure all the necessary steps are successfully performed. The process of offboarding technology devices can be automated through the use of mobile device management (MDM) solutions to ensure that the process is properly followed and correctly executed without reliance on a manual checklist.

Employee Terminations and Mobile Technology

Depending on the specific situation, employee termination may not be a pleasant task. A mutual or voluntary departure is much easier to deal with than a one-sided departure. With respect to information technology, companies must have a detailed policy as to how to handle employee terminations.

As the use of mobile technology, both company-owned and personal devices, continues to increase, so does the amount of company data that is contained within these devices. If not handled correctly, employee terminations could put company data such as intellectual property and trade secrets at risk. In addition to access to data, other considerations such as remote access to company network resources must also be considered. Some statistics show that more than 50 percent of employees who are no longer associated with their former company may still have access to the company's private information. Also, more than 25 percent of former employees may still have access to the company's resources through remote access technologies. A well-defined corporate security policy will help prevent this type of issue and will help to keep company data and resources secure after employee departures.

Mobile Device Deactivation

Deactivating a mobile device can easily be accomplished using a mobile device management (MDM) solution. This can be done remotely in the event that the device is lost or stolen, or an employee leaves the company involuntarily. Performing a remote wipe or remote lock will help to prevent someone from using the device and potentially accessing company resources. Remote wipe and remote lock are discussed further in Chapter 13, "Mobile Device Operation and Management Concepts."

Mobile Device Migrations

With mobile device technology advancing at a fast pace, it is not uncommon for an organization to change which devices and supporting technology they use as a standard. One example is if a company decides to move from a BlackBerry platform to an Android platform, or from Android to Apple iOS devices. Migrations are also likely in organizations that allow employee-owned (BYOD) devices, because people may upgrade to newer, more feature-rich devices at a quicker pace than they do with company-owned devices.

Removing Applications and Corporate Data

When someone who has been allowed to use an employee-owned device to access corporate resources leaves a company, it may be necessary to remove specific mobile applications. These types of applications are typically proprietary or customized for the organization, but in some cases they may be commercial off-the-shelf applications that are owned and licensed by the company. If the device is employee owned, the personal applications that do not belong to or are not licensed by the company should remain intact on the device. In addition to applications, corporate data will more than likely also need to be removed, or "wiped," from the device. This may include company proprietary information, intellectual property data, trade secrets, and any other data that could be a security risk if obtained by individuals who are not employed by the company and bound by company confidentially agreements.

Asset Disposal and Recycling

Statistics show that technology devices that are decommissioned from use with the intent of being destroyed or recycled often still contain personal and company data that is either readily available or easily recoverable. These devices include notebook computers, tablets, and smartphones. Many computers still contain mechanical hard drives with a physical disk that is used to store the data. If you were to delete all of the data, there is still a possibility that some or all of the data could still be recovered with off-the-shelf software. Some newer computers are equipped with electronic or solid-state drives (SSDs), and even though there are no moving mechanical parts or physical disks, the same potential issue of recovering data still exists. In fact, it has been reported that in some cases it is not possible to reliably erase all data from a solid-state drive.

Organizations should take proper measures to ensure that data is not accessible or recoverable from devices that are decommissioned. This should be defined in the corporate information technology security policy with an outline of the process and steps that should be taken to delete data from a device. Home users should follow best practices that are available from the manufacturer of the device. This usually involves performing a factory reset or performing a "wipe" function on the device. The physical disk is only one concern. Other potential media types that can store data or personal information include removable SIM cards, micro SD cards, and other types of internal storage. The removable cards are typically inexpensive, and it may be best to keep them for other uses or arrange for them to be physically destroyed for security purposes.

Summary

In this chapter I explained the System Development Life Cycle. We also explored pilot testing, which includes selecting the individuals or groups to be part of the test, soliciting feedback, monitoring the testing, making any necessary changes, getting proper approvals, and finally performing the full deployment. I explained infrastructure and wireless client devices used with over-the-air technology such as wireless LAN and cellular. We explored the process used for bringing client devices onto a network for the first time, which is known as onboarding, and taking devices offline or off the company network, which is known as offboarding. We explored both manual and automatic onboarding. We also looked at the offboarding process and the proper procedures to ensure that company data is secure after an employee leaves an organization. Finally, I explained the importance of preparing mobile technology devices for recycling or disposal after they have been decommissioned. You saw that deleting data from a device may not really prevent it from being seen or accessed again, and that certain technology factors must be considered.

Chapter Essentials

Know the basic System Development Life Cycle (SDLC) concept.   The phases of technology deployment commonly include plan, design, test, build, deploy, and maintain.

Understand the concepts of a pilot program, testing, evaluation, approval, and launching.   Know the common areas of a pilot test program. These include (but are not limited to) selecting people to participate, receiving feedback and monitoring, making necessary changes, getting approval from management, and deploying the technology.

Understand the importance of proper and complete documentation.   Creating clear and concise documentation will help in providing and maintaining a well-planned deployment. Understand that technical documentation is not static and will need to be maintained and updated due to technology, the corporate environment, and other changes that may occur.

Know that activation of mobile devices will differ depending on the device type.   Mobile devices that use wireless LAN technology such as IEEE 802.11 require only the proper credentials to connect to a network. Unless the user is at a paid hotspot, no fees are required for this type of connection, and no activation is required. Bluetooth technology is used to create a personal area network within the user's immediate area, which requires only the proper credentials, such as a personal identification number. Cellular devices such as smartphones require activation from a wireless carrier, and fees are associated with their use.

Understand the different device types and adapters used to facilitate over-the-air access.   Know the different device types used based on the technology. This includes devices used with wireless LAN and cellular data. Universal Serial Bus (USB) adapters allow for access and Secure Digital (SD) adapters allow for information storage.

Understand the meaning of on-boarding and provisioning.   Be familiar with the different methods available and used to bring a device onto a network for the first time. This process is known as onboarding or provisioning. Know the different methods used, which include manual methods and automatic methods that will be transparent to the user.

Understand the meaning of offboarding and deprovisioning.   Be familiar with the importance of offboarding a device, or removing it from service. This process will help to ensure the integrity of a company's proprietary information, intellectual property data, trade secrets, and other data. Understand the importance of proper mobile device disposal and recycling.

Chapter 13

Mobile Device Operation and Management Concepts

TOPICS COVERED IN THIS CHAPTER:

- ✓ Centralized Content, Application Distribution, and Content Management System
- ✓ Content Permissions, Data Encryption, and Version Control
- ✓ Mobile Device Remote Management Capabilities
- ✓ Lifecycle Operations
- ✓ Information Technology Change Management
- ✓ The Technology End-of-Life Model
- ✓ Deployment Best Practices

It is important to understand available content management options and evaluate what would best fit your organization.

In this chapter, we will look at the differences between some available solutions for mobile device content management and distribution, including enterprise server-based and cloud-based solutions. We will also explore factors related to content management, such as changes, permissions, data encryption, and version control.

Understanding the remote management capabilities of mobile devices is important, particularly to ensure the integrity of company data and resources on a lost or stolen device. We will survey remote management capabilities such as location services, locking a device remotely, and wiping a device remotely, among others. We will also explore life cycle operations such as renewing expired mobile device certificates, software updates, and hardware and firmware upgrades.

I will explain the importance of technology change management and why an organization must control any and all changes to the current information technology infrastructure to ensure proper operation, compliance, and security. We will look at the technology end-of-life model and how it may have an effect on mobile device and networking technology. Finally, I will explain some of the common best practices used to help ensure a successful mobile device deployment.

Centralized Content, Application Distribution, and Content Management Systems

With a centralized location for company data, users can access content from anywhere they have an active Internet connection. Keeping the corporate content centralized also provides better collaboration opportunities, increased employee productivity, and more robust security management compared to the public file sharing services that are readily available.

Centralized content repositories and their associated *content distribution* methods can be either corporate server-based (on-premise) or cloud-based solutions. Both of these options come with their own unique set of challenges. In Chapter 10, "Introduction to Mobile Device Management," we explored the differences between on-premise and Software as a Service (SaaS) mobile device management solutions. The same principles apply to content and application management and distribution methods. Using a corporate server-based on-premise solution may require extra management overhead for the IT department as well as an investment in additional hardware solutions. A cloud-based solution may incur recurring monthly fees and an Internet connection will always be required for all management and control functions. It is important to weigh the pros and cons of each solution to ensure that the best and most cost-effective solution is selected based on your organization's needs.

In addition to static content such as company data and mobile device apps, real-time content such as streaming video needs to be taken into consideration. One method for delivering real-time content is multicast technology. Multicast technology allows for the same content to be delivered simultaneously to a large number of devices compared to unicast technology, which is one-to-one communication. With the continued development of mobile technology, the need for multicast real-time communications is becoming more prevalent.

Server-Based Distribution Models

Server-based content distribution gives an organization the ability to control the entire process of content storage and distribution, which includes physical access, permissions, and other administrative tasks. Like any on-premise technology solution, this type of distribution has both advantages and disadvantages, as mentioned in the preceding section and explained in Chapter 10 with respect to MDM solutions.

Whether the content repositories are centralized or decentralized will also have an impact on the overall performance of the process. Depending on the size of the network, number of devices, and type of content, using a centralized content storage and distribution method comes with its share of challenges. These include potential bottlenecks due to infrastructure connectivity, network latency, and server availability, potentially resulting in a below-par performance for the end user. However, it really depends on the infrastructure that is in place in addition to the connections to the outside wide area network or the Internet. If implemented correctly, the centralized method could be a viable solution, but all of the components must be closely evaluated to reduce the chances of poor results for the end user.

A decentralized solution will lessen the chances of some of the issues with a centralized repository. However, a decentralized solution comes with its set of challenges as well, which include cost and potential administration issues. A decentralized solution will still be centrally managed but will have physical servers at various locations for content distribution to limit the negative effects of a single source location.

Cloud-Based Distribution Models

Another option for mobile device content storage and distribution is using the *cloud-based* model. This operates in a similar manner to other cloud-based networking solutions, where a third-party provider handles the content storage mechanisms. The cloud-based model will allow a user to access information from anywhere in the world an Internet connection is available. This offers a convenient and easy way to access information, data, and applications for which the user has been granted the adequate permissions.

As with the corporate server-based model, the cloud-based model has its own set of challenges and concerns, one of which may be security. If implemented correctly, security should not be an issue, but the key words here are implemented correctly. It will also require end user cooperation to help ensure that they are following policy and are using the services correctly. If not used correctly, cloud-based storage solutions may result in an effect known as "shadow IT," which occurs when information technology solutions are built and used without company approval. These include publicly available file sharing and storage services.

Another disadvantage is that you are at the mercy of the cloud provider when it comes to bandwidth and the data backup and restore solutions.

Selecting the correct cloud service provider will take time, and statistics such as available bandwidth and server availability must be analyzed to determine if they are a good fit for the organization.

Content Permissions, Data Encryption, and Version Control

As you saw in earlier chapters, wireless and mobile device technology is a moving target, and changes are occurring constantly. These changes can be the result of advancements with the actual technology used or due to device and software upgrades. Either way, it's critical to have the correct plan in place to ensure consistency throughout the organization. Basic network administration tasks include assigning permissions or rights to network resources and making certain data is safe in transit by using secure encryption methods. With mobile device technology all information is transmitted across the open air, so proper permissions and encryption techniques are even more critical. Maintaining a version control mechanism will help to guarantee that the user base is consistent with operating systems, applications, and device updates.

Content Permissions and Data Encryption

A well-planned and executed security strategy will help to ensure that company policy is implemented and enforced as intended with every technology deployment. This strategy will include access control methods and encryption policies to protect all company information and available resources. Network administration tasks include validating that access to company resources is available only to those who need it. This is where permissions or rights (depending on the operating system) come into play. Because of the fact mobile device technology is wireless, encryption of data plays a major role.

Permissions *Permissions*, or what some operating systems call rights, gives the administrator the ability to restrict access to a resource by using an access control list. The specific categorization of each user or the group they may belong to determines the type of access they have to a network resource. One example would be access to a shared folder for the sales department. People in the sales department would be added to a group called sales.

Only the sales group would then be granted permissions to access the sales shared folder. Any other users who required access would have to be granted access by the network administrator.

It is important to understand that different levels of permissions exist, depending on the user or group needs. For example, a group could be granted read only permissions, which would allow them to see the data in a shared folder but not make any changes to it.

Permissions can be very strict and will be set based on the specific needs of your organization. The technical details of assigning permissions and the levels of these permissions are beyond the scope of this book.

Data encryption   In the most basic sense, *data encryption* is scrambling data by using some type of encryption algorithm so that only the sender and the intended recipients that know the algorithm can decrypt the data to see the contents. The only way mobile wireless devices can send and receive data is by using radio frequency and over-the-air technology. Therefore, the concept of data encryption is of the utmost importance. Data encryption will ensure that the information sent using over-the-air technologies is secure while in transit. Data encryption methods for data at rest (stationary) and data in transit (traversing a network) are discussed in more detail in Chapter 16, "Device Authentication and Data Encryption."

Version Control

In a general sense, *version control* or revision control is a way to ensure that all software and hardware technology used across an organization is consistent. This is a process that is also used with creating and maintaining documentation, software programs, and computer code that is used to create software programs, web pages, and the like. With mobile device technology and employee-owned or bring your own device (BYOD) adoption, this is even more of a concern due to the diverse array of devices that can be used within an organization. A well-documented plan that is followed and enforced will help to ensure that consistent hardware and software solutions are in use and help to lessen the potential for technical support issues that could have been avoided.

Mobile Device Remote Management Capabilities

If you have ever not been able to locate your smartphone or other mobile device, you have experienced a terrible feeling. The first thought that runs through your mind is that the data on the device is in the hands of someone else. Knowing that you have the ability to remotely lock your device provides some level of comfort if this were to happen. Most of the enterprise mobile device management solutions listed in Chapter 10 provide a feature that enables you to remotely lock, unlock, and wipe the device. There are also some apps available that you can install to provide the same features if a device is for personal use only. One example of such a product is Android Device Manager from Google (used with Android devices).

In addition, an organization's mobile device management solution and security policy will provide information on how to properly handle reporting procedures. This is true regardless of whether the device is employee owned or company owned.

Another remote management option is remote control. A company network administrator or support person can take control of a device and assist with technical issues the user may be experiencing. There are also apps available that will allow you to remotely control your home or office computer from a mobile device such as an Android device or iPhone.

The ability to perform remote management on mobile devices is an important component of mobile device management. These features will allow a network administrator to perform specific tasks on a device to help ensure the security and integrity of an organization's data and to limit unauthorized access to company resources.

Mobile Device Location Services

In Chapter 10, I explained location services as they pertain to geo-fencing and geo-location. Here we will explore how location services can be enabled or disabled on a mobile device and the ability to track a device's location within a few meters if it needs to be located as the result of device loss or theft. Depending on the solution in use, *location services* may need to be enabled on the physical device in order to physically locate it. For example, Android Device Manager from Google will require location services to be enabled on a device in order to track its physical location within a close proximity. Figure 13.1 shows how to enable location services on an Android tablet device.

FIGURE 13.1 Enabling locations services using the settings shown on an Android mobile device

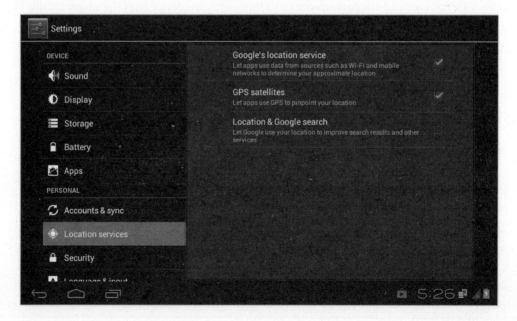

Location services can allow you to pinpoint the exact location of a device within a very close range, in some cases less than 10 meters. There are different methods used for location services:

- Global Positioning Satellite (GPS)
- Cellular provider network
- Wireless LAN technology
- Applications
- Internet Protocol (IP) address

In most cases, these tracking options must be available and enabled with the device's configuration or Internet browser in order to use the location services.

Remote Lock and Unlock

Remote lock allows you to lock the screen of a device that may or may not be in your possession. This will prevent the device from being used by anyone who does not know the correct passcode to unlock the device. Depending on the product used for the remote lock, you may also be able to change the unlock password and provide a message that will appear on the locked screen with instructions or some other type of message. Figure 13.2 shows a screen capture of an Android device that was remotely locked with a message that was displayed.

FIGURE 13.2 Result of a remote lock command by Android Device Manager

Notice in Figure 13.2 that a message was sent with the remote lock command providing instructions to the user of the device. To unlock a device, the user will need to enter the unlock passcode. A user may need a remote unlock, for example, if they forgot the passcode. Most mobile device management solutions offer the capability to clear the passcode so the user can get access to the device. The user will then be able to enter a new passcode to unlock the device.

Device Remote Wipe

If a device has been lost or stolen device and it contains data for "your eyes only," such as corporate information, you may need to wipe the data and reset the device to the factory default settings. This pertains not only to personal devices but also to company-owned devices or devices that are employee owned and allowed to be used on the corporate network.

However, before any *remote wipe* action is implemented, it is critical to follow corporate security policy and determine the appropriate type of response. If *data containerization* is utilized, only the data in the corporate container may need to be wiped. Containerization is the separation of corporate data and apps from personal data and apps. This concept is more prevalent with employee-owned devices than with company-owned devices. It is similar to partitioning a hard disk drive in a computer. You create different partitions on a physical device, keeping content separate regardless of whether it is operating systems and apps or data. Container technology allows an organization to control only what is in the corporate container rather than the entire device. This provides some peace of mind for the user of the device, who will know their personal data is separate and cannot be accessed, deleted, or remotely wiped.

In addition to enterprise-based mobile device management (MDM) products and other web-based products such as Android Device Manager, there are some specific products that will allow a user to perform a remote wipe. One such product is Outlook Web App (OWA, formerly known as Microsoft Outlook Web Access), which is a web-based email interface that is used with Microsoft Exchange Server. This product allows a user to access their email from any place they have an active Internet connection. With only a few keystrokes, a user can remotely wipe a device if it is lost or stolen. This particular product is not limited to only one specific mobile operating system; it works for various device types, such as Android, iOS, and Windows Phone.

Exercise 13.1 illustrates the steps to enable and use Android Device Manager provided by Google.

EXERCISE 13.1

Enable Android Device Manager

In this exercise you will enable Android Device Manager, provided by Google. To perform this exercise, you will need an Android-capable device, or you can use the Android emulator program you installed in Exercise 10.1, "Installing Android Emulator Software in Microsoft Windows," in Chapter 10. To use this feature to remotely locate, lock, or wipe an Android device, you will need to enable it on the device and associate the device with a Google account. If more than one person uses a tablet, only the tablet owner can turn on Android Device Manager.

Mobile Device Remote Management Capabilities

1. Download and install the Android Device Manager app from the Google Play Store.
2. Open the Google Settings screen located on the device apps menu by tapping or clicking the Google Settings icon.

3. The Settings screen will appear.

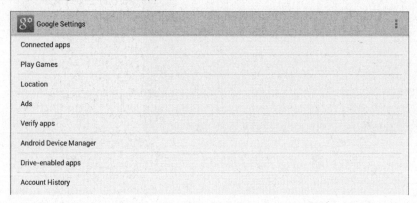

4. Tap or click Android Device Manager. You will see a screen with two options: Remotely Locate This Device and Allow Remote Lock And Erase.

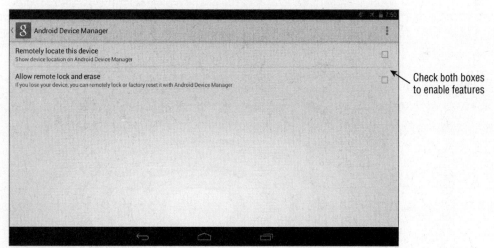

Check both boxes to enable features

EXERCISE 13.1 *(continued)*

5. Verify that both boxes are checked if you wish to enable these features. After you select the checkbox next to "Allow remote lock and erase", the Activate Device Administrator? dialog box will appear.

6. Tap or click the Activate button to enable these features. The device is now ready to be managed using Android Device Manager.

7. Access Google Device Manager by pointing your browser to www.google.com/android/devicemanager. The Android Device Manager Login screen will appear.

8. Enter your account credentials to access the online manager software. The Google Android Device Manager online interface will appear.

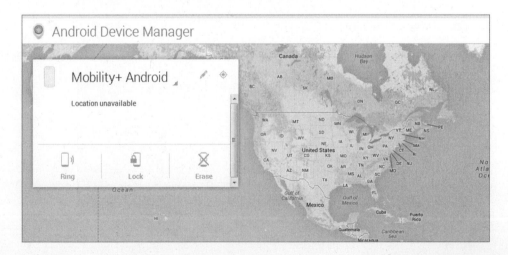

You can now set up your device to be managed. The features available with this service are ring, lock, and erase. Keep in mind that once you associate this device to your Google account, there is no way to remove it. It will remain connected to that Google account. However, the device could be associated to different Google accounts if needed.

If you decide to use Android Device Manager or any service that provides location tracking and feel your device has been stolen, do not attempt to retrieve the device yourself. You should immediately contact the appropriate law enforcement agency and let them handle it based on the laws that are in place.

Remote Control

Mobile device *remote control* allows technical support to access a device remotely and make changes to it as if it were physically in their possession. Many of the MDM solutions listed in Chapter 10 provide this feature. However, stand-alone solutions are also available. Remote control capabilities give support personnel the power to be able to assist a user with technical issues and help resolve problems from any location as long as the user is connected.

Reporting Features

Many of the products that provide location services also provide *reporting features* that allow the user of a device to store and utilize information about the device's location. The ability to reuse this information provides a convenience to the user should they want to revisit the location. This feature will also allow for the device's location history to be stored within the program or the user account database.

Sometimes a user might not want this history to be stored. For example, some users travel constantly, rarely using the same location, and may not want to clutter the database with a lot of irrelevant information. Other users might want to protect their privacy or security. The steps to disable location reporting on an Android device are as follows:

1. From the home screen click or tap the Settings icon.
2. Click or tap the Google Settings icon.
3. Click or tap the Maps & Latitude option icon.
4. Select the Location Reporting option from the Location Settings screen.
5. Click or tap the Do Not Update Your Location option.
6. Uncheck the Enable Location History option to turn the feature off.

Life Cycle Operations

In Chapter 12, "Implementation of Mobile Device Technology," you learned about the System Development Life Cycle (SDLC). With respect to information technology (IT) and especially mobile device technology, the SDLC consists of several different phases and processes, one of which is maintaining the system that was deployed. This part of the SDLC includes several areas that need to be dealt with:

- Certificate expiration and renewal
- Software updates
- Software patches
- Device hardware upgrades
- Device firmware updates

Certificate Expiration and Renewal

In Chapter 12, we explored mobile device onboarding, including the certificate enrollment method. Using certificates is one of the most secure ways to validate a user's identity. Onboarding using certificates is a straightforward process. Figure 13.3 shows one example of an onboarding solution, the certificate settings in Cloudpath XpressConnect.

FIGURE 13.3: The certificate settings on the Cloudpath XpressConnect enrollment page where you can set the validity period of the certificate

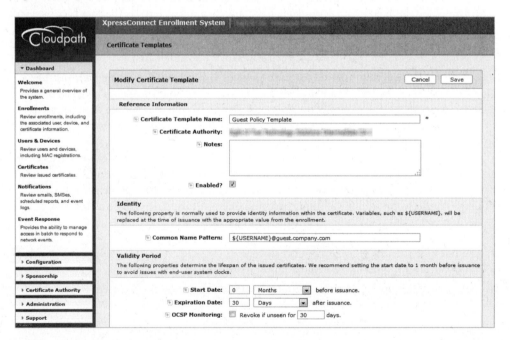

Working with certificates can also mean extra administration overhead. Just as a driver's license will expire and must be renewed, a device certificate has an expiration date and will need to be renewed. In addition to expiring, certificates can be revoked for various reasons. Among MDM administration tasks is the administration of user and device certificates. In most cases, MDM solutions can be set up to automatically renew a certificate as the expiration date approaches. This will ensure that user access is not affected by an expired certificate. If corporate security policy determines that certificates are not to be automatically renewed prior to the expiration date, then another process will need to be implemented to provide uninterrupted service.

Certificates may need to be revoked for various reasons, such as, for example, if a device is lost or if an employee is terminated. Both of these situations would justify immediately revoking a digital certificate. However, proper procedures and policy must be followed prior to the revocation. The task of revoking a certificate is no more difficult than the task to issue or renew a certificate. It can be done within the MDM administration console with a few clicks of the mouse. When a certificate is revoked, the device and user will no longer have access to corporate data, applications, and other resources. This will take effect immediately after the certificate has been revoked.

Other life cycle maintenance operations involve keeping all aspects of the technology up-to-date. This include software updates and patches, device upgrades, and firmware updates, which are discussed in the upcoming sections.

Software Updates

Updating operating systems and software applications will help to enhance the end user experience. Mobile devices require updates for various reasons. Many times updates become available to fix documented issues or to provide security enhancements. In other cases, updates provide new features. There are several possible ways that a mobile device can be updated:

- Software update tools
- Through an Internet connection to the provider or carrier network
- From the device itself

Device manufacturers try to make updates easy to help ensure the best and most secure device operation possible for the end user. Many updates are created because of experiences of the device user population. Therefore, providing software updates will help to lessen potential future technical support or help desk calls.

Although most applications provide backward compatibility to previous versions, it is preferred to have all devices on the same version to maintain consistent usage throughout the organization.

Software Patches

People are sometimes confused about the difference between a software upgrade and a *software patch*. In some cases, the difference may be a matter of interpretation. In most cases a patch, or hotfix, will fix a problem or a bug in the software that is usually the result of technical support issues or engineering and programming changes. In many cases, a patch also provides security fixes that have been created due to security holes that were discovered after the software was released to the general public. In most cases, patches are identified by the same version number as the software, whereas a completely new version number is used for a software update. For example, v7.0 of a software application might have a patch named v7.0.2315, showing the original version number in addition to the version number of the patch that was applied.

Device Hardware Upgrades

This type of upgrade is usually the result of changes to the technology that is in use. With respect to mobile devices such as smartphones or tablets, this occurs in many cases much more often than with computer technology. With BYOD implementations, it can occur much more frequently than with company-owned devices because a single user is much more inclined to upgrade their own personal devices within a shorter time frame than an organization's technology refresh policy will allow.

Device Firmware Updates

Device firmware is the instruction set that allows hardware to operate based on its design. *Firmware* is simply software that remains in memory when the power is removed. It tends to be hardware oriented, but not all firmware involves the control of hardware. Many Layer 7 user interfaces and agents are implemented as firmware burned into read-only memory (ROM) devices.

Conversely, there are many examples of software directly controlling hardware. Most drivers for hardware devices are software (loaded into volatile memory only when power is present) and not firmware (always resident in static memory regardless of power). Firmware upgrades are common support tasks that need to be done periodically, either to fix issues with the way the hardware is operating or to provide new features for the hardware.

All wireless infrastructure devices, either SOHO grade or enterprise grade, allow you to upgrade the firmware. In enterprise environments, firmware upgrades can be performed either manually or automatically with the aid of wireless LAN management system software. For mobile device technology, it is typically accomplished from the device itself and can be performed automatically.

> It is imperative to follow the manufacturer's instructions when performing a firmware upgrade. Failure to do so may result in the device becoming unusable or needing to be returned to the manufacturer for repair. Several items are important when you are performing a firmware upgrade on any device:

- Always verify that you have the correct firmware file for the device to be upgraded prior to performing the upgrade.
- Always read the release notes that come with the firmware.
- Never upgrade firmware on a device that is running on battery power.
- Only upgrade devices that are physically plugged into a reliable power source.
- Never power down the device while the firmware upgrade is in process.

Life Cycle Operations

With respect to wireless LAN infrastructure devices, the firmware in enterprise-grade access points and wireless controllers can usually be upgraded using the command-line interface instead of the GUI if desired.

In Exercise 13.2, you will check the version number of the firmware on an Android device and perform an update if necessary.

EXERCISE 13.2

Update an Android Device to the Latest Firmware Version

This exercise shows the basic steps to upgrade the firmware to the latest version on an Android tablet device. Prior to performing any type of firmware upgrade, I highly recommend backing up all data and plugging the device into a power source. Even though chances are slim that you will experience a power outage or crash, it falls under best practices to maintain a reliable power source.

1. Tap or click the icon to open the device settings screen. The Android Apps and Widgets screen will appear.

2. Tap or click the Settings icon. The Settings screen will appear.
3. Scroll down to the bottom of the Settings screen.
4. Tap or click the About Tablet or About Phone button. The device's Information screen will appear.

EXERCISE 13.2 *(continued)*

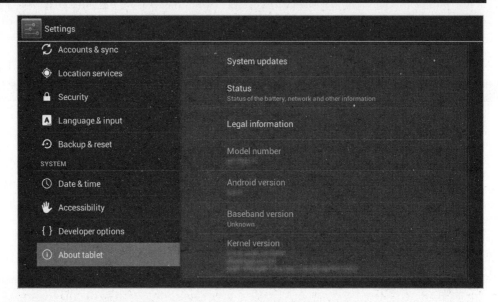

5. Observe the information that is displayed regarding the device.
6. Click on the System Updates button. The device update screen will appear.

7. Tap or click the appropriate button to update or check for updates.

In this exercise, the system was up-to-date, as shown in the preceding screen shot. Keep in mind that this may not always be the case and you may need to update the device.

Information Technology Change Management

The purpose of information technology (IT) *change management* is to make sure any and all changes to the IT infrastructure will meet company requirements, conform to company policy, and maintain a high standard of functionality and reliability. By controlling, managing, and approving changes, the change management group/department will help to reduce potential issues resulting from changes that may not have been adequately tested or proven. The implementation of mobile device technology within the enterprise environment creates many new challenges, especially with the adoption of bring your own device (BYOD) technology. Mobile device technology changes constantly; therefore, all parts must be considered prior to a deployment. For example, if a network administration team was to take it upon themselves to use remote wipe technology on mobile devices without proper analysis and authority, the result could be catastrophic for the organization. Therefore, any changes that might have an impact on the operation of an organization must be carefully considered, validated, and approved prior to implementation. Proper research, assessment, and pilot programs will help to provide a positive experience.

The Technology End-of-Life Model

With respect to information technology, the *end of life (EOL)* is when a hardware or software product is no longer useful. Often this is the result of advancements within the technology sector. Once the EOL has occurred, both the product and support for the product will no longer exist. The EOL is preceded by what is known as the end of sale (EOS). The EOS is when a manufacturer no longer produces the product or makes it available for sale.

Looking at this from the perspective of a wireless LAN infrastructure device manufacturer, as WLAN technology changes and new standards are created, companies must design and manufacture new devices that will provide the newer technology to their current and potential new customer base. For example, the newest 802.11 technology used with enterprise organizations is 802.11ac. However, the availability of IEEE 802.11ac technology does not mean that every organization will be upgrading all of their infrastructure devices immediately, nor does it mean that all WLAN manufacturers will stop manufacturing their IEEE 802.11n products. However, it is almost guaranteed that manufacturers no longer produce IEEE 802.11b–only access points, which makes them end-of-life devices.

When it comes to mobile device technology, operating systems, hardware, and applications can all experience end of life.

Device Operating Systems

In Chapter 10 we explored some of the common operating systems that are used with mobile device technology, including Android, BlackBerry, Apple iOS, and Windows Phone. In many cases if an operating system is upgraded, an existing device will still be capable of supporting the new operating system. An upgraded operating system will commonly provide new features or enhancements to the existing features.

Device Hardware

The technology refresh time frame for an organization really depends on the type of organization, the number of devices it has, the need to upgrade, and the budget it has to work with. In some cases, the idea of "if it isn't broken, don't fix it" may apply. For example, there are companies that use wireless bar code scanners that are considered legacy technology, but these devices are serving their immediate purpose and therefore companies are reluctant to upgrade to more modern technology.

Device Applications

Just as hardware has a specific life span, so do many of the available device applications. Device applications usually fall into one of two different categories: commercial, off-the-shelf applications and proprietary or custom applications.

Commercial, Off-the-Shelf Applications Chances are commercially available applications will be upgraded by the software manufacturer as technology advances. In most cases, the newer version of the applications will provide additional features and enhancements for the product. This is nothing new; we have been dealing with this on the computer side for decades.

Proprietary or Custom Applications An application may be developed that will provide specific functions based on an organization's needs. In many cases, applications of this nature may not be updated as often. Sometimes it is cost prohibitive to update them except when absolutely necessary or when the budget allows.

Deployment Best Practices

Not every mobile device deployment scenario will be the same. One of the factors to be considered is the type of organization, such as whether it is a corporate, government, manufacturing, warehousing, education, or some other type of entity. Other factors include the number and types of devices and whether they are employee owned or company owned. Also, you must consider the type of applications. Following are some best practices that should be taken into consideration when planning a mobile device deployment.

The User Base and the Device Population Understand the type of applications that will be used based on the type of user, either a general user or a power user. Also, determine the number of devices that will be used. Keep in mind most organizations plan for three to five devices per user.

Device Enrollment and Onboarding Plan the device enrollment process carefully, and verify all potential situations that need to be covered for the devices that will be used on the network. This includes but is not limited to self-service methods, certificate enrollment methods, and role-based enrollment methods.

Validation of Device and User Profiles Device and user profiles should be closely evaluated and tested to ensure that they meet the needs of the organization and perform as intended. Profiles should be pilot tested prior to deployment.

Security Validate all processes so that the intended level of security for corporate data and resources is met and maintained. This includes using the correct access control methods, the strongest data encryption, and the correct permissions.

Device Containers For organizations that have accepted the use of employee-owned devices, containerization is a recommended way to keep corporate data and personal data separate. Determine whether the container model is a feasible solution based on use and organization type.

Device and Application Compatibility Make sure all devices and applications are compatible with the organization's specifications and needs. This is especially important with organizations that allow employee-owned devices on the corporate network. Many organizations will use company-owned devices from the same wireless carrier to maintain consistency throughout.

Updates for Operating Systems and Applications Keep all device software current by maintaining updates for the operating systems that are in use as well as any installed applications. Validate a process for testing and installing the latest patches and fixes and security updates. I recommend registering with the device manufacturer or software developer to receive updates that may apply to the technology in use.

Push Technologies Utilize the technology that is available to help streamline the deployment and support of mobile devices. For example, you can use push technology for notifications, content, configurations, and updates.

Feedback from the User Population Always solicit feedback from the user population to ensure that the deployment is operating as intended and meets the needs of the organization.

Summary

In this chapter, we explored the concept of content management, which includes how company information, including applications and data, is stored and distributed. We looked at enterprise server-based and cloud-based solutions that are available to provide mobile device content management. We explored factors that are related to content management, such as changes, version control, permissions, and more. We looked at some of the different remote management capabilities that allow an administrator to remotely track, lock, unlock, and wipe part of or an entire device with the understanding this will help to ensure the integrity of company data and resources. Enabling location services will allow you to locate a lost or stolen device. This chapter covered life cycle operations such as expired device certificates, upgrades, patches, and the importance of information technology change management. We explored the concept of end of life for devices and the understanding that operating systems, applications, and hardware all will need to be upgraded due to advancements in technology. Finally, I shared some of the best practices you can use for mobile device deployment to provide a quality user experience.

Chapter Essentials

Understand the concepts associated with the storage and distribution of corporate content. With mobile device technology, different methods are available for the distribution of applications and company data. These include corporate server-based and cloud-based solutions.

Know the difference between server-based and cloud-based content distribution. Be familiar with the differences between how company content can be stored and distributed from servers that are managed by the organization, which can be centralized or decentralized. Conversely, cloud-based solutions are available for organizations that wish to use external services and providers.

Know that centralized content distribution comes with its own set of challenges. Using a centralized content storage and distribution method may cause some specific problems. These include potential bottlenecks, network latency, and server availability. Using a decentralized or distributed model can help to overcome some of these challenges.

Understand mobile device remote management capabilities. Be familiar with the different ways mobile devices can be managed remotely. These include utilizing location services, remote lock and unlock features, remote wipe, and remote control capabilities.

Understand the potential use of mobile device location services. Know how to enable location services on a device so you can track its location in the event of loss or theft. Be familiar with the different ways devices use location services, including Global Positioning Satellite (GPS), the wireless carrier network, and wireless LAN technology.

Know the potential benefits of remote lock and unlock features of mobile devices. Understand the methods used to remotely lock and unlock devices in the event of loss or theft of a device. Mobile device management (MDM) solutions, along with some third-party applications and services, provide this capability. This technology will also allow an administrator or user to change the passcode or password that locks the device.

Be familiar with remote wipe technology. With the proper management technology, a device can be remotely wiped out and reset to factory default settings to ensure the protection of corporate data. Understand the concept of container technology and that partitioning a device allows an organization to control only what is in the corporate container rather than the entire device. Know that wiping a device may be a last resort and be sure to follow company policy.

Understand the basic life cycle operations. Know the common life cycle maintenance operations, which include device certificate expiration, renewal and revocation, software updates and patches, and device upgrades.

Know the importance of change management as it pertains to information technology. The change management group in an organization is used to control, manage, and approve changes to the information technology infrastructure. Change management will help lessen potential issues that may result from changes that may not have been adequately tested or proven.

Understand the technology end-of-life model. Due to advancements in technology, it is only a matter of time until aspects of mobile technology such as operating systems, device hardware, and applications reach the end of their life cycle.

Know the best practices for mobile device deployment. Become familiar with mobile device deployment best practices, including secure onboarding with certificates, staying in touch with manufacturers for updates, and soliciting feedback from the user population.

Chapter 14

Mobile Device Technology Advancements, Requirements, and Application Configuration

TOPICS COVERED IN THIS CHAPTER:

- ✓ Maintaining Awareness of Mobile Technology Advancements
- ✓ Configuration of Mobile Applications and Associated Technologies
- ✓ In-House Application Requirements
- ✓ Mobile Application Types
- ✓ Configuring Messaging Protocols and Applications
- ✓ Configuring Network Gateway and Proxy Server Settings
- ✓ Push Notification Technology
- ✓ Information Traffic Topology

We will start this chapter by exploring some of the topics that pertain to mobile device technology advancements. This includes understanding the importance of changes to the actual hardware devices (such as computers, smartphones, and tablets) and the mobile operating systems that are used on the devices. I will explain the concept of third-party application vendors for online mobile app stores and the actual apps themselves.

New risks and threats are also an unfortunate byproduct of technology advancements. We will look at the requirements for application (app) types that may be used within an organization. These include in-house, custom, and purpose-built apps. It will be beneficial for you to become familiar with the three types of mobile apps, which are the native app, the web app, and the hybrid app. You will learn how to configure different applications and protocols (SMTP, IMAP, POP3, and MAPI) that work with electronic messaging and allow you to send and receive email messages. You will gain a better understanding of how to configure certain network parameters, such as network gateway and proxy server settings, that may be required for your devices to perform as expected in the environment in which they are installed. Finally, in addition to messaging, we will explore the process and configuration of common push notification services and network operations center (NOC) options.

Maintaining Awareness of Mobile Technology Advancements

Technology changes rapidly, which can have an impact on the enterprise deployment of the mobile devices used within your network. As telecommunications technology and computer networking technology have advanced over the past few decades, so has the distinction between the two. Not too long ago a telecommunications technician would be responsible for technology within the radio and the telephone industries. A computer network engineer's responsibility would be more along the lines of working with the computer network infrastructure, including LANs, WANs, and the associated network devices. The two teams were generally separate. The boundary between computer and mobile technology has changed as the technologies have evolved. Telecommunication and networking equipment manufacturers that were once separate entities now have crossed over into each other's sectors. It is not uncommon for engineers and technicians to design, deploy, and support telephone, voice, video, and computer networking technology. This is becoming more commonplace with the increase of smartphones and other mobile devices that support voice and data technologies.

Faster, more-intelligent devices will require a stronger and more robust network infrastructure that will be able to support the advanced features and capabilities of these newer devices. The following items should be considered when it comes to mobile device technology advancements:

- Mobile device hardware advancements
- mobile operating system(s)
- Third-party mobile application vendors
- New security risks and threat awareness

Mobile Device Hardware Advancements

As computing technology in general continues to advance, the mobile device hardware we use also evolves. This includes smaller notebook computers, smartphones, tablets, and other devices that are used to connect to a network of some sort and to the Internet. In addition to becoming more powerful, these devices are becoming more compact and much more affordable. With physical QWERTY keyboards, touch screens, styluses, and voice recognition software, the capabilities are astounding. Mobile devices are equipped with an operating system and allow the installation of applications (apps) with more and more features and are becoming self-contained computing and voice devices. Features such as cameras, music players, streaming video, live television, Global Positioning System (GPS) technology, mapping applications, location services, games, calendar services, social media, and Internet web access are common in many mobile devices. Using these devices for advanced cellular voice communications such as next-generation LTE, social media, and online meetings must also be considered as the technology evolves.

Mobile Operating Systems

Mobile device hardware advancements usually go hand in hand with software advancements, including advancements to the operating system that are required to control the device. The mobile operating system(s) is what allows the keyboard, physical screen, storage, wireless networks, and other physical components to operate with the graphical user interface and the installed applications. Common mobile operating system(s) include Apple iOS, Android, BlackBerry, and Windows Phone. These were explained earlier in Chapter 10, "Introduction to Mobile Device Management." Keeping the mobile operating system(s) up-to-date will allow for fixes or patches to be installed, security updates to be enabled, and sometimes new features for the user to be added. In some cases, advancements with the physical hardware will also require the operating system to be updated in order to effectively utilize the new hardware to its fullest potential.

Third-Party Mobile Application Vendors

Mobile apps are programs that run on a mobile device and allow the user to perform a variety of tasks that relate to Internet access, web browsing, email, personal calendars, music, games, and more. Mobile operating system(s) apps may be produced by the same large companies that create software programs used on their larger counterparts (desktop and notebook computers), or apps may be created by third-party companies or even individuals. As you saw in Chapter 10, apps are typically available from two different online store types, the enterprise application store or consumer application store. These online stores will allow an individual to download apps for business-related activities or for personal use. Some apps are fee based, while others are available for free. The big players in the mobile app store marketplace are Google Play (for Android devices), the Apple App Store (for iOS devices), and BlackBerry World (for BlackBerry devices). There are other online app stores available. Some may consider this store type to be a third-party mobile application store. One example is the Amazon Appstore for Android available at www.amazon.com. Using these stores or installing third-party apps may require changes to settings and configuration parameters. This can include enabling permissions to the hardware itself. Figure 14.1 shows specific permissions that were required for an app to be installed on an Android device.

FIGURE 14.1 Settings for an app on an Android device requiring specific permissions

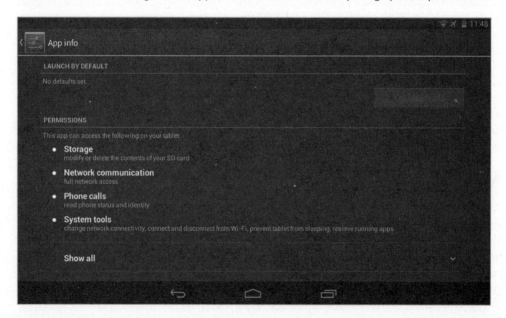

Notice in Figure 14.1 that the app will be able to access the device's storage and have access to full network communications, phone calls, and system tools. There may be some risks associated with installing third-party apps from online stores that may not be well known, including malware, viruses, and other common security risks. You should therefore be cautious about what you download and any configuration changes that may be necessary to install and use a mobile app. In most cases, after the app is installed, you can reverse any configuration changes that may have been made but are not necessary for the app to operate as intended.

New Security Risks and Threat Awareness

Unfortunately, like other types of technology, mobile devices have their share of security risks and threats. These include viruses, Trojans, spyware, and just malware in general. Keeping your devices up-to-date with the most current software protection is critical. Performing tasks on your devices that involve access to personal information, such as online banking with the ability to make payments, transfer money, and even scan and deposit checks, can lead to catastrophic results in the event that personal information is compromised. Subscribing to online forums, newsletters, and email services from software providers will help you stay in touch so you are aware of the latest security threats and can prevent security-related issues. Technology viruses, Trojans, spyware, and other issues related to device security are discussed further in Chapter 15, "Mobile Device Security Threats and Risks," and Chapter 17, "Security Requirements, Monitoring, and Reporting."

Configuration of Mobile Applications and Associated Technologies

There are many mobile device apps that require the user to change or enable settings on the device during the installation process. In some cases these changes are necessary for the app to function as intended, and in other cases the changes aren't required but are needed due to lack of quality application development and programming. When a user understands the changes that are required for an app to be successfully installed, there's a better chance they can ensure that the app will not impact the performance of the device or create any potential security issues. It is best for a user to read all dialog boxes and messages that may appear during the application installation process. If some features or settings have been changed and are not necessary, it is important for the user of the device to reverse any unnecessary changes to ensure the overall integrity of the device and its operation.

In-House Application Requirements

As mobile device technology continues to develop, so does the need for specific app types. In Chapter 10 we took a look at the two different application store models: the enterprise app store and the consumer app store. The enterprise app store is one that is managed and controlled by the organization that hosts the online app website. On the other hand, the consumer app store is one that is publicly accessible to anyone who has access to a device and an active Internet connection.

Both store models can provide many different apps that fall under a variety of different categories. In addition to standard apps, many organizations require apps that are custom or purpose built to serve a specific function for their business model. In cases such as this, there may be specific requirements that must be met for the deployment of the app to be successful. The enterprise app store would be the best method for distribution of the *in-house developed application*. Understanding these app requirements will provide the best possible user experience. When you are considering the requirements, you should keep the following in mind:

- Application (app) publishing
- Mobile device platforms
- Device digital certificates

Application (App) Publishing

Generally, developers can publish apps to an online app store such as Google Play or the Apple App Store. In-house and custom apps that are developed by an organization can be published to an enterprise app store that is run by the company. This makes the most sense because such apps are usually not available for public use and are instead meant for a company-specific purpose.

Prior to publishing any app, it is important to verify that the app works correctly and does not contain any programming flaws or errors. This can be done by choosing a select group of individuals to test the app before it is published and launched.

The Android developer website provides a launch checklist you can refer to before an app is published. The following are some highlights from the checklist you can use to make sure everything is covered for publishing and launching an app used with mobile devices.

- Understand the publishing process.
- Read and understand the store's policies and agreements.
- Validate the app for quality assurance.
- Confirm the app's total size.
- Confirm compatibility with the device platform and the screen ranges.
- Prepare for promotion of the app.
- Upload the app to the online store.
- Perform the trial or beta release of the app.
- List the app in the online store.
- Finalize and launch the app.
- Provide product support for the app.

Keep in mind that this is not a complete list and consists of only key parts of the publishing process. This process can be applicable to both commercially available apps and in-house or custom-created apps not readily available to the general public. Some such apps in an enterprise app store may be available only to the organization for which it was created. Organizations can subscribe to developer services that provide the tools and resources needed to develop apps on certain mobile operating system(s). One such service is the iOS Developer Enterprise Program.

Mobile Device Platforms

With the number of diverse mobile client devices that are available for use in the workplace, understanding device platform support is an important component of a successful app development and deployment. There are three different app types that can be created: the native app, the web app, and the hybrid app. Some app types are platform dependent, and the type used will depend on the environment in which the app is deployed. These app types each have their own advantages and disadvantages and are discussed later in this chapter.

If an organization allows employees to use their own devices on the corporate network, this creates challenges for in-house app development and deployment. This is mostly because of the diversity of platforms that are available. In such instances, an app needs to be developed to work on multiple platforms or to be platform independent. The number and types of devices that will be used on the network must be considered with application development and publishing.

Device Digital Certificates

Using digital certificates is the most secure method available to confirm a user identity. With the continued increase of mobile device technology within the networking environment, use of digital certificates is becoming more commonplace for many different organization types.

At one time, certificate management required a strong knowledge of a public key infrastructure (PKI) technology, extra management overhead, and more complex network administration. Certificate-based implementations were more common in areas that required higher-than-average security. Now, the concept of using digital certificates is much more widely accepted, mostly because technology advancements have made the certificate management process easier to design, deploy, and support. You will see more about PKI in Chapter 16, "Device Authentication and Data Encryption."

Certificates are also used with some push notifications technologies such as Apple Push Notification Service (APNS). With this service, an encrypted Transport Layer Security (TLS) session is established between the mobile device and the APNS using certificate technology. You will see more about configuring APNS later in this chapter.

Mobile Application Types

Mobile apps fall into three different categories: native apps, web apps, and hybrid apps. Each app type has a unique set of advantages and disadvantages, and their use will depend on several factors. These factors include the device platform, the mobile operating system(s) in use, the cost, the purpose, and the amount of time available to the developer. The mobile device types that will be used and the mobile operating system(s) will have an impact on the app type that will be implemented.

Native Apps

Native apps built into the device or installed to the device from Google Play, Apple's App Store, BlackBerry's App World, or a similar app store. The *native app* is developed specifically for the device platform in which it operates on, such as Android, Apple iOS, BlackBerry, and others. Because there is a program installed on the actual device, some of the processing takes place locally, which will help with the overall performance of the app. The native app has its share of advantages and disadvantages:

- Advantages of native apps
  - Good performance
  - Active Internet connection not required for use
  - Widely available through public consumer app stores
- Disadvantages of native apps
  - Platform dependent
  - Must be installed on the device
  - Need to be updated when changed

Web Apps

As the name implies, a *web app* is one that is available over an Internet connection using a web browser that is installed on the mobile device. The browser acts as the interface between the mobile device and the app code located on a web server. This app type is not necessarily an app; it can be a website that the mobile device is accessing. The driving force behind mobile web apps is HTML5, which allows the app to run within a standard Internet browser. The web app has unique advantages and disadvantages:

- Advantages of web apps
  - Web browser based and platform independent
  - Very limited or no client-side updates
  - Centralized web-side updates
- Disadvantages of web apps
  - An active Internet connection is required
  - They affect performance due to no client-side software footprint installed on the device
  - HTML5 integration may cause some known issues

Hybrid Apps

In any sense of the word, *hybrid* means a combination of items. A hybrid vehicle is one that will work with more than one type of energy source, such as gas and electric. With regard to mobile app technology, a *hybrid app* is a combination of a native app and a web app, containing parts or functionality of both. A benefit to hybrid apps is that they can be designed to be platform independent. Here are some of the advantages and disadvantages of the hybrid app:

- Advantages of hybrid apps
  - Can be built quicker than native apps with less client-side coding
  - More functionality than web apps but may have some of the same benefits
  - Can be made available from app stores
- Disadvantages of hybrid apps
  - May lack some of the native or web app functionality
  - Performance is somewhere between that of web apps and native apps, and it may not be as good as native app performance.

Configuring Messaging Protocols and Applications

There are many options available when it comes to configuring your device for messaging services, regardless of whether it is a mobile device such as a smartphone or a tablet or notebook computer. In Chapter 2, "Common Network Protocols and Ports," we explored several of the networking protocols that are used for messaging:

- Simple Mail Transfer Protocol (SMTP)
- Internet Message Access Protocol (IMAP)
- Post Office Protocol 3 (POP3)
- Messaging Application Programming Interface (MAPI)

In Chapter 2, you learned that SMTP, which uses port 25, is used to send email messages from client device to server and to send and receive email messages between email servers. This is one of the most popular Internet protocols in use today. IMAP, which uses port 143, allows a client device to access email that is stored on a remote email server. Keep in mind that IMAP over SSL (IMAPS) uses port 993. The messages are not downloaded locally to the client device's email software program. Instead, the email remains on the email server for retrieval from any device with an Internet connection and the proper credentials to authenticate to the email server. POP3, which uses port 110, is used by client devices to retrieve email from a remote email server. For secure sessions, you can use POP3 over SSL (POP3S), which uses TCP port 995. Currently, though, support for this port from Internet Service Providers is unlikely. When a user connects to their email server using POP3, the messages are downloaded to the local client device. The MAPI protocol, which uses port 135, is a proprietary protocol that was created by Microsoft. If you are using Microsoft's Outlook email client, you can use the MAPI protocol to connect to Microsoft Exchange email servers. In the following sections, we will explore how to configure these commonly used protocols and applications for designed for electronic messaging.

Simple Mail Transfer Protocol (SMTP)

Simple Mail Transfer Protocol (SMTP) is a messaging protocol that serves two purposes. It can be used for client-device-to-server email delivery and for server-to-server email delivery. SMTP communicates using TCP/IP and is the most common messaging protocol used to send messages from a client device to an email server. Other messaging protocols, such as POP3 and IMAP, are used to receive email messages from an email server. Figure 14.2 illustrates the mail delivery process using SMTP.

FIGURE 14.2 Messaging and SMTP

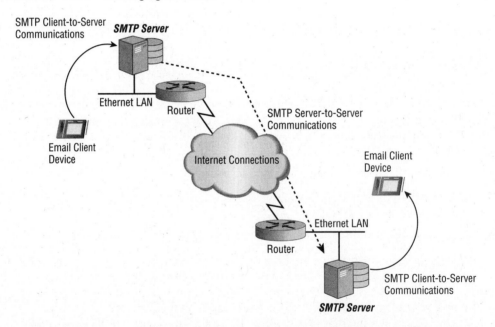

Configuring your device to use SMTP is straightforward. It is a common configuration parameter set within the email client software and is used to access the email server to in turn access the email database. When you configure email client software, you will need to know several parameters, including the login username and password, the outgoing and incoming mail server names or IP addresses, and the protocols and port numbers used. In the case of SMTP, this would be the outgoing mail server, which uses port 25 by default. Figure 14.3 shows an example of an email client configuration screen for the Android operating system.

FIGURE 14.3 Android email app outgoing mail server settings

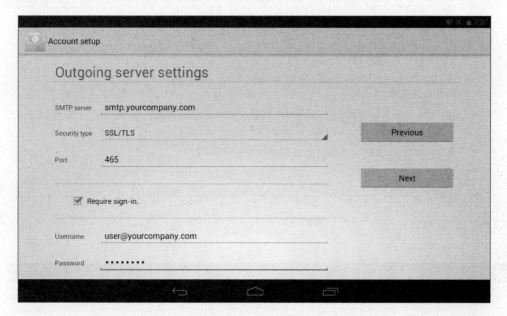

Notice in Figure 14.3 that the port for SMTP is 465 and not 25. This is because this specific application is using SMTP over Secure Sockets Layer (SSL), or SMTPS, which uses port 465.

Configuring Internet Message Access Protocol (IMAP)

Configuring your client device for *IMAP* is also a straightforward process. One main benefit to using this protocol to access email is that the messages are not downloaded locally to a device and they remain on the email server until the user deletes them. The messages may be cached temporarily on the local client device while the email program is in use. This allows a user to retrieve email from any device with an Internet connection and the proper credentials to authenticate to and synchronize the messages with the email server. Here are some of the advantages and disadvantages of using IMAP:

- Advantages of IMAP
  - Email client software does not need to be installed on the local device.
  - Email can be accessed from any device with an active Internet connection.
  - Because the email messages are stored on a server and not locally, they will still be available in the event of a device failure.

- Email stored on the server will be scanned for computer viruses, which lessens the chance of infecting the local device.
- If you change devices, you don't need to transfer the email messages to the new device.

- Disadvantages of IMAP
  - An active Internet connection is required for users to read, create, and send email messages.
  - Because the email is created and read online, it uses data from a data plan if the user is on a cellular network connection.
  - Not all devices or operating systems support IMAP.
  - It can potentially be slow, depending on the Internet connection speed.
  - There may be limited storage space on the hosting email server.

In Exercise 14.1, you will configure an email account on an Android device to use IMAP.

EXERCISE 14.1

Configure an Android Device to Access Email Using IMAP

In this exercise, you will configure an Android device to access email using IMAP. The steps in this exercise were performed using the Android emulator program you installed in Exercise 10.1, "Installing Android Emulator Software in Microsoft Windows," in Chapter 10. Keep in mind that the steps and screen shots in this exercise may vary slightly based on the device and version of the Android operating system used.

1. Tap or click the application's menu icon. The Apps and Widgets screen will appear.
2. Tap or click the settings icon. The settings screen will appear.

3. Scroll down to the Accounts section and tap or click Add Account. The Add An Account screen will appear.

4. Tap or click the Email icon. The Account Setup screen will appear.

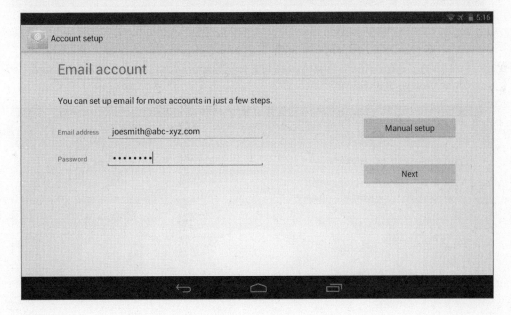

EXERCISE 14.1 *(continued)*

5. Enter your email address and password and tap or click on the Manual Setup button to advance to the Account Type screen.

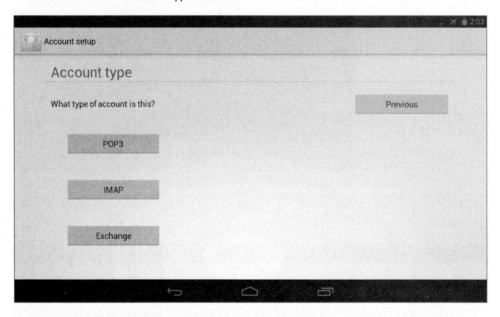

6. Tap or click the IMAP button. The Incoming Server Settings screen will appear.
7. Enter the correct settings for your email account on the Incoming Server Settings screen. The default port is set to port 143. It is recommneded that you use a secure connection such as SSL with port 993.

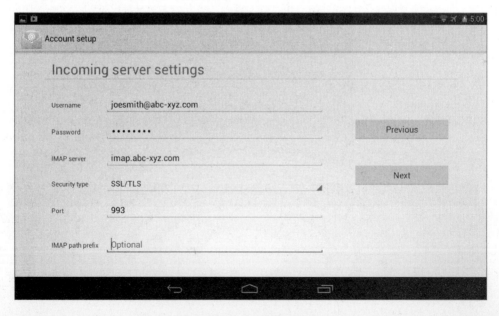

Configuring Messaging Protocols and Applications

8. Tap or click Next. On the Outgoing Server Settings screen, enter the correct settings for your email account and server.

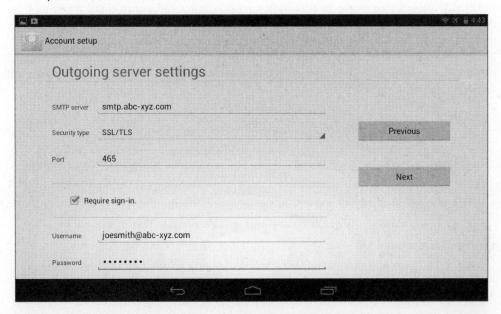

9. Tap or click Next. The Account Options screen will appear.

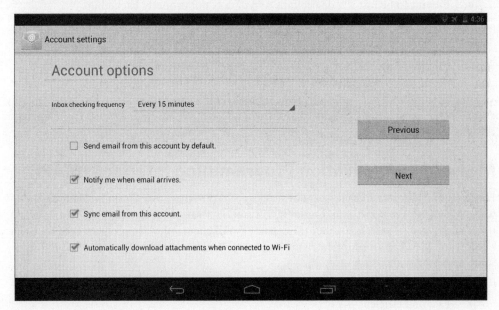

10. Select the desired options and click Next. Your account is now set up.

Post Office Protocol 3 (POP3)

POP3 will allow a user to use an email software program to access an email server and download messages. The user will then have access to the email messages that are stored on the local device. This is a message-retrieval protocol that is quite common and popular because the user is not required to have Internet access to read and create email messages, so the user can work with the email in an offline state. Here are some of the advantages and disadvantages of using POP3:

- Advantages of POP3
  - Many software programs that support POP3 are available.
  - An active Internet connection is not required to access email that is stored on your device.
  - A connection to the Internet is required only to download messages and to send messages that are queued within the software program.
  - Large attachment files can be accessed quickly because they are stored locally.
- Disadvantages of POP3
  - Email messages that include attachments are downloaded locally and can use up hard drive space unless properly maintained.
  - Downloaded attachments may contain viruses, causing problems on a system that is not properly protected.
  - Email can be lost if a device fails and data is not properly backed up.
  - If someone else has access to your device, they may be able to read your email data.
  - If POP3 is not used with a virtual private network (VPN), SSL, or other secure Internet connection, all user information, including username, password, and server configuration, may be seen by hackers on a wireless network.

Email client device software programs that use POP3 can be configured to automatically check the server at specified intervals. This will allow the client to download the email to the local device.

Messaging Application Programming Interface (MAPI)

For users of Microsoft Exchange email services and the Microsoft Outlook client software, *MAPI* is a great solution. Outlook will run on a variety of device types and mobile operating system(s) systems. Other third-party email client software programs may provide support for MAPI. As a proprietary protocol that was developed by Microsoft, it has its share advantages and disadvantages:

- Advantages of MAPI
  - Fast and efficient
  - Easy to configure on the client side

- Disadvantages of MAPI
    - Proprietary and requires Microsoft Exchange Server
    - May require special configuration to operate correctly when crossing a firewall

 It is important to note that there are many web-based email programs that use Hypertext Transfer Protocol (HTTP) via a web browser. Many of these email services allow the user to configure their email to be delivered or viewed using protocols such as POP3 and IMAP. HTTP is an application layer protocol that provides a graphical user interface to the end user and is not considered a messaging protocol. Configuration of the messaging protocols used with this type of email will vary based on the service provider.

Configuring Network Gateway and Proxy Server Settings

In Chapter 6, "Computer Network Infrastructure Devices," we explored the functionality and features of various infrastructure devices for wired and wireless networking, which included network gateways and proxy servers. In the following sections, you will learn how to configure settings for both network gateways and proxy servers.

Configuring Network Gateway Settings

The term *network gateway* can be interpreted in different ways, often depending on the context in which it is used. A network gateway is an interface that allows computer networks using different protocol sets to connect together and share information. In some cases it is simply the IP address of the router port that allows your device to access resources outside of the LAN you are connected to. In the context of this section, the network gateway is the point that allows your mobile device to connect to external resources outside of the local area. If your device is using Dynamic Host Configuration Protocol (DHCP), which enables the device to receive IP settings automatically, there is no configuration because it is provided as part of the IP settings issued by the DHCP server. If for some reason the device is not using DHCP and static IP settings are required, the gateway will need to be manually entered.

FIGURE 14.4 Microsoft Windows computer default gateway settings entered manually

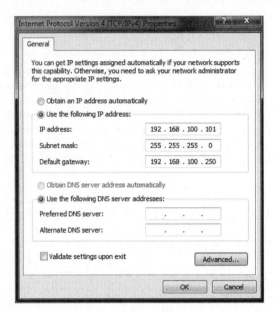

Figure 14.4 shows where the default gateway settings are entered manually in the Microsoft Windows operating system. The same concept holds true with devices such as smartphones and tablets. If DHCP is used, the gateway settings are automatically provided by the DHCP server. If static IP settings are used, the gateway address will need to be entered manually. The following steps show the process of manually entering the default gateway information into an Android client device:

1. Tap the applications menu icon to show the Apps and Widgets screen.
2. Tap the settings icon.
3. Tap the Wi-Fi setting.
4. Tap the + sign to add a wireless network.
5. Enter the wireless network settings, network SSID, and security settings.
6. Select the check box labeled Show Advanced Options.
7. Tap DHCP next to the IP Settings label and choose Static. The window will expand as shown in Figure 14.5.

FIGURE 14.5 Android device showing where to enter the default gateway manually in the IP static settings

8. Enter the settings based on your network and proxy server, then tap or click Save.

The following steps show the process of entering the static IP information including the default gateway into an Apple iOS client device:

1. Navigate to the Settings app and tap Wi-Fi.
2. Tap the Wi-Fi connection for which you wish to set the static IP information, including the default gateway.
3. Tap Static and then tap Router.
4. Tap IP Address and enter the address.
5. Tap Subnet Mask and enter the appropriate subnet mask.
6. Tap Router and enter the IP address of the default gateway.

Configuring Proxy Server Settings

A *proxy server* can be either a hardware appliance or software server solution. It is the "middleman" between a device on a private network and a destination server across another network, which can be a WAN connection or the Internet. A proxy server provides features such as web page caching, content filtering, and more. If a proxy server is used to gain access to the Internet, the client devices, computers, smartphones, and tablets will need to be configured with the correct information. The following steps show the process of entering the proxy server information into an Android client device:

1. Tap the applications menu icon to show the Apps and Widgets screen.
2. Tap the Settings icon.
3. Tap the Wi-Fi setting.
4. Tap the + sign to add a wireless network.
5. Enter the wireless network settings, network SSID, and security settings.
6. Select the check box labeled Show Advanced Options.
7. Tap None next to the Proxy Settings label and choose Manual. The window will expand as shown in Figure 14.6.

FIGURE 14.6 Android device proxy server manual settings screen

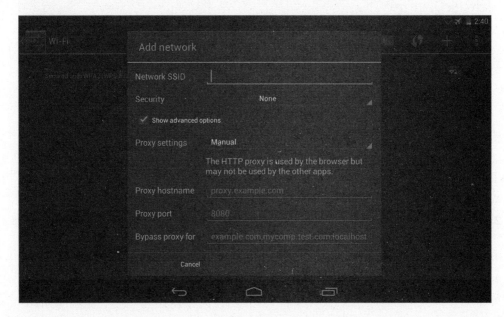

8. You can enter the settings based on your network and proxy server.

The following steps show the process of entering the proxy server information into an Apple iOS client device such as an iPhone or iPad:

1. Navigate to the Settings app and tap "Wi-Fi."
2. Tap the Wi-Fi connection for which you wish to add the proxy information.
3. Tap Static and scroll to the bottom of the screen.
4. Tap Manual.
5. Enter the proxy server address and port if required.

6. Slide the Authentication icon to the right to turn on authentication.
7. Enter the username and password supplied by the network administrator.

Push Notification Technology

Push notification technology will allow an application to provide an announcement of some type to alert a user of events and messages that have been received by a device. These notifications can be delivered to the device through an app that is not even active at the time the alert is delivered. In Chapter 2, we explored some of the protocols and ports that are used with messaging and notification services. Here you will see an overview of the configuration of two of the most common notification technologies, Apple Push Notification Service (APNS) and Google Cloud Messaging (GSM). The device type and the mobile operating system(s) in use will determine how the notifications are delivered to the device for the user. For example, with Android devices, the application's icon and a message will appear in the status bar, providing an alert to the user. Apple iOS devices can alert the user with sounds and using other graphical methods.

Configuring Apple Push Notification Service (APNS)

Apple Push Notification Service (APNS) is used to enroll an Apple iOS device within a mobile device management (MDM) solution. APNS uses a native push notification technology that is available for Apple iOS devices and allows the iPad, iPhone, and other devices to receive information even when the destination app is not open or running. The iOS notifications include sounds, graphical alerts/banners, and badges. Configuring notifications on an iOS device is straightforward, and each app will need to be configured independently.

Exercise 14.2 shows you how to configure APNS on an Apple iPhone.

EXERCISE 14.2

Enabling APNS on an Apple iPhone

Here you will look at how to configure APNS on an iPhone.

EXERCISE 14.2 *(continued)*

1. Tap the Settings app.

2. Tap the Notification Center icon.

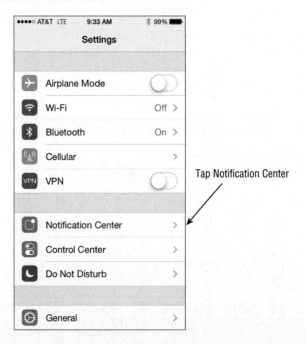

3. You can see apps that are currently in the Notification Center.

In the Notification Center

4. Tap the app to enable the notification. In this case, the Messages app was chosen.
5. Toggle the switch to enable the notification.
6. Choose an alert style from one of three options: None, Banners, or Alerts.

Notice Three Different Alert Types

Keep in mind that a valid encrypted IP session is established between the mobile device and APNS. The device will receive all notifications over the encrypted TLS communications tunnel that was created. The device must obtain a token that contains the required information that will allow the notification service to locate the device where the app is installed and deliver the alert.

Google Cloud Messaging (GCM)

Google Cloud Messaging (GCM) allows for data to be sent from a server to Android-based devices, and also to receive messages from devices that are on the same connection. Like the APNS for iOS devices, GCM provides a push notification service for Android devices. One benefit to a push notification service is that it will lessen the amount of network bandwidth used because the device will receive information that is pushed, therefore, it will not have to poll a server for any of that type of data. The following steps show the process of enabling or disabling notifications on an Android client device:

1. Tap the applications menu icon to show the Apps and Widgets screen.
2. Tap the Settings icon.
3. Scroll down and tap the Apps field.
4. Tap the app in which you want to configure the notifications.
5. Enter the wireless network settings, network SSID, and security settings.
6. Select the Show Notifications check box.

If the check box is selected, the device will receive notifications for the selected app. If it is not selected, the device will not receive notifications for the selected app. Figure 14.7 shows how to enable the notifications setting on an Android tablet.

FIGURE 14.7 Enable or disable notifications on an Android tablet

The preceding steps show only one way to change the notification settings on an Android device. Depending on the version of the mobile operating system(s) used, other methods may be available.

Information Traffic Topology

The *network operations center (NOC)* is where a computer network is managed, controlled, and monitored. Networks of all sizes will contain some sort of operations center, and they will vary in design, size, and complexity. The NOC engineer's responsibilities include infrastructure and device monitoring, handling reported incidents, and notifying the appropriate person or department when problems arise. NOC engineers will be able to accommodate necessary changes to help ensure overall network health, maintain security compliance, and verify that the network meets the organization's requirements.

Many NOCs operate 24 hours a day, 7 days a week, and 365 days a year. An NOC may have support personnel organized into groups with varying levels of knowledge and hours of operations, which can have an effect on the overall costs. For this reason and depending on the size of the network, in some cases it may be more cost effective to outsource this responsibility. Rather than managing their own network, some organizations now use a managed service provider (MSP) to handle the responsibilities of the network.

Basically, there are three different NOC types:

- On-premise network operations center
- Third-party network operations center
- Hosted network operations center

On-Premise Network Operations Center

The on-premise network operations center model is the traditional type of NOC. This model has been around for decades and is a common implementation in organizations from all types of businesses. With an on-premise NOC, the organization is responsible for all aspects and day-to-day responsibilities of managing and supporting the network infrastructure. In most cases, this will require 24/7 support and maintenance 365 days a year. Depending on the size of the organization and network, this model can be a costly endeavor. It will be up to the organization to supply the manpower, hardware, software, and other necessary resources. Although it is beneficial to have all network operations handled in-house, in some cases this model may be cost prohibitive; therefore, the task may be outsourced to a third-party organization.

Third-Party Network Operations Center

The third-party network operations center is becoming a more common scenario for many organizations. In this model, network operations are completely outsourced to an outside independent company that specializes in the process. In some cases, the organization that is outsourcing the NOC will still provide the facilities and equipment, but the manpower is provided by a different company. In other cases, the entire process is handled by the third-party NOC. This managed services model can provide a substantial cost savings to an organization because the around-the-clock responsibilities of dealing with a NOC are minimized. Some organizations choose a third-party solution so they can get on with their business and not have the burden of directly managing the technology.

Hosted Network Operations Center

Some organizations choose to have some or all of their day-to-day network operations hosted by what would be the equivalent of a cloud-based service provider. This solution will allow a company to offload some or all of the network operations, such as web services and application and data servers, to an offsite provider. The hosting service provider will be able to provide uninterruptible power supply (UPS) services, allowing for generators and battery backup power sources to supply the required power and the necessary actions in the event of a power outage. Also, network monitoring tasks will be handled by a staff trained to respond to issues that may have an impact on the performance of the network. Hosted network operations providers can usually provide a guaranteed amount of bandwidth and the ability to dynamically allow additional bandwidth. Finally, the offsite responsibility of data backup, storage, and restore will relieve the organization of the burden to maintain its data integrity.

Summary

In this chapter you learned about the importance of being aware of mobile device technology advancements, including those that are related to device hardware and the mobile operating system(s). We also explored the concept of third-party application vendors for online app stores and the potential risks and threats from changes to apps and from technology advancements. I explained the three different app types: native apps, web apps, and hybrid apps. You also saw the advantages and disadvantages of each app type. You were given the opportunity to see how to configure various messaging protocols on different platforms, including Android and Apple iOS devices. We also explored the configuration of specific network parameters, such as network gateway and proxy server settings that may require manual configuration, including where they are located within mobile operating system(s). Finally, we took a quick survey of two popular push notification services and explored the concept of different network operations center (NOC) options.

Chapter Essentials

Understand the importance of maintaining awareness of advancements in mobile technology. Staying informed of technology updates and enhancements will help to provide a quality user experience. Changes to device hardware, mobile operating system(s), and third-party apps can have an adverse effect on a deployment, as can new risks and threats. One of the best ways to stay informed is to stay in touch with hardware and software vendors by connecting with them through email updates and social media outlets.

Be familiar with in-house application requirements. Some organizations use apps that are custom or purpose built to serve a specific function for their business model. These apps are best acquired and deployed from an enterprise app store rather than a consumer or publicly available app store.

Understand the different app types. Know the advantages and disadvantages of the different available app types, which include native apps, web apps, and hybrid apps. Understand that native apps offer good performance but are platform dependent. Web apps are platform independent but require an active Internet connection. Hybrid apps can be built more quickly than native apps but may lack some common functionality.

Be familiar with the configuration of some specific messaging protocols. Understand how to configure various messaging protocols, such as SMTP, IMAP, POP3, and MAPI. The configuration process will vary based on the device hardware type and the operating system that is used. If IMAP is used, messages stay on the email server and any work done while not actively connected to the Internet will require synchronization when a connection becomes available. Conversely, using POP3 will allow email messages to be downloaded to the client device, and email can be read and created while offline.

Understand the configuration of additional network settings such as those for gateway and proxy servers. In some cases, you may need to manually configure network settings such as network gateway and proxy server parameters. Although these settings can be delivered automatically, it is beneficial to understand the manual configuration process.

Be familiar with push notification services. Understand how common push notification services such as Apple Push Notification Service (APNS) and Google Cloud Messaging (GCM) operate. Know how to enable and disable push notification services on Android and iOS devices.

Chapter 15

Mobile Device Security Threats and Risks

TOPICS COVERED IN THIS CHAPTER:

- ✓ Understanding Risks with Wireless Technologies
- ✓ Understanding Risks Associated with Software
- ✓ Understanding Risks with Mobile Device Hardware
- ✓ Understanding Risks Affecting Organizations

Mobile devices have their share of risks and threats associated with the technology. The risks can be with wireless (radio frequency) technology, software, hardware, and within the organization itself. This chapter will explore some of the threats that fall under each of these risk categories. Keep in mind that many of the threats discussed here are not limited to wireless technology; they are threats that you may already be familiar with from other networking technologies. Because public Wi-Fi hotspots are easily accessible to virtually anybody, they can be a target for various types of intrusion attacks. Rogue access points (APs) are also wireless threats because they can result in security breaches within the organization. These and other wireless concerns will be discussed. You may already be familiar with risks associated with software, which includes threats such as malware, viruses, and spyware. These are also a concern with mobile device technology. In addition to software, hardware risks such as lost or stolen devices are a concern. In this chapter, we will explore different software and hardware risks and the risks that have an effect on organizations in general.

Understanding Risks with Wireless Technologies

In previous chapters you learned that wireless technology uses an unbounded medium (the air) for communications. That fact itself makes this type of communication fall into a high-risk category and therefore be vulnerable to various types of intrusion attacks. Understanding these risks is essential to provide an adequate security solution for wireless communications. Here are some common risks associated with wireless technology in general:

- Public Wi-Fi hotspots
- Rogue access points
- Denial of service attacks
- Radio frequency jamming
- Cell tower spoofing
- Wireless hijack attacks
- Wireless man-in-the-middle attacks
- Weak security keys
- Warpathing

The following sections provide a basic understanding of these risks and how they may be implemented. Figure 15.1 shows some of the risks associated with wireless technology.

FIGURE 15.1 Wireless-enabled devices are subject to many potential security threats.

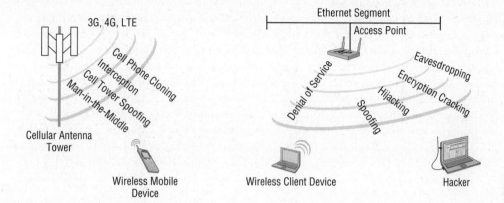

Public Wi-Fi Hotspots

Anytime you access a wireless network from a public Wi-Fi hotspot, extra care must be taken to ensure a secure connection. Many public guest networks require the user to authenticate through a captive web portal. A captive web portal will intercept a user's attempt to access the network by redirecting them to a web page for authorization of some sort. This web page may request account credentials, payment information from a user, or a simple agreement to terms and conditions before granting access to the wireless network. Some assume that because they were required to "authenticate," the wireless connection will be secure; however, this is not the case. Once the captive web portal authentication has completed, all communications will be in clear-text unless other security measures are implemented by the user, such as using secure protocols such as Hypertext Transfer Protocol with Secure Sockets Layer (HTTPS) to create a secure connection or connecting through a virtual private network (VPN). Many people do not realize that a captive web portal is used only to restrict access to the wireless network and does not provide any additional encryption or security. In addition to ensuring a secure connection, additional precautions should be in place, including but not limited to the following:

- Secure personal firewall
- Current antivirus software
- Client-side software/agents

Rogue Access Points

In order for IEEE 802.11 wireless client devices to connect to an infrastructure network, they must connect to a wireless access point (AP). The AP will act as a portal for the devices using IEEE 802.11 wireless technology to connect to IEEE 802.3 wired technology (Ethernet). We explored this and other AP fundamentals in Chapter 5, "IEEE 802.11 Terminology and Technology."

It is important to understand that an AP to which you are connecting may not necessarily be what is considered an authorized AP. Connecting to an unauthorized AP comes with its share of risks, including various types of wireless intrusion attacks. The term *rogue AP* does not necessarily mean that the AP is a specific threat, but it very well could be depending on the specific situation in which it is placed. With respect to wireless security, typically the term *rogue* means that the AP is not an authorized AP and has the potential to be a threat. One example is when an employee of a company brings a personal wireless AP from home for use in their office and connects it to the corporate LAN. The user unknowingly has connected an unauthorized device that is not managed by the corporate IT staff, and the device is now considered a threat to the corporate network. This would fall under the unintentional rogue AP category. Although the employee meant no harm, it could compromise the security of the entire corporate LAN. Conversely, when someone knowingly connects an unauthorized AP to a corporate network for malicious intent, it would fall under the category of an intentional rogue AP.

Proper monitoring of the network for rogue devices can be a time-consuming task, but it must be considered as part of the corporate security policy. A couple of particular concerns to watch for are ad hoc networks and mobile wireless hotspots. If these are not allowed by corporate security policy and properly secured, they could be considered rogue AP and a security threat to the organization.

Ad Hoc Wireless Networks   The ad hoc, independent basic service set (IBSS) and peer-to-peer networks can also be considered threats and may fall under the category of an AP point. The IBSS network explained in Chapter 5 is typically noncompliant against most corporate security policies. One reason is that if a network of this type is connected to a corporate LAN, it basically becomes a rogue AP. The ad hoc network is an independent unmanaged network that can pose a serious threat to a corporate LAN.

Mobile Wireless Hotspots   Many mobile devices such as portable wireless hotspots and smartphones have the capability to become a "wireless hotspot." Although these devices are not physically connected to a corporate network, they can also potentially become a threat to corporate security. This is because a user connected wirelessly to a device of this nature may inadvertently send a company's private information over a network that is not managed by the enterprise network staff, thereby compromising corporate security. Although most of these devices have the capability to be secured, there is no telling whether they are secured or the extent to which they are secured.

Denial of Service (DoS) Attacks

A *denial of service (DoS)* can fall under two different categories, unintentional or intentional. An intentional DoS is caused by a hacker or intruder implementing a technology of some sort that will prevent authorized users from connecting to, accessing, or maintaining a connection to their wireless network. With IEEE 802.11 networks, a DoS attack could be performed at the Physical layer (Layer 1) or the Data Link layer (Layer 2) of the Open Systems Interconnection (OSI) model. As shown in Figure 15.2, a Physical layer DoS attack typically entails the use of a malicious radio frequency (RF) jamming device operating on the same RF channel as the wireless AP. This will prevent users from maintaining a connection to the AP, thereby denying access to the network. Jamming is discussed in more detail in the next section of this chapter.

FIGURE 15.2 The MetaGeek Chanalyzer spectrum analyzer shows an RF DoS attack using a narrow-band jamming device.

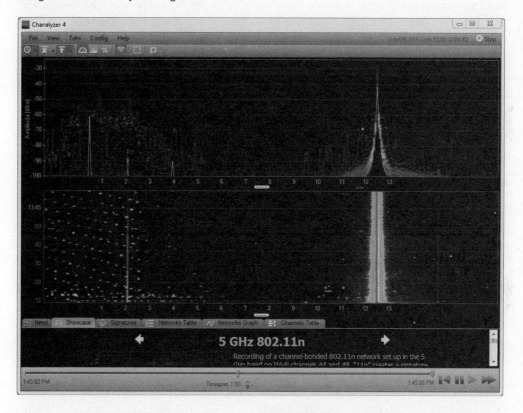

A Layer 2 DoS attack is more sophisticated, and adequate knowledge and the proper tools are required to perform this type of attack. With one type of Layer 2 DoS attack, the intruder will send either IEEE 802.11 deauthentication or disassociation frames that prevent an authorized device from maintaining a connection to the wireless network. These frames are sent by spoofing the Media Access Control (MAC) address of the wireless AP. The connected wireless devices do not realize that these are unauthorized frames and therefore will not be able to maintain a connection to the AP. Remember that both IEEE 802.11 deauthentication and disassociation frames are notifications and not requests; therefore, they will not be refused by the client device.

Radio Frequency (RF) Jamming

Radio frequency (RF) jamming can be the result of either intentional or unintentional interference. Intentional jamming occurs when an illegal radio transmitter is operating on the same RF channel as an authorized transmitter, preventing the authorized legitimate transmission from occurring. With regard to IEEE 802.11 wireless networking, this is known as a Layer 1 DoS attack.

Due to the unbounded nature of RF communications, there is no way to prevent this type of attack other than physical security. However, there are many tools available that will help you to locate the jamming device and therefore allow you to eliminate the source once it has been identified as a threat. One example is an RF spectrum analyzer with a directional antenna. This device will allow a network security specialist to track down the exact location of the jamming device and eliminate the threat. In IEEE 802.11 wireless networking, a *wireless intrusion prevention system (WIPS)* is a software/hardware solution that monitors the radio waves and, using a wireless hardware sensor, can report captured information to software to be recorded in a server database. The WIPS solution will then be able to take the appropriate countermeasures to prevent wireless network intrusions. These countermeasures are based on identifying the intrusion by comparing the captured information to an intrusion signature database within the WIPS server. Figure 15.3 shows a WIPS dashboard that displays a Layer 1 RF jamming DoS attack that was recorded.

FIGURE 15.3 AirMagnet Enterprise dashboard displays a recorded RF DoS attack.

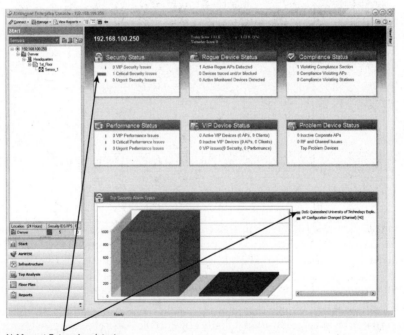

AirMagnet Enterprise detects an RF denial of service (DoS) attack that was reported.

Unintentional RF jamming is the result of devices that are operating in the same frequency space. Devices that can be the cause of this include microwave ovens, RF cameras, cordless phones, and others. Although these are not intended to be the cause of a jamming DoS, in some cases they will have the same overall effect. Identifying, locating, and removing these interference sources will help improve the performance of the wireless network.

Cell Tower Spoofing

Cell tower spoofing is a newer form of mobile device wireless hacking, intrusion, and eavesdropping. Smartphones and mobile devices of all types may be vulnerable to cell tower spoofing. With the proper hardware and software, someone can transmit what appears to be a legitimate cellular signal and, acting as a provider or carrier, force the unsuspecting user to connect to their device. Devices do not require two-way authentication, and from the mobile device perspective a smartphone does not realize the transmitting "station" is not provided by the wireless carrier to which it is subscribed.

Mobile devices that are cellular capable use what is known as an international mobile subscriber identity (IMSI) number to uniquely identify the device. Typically about 15 digits in length, this number allows mobile devices to interconnect with various cellular networks. The IMSI may be embedded within the device itself or contained within the Subscriber Identity Module (SIM) card. If the IMSI is contained within the SIM card, a user can move the card from one device to another and maintain the same identifier.

An IMSI catcher is a system that is used to connect unsuspecting users to an imposter cell tower. Once the mobile device has connected to the "fake" tower, it has essentially become the victim of a man-in-the-middle attack. This would allow the operator of the IMSI catcher to eavesdrop on all transmissions between the user and the legitimate provider's network and potentially gain access to sensitive and personal information.

Wireless Hijack Attacks

A *wireless hijack* is caused by a hacker creating a situation that will allow or even force a wireless client device to connect to an unauthorized wireless AP or base station. Figure 15.4 shows the steps a hacker could use to perform a hijack using an RF DoS attack. The hacker would set up a rogue software AP and then perform an RF DoS attack on the channel in use. This will cause the authorized users to seek another AP to connect to.

This is something that can go unrecognized by a user for some time, especially if the hacker then follows through with a wireless man-in-the-middle attack. The AP used by the hacker in this specific situation can be software based or hardware based. There are many ways that a wireless mobile device using Wi-Fi can be hijacked:

- The client issues IEEE 802.11 probe request frames to connect to a network that is set to automatic (normal behavior).

FIGURE 15.4 Wireless hijack using an RF DoS attack with a wireless jammer

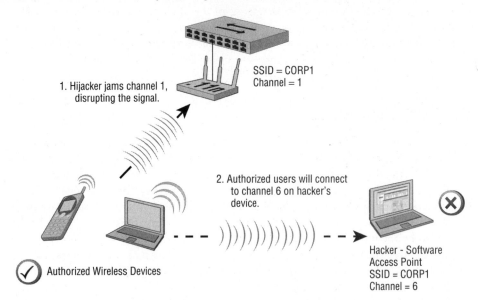

- The hacker creates a DoS attack against the authorized AP. This will force the device to roam.
- The hacker sends deauthentication frames to the wireless mobile device by spoofing the address of the authorized AP.

Any one (or any combination) of these methods will allow for a wireless client device to connect to a rogue AP and become a victim of a hijack situation.

Wireless Man-in-the-Middle Attacks

Once a device has been hijacked, the hacker can follow through with a wireless *man-in-the-middle attack*. This type of attack will enable a hacker to provide an unsuspecting user with a connection to a legitimate corporate network or to the Internet. With the hijacked user connected to the rogue AP, the hacker will retransmit all data to the authorized AP so the hijacked user does not even realize there is a problem. This is accomplished by the hacker having a second wireless network adapter and connecting that adapter to the authorized AP. All information that the user sends to the authorized AP is captured by the hacker, potentially exposing personal or sensitive information. Figure 15.5 shows a graphical representation of a wireless hijack and wireless man-in-the-middle attack. After the hacker performs the wireless hijack and forces the authorized users to connect to the rogue software AP, they will retransmit to the real AP, exploiting all traffic that is sent to the network.

FIGURE 15.5 Wireless RF hijacking, progressing to a man-in-the-middle attack

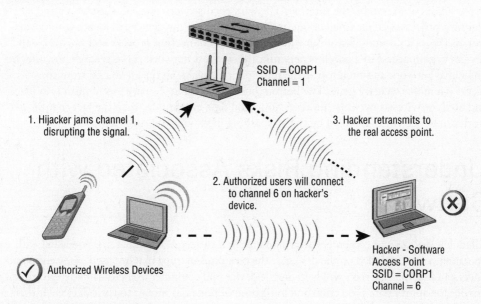

Weak Security Keys

A key in any sense of the word is something that is used to gain access to a secured environment such as an automobile, home, or office. With respect to technology, a *key* is used for cryptography purposes and may be used for authentication purposes or to gain access to encrypted information. Software programs that will allow a hacker to break methods that use weak keys are freely available on the Internet.

One example involves IEEE 802.11 wireless LAN technology that uses passphrases and pre-shared keys. A passphrase (basically a password) is entered into all devices participating in the secured wireless network connection. The IEEE 802.11 standard defines an algorithm that will create a 256-bit pre-shared key from the passphrase that was entered. This algorithm is published in the IEEE 802.11 standard, and freely available software programs exist that will allow a hacker to obtain a weak passphrase. This is done by capturing specific frames used during the authentication process. With the software and use of a dictionary file, a weak passphrase can be quickly determined. Once the passphrase is known, the hacker can enter the passphrase into a device and gain access to the network. In addition to network access, the experienced hacker will be able to re-create the encryption keys created on a per-user basis and will be able to view their encrypted information.

Warpathing

Wireless *warpathing* is the process of seeking and discovering wireless networks. *Wardriving* is another term used to refer to the same process. In the early days of wireless networking, this was considered a popular way to find and document the locations of open wireless networks.

These networks could then be used by the casual hacker to gain personal information or just to provide a connection to the Internet. The wireless discovery phase is the passive and active scanning processes you learned about in Chapter 5 that allow devices to locate wireless networks. This is a normal part of the wireless network connection process and is used by all devices to gain access to a wireless network. If a wireless network is not secured, locating the unsecured network and monitoring the information traversing the air to and from the wireless access is a major security issue. The bottom line is if a strong security mechanism is in place and used for wireless network authentication and data encryption, then the fact that the network can be seen or discovered by the warpathing method is not really an issue.

Understanding Risks Associated with Software

All mobile devices and computers are vulnerable to a variety of risks that are associated with the operating system installed on the device or the programs or apps that are used. Awareness and understanding of these risks will help mobile device and computer users to be better prepared to provide the appropriate protection and mitigation of potential threats to their devices. It is best for users to have the appropriate software protection installed, such as firewall and antivirus programs, to help protect against and prevent any damage software threats might cause.

Protection against software threats such as computer viruses, malware, and spyware are discussed further in Chapter 17, "Security Requirements, Monitoring, and Reporting." The intention of this chapter is to provide you with an understanding of the different potential software threats to computers and mobile devices.

Malware

Technology devices are subject to many attacks in which software programs or applications designed for malicious intent are used. These programs are known as *malware*, which is an abbreviated term for *malicious software*. Malware is a general term used to describe a variety of different software-based threats:

- Information technology viruses
- Trojans
- Computer worms
- Spyware
- Keylogging

Although this is not a complete list, these are some of the common software threats that you should be aware of in order to protect your technology devices and the data contained within.

Information Technology Viruses

I like to refer to this threat as an information technology *virus* rather than a computer virus because viruses can affect many devices other than computers, including mobile devices such as tablets or smartphones. The Android operating system may be more vulnerable to viruses than its proprietary counterparts such as Apple iOS, BlackBerry, and Windows Phone. This is because the Android operating system is open source.

A virus can be any code that can have adverse effects on the device that it has infected. The effects of information technology viruses can range from being a minor inconvenience for users to being serious security breaches for a corporation. Some viruses are programmed to replicate and reproduce themselves to other devices they connect to and even across entire corporate networks and the Internet. This can result in wasted time for the information technology staff as well as the users of the infected devices.

Jailbreaking and rooting a mobile device, both explained later in this chapter, may make a device more susceptible to viruses and other technology threats.

Trojans

A *Trojan*, also known as a Trojan horse, is a type of a technology virus, usually in the form of an executable program that is designed to perform malicious tasks. These will often create a "back door" to the operating system that is installed on the device, allowing for unauthorized access to the device and the data it contains. Trojans don't infect only computers. Mobile devices like smartphones and tablets are also vulnerable to these types of attacks. Trojans are sometimes programmed to perform the following:

- Modify or delete data
- Steal personal information such as passwords
- Steal money electronically through transfers
- Control cameras installed on the device
- Install software such as keyloggers

This list is just a small sample of the types of security issues a Trojan can cause.

Computer Worms

A *computer worm* is a malicious software program that is designed to replicate and spread to other devices. A computer worm can be spread from data that is transferred manually (such as on a removable disk drive) or from the connection to a computer network. Some worms are designed to be a nuisance and cause a slowdown on the network by increasing network traffic and utilizing the bandwidth, while others can create more serious security issues.

Spyware

Spyware consists of computer programs that are used to gather information about the device and the information (data) that is stored on the device. This type of program is installed without the user's knowledge or consent and can cause a variety of privacy- and security-related issues. In many cases, spyware programs are installed as a result of the user clicking a hyperlink contained within a web page, causing embedded code to run and install the unwanted software. Spyware can report back to other servers and provide specific information regarding the infected device's use. For example, spyware can provide information on websites the user visits and online shopping habits. Spyware can also include software programs that are manually installed on a smartphone or other mobile device and used to monitor, track, and log everything the user does.

Keylogging

Keylogging is a method by which someone with the use of software can "record" keystrokes that are executed on computers and mobile devices. This is usually done unbeknownst to the user. The data that is collected from the keylogging process can then be viewed, possibly exposing sensitive information. This can lead to serious security issues such as stealing from bank accounts, sending fraudulent information to unsuspecting users, and even identity theft.

Mobile App Store Usage

When you download and install a mobile application from an online store, you may be authorizing the use of certain services regardless of whether the app requires them. These services often include contacts, email, and location services, among others. Therefore, the apps may be somewhat of a security threat and present privacy concerns. If the devices are allowed to be used on a corporate network, the security of both personal and company information can be an issue.

Mobile Device Jailbreaking

From a mobile device perspective, *jailbreaking* is a way to remove manufacturer or carrier restrictions from an Apple mobile device. Although jailbreaking is performed on Apple devices that run the iOS operating system, such as the iPad and iPhone, the term is sometimes used with other devices, including Android devices. (However, a similar concept that is known as *rooting* applies to Android devices. See the next section.) Jailbreaking will allow a user to modify the device's operating system for various reasons. In addition to voiding the manufacturer's warranty, jailbreaking a phone can make the device more susceptible to viruses, malware, and Trojans in addition to other security issues. Jailbreaking applications can damage your mobile device, rendering it useless, and once jailbreaking has occurred, the manufacturer or carrier where you purchased the device is not responsible for repairing or replacing it.

Android Rooting

Android *rooting* is a method of modifying an Android-enabled device to allow the owner to obtain administrative privileges or control over the operating system. This will allow the user to modify the actual operating system code and perform tasks that are not intended for the average user to do. The following list includes some of the disadvantages of rooting your device:

- Voiding the manufacturer's warranty
- Rendering the device useless
- Allowing for viruses, malware, and Trojans

Keep in mind that the Android operating system is designed with several limitations and restrictions by carriers or manufacturers for commercial software usage and for security reasons. Therefore, you should consider the potential drawbacks prior to rooting a device.

Understanding Risks Associated with Mobile Device Hardware

In addition the potential software risks associated with mobile devices, there are certain hardware risks. They include loss or theft of devices and device cloning. A well-defined security policy will help to lessen the impact of lost or stolen devices. Becoming aware of newer technology such as femtocell technology and the threats associated with it will also help to lessen other potential security risks.

Mobile Device Loss or Theft

If a mobile device or computer is lost or stolen, this could have serious consequences for the security of an organization and its users. Many mobile devices are configured to automatically connect to and log into a corporate wireless network. Once logged in, the device will have access to any data or resources that are assigned to the user. It is imperative that corporate policy or company best practices educate users on the potential implications of a mobile device configured as such if it is lost or stolen. Asset management will help to identify whether a device is lost or stolen. However, in some cases it may take time to determine that this is the case. The user of the device must take some responsibility and report any missing or stolen device that may result in a security threat to the organization. Here are some recommended practices to help protect your information in the event of a lost or stolen mobile device:

- Password protect the device.
- Always keep the device in your possession.
- Never leave the device unattended.
- Refrain from having the device cache user credentials for sites capable of financial transactions and other sensitive information.
- Keep all information backed up in a secure location (for example, a trustworthy online service).

Mobile Device Cloning

Cloning is a generic term that means to make an exact copy or replica of something using a part of the original item. The concept of cloning a mobile device is no different. The technology used will determine how difficult the process will be. Some mobile devices, such as smartphones, contain numbers to uniquely identify them:

- The electronic serial number (ESN)
- The mobile equipment identifier (MEID)
- The mobile identification number (MIN)

One way of cloning a CDMA device is with the use of a femtocell, which basically is a mini or personal cell tower. The characters of a femtocell are similar to those of a home broadband router. These devices typically have a cellular radio to connect to the mobile devices and a connection to the Internet through the use of a DSL or a cable modem. A mobile device such a smartphone will connect to a femtocell and then the data will be transmitted over the Internet to the cellular carrier. A modified femtocell will be able to see specific information such as the unique identifiers. The identifiers can be entered into a different device, which is then made into an exact replica or clone of the original device. There are some limitations with this method and the details are beyond the scope of this book. One reason for cloning is to make free (fraudulent) phone calls from the cloned device. The bill for these calls would be the responsibility of the subscriber.

Understanding Risks Affecting Organizations

In addition to risks associated with software that is installed on computers and mobile devices, other situations must be taken into consideration, such as BYOD and proper use of removable media. These have the potential to affect an organization and the technology devices that are in use.

Bring Your Own Device (BYOD) Ramifications

Bring your own device (BYOD) is becoming an important but debatable topic in the technology sector. Many organizations are working to determine the pros and cons of employees bringing their own personal devices to use on the corporate network. Several key points must be considered with BYOD, including network capacity and security. Depending on when the network was installed, it may not have the capacity to handle this type of arrangement. Additionally, concerns about network security are a factor. One benefit of BYOD is that the devices used on the corporate network may be leveraged for company use, therefore providing a cost savings to the organization. Although the device is a personal asset, it can be managed by the organization using mobile device management (MDM) control, mobile security solutions, and application management.

Risks with Removable Media

Securing removable media is an important consideration when it comes to mobile device security. Removable media includes Universal Serial Bus (USB) drives, SIMs, Secure Digital (SD) cards, and other portable drive sources. These types of removable media are convenient, are readily available online and in many retail stores, and can be considered dangerous. Some corporate policies do not allow the use of removable media under any circumstances. In some cases, the penalty for violating this policy could mean dismissal by the organization. This is especially true in government organizations and those that have high security concerns. To help lessen the temptation to use these devices, some corporate hardware device policies will require that the devices be locked, not allowing the removable media to be recognized by the operating system and therefore rendering it unusable.

Wiping Personal Data

When the time comes to dispose of your mobile device, I highly recommend that you delete all personal information to protect yourself from someone who might gain possession of the device. Mobile devices are often disposed of, upgraded, donated to a charity, or recycled. You may be surprised at the amount of personal data that is retained within a personal device. This includes but is not limited to the following information:

- Your contact list
- Your call history
- Your text messages
- Your personal photos
- Your email account credentials
- Your emails stored locally
- Your calendar history and appointments

There are different ways to remove personal data from a device. There may be a built-in wipe feature or you may be able to reset the device to factory defaults. Before activating any feature that will wipe out any personal information, perform a complete backup of all data in case it needs to be recovered in the future. Exercise 15.1 shows how to perform a wipe of mobile devices.

The following steps are intended for informational purposes only. Executing these steps will erase all data and settings from the mobile device, and the process is not reversible.

EXERCISE 15.1

Steps to Wipe Various Mobile Devices

This exercise will show you how to wipe all data from three popular mobile devices: Android, BlackBerry, and the iPhone. It is important to understand that this is for informational purposes only and once a wipe is executed, it cannot be reversed. The following steps may vary slightly depending on the version of the device and operating system in use.

Android

1. Perform a complete backup of your device.
2. Select the gear icon.
3. Select the Privacy icon.
4. Select the Factory Data Reset option.
5. Scroll to the bottom of the screen.
6. Select the Erase Internal Storage check box.
7. Select the Reset Phone button.
8. Draw your security pattern or enter your passcode.
9. Select the Erase Everything button.

The amount of time it takes will depend on the amount of information that is deleted.

BlackBerry

1. Perform a complete backup of your device.
2. Open the device's Options menu icon.
3. Select the Security menu.
4. Select the Security Wipe icon.
5. Check the appropriate boxes for emails, contacts, user-installed applications, and others if available.
6. Type **blackberry** in lowercase in the confirmation field.
7. Select data to delete and type **blackberry** to confirm wiping of the device.
8. Select the Wipe Data box to proceed.

All selected information will now be wiped from the device.

The amount of time it takes will depend on the amount of information that is deleted.

iPhone 4

1. Perform a complete backup of your device.
2. Select the Settings menu icon.

3. Scroll down and tap the General icon.
4. Scroll to the bottom to the Reset option.
5. Select this option to continue.
6. Select the Erase All Content And Settings option.
7. Enter your passcode. A warning dialog box will appear on the bottom of the screen.
8. Select the Erase iPhone red button to continue.

The amount of time it takes will depend on the amount of information that is deleted.

With Apple mobile devices using iOS 5 or later, the Erase All Content And Settings option erases user settings and information by removing the encryption key that protects the data. This can be a fast process and applies to iPhone 3GS and later and all models of the iPad and the iPod Touch (third generation) or later. Any other Apple device, such as the original iPhone, iPhone 3G, iPad Touch, and iPod Touch (second generation) are devices that can be "erased" by overwriting memory, which is a more involved process. If you choose the Erase All Content And Settings option, the entire process can take minutes or even several hours, depending on the device. I recommend that you connect your device to a charger and leave it connected until the entire process has completed.

Unknown Devices on Network or Server

Unknown devices on a network or connected to a corporate server can create a number of issues and concerns, such as how unknown devices might interact with available resources, network capacity, IP address management, bandwidth, and others. An unknown device is not necessarily a rogue device; it just may not be registered with the company as a company-owned asset and could be the result of BYOD. If users are permitted to use personal devices on a corporate network and an MDM solution is not in place or a policy is not properly set, legitimate devices with the correct access credentials will be able to connect and utilize the network. MDM solutions can require that devices are properly enrolled, data is encrypted, and the device is compliant with corporate policy. This will lessen the number of "unknown" devices and relieve the network of the impact they may have on the resources.

Summary

In this chapter we explored the various risks that are associated with mobile device technology, including risks related to wireless technology, software, hardware, and organizations. Many of these risks are not limited to computers but also to mobile devices, including smartphones and tablets. You saw the risks that could have an effect on a wireless connection as well as how a hacker can hijack a connection to gain control and then retransmit to a legitimate AP, thereby exploiting all traffic that passes through. We looked at various software risks that could have a negative impact on all devices connected to a network. These include threats related to malware, viruses, and spyware, to name a few. Finally, the chapter covered hardware risks, including lost or stolen devices and cloning. Risks affecting an organization in general were also discussed.

Chapter Essentials

Understand the risks that are associated with wireless technology. Be familiar with the general risks associated with wireless technology, including public Wi-Fi hotspots, rogue AP, cell tower spoofing, weak security keys, and others.

Understand that the only protection against intentional RF jamming is physical security. RF jamming can be intentional or unintentional, and intentional jamming devices must be physically located and removed to prevent potential security threats.

Know how an RF DoS attack can affect wireless mobile devices. Understand that a hacker can interrupt a valid RF signal and perform a wireless hijack attack. This will then provide the opportunity to create a wireless man-in-the-middle attack, potentially causing serious security breaches.

Understand the meaning of malware. *Malware* is a general term used to describe a variety of different software-based threats, not any one specific threat.

Understand the risks associated with software. The risks associated with software include information technology viruses, computer worms, Trojans, keyloggers, and spyware. Understand the differences between these threats and that they are potentially damaging to both computers and mobile devices.

Be familiar with the risks associated with mobile device hardware. Mobile device loss or theft may have serious adverse security effects. Understand that someone can clone a mobile device even without gaining physical possession of it.

Be familiar with the risks associated with an organization. Understand the effects of BYOD, using removable media, wiping personal data, and other behavior that could have an impact on an organization.

Chapter 16

Device Authentication and Data Encryption

TOPICS COVERED IN THIS CHAPTER:

- ✓ Access Control
- ✓ Authentication Processes
- ✓ Encryption Methods for Wireless Networking
- ✓ Cipher Methods
- ✓ Secure Tunneling
- ✓ Public Key Infrastructure (PKI) Concepts
- ✓ Physical Media Encryption

In computer networking, authentication is defined as a way of confirming an identity; basically, it determines that you are who you say you are. Data privacy is ensuring that information or data is understandable only by the individuals or groups it is intended for: the sender and the intended receiver. In this chapter you will learn about various methods of access control, the authentication process, and encryption types used with mobile devices and wireless networking. Information that is either stored on a device (at rest) or traversing a network (in transit) should be secured to ensure that only those authorized to view it are able to do so. This chapter will explore various types of encryption and the encryption processes (ciphers) they work with. Finally, you will see that physical media such as hard disks and removable media are vulnerable to security breaches if they are not properly protected.

Access Control

Restricting access to any type of computer network system is an important part of network security. *Access control* can be accomplished in several ways using different authentication methods. One way people may think of authentication is to simply supply a username and password to log onto a computer. Another would be to enter the appropriate logon information before performing an activity like Internet banking. Here are several common authentication concepts that provide access control:

- Username and password
- Personal identification numbers (PINs)
- Security tokens
- Certificate authentication
- Biometrics
- Multifactor authentication

Username and Password

One very common way to validate a person's identity is by supplying them with a username and password pair. This is a method that most users of technology can relate to. When a user is attempting to log on to a system, the information they supply will be challenged against a database for validity. If the information supplied provides a match, then the user will gain access to the system. One potential issue with this type of authentication is the fact that anyone with access to this logon information will be able to gain access to the system with the information entered. Therefore, this access control method is only as secure as those who have possession of the username and password. Unfortunately, many people do not understand the importance of keeping this type of information confidential and can make this a weak access control method.

Personal Identification Numbers (PINs)

A *personal identification number (PIN)* is a password that usually consists of numeric digits. This is a common authentication method used with ATM cards or debit cards. The most common length for a PIN is 4 digits. However, the international standard (ISO-9564-1) allows from 4 to 12 digits. Just as you would keep a password private, a PIN should be treated the same way to ensure security. A PIN is considered to be one of the best methods to restrict access to a mobile device such as a mobile phone or a tablet and is a common method used to lock the device from unauthorized access.

Security Tokens

Security tokens consist of an electronic device of some type, commonly part of a card and used for authentication in smart cards and other physical devices. Like some of the other authentication methods mentioned, security tokens are used to validate a user's identity. A security token may also be called an authentication token and is used with multifactor authentication. A key fob is another physical device that can be used as a security token.

Certificate Authentication

Using digital certificates (*certificate authentication*) as an access control method for mobile devices, computers, and other devices will ensure that only valid authorized users will gain access to the corporate network infrastructure. Digital certificates are used to validate a user's identity and help protect network resources. Certificates, and the public key infrastructure are discussed in more detail later in this chapter.

Biometrics

Biometrics is something unique to you as an individual. This may include a *fingerprint*, *palm print*, *facial scan*, *retinal scan*, or some other human characteristic. Since these are unique to a specific person, they cannot be duplicated or stolen and therefore provide a high level of security. Biometrics is often used in conjunction with something you know, such as a PIN, or something you have, such as a card, and therefore will provide multifactor authentication.

Multifactor Authentication

This type of authentication provides a more secure level of access control then the previously mentioned username and password pair. *Multifactor authentication*, also known as *two-factor authentication*, requires at least two components of authentication criteria for a login to be validated and the user to gain access to a system. The information types are as follows:

Something You Know   This includes a username/password pair or personal identification number (PIN).

Something You Have   This is something physical you have in your possession such as an automated teller machine (ATM) card, common access card (CAC), digital certificate, or token smart card.

Something Unique to You   This includes a biometric characteristic such as a facial scan, retinal scan, palm print, or a fingerprint.

One familiar method of multifactor authentication is an automated teller machine (ATM) card. The ATM card cannot be used without the physical card and knowledge of a personal identification number (PIN). The card is something you have and the PIN is something you know. To use this type of multifactor authentication, you would insert the ATM card into a banking machine and enter a PIN. This type of access control is becoming a much more common method of authentication because of fraud and identity theft concerns.

Authentication Processes

There are many different authentication processes available, and the one you choose to use will depend on the type of system you are planning to secure. In the following sections, we will focus on authentication types that are used with wireless networking and mobile device security. However, keep in mind that some of the authentication processes discussed here will also be valid with other types of systems, including wired networking systems.

Wireless Passphrase Security

Wireless *passphrase-based security* was designed with the small office, home office (SOHO) or home-based wireless network user in mind. This type of security allows a user to create a secure wireless LAN solution without the experience or knowledge necessary to configure enterprise-level components such as an 802.1X/EAP and a RADIUS server. Passphrase-based security requires all wireless devices that are part of the same wireless network to have the same 256-bit pre-shared key (PSK) in order to securely communicate with each other. Deriving a secure key of this length would be a daunting task; to ease the burden of having to create a long secure key, the passphrase was introduced. This works by requiring the user to enter a strong passphrase on all wireless devices that are part of the same wireless network to be secured. The benefit of a passphrase is that it can be a sequence of words or other text that is memorable only to the user who created it. After the passphrase

Authentication Processes

is entered into the device, with the help of an electronic algorithm from the IEEE 802.11i amendment, a secure 256-bit pre-shared key will be created.

The Wi-Fi Protected Access certifications (WPA and WPA2) refer to the passphrase method of security as personal mode.

The characteristics of the passphrase used with IEEE 802.11 devices are as follows:

- It consists of 8 to 63 ASCII (case-sensitive) or 64 hexadecimal characters.
- It creates a 256-bit pre-shared key.
- The longer and more random the passphrase and the more mixed-case and special characters and such it includes, the more secure it will be.
- Weak passphrases can be compromised.

Several software programs and websites have random password generators to aid in the generation of strong passphrases. One such website is Gibson Research Corporation. GRC's Ultra High Security Password Generator can be found at www.grc.com/passwords.htm.

Figure 16.1 shows how to configure a passphrase on a SOHO wireless LAN access point.

FIGURE 16.1 D-Link wireless access point pre-shared key passphrase settings

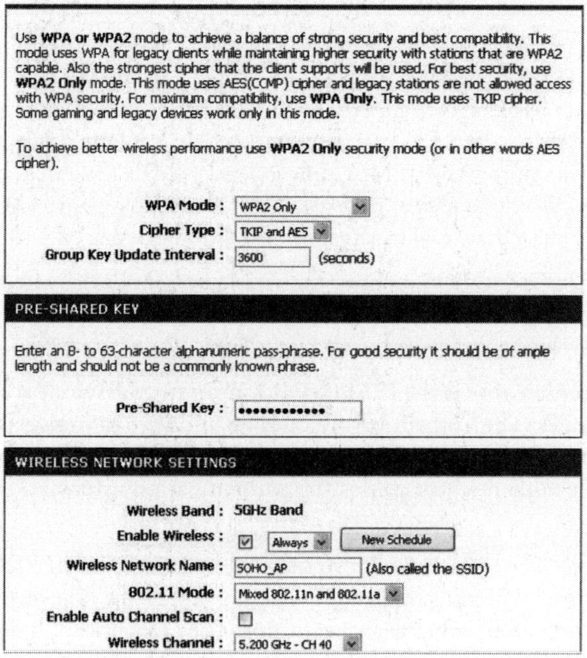

Wireless Networking WPA and WPA2 Enterprise Security

Concerned about problems connected with MAC address filtering and WEP, the industry drove the development of additional, improved wireless security solutions. One of these solutions also operates at Layer 2 and is an IEEE standard. This advanced enterprise-level solution is known as IEEE 802.1X, which addresses port-based access control and, used in conjunction with Extensible Authentication Protocol (EAP), allows for user-based authentication. You will learn more about IEEE 802.1X and EAP next. User-based security allows an administrator to restrict access to a wireless network and its resources by creating users in a centralized database. Anyone trying to join the network will be required to authenticate as one of the users by supplying a valid username and password. After successful authentication, the user will be able to gain access to resources for which they have permissions. This type of mutual authentication is more secure than the previously mentioned passphrase security method.

IEEE 802.1X/EAP

IEEE 802.1X/EAP consists of two different components used together to form an enterprise network security solution. In the standards-based wireless networking world, IEEE 802.1X/EAP is defined in the IEEE 802.11-2012 standard but was originally part of the IEEE 802.11i amendment. We'll first discuss the IEEE 802.1X standard and then Extensible Authentication Protocol (EAP). Then we'll combine the technology and terms to form IEEE 802.1X/EAP.

IEEE 802.1X

IEEE 802.1X is a port-based access control method and was designed to work with IEEE 802.3 Ethernet wired networks. However, this standard was adapted into the wireless world as an alternative, more powerful solution to legacy 802.11 wireless networking security technologies. Wireless devices that use 802.1X technology are identified using different terminology than that used in IEEE 802.11 standards-based wireless networking:

Supplicant   *Supplicant* is another name for the wireless client device attempting to connect to the wireless network. This is the IEEE 802.1X terminology for the software security component of the wireless client device.

Authenticator   *Authenticator* is the IEEE 802.1X term for the wireless access point or wireless LAN controller. The authenticator acts as a middleman between the wireless supplicant and the authentication server. When the supplicant requests to join the wireless network, the authenticator passes the authentication information between the two devices.

Authentication Server   The term *authentication server* is used by the IEEE 802.1X standard to identify the server that will authenticate the wireless supplicant. The authentication server receives all information from the authenticator. The authentication server may use authentication, authorization, and accounting (AAA) services and perform RADIUS server functionality; both AAA and RADIUS are explained later in this chapter.

Figure 16.2 illustrates the 802.1X/EAP process for a wireless LAN.

FIGURE 16.2 Wireless LAN client authenticating to a RADIUS server using IEEE 802.1X/EAP

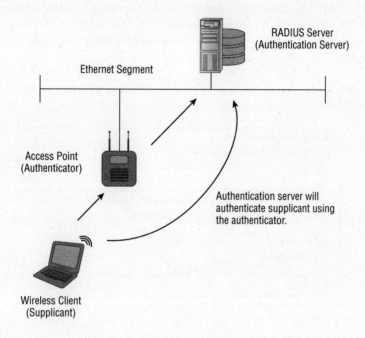

Extensible Authentication Protocol (EAP)

IEEE 802.1X is a framework that allows for an authentication process. The authentication process used with IEEE 802.1X is *Extensible Authentication Protocol (EAP)*, specified under RFC 3748. The IEEE 802.1X standard will employ some EAP type to complete this process. Many types of EAP are available in the industry that can be used with wireless networking technologies. They vary from proprietary solutions to very secure standard solutions. Here are some examples of some popular EAP types:

- EAP-TLS (certificates required)
- TTLS (EAP-MSCHAPv2)
- PEAP (EAP-MSCHAPv2)
- EAP-SIM for GSM Subscriber Identity Module (SIM)
- EAP-AKA for Universal Mobile Telecommunications System (UMTS)

These and other EAP types allow a user of mobile devices to authenticate to a network in several ways, including using credentials such as username/password and digital certificate–based authentication.

IEEE 802.1X and EAP Together: IEEE 802.1X/EAP

Now it is time to put these two parts together to form the IEEE 802.1X/EAP authentication process. This authentication process is typically used for enterprise-level security, but it's sometimes used in smaller wireless installations. As mentioned earlier, a variety of EAP types are available in the industry that work very well with IEEE 802.11 wireless networking. The EAP type chosen will depend on the environment in which the wireless network is used. EAP types vary in specifications, costs, and complexity. Figure 16.3 shows IEEE 802.1X/EAP configuration on a cloud-based wireless network management system.

FIGURE 16.3 Aerohive HiveManager Online cloud-based IEEE 802.1X/EAP configuration screen

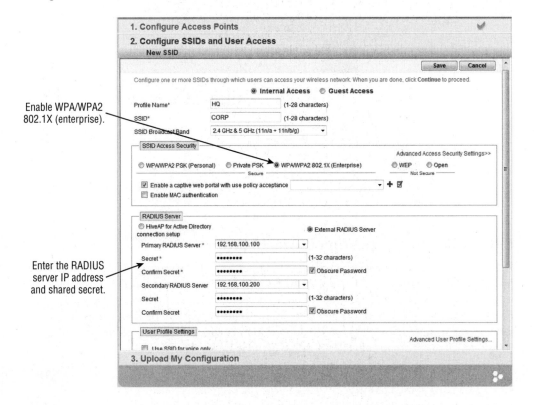

 The Wi-Fi Protected Access certifications (WPA and WPA2) refer to the IEEE 802.1X/EAP security method as *enterprise mode*.

Remote Authentication Dial-In User Service (RADIUS)

Remote Authentication Dial-In User Service (RADIUS) is a networking protocol that provides centralized authentication and administration of user login accounts. RADIUS started years ago as a way to authenticate and authorize dial-up networking users. A network user would dial up to a network from a remote location using the public switched telephone network (PSTN) and a computer modem. A modem from a modem pool on the receiver side would answer the call. The user would then be prompted by a remote access server to enter a username and password in order to authenticate and gain access to the network. Once the credentials were validated, the user would have access to any resources for which they had permissions. Figure 16.4 illustrates the remote access service authentication mechanism.

FIGURE 16.4 Remote user authenticating to a remote access server (RAS)

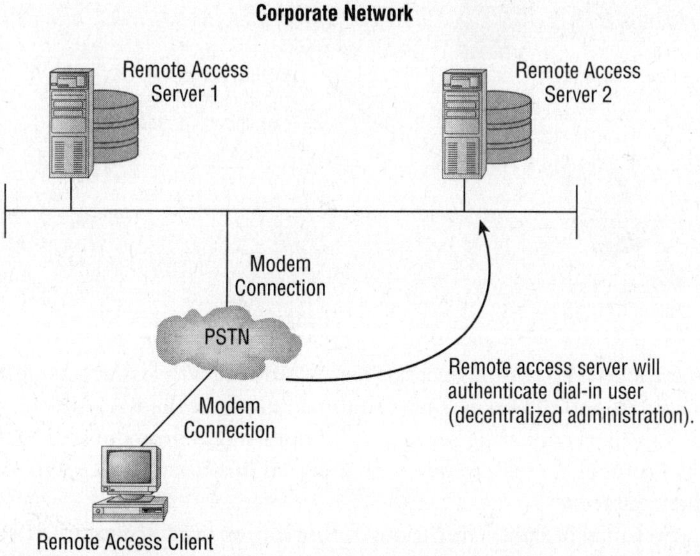

In this example, the remote access client would be the computer dialing into the network, and the remote access server would be the one performing the authentication for the dial-up user. As computer networks grew in size and complexity and remote access technology improved, there was a need to optimize the process on the remote access server side. This is where RADIUS provides a solution. RADIUS took decentralized remote access services databases and combined them into one central location, allowing for centralized user administration and centralized management. It eased the burden of having to manage several databases and optimized administration of remote access services.

A company does not need a large number of RADIUS servers. For a small to medium-sized company, one RADIUS server should be sufficient (with a backup if possible). Larger enterprise organizations may need several RADIUS servers across the entire wide area network. In wireless networking, the wireless access point can act as a RADIUS client, which means it will have the capability to accept requests from wireless client devices and forward them to the RADIUS server for authentication. Figure 16.5 shows an example of this configuration.

FIGURE 16.5 Wireless access point configured as a RADIUS client device

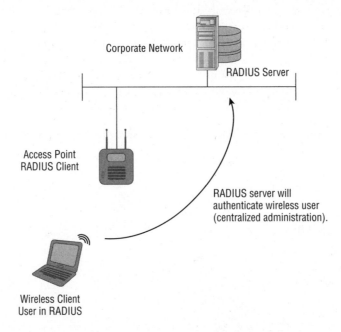

As Figure 16.5 shows, the remote access client is now the wireless access point. The wireless client device is authorized as a user in the database of the RADIUS server. The RADIUS server is the authenticating server, and if not using a protocol such as Lightweight Directory Access Protocol (LDAP) to query an external directory service will also contain the RADIUS client database.

A RADIUS server may use authentication, authorization, and accounting (AAA) services. In this configuration, the RADIUS server will authenticate users (validate credentials) and provide access to the resources for which they have permissions, which is also known as access control. In addition, it will keep track of all transactions using accounting methods by logging all information and activity.

 Terminal Access Controller Access Control System (TACACS) is a legacy authentication protocol that was available prior to the adoption of RADIUS and was used to forward an encrypted user password to an external authentication database. TACACS would then validate the user credentials prior to allowing access to a remote access server (RAS). TACACS+ is a newer, proprietary protocol designed by Cisco Systems that uses Transmission Control Protocol (TCP) (as compared to UDP used by RADIUS) and will encrypt both a username and password (the entire packet). Network administrators may choose to use TACACS+ rather than RADIUS as a network access protocol.

Encryption Methods for Wireless Networking

Computer data and information can be encrypted using two different methods, symmetric encryption and asymmetric encryption. Symmetric encryption uses a private key process and asymmetric encryption uses a public key process. Symmetric key encryption is used with wireless networking to encrypt the data payload of the wireless frame.

The type of security in use (either passphrase or enterprise) will determine the information that is used with the encryption key generation process. Note that the data encryption keys are never transmitted across the wireless medium. However, some information (keying material) shared between each pair of devices will generate or derive the same data encryption key. Legacy WEP encryption uses static keys between all devices on the common wireless network. Conversely, TKIP and CCMP use a dynamic process to derive the keys based on the keying material that is used.

Wired Equivalent Privacy (WEP)

From a security perspective, one major drawback to any wireless network is the fact that all information, including data payload, travels through the open air from one device to another. This makes wireless networks vulnerable to eavesdropping and inherently less secure than bounded networking using other mediums such as Ethernet cabling. With IEEE 802.11 open system authentication, all information is broadcast through the air in plain text. This means anyone with knowledge of how to use a wireless packet analyzer or other wireless scanning software program can easily see all the information that is passing between devices through the air.

Wired Equivalent Privacy (WEP) was designed as a way to protect wireless networking from casual eavesdropping. The original IEEE 802.11 standard states that the use of WEP is optional. The manufacturer supplies the capability, but it is up to the user of the wireless device to implement it.

With wireless networks, WEP can be used in one of two ways: with IEEE 802.11 open system authentication to encrypt the data only or with legacy IEEE 802.11 shared-key authentication, which is used for both wireless device authentication and for data encryption. The original standard specified only 64-bit WEP, which consists of a 40-bit key plus a 24-bit WEP initialization vector (IV).

 It is important not to confuse IEEE 802.11 legacy shared-key authentication with the pre-shared key that is used currently with WPA or WPA2 passphrase mode. IEEE 802.11 shared-key authentication requires legacy WEP encryption, which is not secure and should not be used.

WEP is fairly simple to implement. It requires all wireless devices on a common network to have the same WEP key. The WEP key can be either 64-bit or 128-bit; however, the standard requires only 64-bit WEP.

As mentioned earlier, one disadvantage to WEP is that it uses static keys, which means all wireless devices—access points, bridges, and client stations—must have the same key and the key must be manually entered into them. Anytime the key is changed, it must be changed on all of the wireless devices that are on the same SSID. No matter which you use, 64-bit WEP or 128-bit WEP, you are still only using a 24-bit IV that is transmitted in cleartext. 64-bit WEP uses a 40-bit secret key and a 24-bit IV. 128-bit WEP uses a 104-bit secret key but still only the 24-bit IV, making both 64-bit WEP and 129-bit WEP vulnerable to the same attacks. Gathering a number of frames containing these IVs allows a hacker to crack your WEP key very quickly.

Temporal Key Integrity Protocol (TKIP)

Temporal Key Integrity Protocol (TKIP) was designed as a firmware upgrade to WEP. This provided a fix for some of the inherent flaws with WEP and an interim solution pending the release of the IEEE 802.11i amendment, which specified CCMP/AES to provide a strong security solution. TKIP added several enhancements to the WEP algorithm and was the foundation for the Wi-Fi Protected Access (WPA) certification from the Wi-Fi Alliance. These enhancements are as follows:

- Per-packet key mixing of the IV to separate IVs from weak keys
- A dynamic rekeying mechanism to change encryption and integrity keys
- 48-bit IV and IV sequence counter to prevent replay attacks
- Message integrity check (MIC) to prevent forgery attacks
- Use of the RC4 stream cipher, thereby allowing backward compatibility with WEP

Configuring a wireless network to use TKIP is a fairly straightforward process. It can be accomplished either by using the web interface available on most SOHO access points or by using the web interface or command-line interface for enterprise-level access points. For the wireless client devices, TKIP will be configured through the client software utility. Some older wireless hardware devices may not support TKIP. If this is the case, replacement of the hardware will be necessary in order to take advantage of newer security solutions. Figure 16.6 shows a block diagram of the TKIP process.

FIGURE 16.6 Block diagram of TKIP encapsulation process

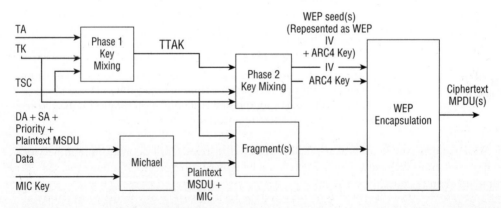

Image provided by IEEE Std 802.11-2012.

Counter Mode with Cipher Block Chaining Message Authentication Code Protocol (CCMP)

Counter Mode with Cipher-Block Chaining Message Authentication Code Protocol (CCMP) is a mandatory part of the IEEE 801.11i amendment, now in the IEEE 802.11-2012 standard and part of Wi-Fi Protected Access 2.0 (WPA2) certification from the Wi-Fi Alliance. CCMP uses the *Advanced Encryption Standard (AES)* algorithm block cipher. CCMP capability is mandatory for robust security network (RSN) compliance. If an RSN is required to comply with an industry or government regulation, CCMP must be used. CCMP is also intended as a replacement to TKIP. Because of the strong encryption CCMP provides, it may require replacement of legacy wireless hardware devices that are not capable of the newer technology. In some cases, it may use a separate chip to perform computation-intensive AES ciphering.

Configuration of CCMP is similar to that of TKIP, discussed earlier. The main difference with CCMP is that legacy hardware devices may not support it and it is a stronger encryption solution. Figure 16.7 illustrates the CCMP encryption process. CCMP is the most secure encryption method to use to secure a wireless network.

FIGURE 16.7 Block diagram of TKIP encapsulation process

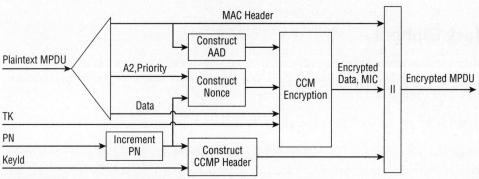

Image provided by IEEE Std 802.11™-2012.

Cipher Methods

The term *cipher* is defined as a method used to encrypt and decrypt information. There are many different cipher technologies available that are derived from two common processes, stream cipher and block cipher. Both processes use an encryption key and an encryption method to encrypt (scramble) cleartext information.

Stream Ciphers

A stream cipher will encrypt information on a per-bit basis. This means that each bit in a data stream will be encrypted individually. Using an exclusive OR (XOR) process, the data stream is combined with what appears to be a random code, but it's really not random and is known as a pseudorandom stream. Figure 16.8 illustrates the bit-level stream cipher process.

FIGURE 16.8 Stream cipher encryption process

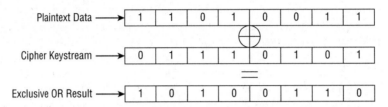

One common stream cipher is RC4. This cipher was created in 1987 and developed by Ron Rivest of RSA Security. The *RC* stands for *Ron's code*, and RC4 may also be known as ARC4 or ARCFOUR. This cipher is used with WEP and TKIP, both discussed earlier in this chapter. The vulnerability with WEP was not necessarily RC4 but the plaintext 24-bit initialization vector (IV) used to create the encrypted ciphertext. TKIP also uses the RC4 stream cipher, but enhancements such as a message integrity check (MIC) help to secure the weaknesses that were in WEP.

Block Ciphers

Unlike a stream cipher that encrypts each data bit individually, a block cipher will encrypt information in specific block sizes. A common block size is 64 bits of information. Figure 16.9 illustrates the block cipher encryption process.

FIGURE 16.9 Block Cipher Encryption Process

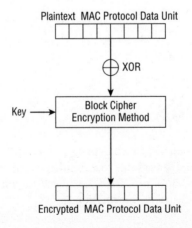

Following are some block ciphers:

Data Encryption Algorithm (DES)   *DES* was developed in the early 1970s by IBM and uses a block size of 64 bits with a key length of 56 bits. In the late 1970s, a modified version of DES became Federal Information Processing Standard (FIPS) compliant. DES was compromised in early 1999, therefore making it insecure in many cases. As technology advanced, so did the capability to break encryption algorithms requiring more sophisticated methods.

Triple Data Encryption Algorithm (3DES)   *3DES* is an enhanced version of DES that basically encrypts three times, hence the term *triple DES* or *3DES*. This stronger version of DES used the 56-bit key length but was used three times with each 64-bit block of data, thereby increasing the key size to 168 bits. This stronger algorithm decreased the possibility of security weaknesses such as brute-force attacks on the transmission of data.

Advanced Encryption Standard (AES)   *AES* is a strong encryption algorithm that is widely used in modern day wireless networks. In conjunction with CCMP encryption, it is considered unbreakable and is the required cipher suite for IEEE 802.11i compliance and WPA2-certified devices. AES uses a larger block size of 128 bits (recall 64 bits with DES and 3DES) and three possible key lengths of 128, 192, and 256 bits.

Twofish   *Twofish* was one of the five finalists for the AES standard. Twofish is another block cipher and uses a block size of 128 bits with a key length of 128 bits up to 256 bits. It's optimized for 32-bit central processing units (CPUs). Twofish is not under patent protection and is free for anyone to use.

Elliptical Curve Cryptography (ECC)   Based on public-key cryptography, *Elliptical Curve Cryptography (ECC)* is used widely across the Internet to secure web-based communications and data transfers. ECC is a complex method based on elliptical mathematical curves referenced to an x- and y-axis providing an equation and difficult-to-replicate results. This provides a very strong, if not the strongest, method for securing data on mobile devices.

The CompTIA exam objectives categorize these block cipher encryption algorithms as "data at rest" encryption methods for the Mobility+ (MB0-001) certification exam. *Data at rest* is a term that is commonly used to refer to information that is physically stored on a device.

Secure Tunneling

Whether using remote or local connectivity, securing access to computer networks is a critical component with today's computing technology. Connecting from a remote location usually involves a public Internet connection. By using a tunneling and encryption mechanism, you can securely exchange information on a public or private network infrastructure. In the following sections, you will learn about the virtual private network (VPN) technology and tunneling and encryption methods that are commonly used to secure data transmissions.

The CompTIA exam objectives categorize the data for which some of the encryption techniques discussed in the following sections are used as data in transit for the Mobility+ (MB0-001) certification exam. *Data in transit* is a term that is commonly used to refer to information (data) that is encrypted as it traverses a corporate network, an intranet, or the Internet. This includes websites and mobile device apps. These methods will protect the information from hackers who know how to intercept the information as it traverses the infrastructure.

Virtual Private Networking (VPN)

Virtual private networking (VPN) provides the capability to create private communications over a public network infrastructure such as the Internet. The security solutions discussed earlier in this chapter are Layer 2 security solutions; that is, they all work at Layer 2 of the OSI model. By contrast, VPNs are based on the Internet Protocol; they typically operate at several layers of the OSI model, depending on the encryption/protocol used, and there are some that operate at Layer 2. Figure 16.10 illustrates VPN technology in relation to the OSI model.

FIGURE 16.10 OSI model representation of a Layer 3 (IPsec) and Layer 4 (SSL/TLS) VPN security solution

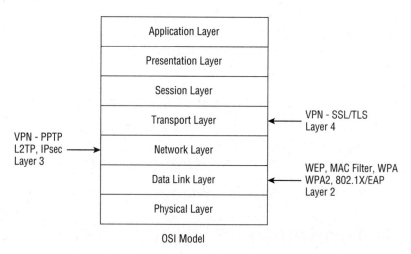

Prior to the ratification of the IEEE 802.11i amendment to the standard, VPN technology was prevalent in enterprise deployments as well as in remote access security. Since Layer 2 security solutions have become stronger (mostly thanks to the 802.11i amendment and the Wi-Fi Alliance WPA and WPA2 certifications), VPN technology is not as widely used within enterprise LANs. However, VPN still remains a powerful security solution for remote access in both wired and wireless networking.

VPNs consist of two parts—tunneling and encryption. Figure 16.11 illustrates a VPN tunnel using the Internet. A stand-alone VPN tunnel does not provide data encryption, and VPN tunnels are created across Internet Protocol (IP) networks. In a very basic sense, VPNs use encapsulation methods where one IP frame is encapsulated within a second IP frame. The encryption of VPNs is performed as a separate function.

FIGURE 16.11 Representation of a VPN tunnel

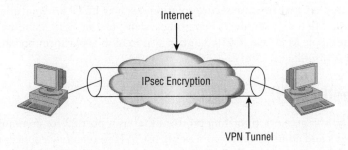

Two common types of VPN protocols are Point-to-Point Tunneling Protocol (PPTP) and Layer 2 Tunneling Protocol (L2TP).

Point-to-Point Tunneling Protocol

Point-to-Point Tunneling Protocol (PPTP) is mentioned in this book because at one time it was a popular VPN technology due to the ease of configuration, and it is contained within the Microsoft Windows operating system starting with Windows 95. You saw in Chapter 6, "Computer Network Infrastructure Devices" that one benefit of PPTP is that it provides both tunneling and encryption capabilities for the user's data. The encryption is achieved by utilizing the Microsoft Point-to-Point Encryption (MPPE-128) protocol. I recommend you not use PPTP because potential serious security vulnerabilities have been found within the protocol. The next topic discussed, Layer 2 Tunneling Protocol (L2TP), is the recommended tunneling protocol that should be used today.

 If the PPTP configuration uses MS-CHAP version 2 for user authentication on a wireless network, it can be a security issue. This authentication process can be captured using a wireless protocol analyzer or other scanning software program and potentially allow someone to perform a dictionary attack, allowing them to acquire a user's credentials and eventually giving them the capability to log on to the network. A dictionary attack is performed by software that challenges the encrypted password against common words or phrases in a text file (dictionary). Therefore, using PPTP on a wireless network should be avoided. Keep in mind that the security vulnerability is not PPTP itself; it is that the authentication frames on a wireless LAN can be captured by an intruder, who can then acquire user credentials (username and password) and be able to gain access to the VPN.

Layer 2 Tunneling Protocol

You also learned in Chapter 6 that Layer 2 Tunneling Protocol (L2TP) is the combination of two different tunneling protocols: Cisco's Layer 2 Forwarding (L2F) and Microsoft's Point-to-Point Tunneling Protocol (PPTP). L2TP defines the tunneling process and requires some level of encryption in order to provide the needed security which will protect the data in-transit. L2TP provides much improvement including enhanced security mechanism over a PPTP VPN solution. It is important to note that PPTP should not be used unless absolutely necessary and only when more secure solutions such as L2TP are unavailable. L2TP became part of the Microsoft Windows operating system starting with Windows XP, is included with the Apple MAC OS X v10.3 and is also in mobile platform operating systems such as iOS and Android.

Secure Shell (SSH)

Secure Shell (SSH) is a network tunneling protocol that uses port 22 for communications. Network managers and administrators commonly use SSH to remotely connect to and administer network infrastructure devices and provide data or file transfers. This is often referred to as a replacement for Telnet, which was used for remote terminal sessions and used plaintext communications, making them vulnerable to interception by an unauthorized user. SSH can be used within the local area network or across a wide area network to secure these transmissions. SSH resides on a server and will provide a redirection to the appropriate networking service over port 22. For example, if a network administrator wanted to manage computer services such as email or web servers or even configure network infrastructure devices such as a wireless access point, SSH would provide secure data transmissions for the session. A benefit to using SSH is that it is platform independent and server and client software are both available commercially and open source. SSH2 is a stronger version of SSH and should be used whenever possible.

Web-Based Security

You learned in Chapter 2, "Common Network Protocols and Ports," about different protocols that are used to secure communications over the Internet. The native protocols of the Internet (TCP/IP) alone do not provide any security. Therefore, additional mechanisms to secure information are necessary. The methods that include securing Internet protocols include Secure Sockets Layer (SSL) and Hypertext Transfer Protocol Secure (HTTPS).

Secure Socket Layer (SSL)

Secure Socket Layer (SSL) technology can be used with a variety of protocols to secure Internet-based communications. This includes but is not limited to HTTP, POP3, and FTP. As the name implies, it is a security protocol that creates an encrypted link between a client device and a server. Using SSL in conjunction with the protocols just mentioned will basically add the SSL functionality, resulting in HTTPS, POP3S, and FTPS. This will secure what were otherwise unsecured networking protocols and provide some level of privacy to the communications.

SSL was succeeded by Transport Layer Security (TLS), which is based on SSL technology. SSL is native technology to most if not all modern Internet web browsers. Web servers that are SSL enabled will be issued a digital certificate from a valid certificate authority (CA). This will guarantee to the user that the server they are connecting with is in possession of a valid certificate issued from a CA and that transactions made will be secure. Although not the case 100 percent of the time, if a user receives a certificate error, it may be a red flag that the server they are connecting to is not a legitimate source.

Hypertext Transfer Protocol Secure (HTTPS)

Browser-based Internet communication has grown by leaps and bounds over the last 20 years. With this growth, the need for secure web communications has also increased. Browsers are now available on any device that has a connection to the Internet, including computers, smartphones, and tablets, for example. These devices are now used for applications such as online banking, online purchases and paying bills. Therefore, secure communications are a must to ensure that these exchanges cannot be compromised.

You learned in Chapter 2 that the Hypertext Transfer Protocol (HTTP) uses TCP port 80 for communications, and TCP port 80 provides cleartext communications. Adding a secure socket layer to the HTTP connection ensures a secure data exchange. HTTPS uses TCP port 443, providing the needed SSL secure connection. It is important to understand that in some cases the SSL connection is used only for the authentication process to secure the user's logon credentials. Once the logon is complete, the SSL is dropped and the communications are back to unsecured HTTP port 80 communication in cleartext. Users need to be educated to always be aware of this type of web-based and Internet connectivity because they sometimes assume that security stays in place after the logon process is complete.

Public Key Infrastructure (PKI) Concepts

Another method used to secure data transmissions is *public key infrastructure (PKI)* technology. PKI is used in an asymmetric key encryption process that uses both public and private encryption keys that will secure the data communications. Each device possesses a pair of these keys. The public key is distributed to other devices, and the private key is kept private with the source device.

PKI is a secure method but comes with extra administration overhead, which can get expensive. PKI implementations are typically cost prohibitive for small networks and therefore more common with enterprise network installations.

The components of a PKI infrastructure are as follows:

Certificate Authority (CA) A certificate authority is responsible for issuing and managing digital certificates. Two types of CAs exist: commercial trusted certificate authorities or private authorities. Private CAs are typically used for internal use and vendor access.

Digital Certificates A *digital certificate* is an electronic method of validating a user's identity. The certificate is the public key component in a PKI infrastructure and is electronically signed by the authority from which it was issued.

Certificate Revocation lists A *certificate revocation* list contains a list of certificates that have been revoked for a variety of reasons. The reasons for revocation include a terminated employee or devices that are lost or stolen. If a certificate is revoked, the authentication process will fail.

Entity In PKI terms, an *entity* is a server, computer, mobile device, user, router, or other network component that is part of the PKI. The client device that receives the certificate is known as the end entity.

Physical Media Encryption

Security of the physical media in which information is stored, such as a computer hard drive or a removable electronic disk drive, is just as important as other types of network security, such as authentication. Physical media encryption may consist of encrypting entire hard drives, folders, or even specific files. This also applies to removable media such as Universal Serial Bus (USB) flash/portable drives, mobile devices, smartphones, and tablets, which also have data storage capabilities. In the following sections, we will explore how media encryption should be considered to protect stored information.

Full Disk Encryption

Full disk encryption (FDE) is a method of protecting the contents of an entire hard disk from anyone with physical access to the device that contains the hard disk. FDE will ensure that only users who are authorized to access the information on the device will be able to view the contents on the disk. This is accomplished by encrypting all of the content on the hard disk using third-party software programs or available hardware encryption mechanisms. Depending on the method used, the disk partitions, including the master boot record (MBR), can be encrypted. Some versions of operating systems such as Microsoft Windows may include disk encryption capabilities that are native to the operating system.

One thing to keep in mind is the trade-off of security with system performance. Depending on the software or hardware and the devices used, FDE may have an adverse affect on system performance. However, this would be the best way to protect the entire contents of a disk and the storage capabilities of a mobile device.

Block-Level Encryption

Understanding the mechanics of how disk storage works is beyond the objectives covered in this book. However, it is important to note that physical hard disks store information on sections, or blocks, within the physical hard disk. In other words, the area reserved for the information may be larger then what is required for the data to be stored. Full disk encryption software interacts directly with the device's operating system architecture and the file system itself. The block-level encryption process will encrypt each individual block of information that is stored on the disk using a *block-level encryption* process. When the user wants to view the encrypted information, the blocks containing the data will decrypt the data into the device's electronic memory (RAM), allowing the authorized user to see and use the encrypted information.

Folder and File-Level Encryption

Two alternatives are *folder encryption* and *file-level encryption*. These types of encryption will allow the user of a device to encrypt specific folder or files that are contained on the physical disk. They may be beneficial where the user wishes to protect only specific information such as a database or files that contain personal information about the user. One benefit to folder and file-level encryption is that there may be less overhead then full disk encryption, depending on the system and the information that is encrypted.

Exercise 16.1 demonstrates the basic folder and file encryption process on a computer using the Microsoft Windows 7 operating system.

EXERCISE 16.1

Encrypting a folder in Microsoft Windows

In this exercise, you will encrypt a folder and its contents in Microsoft Windows 7 Professional using the built-in Microsoft encryption technology. To encrypt a folder and its contents, follow these steps:

1. Open the C:\ drive using Windows Explorer.
2. Create a test folder.
3. Save a file to the test folder you created.
4. Right-click the test folder you created and click Properties.
5. Click the General tab, and then click the Advanced button.
6. Select the Encrypt Contents To Secure Data check box, and click OK. A Confirm Attributes Changes dialog box will appear.
7. Select the Apply Changes To This Folder, Subfolder And Files radio button and click OK.

The contents of this folder and subfolder are now encrypted. You will notice text representing the name of the folder is a lighter shade, showing that the encryption attributes have been applied for that folder.

Removable Media Encryption

The advancements in device storage technology can create a big problem within the corporate security environment. Removable media is getting less expensive, smaller with larger storage capabilities, and available in a variety of form factors. Removable media can be in the form of a USB flash drive, portable hard disk, digital camera, smartphone, tablet, or other device. The same principles of encryption apply here as previously mentioned for software or hardware solutions. However, the *removable media encryption* aspect cannot be overlooked.

How removable media is handled is usually indicated in a corporate security policy. Regardless of whether the media/device is corporate owned or a personal device, it is important to understand that removable media and other devices capable of data storage can be lost or stolen. This can be a large security concern, which is why some corporate security policies prohibit the use of removable media and restrict the use of USB or other external connected devices on corporate-owned devices.

Summary

In this chapter we explored different methods used to restrict access to a system such as a computer and devices such as mobile phones and tablets. You saw that this can be accomplished using several methods, including but not limited to username and password pair, biometrics, and digital certificates. You also learned about the importance of data encryption and the methods used for encrypting data at rest (stationary) and data in transit (traversing a network). We also explored encrypting information that is stored on both physical disks and removable media.

Chapter Essentials

Be familiar with the various access control methods. Understand the different methods used to restrict access to a system and that some are more secure than others. Multifactor authentication and digital certificates are secure methods and are used with mobile device authentication.

Know the different authentication processes used with networking. Wireless networks use different authentication processes for user authentication, including passphrases, IEEE 802.1X/EAP with various types of EAP, and RADIUS. Strong passphrases will provide more secure authentication than weak passphrases. IEEE 802.1X/EAP is more common in larger enterprise networks but is becoming more available to mid-size network deployments.

Understand the different encryption methods used with wireless communications. WEP is a legacy encryption method and should be avoided. TKIP was a fix for WEP and provides better security than WEP. CCMP is currently the strongest encryption method available for wireless communications.

Understand the different cipher types that are used to encrypt data. Know that in addition to an encryption type, a method to encrypt that information is required. Stream ciphers such as RC4 encrypt at the bit level, and block ciphers such as AES encrypt chunks of information. A common block size is 64 bytes. Understand that different methods use various key sizes. AES, for example, uses three possible key lengths of 128, 192, and 256 bits. ECC provides a very strong, if not the strongest, method for securing data on mobile devices.

Know that VPN technology will create private secure communications over a public network infrastructure. Be familiar with the different methods of VPN technology, including tunneling and other protocols or encryption methods such as IPsec. These protocols can help secure data in transit. SSH is a port redirection protocol and is often used by network administrators to manage infrastructure devices and transfer files using secure methods.

Understand the different secure protocols commonly used for secure web communications. Be familiar with the basics of SSL and HTTPS and how these protocols can aid with securing browser-based communications. Realize that these protocols are commonly used to secure login to websites and other activities on the Internet, such as email access.

Be familiar with public key infrastructure (PKI) concepts. Although PKI increases administration overhead compared to other network security methods, it provides a very secure network infrastructure. Know the basic components that make up the PKI, including the certificate authority (CA), digital certificates, revocation lists, and entities (servers, computers, mobile devices, users, routers, and so on). A PKI infrastructure will build secure communications with devices that were previously not authorized.

Understand physical media encryption. Know the differences between the different types of media encryption, including full disk encryption, block-level encryption, folder and file-level encryption, and removable media encryption. Understand the security risks of unencrypted removable media such as USB flash drives.

Chapter 17

Security Requirements, Monitoring, and Reporting

TOPICS COVERED IN THIS CHAPTER:

- ✓ Device Compliance and Report Audit Information
- ✓ Third Party Device Monitoring Applications (SIEM)
- ✓ Understanding Mobile Device Log Files
- ✓ Mitigation Strategies

In modern day business operations, IT security is no longer optional. Businesses that do not make a reasonable effort to secure their own and their customers' data are considered to be negligent and can be liable for damages to third parties.

Not only must business decision makers take ownership of external threats to the organization, they must also ensure that authorized parties within the business are not using confidential financial and personal data for illicit gain, monetarily or otherwise. More than ever, IT security operations teams will be at the forefront of ensuring effective business continuity as the reliance on electronic transmission mechanisms has become almost absolute.

The use of mobile devices has added an additional, highly complex variable to the mix because regulators do not see any difference between data breaches on corporate-owned machines or those on personal devices that employees also use for work purposes. Some organizations have opted to completely lock down and prohibit the use of mobile devices. However, the advantages of being able to support operations regardless of location is driving businesses to find acceptable ways to allow the use of mobile devices while at the same time providing reasonable controls around sensitive data.

In this chapter, we will look at the different options that are currently available to provide security controls for mobile devices and that are in line with longstanding preferences to help secure the corporate network as a whole.

Not all organizations will require every single control mentioned here, nor should they limit themselves to only the options that follow. Yet, in order to maintain high levels of trust with business partners and to maintain a modern security operations team that can cope with a constantly fluctuating external environment, all of these tools and strategies should definitely be considered as solutions to gaps uncovered during regularly recurring security audits.

Device Compliance and Report Audit Information

To ensure device compliance and to be able to report meaningful audit information, the key starting point is to have a robust mobile device asset inventory. This entails having a detailed record of every corporate-issued mobile device that includes (but is not limited to) device characteristics, location, currently entrusted user and permissions, and current security software as well as proof of compliance with federal and state regulations in regulated environments. In some cases this also includes having full device encryption as well as other safeguards that would prevent unauthorized customer or patient data from being accessed by third parties if the device were to be lost.

Asset inventories must be reviewed on a periodic basis to ensure that not only is the data accurate but also the level of device permissions and the distribution of the devices themselves are appropriate and consistent with a valid business case. Internal audits performed by employees that are not part of the immediate security team can help uncover a host of resolvable compliance concerns well before they become eyesores on an external third-party report.

Security and compliance teams should also be aware of specific reporting requirements mandated by Sarbanes-Oxley (SOX), HIPAA/HITECH, GLBA, and similar regulations and organizations. In the case of healthcare data, large breaches or loss of devices that contain patient data must be quickly reported to Health and Human Services's Office of Civil Rights (OCR). Failure to do so has resulted in fines ranging from hundreds of thousands to millions of dollars. Even the loss of a single thumb drive can generate large penalties, despite the fact that it may not be certain that any sensitive data was contained on the device in the first place. Unless an organization can provide documentation that proves compliance, it will have to be prepared for some form of punitive action if it loses sensitive data within a regulated environment.

With so many regulations affecting handling of confidential information at both the federal and state levels, security teams can quickly feel overwhelmed, especially when mobile devices are thrown in the mix. While it is okay to provide a level of data security that surpasses what is required in a particular state, a lower level is not acceptable, even if it is in line with federal guidelines. This can be especially cumbersome for organizations that have operations in different states and are handling sensitive data across state lines. To simplify security operations, a good rule of thumb is to implement security and compliance initiatives based on what is required by the federal government in conjunction with the strictest state laws. While many states do not currently have laws on the books that protect personally identifiable information (PII) or other sensitive data, many do or are in the process of developing such requirements. A proactive approach will help avoid being blindsided by ever more stringent expectations around the handling of data, even when it's being accessed on corporate mobile devices.

Third-Party Device Monitoring Applications (SIEM)

Security Information and Event Management (SIEM) tools are a combination of two older technologies known as Security Event Management (SEM) and Security Information Management (SIM). A SIEM system is designed to digest the sea of log and network data that an organization produces on a daily basis into actionable information that can be used to improve an organization's security posture. It is also used to gain awareness of emerging threats in the enterprise environment.

The amount of log and network data produced by most organizations is beyond what most security teams would be able to review and analyze on a daily basis. Also, many security threats today can be detected only when multiple events are occurring in a certain combination, in separate locations, which is impossible to correlate in real time on a manual basis. The role of SIEM is to analyze large sets of network data and security logs to find security-relevant events that otherwise would not be detected.

Modern SIEM systems rely on the same tools and techniques used by big data analytical warehouses. Therefore, many are simply add-ons to broader searching and reporting devices that can provide much more business intelligence than simply tracking security events and correlations.

Before SIEM technology was widely available, it would take a security team months to find out how their organization was breached, with little ability to even determine for a fact what the actual intent of the attackers was. In most cases, they would not even have known that they had been breached at all unless they had been tipped off by a third party. SIEM technologies allow modern-day security teams to respond in real time and in some cases locate who or what is actually attempting to compromise their network while it is occurring.

Yet, SIEM systems have their downsides as well. Capturing and storing every log file from every device, as well as every packet that traverses the wire, is usually not possible outside of major government entities. Therefore, security analysts must spend a significant amount of time defining rules and correlations that are actually relevant to their environment. Otherwise, they will quickly find that they are unable to capture the data effectively, much less use it effectively in any way, and retention capabilities will be extremely limited.

Some SIEM systems can make use of Layer 2 Berkeley Packet Filters. This greatly enhances the ability to rapidly sift through large amounts of data and allows inspection points to handle much higher traffic loads. Also, be aware of the advantages of SIEM systems that have a Hadoop-based, big data architecture over those that still rely on a traditional database model. The former will provide much faster query speeds and will allow the security team to investigate and respond in a near real-time fashion.

Security analysts must have some understanding of what the specific alerts actually mean and the possibility of false positives. SIEM technology is still not Plug and Play and requires extensive user involvement. Poorly deployed and operationalized SIEM systems produce an unwieldy amount of security alerts that are eventually ignored, giving the organization no better security footing than before.

 Provisioning adequate storage is an integral part of a successful SIEM deployment. During the vendor selection process, it is important to discuss what the data retention requirements are from the business and the ability of the solution to actually provide functionality within those requirements.

Many SIEM solutions cannot adequately handle extremely large amounts of data across extended time frames (over 90 days). Being able to store five years of security log events is of no use if it takes months to query the data.

Corporate legal teams can become extremely frustrated when they realize that the effort to extract relevant data from the SIEM solution can be in excess of what's at stake in ongoing litigation.

Understanding Mobile Device Log Files

As with any computing resource, *log files* can provide a wealth of information for troubleshooting purposes as well as aid in security investigations. While it is definitely unwieldy and unnecessary to manipulate mobile device log files on the actual mobile device, there are excellent tools on the market that can aid in collecting and working with these log files in a desktop environment.

It is important to realize that log collection will differ between Android and iOS devices. Apple provides the iPhone Configuration Utility for free on its support website. There are similar tools for Android that allow for remote log collection by emailing the logs to a specified mailbox, among other options. While no specific tool will be endorsed here, new apps that provide Android log analysis are popping up all the time. As with any application, the tool should be vetted and researched before it's adopted to avoid introducing a potential security risk. Special caution should be exercised with any security tool being distributed for free.

At the moment, most logging for mobile devices is oriented around application debugging and not security use cases. When mobile devices are interacting with the corporate network, the security logs from the network devices will in many cases provide much more robust and usable data that will allow the security team to extrapolate potential issues emerging from the mobile devices themselves.

For example, while it may be expected that a mobile device might access corporate resources at odd hours, it may not be normal for the device to be downloading large amounts of data from corporate shares or to be doing so from locations far from where the user is based.

Log files produce a wealth of useful data, but it can be extremely time consuming to analyze and take action on it. This difficulty is compounded when your security team does not have any specialized resources for mobile device and application security. One way to alleviate the load is to integrate log events and correlations into a SIEM solution. This will allow real-time alerting when an immediate response is required and permit non–mobile security specialists to stay on top of emerging threats.

Mitigation Strategies

An entire book could be dedicated to just a slice of the threats that are in the wild at the moment. Most enterprises would rather focus their time on what tools and processes can be implemented to mitigate the security risks of corporate use of mobile devices. Thankfully, many of the mitigation strategies available are extensions of options that help secure traditional networks and can be easily understood and operated by most experienced security professionals.

In the following sections, I will go over many different security controls that not only secure your mobile footprint but also strengthen your network security posture. This may seem like a lot to chew on, but many of the security controls have capabilities that overlap because most organizations are driving vendors to develop unified security tools for their most pressing security needs.

Antivirus

Antivirus (AV) technology has become a bread and butter item for nearly every enterprise security team in the world. Very simply, a virus is an executable computer code that produces unwanted and in many cases malicious results for the end user. Although the line between viruses, malware, and spyware is becoming increasingly blurry, AV software is designed to detect signatures or patterns associated with known viruses in the wild. Most vendors today also offer at least some capability to detect morphs of older viruses as well as brand-new ones. This is due to the fact that AV software relies on not only known malicious signatures but also known patterns of malicious behavior.

Antivirus software usually has two main tiers of virus detection and prevention. A real-time component runs in the background to detect and remediate threats as they occur, and a scanning tool can be used on internal and external hard drives, as well as other devices that can be connected to a workstation, for on-demand virus scanning. Additionally, AV will monitor volatile memory (RAM) during normal use as well as the boot sector during system startup. USB keys, SD cards, external hard disks, and similar storage devices are the biggest culprit of transmission of boot sector viruses.

One of the biggest problems with AV is the number of systems that are behind in their virus definition files. While there is some possibility that a virus will be caught even when AV software and associated virus definition files are old, the chance of being infected by an advanced viral attack increases as time passes and no updates are made. At a minimum, virus definitions should be updated daily, with real-time updates being preferred. Many virus outbreaks can be traced to the fact that even though the workstations and servers did have AV software running on them, their virus definitions were well out-of-date.

Software Firewalls

Software firewalls have long been included in server and desktop operating systems. They are designed to prevent network-based intrusions coming from the outside network as well as to prevent malicious code from using the person's PC as a platform for launching attacks on other machines in the same LAN or beyond.

While the software firewall included with your OS may be sufficient for home use, most large organizations have preferred more comprehensive managed endpoint protection solutions that combine an enterprise class software firewall with antivirus and anti-malware capabilities. The advantage of this solution is the additional capability to detect end-user threats as they occur as well as the ability to isolate locations that may be serving as a launching pad for attacks to the rest of the network. Blocking the threats as they happen is important, but identifying the root causes and eliminating them before they can materialize again is just as critical.

Software firewalls are also available for mobile devices. They can provide many of the same functions offered for desktop environments as well as some additional options specific to mobile devices. One frequent use case is to use mobile software firewalls to restrict application access to the network unless given explicit authorization. This prevents an application that was installed by the user but is not currently known to be entirely safe from reaching out to other devices or downloading additional code from the Internet. Another option specifically geared toward devices with mobile phone capabilities is to block the device from sending or receiving spam SMS messages.

Before implementing any software-based firewall or other security application that includes a software-based firewall to mobile devices, be sure to perform extensive front-end user acceptance testing (UAT). As with any application, system resources will be consumed, and an excellent solution from the security team's perspective may be untenable from the user's perspective in terms of a degraded user experience and inferior mobile system performance.

Access Levels

Access levels control what users on your network are able to visualize. From a file system perspective, this includes the ability to access and open files on the corporate network. From a device perspective, this can include the ability to view firewall rules, incident data, or any other system parameters associated with a security device. Access levels can restrict which devices can be accessed as well as what data within the device a user can look at.

An example of this would be to allow a security user access to a data loss prevention console that presents incident data only from the particular business unit that the user has responsibility over, such as HR. Incident data from Legal would be accessible only to the security user assigned to that particular team.

Such access-level restrictions allow for a clear separation of duties within and across security tools. In the previous example, the first user would not be viewing sensitive incident data of other business units and would not be able to use their credentials to make changes on the corporate firewalls because access is specific to the data loss prevention toolset.

If the employee left the company under less-than-favorable terms, the exposure to risk has been minimized. If the individual wished to cause harm to the organization, the tools they accessed and the data within them was limited only to what was essential within the person's job function.

Employee access to corporate resources should be audited on a regular basis, and terminated employees should have their access to electronic resources revoked immediately. When a standard process for this does not exist, it is not uncommon to find users with access to sensitive system resources for weeks or even months after they have departed. While in the overwhelming majority of cases, the user will not make use of these privileges once gone, it does not make sense to leave the door open to such risks.

Be wary of users logging in under generic administrator accounts that are not associated with a specific account holder. Not only does this obfuscate who actually was logged into the account, it makes it much harder to perform forensic investigations of who viewed and modified system resources and files when multiple people were using the same username.

Permissions

Permissions govern what actions the user can actually perform on a server file system or a device's resources on the corporate network. For example, a user may have an access level that allows them to view what firewalls are on the network or what rule sets are being run but at the same time not have sufficient permissions to actually modify the rules on them.

File system and device owners usually have the highest level of permissions and have the ability to delegate certain permissions to other users. Some users may be given full administrative access to systems, minus the ability to manage other users, and others may simply be given the ability to manage and monitor certain aspects of these systems.

As with access levels, users should have only the level of permissions necessary to perform their current job role. When users are terminated or change roles within the organization, these permissions should be reviewed and terminated or modified, depending on the specific situation.

It is important to be wary of allowing multiple security and network teams to have administrative permissions over security devices. This may lead to system configuration changes that are not wholly beneficial for all parties involved. Each security device should have clear ownership, and changes should be made by following a standardized change control process.

Similar precautions should be taken with files and file systems. Critical information should be locked from being deleted or modified, and users should not be able to edit sensitive information without prior explicit authorization. For example, employees may be given permission to view their own HR files but not be given permission to modify them.

Auditing user access levels and permissions manually on file systems is all but impossible in a large organization. Thankfully, there are tools on the market that can help automate the process and even help identify users that may have permissions that go well beyond their job role. More advanced products can even tie in to data loss prevention tools to help security teams determine if individuals have accessed or modified sensitive data when they should not have.

Host-Based and Network-Based IDS/IPS

IDS and IPS stand for intrusion detection system and intrusion prevention systems, respectively. They can be located as network devices within the corporate LAN, DMZ, or network perimeter. They can even be located directly on endpoint devices as software-based security tools.

Intrusion detection systems (IDSs) are designed to simply detect possible malicious incursions into a network. While many firewalls today incorporate some form of IDS capabilities, an IDS does not necessarily have to be part of a firewall deployment and can be a separate stand-alone device. As the name implies, IDSs are not expected to block or respond to malicious traffic in any way but rather just monitor and audit suspected and known attack signatures and behavior. One of the biggest difficulties with IDSs is the fact that network traffic across one segment may seem to be innocuous in isolation but can be deemed to be dangerous when taken into context with other points on the network. SIEM software can pick up the slack in this regard, by correlating activity across the network with known bad behavior.

Intrusion prevention systems (IPSs) are usually an add-on or extension to IDSs and provide an actual response to potentially dangerous traffic. This can include closing connections or access to certain exposed ports in addition to alerting the security team in real time of what is going on.

Software-based IDS and IPS systems located on endpoints are usually referred to as HIDS and HIPS, where the *H* stands for *host-based*. These can be installed on desktops, laptops, servers, mobile devices, and virtual devices. They are usually bundled with anti-malware and antivirus tools and serve as the last line of defense against traffic going both to and from the network. They are also especially useful for traffic traveling between endpoints within the same LAN where an intermediary IDS/IPS may not be located.

Installing a network-based IDS and/or IPS will be fraught with issues if done on an unhealthy network. Network instability can overwhelm security administrators with false positives from the monitoring devices that incorrectly flag large numbers of dropped packets, session retransmits, and so on as a potential security risk. Before implementing any technology that will execute signature-based blocking of network traffic, significant threat analysis should be performed by the security team. A hasty implementation may result in the interruption of normal business and user activity.

 Real World Scenario

Integrated Wireless Intrusion Prevention System (WIPS) Sensor Analogy

A wireless intrusion prevention system (WIPS) is used with wireless networks and can consist of either an overlay system (dedicated sensors) or an integrated system (where the sensor is part of a wireless access point). Depending on the sensor/access point configuration, if an integrated WIPS solution is used, it may share radio time with wireless devices that are accessing the wireless network and only passively scan the air part time. Think about this analogy. An after-hours security officer is sitting at a desk watching the security monitors for any unauthorized access or other activities that could be a security violation. The officer hears a noise down the hall and leaves his post to investigate. Because he has stepped away from the desk, he cannot be watching the security monitors. Therefore, an unauthorized person may be able to enter the area without being noticed by the security officer. This is similar to the way a wireless access point that is a part-time integrated WIPS sensor and that is servicing wireless client devices may not be able to see and detect potential WIPS violations. For example, if the wireless access point is providing Voice over Wireless LAN (VoWLAN) access, because of QoS features, the access point will prioritize the wireless voice frames and may forgo the WIPS monitoring. This will result in the WIPS server not being able to capture potential wireless intrusions or other wireless security violations.

Anti-Malware

Malware is computing software that, when run, unbeknownst to the end user, could harm the user's privacy or financial security or could subject the user to participation in damages to third parties without consent, among many other negative purposes. While the broader definition of malware would also include viruses and any software that runs malicious and/or undesired operations on a user's device, a more restricted view will be taken here.

One well-known form of malware is a *key logger*. This kind of software runs in the background, captures user input as it is performed, and transmits the data to a third party. It is an excellent tool for stealing passwords, which can later be used for unauthorized access to personal email, banking, and social media.

Once the bane of the desktop world, key loggers are now making their way to mobile devices. For mobile devices, the easiest way to install a key logger is to first have actual possession of the device. This means that users should be wary about sharing their corporate-issued device with third parties and make sure their device has a complex password for access. Additionally, users may be tricked through email to install a key-logger onto their device along with another seemingly legitimate application. Once it's installed, they will not realize that the program is even running.

Mobile devices have also become prime targets for groups wishing to leverage distributed denial of service (DDoS) attacks. The number of mobile devices worldwide has eclipsed that of traditional laptops and desktops, with computing power that is not too far behind.

Combine this with the fact that most mobile users are not as proactive with their mobile device security as they are with their PCs, yet at the same time they have these mobile devices connected to the Internet on an "always on" basis, resulting in billions of potentially victimized devices for use as part of botnets and groups of zombies.

Another fairly recent addition to the world of mobile malware are executables that consume mobile resources for the purpose of virtual currency mining. This puts an enormous strain on the device's CPU and memory while simultaneously draining battery life at a higher than usual rate.

With these and other forms of malware, the so-called "human firewall" is the first line of defense. Users should be made aware of the inherent risks of downloading mobile applications from untrusted publishers. If possible, corporate devices should be locked down so that applications that have not been tested and approved by IT security cannot be installed onto the device.

At the moment, there are also software packages that can be purchased for mobile devices that address a wide breadth of malware as well as virus issues. They can ensure that the user can browse in relative safety while alerting the user of any unusual attempts for access to personal data on the mobile device. While any constantly running application will affect device performance, the state of currently marketed mobile devices may make such an impact seem negligible to the end user. Along with user education, anti-malware and antivirus software are as much a must for mobile devices today as they are for desktops and laptops.

Data Loss Prevention (DLP)

Data loss prevention (DLP), also known as data leak prevention, is based on the functionality of content-aware reverse firewalls. In other words, instead of focusing on external threats that can be mitigated with antivirus software, traditional firewalls, and so on, DLP is designed to prevent the loss of sensitive data from within an enterprise division or from the enterprise itself.

DLP has been of great interest in recent years as organizations are now maturing from ensuring the secure transit of data to also ensuring that the recipient of the data is authorized. For example, a file containing credit-card data can be delivered securely by email with TLS encryption. But if the file is going to a member of organized crime for later unauthorized use and resale, it is unlikely that the enterprise that had custody of the data would be comfortable knowing that such an event had occurred.

Regulatory compliance has also been a key driver of DLP initiatives. Large, publicly traded organizations are required to do everything in their power to prevent the leakage of nonpublic financial data that could be used for illicit financial gain. Healthcare entities are now required to ensure that private health information is not disclosed to unauthorized third parties. Manufacturing companies are now on the hook to make sure that no sensitive technological information is shared with certain blacklisted countries and organizations.

Security teams must be aware of local privacy laws when deploying DLP in foreign locations. While acceptable in the United States, DLP usage is illegal in some countries, even when the data is traversing the corporate network using corporately owned devices. Worldwide deployments require a country-by-country analysis to ensure that implemented controls do not exceed what local law allows.

Three types of components are particularly useful in dealing with data loss prevention: data in motion (DIM), data in use (DIU), and data at rest (DAR).

Data in motion (DIM) considers all data that is traversing network components, regardless of whether the components are located internally or are external to the network. Commonly monitored protocols include those that center on email, web traffic, FTP, instant messaging (IM), and peer-to-peer file sharing.

Data in use (DIU) monitors every action a user is taking as they view and manipulate data on a workstation, regardless of whether it is a physical or virtual instance. Normally, DIU requires that an agent, or software tool that is running in the background, be enabled for monitoring. Most organizations want DLP to be a nontransparent control, or in other words they want DLP monitoring to be known by the end users. This alone reduces potential incidents by over 90 percent because users modify their behavior once they know they are being monitored. While it is possible for an end user to completely disable an endpoint agent, most vendors have implemented controls that prevent this from occurring or that re-enable the agent within a brief time.

The endpoint agent will monitor and/or prevent users from moving sensitive data to file shares, DVD/CD drives, SD cards, USB keys and hard drives, printers, and almost any device that physically connects to a workstation. It will additionally monitor all the protocols that are being monitored by DIM components, providing a kind of data loss defense in depth. More granular controls can even monitor and block users from copying sensitive data from one file and moving it to another. Some solutions also have the ability to provide DAR scanning at the workstation level. DAR is discussed later in this section.

Android phones have been a sticking point for most DLP vendors because the way Android mounts its storage drive when plugged into a workstation (it appears to be an internal drive) confuses the DLP agent. Although DLP vendors are aware of the issue and are working toward a solution, this should be a consideration taken by the security team when deploying a DLP solution in an environment where users may be physically moving data to their phones.

Data at rest (DAR) is an asynchronous monitoring component that scans internal file shares, databases, and other systems that provide data warehousing and storage for sensitive data that may be located in unsecure locations. DAR scanning tools can go beyond simple auditing and can be set up to automatically delete or move files to a secure location upon discovery. They can even leave a "ransom note" placeholder file that alerts the user as to what action was taken with the now-moved file.

Setting up a large DLP deployment requires taking into account geographic scalability. Servers handling DAR are best located near the actual file servers and databases to be scanned, and endpoint management servers should be near the actual endpoints being managed. While alternate configurations are possible, they come at the cost of higher loads on WAN link traffic.

DLP for mobile devices has been deployed with varying levels of success by current DLP vendors. A mobile device management system is a must, and in many cases some form of VPN access to the corporate environment is required as well. While it is possible to have a DLP agent running on tablets, phones, and other mobile components, this will impact the users' experience as it consumes system resources. The level of noticeable impact depends on a combination of factors, including what programs the user is actively running as well as the CPU, memory, and other hardware specifications. Developing a DLP agent with iOS has proven to be especially difficult due to inherent issues with integrating monitoring software with the iOS operating system.

Figure 17.1 illustrates a simplified DLP architecture for mobile deployments. In this situation, the end user is located off site and is sending emails, browsing the Internet, and performing other online activities while on the corporate VPN. The end user is either directly connected to a nearby Wi-Fi hotspot or accessing data through a mobile phone provider's nearest cell tower and has established an end-to-end VPN SSL tunnel from the mobile device to the corporate VPN concentrators, using the public Internet as a transmission medium.

FIGURE 17.1 Simplified DLP architecture of network monitoring for mobile deployment

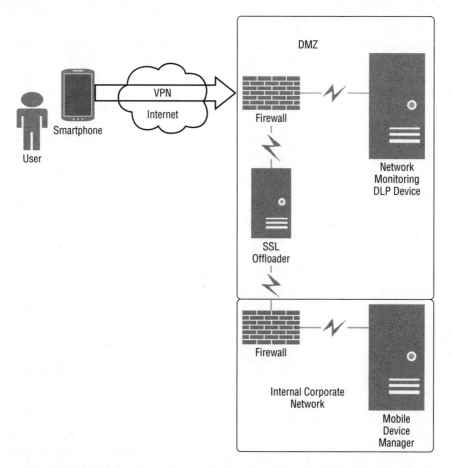

As seen in Figure 17.1, all inbound and outbound traffic passes through a data loss prevention network monitoring server, which usually is sitting as close as possible to the network egress point without actually facing the public Internet (which would overwhelm the server with uninteresting traffic). The monitoring device will audit any instances where users attempt to move sensitive data out to the public Internet. It is usually limited to HTTP, HTTPS, and FTP protocols, although any and all protocols can be audited at the discretion of the local security team.

Additionally, a mobile device management (MDM) server is usually a requirement for successful deployment of mobile DLP. An MDM server is responsible for maintaining the proper configuration of managed mobile devices. In the past MDM servers would sit in the DMZ, but current options include having a dedicated SSL off-loader that permits locating the MDM server within the relative security of the internal corporate network.

While many DLP deployments begin by only monitoring for the movement of sensitive data, over time real-time blocking, notification, and remediation is possible as well. For example, a corporation that allows its users to use Facebook, discussions forums, and other social media tools can not only discover when sensitive data is being posted, they can also block the posting from even occurring in the first place as well as notifying the user and other interested parties of the violating activity.

Although DLP tools have come a long way and have achieved a high level of capability and maturity, most organizations are unable to leverage these capabilities in any significant way. DLP requires a great deal of business buy-in from units outside of enterprise security in terms of identifying what actually is sensitive content as well as how policies should be enforced. Plus, it can be a challenge to even determine which federal and state regulations are the most critical to adhere to because companies that previously did not have DLP are ill-equipped to enforce everything at once.

Many DLP deployments fail because the enterprise decides to monitor a quantity of data that cannot be reasonably analyzed by team members or remediated by relevant parties in the business. It is recommendable to deploy both DLP policy and components in small stages that can help provide sample data that can be used to later adjust subsequent phases of a larger company-wide rollout.

Many vendors claim that their security device (email gateway, mail server security appliance, firewall, and so on) has the added benefit of providing DLP capabilities. Be wary of these claims. Many can only provide so much as "DLP-Lite" or simple pattern matching on strings of numbers and certain phrases, and on only a certain threat vector such as email. In some cases, these add-ons are as expensive as a dedicated DLP product suite, especially when the customer does not take the time to do their research and negotiate an appropriate price. While these tools may be a sufficient substitute when a full-suite DLP solution is not required, or affordable, it is important to ensure that a specific DLP use case has been defined before engaging product vendors. Otherwise, the organization may end up purchasing capabilities that are not needed, or come up short on critical regulatory compliance requirements, along with the additional burden of having to monitor different DLP toolsets located on different devices. If you are planning a mobile deployment that requires DLP monitoring, verify that the DLP vendor being used or the product to be purchased has an actual working solution that addresses mobile. They may claim that it is under development and will be "coming soon," but "soon" may be years away.

Device Hardening

Device hardening is the practice of reducing an attacker's possible approach vectors into a mobile device. This includes not only keeping device applications to a minimum but also turning off communication protocols and ports when they are not explicitly necessary. A good practice is to turn off unnecessary wireless communications such as GPS, Wi-Fi, and Bluetooth when possible. This is a good security measure, and it will extend battery life as well, especially on older devices. Additionally, if Bluetooth is to be kept on, the ability of the device to be discoverable should be turned off. Also, it is important to ensure that any data on the device is in the hands of only authorized users.

Data usage of mobile devices should be monitored on a periodic basis. If there is an unusual spike in network bandwidth consumption after installing a particular mobile application, review what access the application has to the network as well as the functions it should be performing. Corporate IT security should be made aware if any mobile application is behaving erratically because the root cause may be a security as well as an application development issue. Instead of reviewing individual devices one by one, an option would be to export usage metrics to a SIEM or other log consumption and alerting tool.

Other device hardening methods include strong passwords, multifactor authentication, and mobile device encryption. These methods have been discussed in previous sections of the text and will not be extensively reviewed here.

Regulations such as the Health Information Technology for Economic and Clinical Health Act (HITECH) consider data secure if it is either encrypted or destroyed. While throwing all mobile devices into a metal shredder will probably not be a reasonable option, pursuing a device encryption solution for all corporate devices should be. Though this may be an additional burden for users that access data using personal devices, noncompliance is not an option in regulated environments.

Physical Port Disabling

Physical port disabling is the closing of communication channels to external devices that may be connected to corporate computing resources such as desktops, laptops, and servers. This includes shutting down access to external USB storage media, CD/DVD burners, SD cards, mobile devices connected through a physical interface, or any other device that may be used to ingress/egress data into the corporate network through an endpoint location.

Special care should be taken that any physical port lockdown does not affect tools that are used on a daily basis by the business. For example, a full lockdown of external hard-drive access through USB may be feasible, but a full closure of all USB ports is probably not. The latter would probably shut down users' ability to use external mouse and keyboard devices that currently integrate themselves through a USB connection.

As with any security control that affects end users and endpoint devices, a site survey should be conducted before implementation. In modern day healthcare settings, many diagnostic tools connect to physician laptops through a USB or peripheral device port. Other industries have similar use cases as well.

Physical port disabling is usually an option that is included with most enterprise-class endpoint protection software suites. They offer granularity as to the type of ports to be blocked, specific devices to allow/disallow, and the enforcement of using encrypted media at a managed endpoint.

Firewall Settings

One of the most basic but effective firewall controls is to lock down unnecessary network communications ports and protocols, or ones that are associated with unproductive use of network resources. Specifically what these ports actually are depends on an understanding of current corporate policy regarding the use of electronic devices. For example, if the business does not want employees to use an instant messaging (IM) client outside of a certain one, such as Microsoft Lync, it is possible to set the firewall to block all IM traffic on every port except the one allowed.

Modern-day firewalls also have the ability to block traffic based on certain traffic signatures, regardless of what port is being used. In the past, users would attempt to circumvent the firewall by using a nonstandard port as a transmission medium for whatever prohibited tool or software package required network access. Security teams have long ago caught on to this and have at their disposal the tools to prevent such usage.

While on the surface it may sound reasonable to block all IM, peer-to-peer file sharing, social media, and audio and video hosting websites, it is important to investigate the realities of the business before enabling filtering at the firewall level. While access to social media and video hosting websites may seem like a huge productivity drain for the enterprise, they may be a critical part of the firm's marketing activities. Shutting down streaming audio may seem like a no-brainer, but it's definitely not the way to go if the CEO thinks it helps people concentrate on their work. Always make sure to communicate to the business the total impact of any proposed security settings in order to avoid misunderstandings down the road.

While firewalls are excellent security controls in themselves, they do not liberate the enterprise from ensuring that the design of the network itself is inherently secure. This includes making sure sensitive network traffic is segmented from nonsensitive traffic, making use of VLANs to provide an additional layer of isolation between departments, and so forth.

While most modern network security devices that have firewall capabilities come with some level of out-of-the-box capabilities, this does not obviate the need for firewall administrators to review deployed polices in the light of business requirements.

Many business leaders make the mistake of thinking that having a firewall is the equivalent of having a secure corporate network. The security requirements of a segment handling Payment Card Industry (PCI) data is going to be vastly different than a segment that is set aside for public Wi-Fi access.

Even when not required by law, it is advisable to have regular internal and, if possible, third-party audits and reviews of currently enforced policies on firewalls and other network security devices. In some cases it may be found that a firewall is passively passing through all traffic with no enforcement at all!

Port Configuration

Outside of firewalls, other networking devices can also provide an additional layer of security through port configuration. Network switches and routers can be configured to disallow any communication on selected TCP or UDP ports deemed unnecessary or dangerous to the organization.

Port configuration stops potentially harmful traffic well before it even reaches a firewall or other network security device. Even endpoints can be configured to enforce a certain level of port configuration.

Nevertheless, since it is a protocol and not a signature or technology based on security intelligence, security teams may end up blocking traffic that is in no way malicious simply because it is attempting to access a locked-down port.

Unless it has been clearly established that a certain port will have no business use case and its existence is a potential risk to the organization, it is often better to rely on other tools such as traditional firewalls and IDS/IPS technology before implementing port configuration on network communication points.

Application Sandboxing

Sandboxing in the context of a mobile environment usually refers to running applications in an isolated portion of the OS; the entire application and all related files have no interaction with other files and executables on the mobile device.

While such an arrangement is excellent for testing purposes or for applications that will never require interaction with user information and system files, in most cases a completely airtight environment for code execution is not an option because it greatly degrades the user's experience.

Nevertheless, security teams also have the option to execute and test production-ready applications in a sandboxed environment to detect if any malicious code has been attached to the executable content well before it is used in the field.

One caveat to be aware of with application sandboxing is that some types of advanced malware can actually detect if it is being run in a sandboxed environment and withhold from performing any malicious activity while in the sandbox. This can be especially frustrating for security teams, who may unwillingly deem malicious software as being "safe" when actual behavior will be radically different once it is in the user environment.

With newer versions of Android and iOS, application sandboxing capabilities have been improved to only permit applications to access the files and OS processes that are critical for the proper functionality of the application, shutting the door on the possibility that a rogue application may exploit vulnerable system processes and file systems.

A best practice would be to ensure that mobile users have access to updated and more secure mobile operating systems as they are released. For users who refuse or are unable to upgrade personal devices, a good policy is to not allow access to sensitive corporate data through them if they present a risk of data exposure or loss.

Business decision makers should always remember that, regardless of the level of security imposed on mobile, any use of third-party applications has an associated risk, and functionality in many cases will diminish as the level of security imposed increases. Striking an acceptable balance is more of an art than a proven science.

Trusted Platform Modules

Trusted platform modules (TPMs) are physical devices that are located on a motherboard with the specific function of acting as a secure location for encryption keys. As a unique physical device, usually in the form of a secure microprocessor, a TPM can offer a much higher level of security than software-based solutions alone.

Although TPMs are widespread in desktops and laptops, they are a fairly recent arrival in the world of mobile. Recently, Microsoft has incorporated TPMs in its mobile offerings, but other vendors have been slower to bring the technology to market.

TPM technology still has many issues to overcome and is definitely not a "silver bullet" for companies that are coping with employees using personal devices for work. One of the biggest issues is the fact that because it is a physical control attached to the device, the enterprise in essence cannot control it in any way.

Nevertheless, security teams managing mobile devices can expect more vendors to offer TPM capabilities and subsequently the opportunity to leverage them into a potentially more secure handling of corporate data on personal devices through better encryption key technology.

Data Containers

Data containers can mean a host of different things in computing. In the case of securing mobile devices, it represents a school of thought that wishes to avoid the monumental task of ensuring the security of the device as a whole, whether personal or company issued. Instead, it focuses on securing only the corporate data on the device as well as maintaining control and ownership of it.

To this end, numerous security vendors are currently offering tools that allow users to view and manipulate corporate data securely because it is isolated from other data on the device. Additionally, these tools allow security teams to remotely wipe the secure data container when a user leaves the organization or when otherwise instructed by management without affecting other data on the device or requiring physical access to the device itself.

Content Filtering

If a mobile device is managed and connected to the corporate network, there are many existing web and email filtering tools on the market that can be leveraged to provide *content filtering* to them.

Web filtering tools provide the ability to block or restrict access to sites based on a wide variety of factors. Currently, most vendors keep up-to-date categorizations of websites that host malicious code and/or are part of phishing scams. Instead of having security teams worrying about reacting to these threats as they are discovered by end users, many web filtering tools provide website blacklists that are updated on a daily or in some cases a real-time basis. Also, web filtering tools give security teams, in coordination with management, the ability to control access to websites that may be legal but not appropriate for viewing on corporate-issued devices. This includes websites for gambling, pornography, weapons, and hate speech and in some cases can even include access to social media, among many other possible options.

Email filtering tools such as email gateways have the ability to stop malicious or unwanted content from even being received by the mobile device, and they can also block mobile devices on the corporate network from sending undesired content. Email gateways primarily have the role of filtering viruses, spam, and other undesirable content attached within email from entering or leaving the corporate network.

For an extra layer of security, there are also tools that provide these exact same capabilities for the internal mail servers and relays. Email filtering on internal mail servers helps prevent mobile users from accidentally sharing malicious content from their mobile devices to desktop users in the corporate environment and vice versa. Some organizations go as far as to provide content filters on email as well as instant messaging for all managed devices, including mobile. An example of this would be a filter that prevents users from using profane language in internal communication.

Ultimately, while there is a plethora of out-of-the box content filtering controls available for web, email, and IM, it is up to corporate policy makers to determine what rules are appropriate for an organization. While some rules are beneficial from a legal liability standpoint, others are purely advisable from a company-culture point of view. It is not recommendable for security teams to enforce such policies without buy-in and explicit direction from upper management.

In many cases, DLP, web and email filtering tools, and firewalls overlap in capabilities. Instead of "defense in depth," what are often seen are cases of redundancy and inefficient use of technical resources. Security teams, in conjunction with business decision makers, should periodically assess the tools and processes that are in place in order to rationalize existing deployments.

Demilitarized Zone (DMZ)

A *demilitarized zone* (DMZ) is an essential part of securing a corporate network. Connecting a device to the Internet will expose it to some form of attack within minutes. Most servers providing critical IT services would be unnecessarily exposed if they were directly linked to the WAN or Internet without any intermediate filtering and protection from enterprise firewalls.

Currently, there are two basic configurations for setting up a DMZ. The first, which is usually selected due to lower cost and is seen more often in smaller environments, involves setting up a single firewall that has three interfaces (see Figure 17.2). One is connected to the Internet, another to the internal corporate LAN, and another to the DMZ.

FIGURE 17.2 Single firewall configuration for DMZ

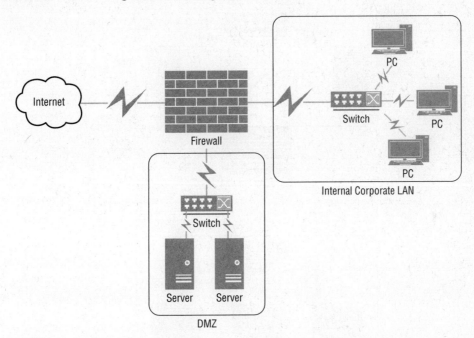

The main disadvantage of a single firewall solution is that if the firewall is compromised in any way, whether by an attack or due to hardware failure, both the DMZ and the internal corporate network may be at risk. Yet, this is definitely better than nothing at all and may suit the needs of a small or medium-sized business with limited ability and resources to set up and maintain the firewall.

The other option is to have two firewalls, one on the front end facing the Internet or WAN and another on the backend that is facing the internal corporate network (see Figure 17.3). This solution can be further augmented by having hot failovers to secondary devices if the front-end or backend firewalls fail for any reason.

FIGURE 17.3 Dual firewall configuration for DMZ

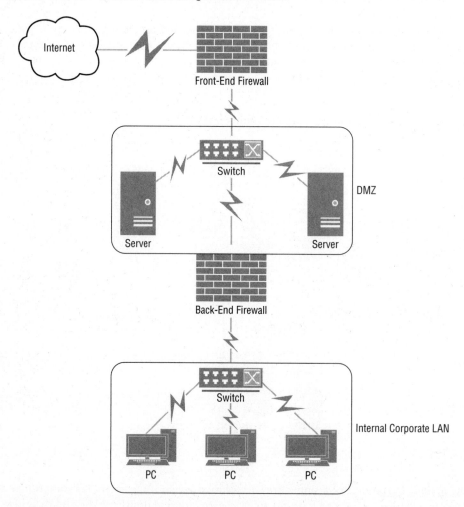

While using the same vendor for the firewalls would save costs and make maintenance simpler, many security professionals prefer to have different brands for the front ends and backends. The thinking is that if the front-end firewall is breached, the same attack vector will not be available on the backend, thus giving the security team enough time to respond and prevent an incursion into the internal network.

Summary

This chapter provided a high-level overview of the security tools and concepts that IT security teams should be aware of, both in traditional deployments and in deployments involving mobile technologies. Device compliance and report audit information ensure that regardless of the level of enterprise IT security (or lack thereof), distribution of electronic assets is well documented and can be readily communicated to third parties when necessary. SIEM technology provides excellent visibility into log events occurring across the network as well as being a source of usable security analytics for business decision makers that may or may not have extensive technical know-how. In this same train of thought, mobile device logs are differentiated from logs that security teams are accustomed to using with other devices. Mobile application developers are aware of the need to provide better event data and mechanisms to relay this information back to the security team and their tools as newer and more sophisticated mobile log management applications are arriving onto the Android and iOS marketplaces.

The bulk of the chapter was dedicated to mitigation strategies, including firewalls, access levels, sandboxing, content filtering, port configuration and disabling, DMZ, and more.

While the security tools and techniques in this chapter do not represent an exhaustive list of everything that is currently available, implementation of any or all of these measures will help drive the organization to a better IT security state. For those who wish to further their knowledge in this realm, a dedicated course in network security or security specific to mobile devices would be highly recommended.

Chapter Essentials

Understand current technical controls that are available to mitigate IT security risks and sensitive data loss. Content-aware controls include DLP and content filtering. Firewalls, both software and hardware based, provide network security. Permissions and access levels are part of a larger identity and access management strategy. Antivirus, anti-malware, host-based IDS, and physical port disabling are generally offered under the same umbrella within an existing enterprise endpoint protection suite. Application sandboxing, trusted platform modules, and data containers are all emerging security technologies being applied to mobile devices.

Implement a secure architecture in your network design. A DMZ can separate perimeter-facing devices from the rest of the corporate network. The DMZ can be built with a single firewall for a basic level of security, or it can be designed using front- and backend firewalls with failovers for enterprise configurations.

Understand your security team's compliance obligations and ability to provide meaningful data and reports to auditors.   Maintain a comprehensive asset inventory that includes mobile devices issued by the enterprise. Internal audits should be performed on a regular basis to avoid any surprises during mandated external audits. Regulated industries must provide compliance information in accordance with federal and/or state laws.

Understand mobile device log files.   Be aware that iOS and Android devices have distinct log management toolsets for working with and maintaining mobile device log files. While generally robust for debugging operations, mobile logs can still be extremely useful when analyzed in conjunction with other network security device logs.

Understand the Concept of SIEM.   SIEM technology allows security teams rapid log collection and aggregation across multiple networks and network security devices. They allow for real-time security alerting and event correlation, well beyond what can be accomplished during manual security log review.

Chapter 18

Data Backup, Restore, and Disaster Recovery

TOPICS COVERED IN THIS CHAPTER:

- ✓ Network Server and Mobile Device Backup
- ✓ Backing Up Server Data
- ✓ Backing Up Client Device Data
- ✓ Disaster Recovery Principles
- ✓ Maintaining High Availability
- ✓ Restoring Data

When it comes to technology, there is no worse feeling than turning on an electronic device such as a computer, tablet, or smartphone and realizing it no longer functions. This is often the time when many people think about their backup, but it may be too late if the backup was not created or does not work. Backing up data and configuration information is an essential part of all aspects of information technology, and mobile devices are no exception. In this chapter we will explore the concept of data backup and recovery solutions for both network servers and client or mobile device technology. This includes understanding common disaster recovery procedures, high availability, and backing up and restoring data for both the server side and the client device side. Think about the influx of bring your own device (BYOD) technology: both corporate and personal data must be taken into consideration when it comes to backup and restore policies. This chapter will also provide an overview of the data restoration process.

Network Server and Mobile Device Backup

Protecting information on a computer network is an important component of computer network administration. Although it may not be a pleasant task, it is something that must be done to ensure the availability and integrity of stored data. In the early days of local area networking, aside from sharing printers, one main reason local area networks were installed was to provide a repository of user data that would be considered secure and always available or recoverable if necessary. For many, backing up data was considered a mundane task that often caused problems because of hardware or software incompatibility and availability of the technology or lack thereof. How the corporate policy was implemented would determine the backup policy and procedures. Backup often occurred after hours at times when data was not accessed to lessen both the impact on the network's resources and performance degradation. For the client device side, usually a software program known as an agent was installed and the user's computer was left powered on for client-specific data to be successfully backed up.

As technology has evolved, so has the backup and user data availability process. We now have the capability of redundant hardware or server solutions to maintain close to a 100 percent uptime or availability in the event of a server, network, or infrastructure malfunction. Other technologies, including high availability, cloud-based backup solutions, dynamic backup, and "hot site" solutions, are available to satisfy corporate disaster recovery requirements. Terms like *service-level agreement (SLA)* come into play. An SLA is basically an agreement between an organization and a service provider, or even between departments within an organization. An example of an SLA from a service provider is shown in Figure 18.1.

FIGURE 18.1 Amazon Web Services (AWS) web page showing the service level agreement (SLA)

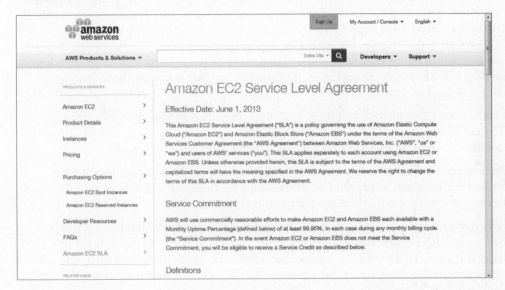

The SLA specifies the terms of the arrangement and consists of items such as the percentage of network, or service uptime and the acceptable amount of network outages. The SLA will also provide some accountability in the event of outages or other related service issues.

Backing Up Server Data

Chances are most information technology departments already have a plan or policy in place that pertains to disaster recovery options for all corporate data. This plan will include catastrophic recovery options, fault tolerance options, and other recovery options such as how or if specific user device data is backed up. There is no absolute "right way" or "wrong way" regarding the topic of data disaster recovery; this will vary based on the specific function of the organization. Rather, there are many *different* ways to accomplish the same task.

The amount of data we create now, compared to 10 years ago, has grown tremendously. Statistics show that companies produce 50 percent more data every year than they did the previous year. For reasons such as this, data backup solutions need to be faster, more reliable, and highly efficient. Over the years, we have seen data backup methods evolve from magnetic tape to optical storage to cloud-based solutions that can include considerations such as high availability, load balancing, and complete redundancy.

Keep in mind that with cloud-based backup solutions, the data does not go to some mysterious, magical place where it is retained forever. With cloud-based solutions, you are offloading the process to a third party that still provides a physical method of data storage and backup. It is really no different than what you would do within your infrastructure; you are just offloading the task to another company. The benefit of a remote backup solution is that it relieves the organization and information technology staff from the burden of the process, which then hopefully will result in a cost savings to the organization.

There are three common traditional backup methods available. Which one you choose will depend on your specific organization, and many different factors based on the business model itself. The following backup methods can be used for either server-side backup or client-side backup.

Full Backup This method will back up all data, regardless of how the files are flagged. The *full backup* method is the most time intensive of the three methods, and requires the most media because every file is backed up, regardless of how it is flagged.

Incremental Backup Any file that has been changed or not backed up since the last backup occurred will be flagged for backup, using the *incremental backup* method. After it has been backed up, it will be flagged accordingly.

Differential Backup All files that have been changed since the last full backup will be backed up with a *differential backup*. However, once backed up, the files will not be flagged as such. This method is quicker than a full backup, but will take longer with each differential backup that occurs, because even if a file was backed up in a prior differential backup, it will be backed up again.

Flagging a file for backup means that an attribute for the file will show that it has been changed or that it is new, and will need to be backed up. The backup program will look at each file to see if this file attribute is set; if it is, the file will then be backed up. The backup method used (full, incremental, or differential) will determine if that attribute is reset after the backup has occurred. The way this is achieved will vary based on the operating system in use. For example, one method Unix systems use is to look at the time stamp of a file. If the file time has changed, then it will be added to the backup. Figure 18.2 shows the backup attribute for a file using the Microsoft Windows 7 operating system.

Notice in Figure 18.2 that the backup attribute in the Microsoft Windows 7 operating system is labeled as File Is Ready For Archiving. If this check box is selected, the file will be backed up. It will be cleared (unchecked) after the file has been backed up, depending on the backup method that is used.

FIGURE 18.2 File backup attribute setting for the Microsoft Windows 7 operating system

 If an incremental or differential backup method is used, it is very important to only use that method for future backups. You cannot mix the differential and incremental backup methods. If you decide to change methods, you must first do a full backup, and then choose either differential or incremental. The restore order is also very important with the incremental or differential methods. Restoring data will be discussed later in this chapter.

Server backup solutions are available for small, medium, and enterprise networks. These solutions are available in standard software formats, cloud-based solutions, physical appliances, or any combination of the three. The cost of these products will vary based on the available feature set and capabilities. The following features are among those available with server-based backup solutions:

- Various operating system platforms
- Virtual machines
- Standard backup types
- Complete disk imaging
- Bare metal support
- Cloud storage
- External drive support

630 Chapter 18 ▪ Data Backup, Restore, and Disaster Recovery

The following list includes some companies that provide enterprise backup solutions for 50+ servers.

Product	Website
Acronis	www.acronis.com
Barracuda	www.barracuda.com
NovaStor	www.novastor.com
Symantec	www.symantec.com

The term *bare metal support* is used in the information technology industry to identify a data restore to a computer or device that basically has no operating system or other data previously installed. The information to be restored will be hosted on an external storage device, and a software program can be used to boot and restore the entire system. The operating system, applications, and user data are included and available from the external data storage source. This method is often used with computer software imaging programs.

In Exercise 18.1 you will view the archive attribute of a file using the Microsoft Windows 7 Professional 64-bit operating system. The archive attribute will signal the backup program that the file is ready to be backed up.

EXERCISE 18.1

Viewing the Archive Attribute

In this exercise, you will view the archive attribute of a file in the Microsoft Windows 7 Professional operating system.

1. Click Start ➢ All Programs ➢ Accessories ➢ Windows Explorer. The Microsoft Windows Explorer window appears.
2. Expand the (C:) folder in the left pane. The contents of your C:\ drive will appear.
3. Left-click the Windows folder in the left pane. The contents of the Windows folder will appear in the right pane of Microsoft Windows Explorer.
4. Scroll through the right pane to find the Notepad.exe file.
5. Right-click the Notepad.exe file in the right pane and then left-click the Properties button. The Notepad.exe file properties page will appear.

6. Click the Advanced button in the Attributes section. The Advanced Attributes window will appear.

7. Notice that the File Is Ready For Archiving box is checked.

The Microsoft Windows 7 Professional 64-bit operating system was used for this exercise. Depending on the operating system and version used, the steps may vary slightly.

Backing Up Client Device Data

As explained earlier in this chapter, *client device backup* for computer data traditionally consisted of software of some sort that would allow a file server to contact the client on the network and retrieve data for the purpose of backup and disaster recovery. This was typically done after hours to lessen the impact on the user and the available network resources. At that time, in many cases the need for dynamic, "on the fly" backup of data was not as urgent as it is today.

As technology has evolved, so has the process of client device backup. Urgency, speed, and efficiency are important factors with modern technology as it relates to data backup. In addition to the "technological" aspects, the type of data (corporate, personal, or both), its frequency of use, and availability need to be considered when selecting the backup solution.

Software imaging programs are now very affordable, and are a great option for performing a backup that provides an exact replica, or "image," of what is installed on the client computer. This will allow an easy restore if the entire drive needs to be recovered because of a device failure, if you are upgrading the client device to a bigger drive with more storage, or even if you are restoring specific files or folders. In addition to the previously mentioned backup strategies, newer backup technology for mobile devices is commonly used. This includes apps that can be downloaded to a device to provide the capability for cloud-managed backup solutions. This will give the user the capability to restore data in the event of a device failure, or to restore data to a replacement device because of loss or theft of their device.

To perform backups, you need backup software. The following list shows some of the products that are available for client device backup.

Product	Website	Free or Fee Based
AceBackup	www.acebackup.com	Freeware
Acronis True Image	www.acronis.com	Fee based
Genie Backup Manager	www.genie9.com	Fee based
NovaStor	www.novastor.com	Fee based
Redo Backup & Recovery	redobackup.org	Freeware
UrBackup	urbackup.sourceforge.net	Open source

One way to get comfortable with backup software is to download freeware and start using it. Exercise 18.2 shows you how to install a freeware backup program.

EXERCISE 18.2

Installing a Backup Program

In this exercise you will install a free computer backup program available from Ace Bit. A computer with the Microsoft Windows 7 Professional 64-bit operating system was used for this exercise.

1. Download the AceBackup 3 software package from the Ace Bit website by pointing your browser to www.acebackup.com.

2. Click the Download button on the home page to download the software program to your computer. At the time of this writing, the file is named AceBackup3.exe.

3. Execute the file to start the installation process. Depending on the version of the Microsoft OS you are using, you may see a User Account Control dialog box appear. Click the Yes button to continue. The Choose Setup Language screen will appear.

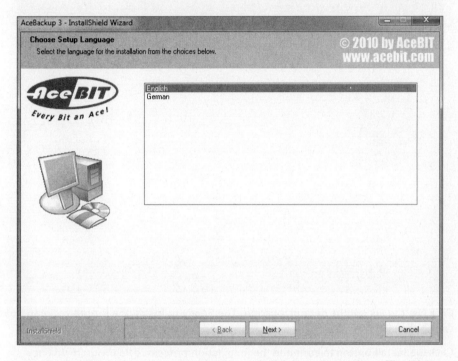

4. Verify that English is the selected language, and click Next to continue. The Welcome To The InstallShield Wizard for AceBackup 3 dialog box will appear.

5. Click Next to continue. The License Agreement dialog box will appear.

6. Review the agreement, verify that the I Accept radio button is selected, and click Next to accept the terms of the agreement. The Choose Destination Location dialog box will appear.

EXERCISE 18.2 *(continued)*

7. Click Next to accept the default location. The Ready to Install the Program dialog box will appear.

8. Click the Install button to continue the installation process. The program will now be installed on your computer. When the installation is complete, the InstallShield Wizard Complete dialog box will appear.

9. Click Finish to end the installation and start the program. The AceBackup 3 home screen will appear.

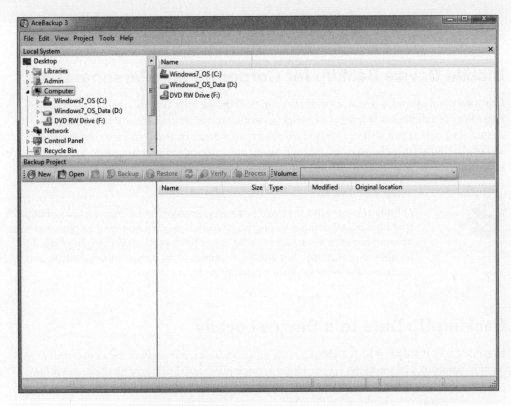

You can now experiment with the AceBackup 3 program. Because this is a freeware program, you can leave this installed on your computer and use it based on the license terms you agreed to during the installation process.

Of course, corporate policy will vary based on organizational needs and requirements for how client device backup solutions will be implemented.

 Just as when backing up servers, you can perform full, incremental, or differential backups with clients. For more information on this, see the section "Backing Up Server Data," earlier in this chapter.

Now you'll take a closer look at some of the considerations for backing up client data, how frequently to do so, and how to test your backups.

Mobile Device Backup for Corporate and Personal Data

Corporate policy and a disaster recovery plan will define how the backup data is handled, regardless of whether it is from a network server, desktop computer, or a mobile client device. The procedure will vary based on the type of organization, and specific business model. Earlier in this chapter you learned about server backups and the common methods and procedures. With mobile devices, many of the same principles apply.

I highly recommend that you *password protect* the backup archive of your mobile devices using a strong password. If an unauthorized person were to obtain access to a backup file that is not password protected, they would be able to restore the file and have access to all data (both corporate and personal) that was on your device prior to the backup.

Backing Up Data to a Device Locally

In some cases it may not be feasible to back up your device remotely or to a corporate backup solution, corporate server, or cloud service provider. This may be the scenario for people who are working on the road remotely, or with limited network or Internet access. In a case such as this, backing up data locally is an acceptable solution.

It is important to be fully prepared, because you never know when you may experience a device failure. For this purpose I recommend that a *local backup* be stored on an external media source such as a *Universal Serial Bus (USB) drive*, *Subscriber Identification Module (SIM) card*, *Secure Digital (SD)* card, or other portable drive source. Using the same physical device or storage drive for data backup and your user data is not recommended. If you were to store the backed-up data on the same device or drive, the backup would be useless in the event of a device failure. Keep in mind that everything on the device, such as the operating system, applications, and other information that can be easily replaced, may not need to be backed up in this situation. Depending on your needs, it is often sufficient to back up only data and other information that cannot be easily replaced.

A SIM card is a small memory device that is used in GSM devices to store personal information about the device's owner, and other information such as contacts, text messages, and address books. Although these cards are available with different memory capacities, you should not expect to be able to store large amounts of data. SIM cards are available in different form factors, and the capacity is relatively small, ranging from 16 KB to 128 KB. The number of entries or amount of information the card can store depends on the capacity.

Frequency of Backups

The frequency of backups is usually specified as part of corporate policy or in the disaster recovery plan. Looking from the client device side, how dynamic or critical your data is will determine the frequency with which it is backed up. Something to think about is the "what if" scenario:

- What if your device failed?
- What data would you be able to live without in the event of a failure?
- How much time or work would it take to re-create the important information that was lost and not recovered?

From the server or network side, it really depends on the size and the complexity of the network, and the number of servers. Some organizations are satisfied with what is considered a more legacy backup solution, such as nightly with a rotational scheme in mind; others may rely on more modern solutions such as cloud-based, server virtualization, and clustering technology. Keep in mind that many networks are now available 24/7/365, and the data can be accessed at any time of the day or night from anywhere. Remote access capability has really changed the landscape of the entire computing process. Some recommendations include using your best judgment, researching industry best practices, contacting your backup solution provider for recommendations, and reading industry white papers written by experts in the field.

Testing Your Backup

What is worse than powering on your device only to see that it's not working? Realizing that the data backup solution you used did not work and you have no way to recover your device's data or the lost configuration information. Testing a backup solution to verify that it works is one component of a disaster recovery plan. Just as you would test a fire escape plan by executing a fire drill, data backups should also be tested to ensure that they are working correctly and will be accessible if needed.

The process of testing includes determining the frequency of testing and the data types to be tested, which will vary based on your organization. I would start by creating files that can be deleted, and use those for the test. Back up, delete, restore, and verify that the data is intact. Another option would be to perform a bare metal restore on an unused server or device, if possible, to verify that the backup is 100 percent valid.

Disaster Recovery Principles

It is important to understand the difference between disaster recovery and restoring computer data. *Disaster recovery* is a plan put into action after an event such as a catastrophic technology failure or natural disaster has interrupted normal business operations, due to a lack of the technology infrastructure. Restoring computer data could be as simple as restoring data for a single user who has experienced a device failure that impacts only that user, or a non–mission-critical server that has experienced a hardware failure or software malfunction.

Definition of a Disaster

Years ago I was working in a subcontractor position that involved a server migration and standardization project for a large enterprise network covering five states and over 20 sites. One of my assignments was to develop and implement a new disaster recovery plan for the organization. This was before some of the newer technologies like cloud-based and virtualization solutions. It involved a traditional hardware and software solution that provided a complete rotational backup of servers and client devices to physical media, with offsite storage of the media to a secure remote location.

This disaster recovery plan was intended for recovery after a natural disaster or catastrophic technology failure, not for restoring a single user's data. On more than one occasion, a user would visit my office and ask me to restore files for them that were located on the file server in their home directory or in the shared folder and had been inadvertently deleted. Depending on the specific situation, this could be a time-consuming task. Many times the deleted data was not of much importance, and the amount of time it took to retrieve the backup and restore the files was not cost effective. Because some users were taking advantage of the process, I had to inform the user base that the disaster recovery procedure was specifically for disaster recovery only, and not to restore individual user data due to user error.

After about a week had passed, I had a user visit my office asking me to recover some files that "somebody else deleted" from their home directory on the file server. I explained to the user that the server backups were intended for disaster recovery purposes only. I was informed by this user that their files needed to be restored because "it was a disaster that they were deleted."

My objective here is to add a little humor and to make a point. I am not suggesting that a backup could not be used to recover files a user deleted. I am suggesting that guidelines and a policy should be implemented that also clarify the differences between disaster recovery and recovering user-specific files that were deleted. So yes, if "someone else" deleted their files, they could be restored, but the policy should specify the priority of this specific situation, and how it is handled.

Disaster Recovery Plan

A network *disaster recovery plan* is a documented procedure with specific sections that include steps and processes that will allow an organization to restore network operations in the event of a disaster or catastrophic failure. Disasters fall under two main categories: natural disasters and human-created disasters.

Natural Disasters A natural disaster is something that is out of our control, and can occur anytime and anyplace. These disasters include events such as floods, fires, earthquakes, and high winds.

Human-Created Disasters These disasters include fire (which also could be considered a natural disaster), power outages, hardware or device failures due to technology issues, terrorism, industrial accidents, and more.

Earlier in this chapter you learned about backing up data and other information from servers and various client devices. The different methods for well-planned backup solutions were also discussed. All of this information is great, but what happens in the event of a network or infrastructure failure? This is where the disaster recovery plan comes into play. Most people know what needs be done, but it's important to identify and document the actual procedures and processes.

The severity of the disaster will determine the course of action that will be taken. A disaster recovery plan will specify the proper procedures that need to be taken, in order to restore network operations to as close as normal as possible. There are many variables to consider. From an overview perspective, there are several main areas to be considered:

- Designing and creating the plan
- Obtaining management buy-in for the plan
- Implementing the plan
- Testing and validating the plan
- Keeping it current
- Planning document storage and accessibility
- Identifying key contacts

Designing and Creating the Plan

To design and create a disaster recovery plan, you must understand the operation and functionality of the entire business and its operation procedures. This will require research and interaction with the entire organization as a whole. Identify the most critical departments or business units within the company, and prioritize accordingly. Different businesses or organizations will have different priorities that will need to be taken into consideration. Heads of all departments need to be involved with this process, in order to ensure that the plan will work as designed. Gather all the necessary information regarding the details and day-to-day operations of each department or business within the company. This will require interviewing department heads and top business unit personnel. The following major departments are among those that must be considered:

- Information Technology
- Security
- Human Resources

- Accounting
- Manufacturing
- Production

Obtaining Management Buy-in for the Plan

Management commitment to a disaster recovery plan is one of the key factors in its success. Surprisingly enough, this may be harder to obtain than you might expect. Let's face it, a project like this will cost money, and depending on the size of the organization, it could be a substantial amount of money. Unless the plan is actually put into action (which you hope it never is), it is very difficult to see the return on investment. A good understanding of what a plan will provide, and the long-term benefits, must be understood by all parties, and you need to get the commitment of the management team.

Implementing the Plan

Once the plan has been approved by management, it should be implemented immediately. This will ensure that you are ready to put the plan into action if the need to do so arises.

Testing and Validating the Plan

Most companies and organizations have plans in place for how to react to emergencies such as fire, tornado, earthquake, and other types of disasters. To make sure these plans work, they are tested. This will include fire drills and other testing of that nature. A disaster recovery plan for technology should be no different. Proper testing procedures should be implemented to ensure that the plan will work as designed.

Keeping It Current

Like many other documents within an organization, the disaster recovery plan must be monitored and updated as needed. Technology and infrastructure services are constantly changing, and the plan can become outdated quickly. This is especially true if the plan has never had to be put into action. Changes should also be approved by a change management team, and documented within the change record pages.

Planning Document Storage and Accessibility

The plan should be stored securely on various media types, possibly in multiple locations, and readily accessible to those requiring access to it. This may include hard copy documents, and electronic files stored on CDs, DVDs, or removable electronic disk drives. Ensuring that those who require access are the only ones to obtain access is also important, especially if the document contains a company's private information. You also may want to consider a cloud-based storage solution that provides only secure access to the document.

Identifying Key Contacts

The main or key contacts responsible for the creation, implementation, and execution of the disaster recovery plan should be noted. This will allow those individuals to be reached if the need arises.

 The hardest part of any type of documentation is figuring out where to start. The objective of the disaster recovery section in this chapter is to provide you with an overview of the process and to stress some key points. There are several websites available with best practices and templates you can use for your disaster recovery plan. One such site is the SANS Technology Institute website at www.sans.org. From the home page, you can click the Infosec Reading Room link and search for a variety of technology-related documents to review.

Disaster Recovery Locations

Disaster recovery locations will depend on the plan that is implemented, and whether the backup is on actual physical media or a dynamic hot site. Backup solutions that are stored on physical media, such as magnetic tape or optical devices, should be stored in a secured offsite location. This will ensure that the media with the backed-up data can be used to rebuild the systems with the data that was lost or destroyed from the live network. Recovery locations are commonly categorized as one of three types of sites: hot sites, cold sites, and warm sites.

Hot Site   A *hot site* is an exact replica of the company's network operations or datacenter. It can be a dynamic site that mirrors all information in real time and would immediately be able to take over in the event of a network failure. Other types of hot sites may need to have the most recent user data restored to get it 100 percent up to date. This is the most expensive solution, but has the shortest recovery time.

Cold Site   The *cold site* location is basically a building space that would have to be brought up to speed with the purchase, delivery, and installation of all equipment, computers, and the infrastructure, and to have all data restored from backup. This is the least expensive type of site, but would take the longest amount of time to be brought up to speed.

Warm Site   This site is the middle ground between the hot and cold sites. The *warm site* will have equipment and computers capable of acting as a temporary replacement of the company's datacenter. A warm site would need to have the latest data restored in order to make it operational.

Choosing the recovery location that will best fit your needs depends on the organization, type of business, urgency of restoration, and, of course, your budget. You should evaluate all options carefully prior to making a decision.

Maintaining High Availability

Providing a physical backup mechanism may not be an adequate solution for some network deployments. Depending on the business model, 100 percent uptime or very close to it may be a requirement for day-to-day business operations to succeed. Any downtime could cost the organization substantial amounts of money, because of the inability to access the network resources or stored data.

This is where server and network clustering may play a role. Providing an exact, real-time replica of all configurations, data, and transactions will ensure the most amount of uptime possible. The process of clustering does not necessarily provide a static backup solution, but more of a *high availability* fault tolerance solution, as well as possible load balancing of services. Backup of servers, user data, and devices is still a factor. For example, a company that sells widgets may have a website running that could become overloaded because it advertised a special promotion on a television commercial, or used social media to promote a product. With *clustering* of services, some of the traffic can be offloaded to additional web servers that are connected together through a common infrastructure, and possibly a *storage area network (SAN)*. This type of configuration provides what is known as an *Active/Active* state, meaning both sites are up and accessible, not just on standby in the event of a server or network failure. A basic framework of this process is shown in Figure 18.3.

FIGURE 18.3 Framework of high-availability infrastructure

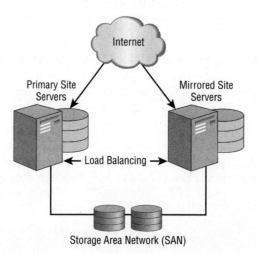

This will also provide a backup solution in the event a site goes down unexpectedly, because the data is theoretically "mirrored" between the sites, and the combined sites act as one. *Virtualization* technology also plays a big role with high availability, allowing for multiple servers at different locations to be managed and accessed as one entity. If needed for *load balancing*, redundancy, fault tolerance, or disaster recovery, it can take only minutes to spin up a new virtual server from a snapshot that was taken of the live network. The snapshot is the state of the server at a specific point in time that can be restored quickly, in most cases.

When it comes to network disaster recovery I like to think of the P-Cubed, or P^3, method.

- Plan – Plan your strategy.
- Prepare – Prepare for the worst.
- Protect – Protect your data and infrastructure.

Restoring Data

With data backup, we must also think about the *data restore process*. Restoration is the part of the data backup and disaster recovery process that ensures that data and configuration information can be restored or rebuilt if necessary. Because of the required accessibility of network resources and data, the restore process is as important as the backup process. Some would say it is even more important. If a restoration is needed, you may not have immediate access to the servers or the data required for the business to operate normally, or for you to do your job. Depending on the type of business, this could be much more than just an inconvenience.

A few items need to be taken into consideration about the restoration process.

- Why is the data restoration needed?
- Does a user need a few files restored that were deleted or corrupted?
- Is it the result of hardware or device failure, or a technological catastrophic event?
- Is it the result of a natural disaster?

The answers to these questions will help determine how or whether the data recovery process proceeds. You hope the day never comes when you will need to put a disaster recovery plan into action, but you need to be prepared in the event that it happens. The disaster recovery plan will document the data restore process for servers and client devices, as well as how corporate and personal data types will be handled.

Restoring Corporate Data

The restoration of data to a device that is used to access *corporate data* will have high priority. The backup method that is used will determine how the restoration is handled. For networks that rely on local backup technology using media that is stored at a remote location, the latest backup media will need to be retrieved and restored. For networks that use high availability, virtualization, and disk imaging technology, the process will be much quicker than those that use the traditional media restore process. The data restore process may include installing the server or device operating system to get a place to start, and then proceeding with the complete restore process. This will include the required applications, and then the user data files.

Verifying the integrity of the restored data is also important. It is not safe to assume that after the restore is complete, everything will be 100 percent operational. Mobile device users may need to synchronize their mobile devices to the servers located on the remote network to ensure that the data is consistent and current, because the local remote user data may have changed since the restore occurred.

Restoring Personal Data

Restoring *personal data* usually has the lowest priority (that is, unless it is the CEO's device that needs to be restored, of course). All kidding aside, how personal data is handled will usually be specified in the company's policy document. It is important that this procedure be documented so the users have a clear understanding of how the process works, and the priority that it will be given. If policy states an employee of the company should not store personal data on a corporate device anyway, you have to determine whether the company should even be concerned about the user's personal data.

Newer concepts like BYOD technology shine some new light on the topic of how personal data is handled. Some corporations, organizations, school districts, and other places that allow people to bring their own devices will need to evaluate policy closely. In many situations, personal mobile devices such as computers, smartphones, and tablets are now allowed to be used on the corporate or organization business network, and access the available resources. Some school districts and businesses are leveraging the fact that people use their own devices on the network, which parlays into a cost savings because they will not need to provide the user with a device to use. Even so, chances are that the company's or organization's network IT staff will be responsible for personal data as well as work-related user data.

Summary

In this chapter you learned about the importance of backing up data, and explored different backup strategies that can be used to ensure the availability and integrity of corporate data. It's important to have a basic understanding of a service-level agreement (SLA), which is an agreement that specifies the details of a service. We explored some common methods used for backup, including traditional storage media types such as magnetic tape and optical storage. You learned about the different types of media backup methods available, which include a full backup, an incremental backup, and a differential backup. We explored server backup software programs and strategies, as well as the features that are available in many commercial off-the-shelf server backup programs. These features include support for various operating systems, cloud-based backups, disk imaging, and bare metal restoration. We also looked at the importance of client and mobile device backup for corporate and personal data. You learned about common disaster recovery procedures, and the importance of maintaining high availability. Concepts for high availability include redundant or mirrored sites and load balancing. Finally, the concept of bring your own device (BYOD) technology and both corporate and personal data were discussed in terms of backup and restore policy and procedures, and the disaster recovery plan.

Chapter Essentials

Be familiar with the different traditional backup methods. Full, incremental, and differential backup methods are used with backup strategies to store backups to physical media such as magnetic tape, optical storage, and external storage devices.

Understand the different methods used to back up servers. Servers can be backed up using physical storage methods. Be familiar with technology that provides high availability, including redundant or mirrored sites, virtualization, and load balancing of network servers.

Know the common methods used for client and mobile device backup. Understand different methods are available for client backups, including traditional storage media, cloud-based storage, and disk imaging.

Understand the concepts of corporate and personal data backups. Corporate policy will play a role in choosing the backup strategy for corporate and personal data. Be familiar with the implications and storage methods for client device local backups.

Be familiar with disaster recovery principles. Understand the framework of a disaster recovery plan. Know the concepts as they pertain to backup and restore procedures.

Understand the types of disaster recovery locations. Be familiar with the difference between a hot site and a cold site. Know that a hot site provides the quickest restore process, and provides a replica of a data center. A cold site is built from the ground up, and the restore process takes longer.

Understand the concept of high availability. Technology such as site mirroring, load balancing, clustering, and storage area networking (SAN) provide high availability for data access, and will provide access until the failed systems are restored.

Be familiar with the data restoration process. Know the options and steps regarding data restoration, and that it can be accomplished by restoring a complete backed-up image. It can also be performed manually by using traditional media methods, which may require restoring an operating system, then the user applications, and finally the user data.

Chapter 19

Mobile Device Problem Analysis and Troubleshooting

TOPICS COVERED IN THIS CHAPTER:

- ✓ The Troubleshooting Process
- ✓ Understanding Radio Frequency Transmitters and Receivers
- ✓ Troubleshooting Specific Problems
- ✓ Common Over-the-Air Connectivity Problems
- ✓ Troubleshooting Common Security Problems

Wireless networks, like most areas of technology, require support and maintenance. The extent depends on the size and complexity of the network. Troubleshooting problems is a key component in the maintenance, reliability, and operation of a wireless network. A wireless technical support engineer will encounter all the same problems that occur with wired networking as well as additional problems because wireless technology uses radio frequency for communication. In this chapter, you will learn about common problems associated with wireless networking, how to identify problems based on the symptoms, and how to determine whether they are global (affecting many) or isolated (affecting an individual). These problems range in complexity and magnitude and many times are associated with wireless connectivity, radio frequency signal strength, and other network-related issues. Troubleshooting is not something that can be learned overnight or from reading a book. Troubleshooting is an acquired skill that requires many hours of hard work. The steps involved in troubleshooting a wireless networking problem vary and depend on the complexity of the network and the environment. We will explore causes and solutions associated with wireless network connectivity, roaming issues, authentication failures, and network saturation.

The Troubleshooting Process

A key step in troubleshooting networking problems is to identify whether an issue is a global problem or an isolated problem. Global problems often include wired and wireless infrastructure devices and components. A global problem usually involves many client devices or groups of devices or a complete network failure. The following network devices can be related to a global problem:

- Wireless access points
- Wireless bridges
- Wireless LAN controllers
- Wired infrastructure devices

In addition to wireless infrastructure devices such as access points, wired infrastructure components like Ethernet switches or Layer 3 routers can be a potential source of global problems. Keep in mind that wireless client and mobile devices nearly always require a wired infrastructure of some sort in order to pass information between infrastructure devices or outside the wireless network and to a cellular carrier network. The following devices and components can contribute to wired infrastructure problems:

- Ethernet cabling
- Ethernet switches

- Layer 3 routers
- Wide area network (WAN) connectivity

As shown in Figure 19.1, there are many components that can be the source of problems associated with wireless networking.

FIGURE 19.1 Many components, whether wired or wireless, can be the source of or contribute to wireless LAN problems.

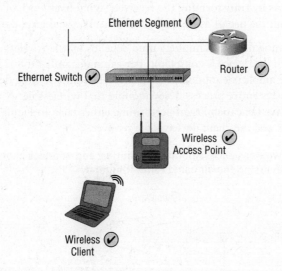

Once it has been determined whether the problem is global or isolated, appropriate steps are necessary to determine the solution. But before we look at common problems associated with wireless networking, it is important to understand a little more detail about how wireless technology functions. This includes understanding the components used to send and receive radio frequency information and is accomplished by two main components:

- Radio frequency transmitter
- Radio frequency receiver

Understanding Radio Frequency Transmitters and Receivers

As discussed in previous chapters, wireless device communications use radio frequency to send and receive data. This is possible because all wireless devices have transmitter and/or receiver capability. Unlike a television or FM radio you may have in your home, both of which are only receivers, all wireless and mobile devices have the capability to act as either a transmitter or a receiver (also known as a *transceiver*). The following explains transmitter and receiver functionality to provide the groundwork necessary to diagnose and resolve wireless communications problems.

Radio Transmitter In wireless communications, a *transmitter* combines binary digital computer data (all of the 1s and 0s) with high-frequency alternating current (AC) signals to prepare the computer data to be sent across the air. This is known as *modulation*. The connected antenna then transforms this signal into radio waves and propagates them through the air. The frequency of the signal depends on the technology in use. With IEEE 802.11 standards–based wireless networking, there are a few select frequency ranges.

Radio Receiver A *receiver* collects the propagated signal from the air using an antenna and reverses the process by transforming the received signal back into an alternating current signal. Through the use of a demodulation process, the digital data is recovered.

The modulation/demodulation technology used depends on the wireless standard or amendment with which the device is compliant. For example, an IEEE 802.11n wireless LAN will use 64 quadrature amplitude modulation (QAM) when transmitting data at 300 Mbps. Figure 19.2 illustrates this entire process. Keep in mind that wireless network devices are transceivers, so they have the capability of performing either task. In Figure 19.2, the access point is the transmitter and the client device is the receiver.

FIGURE 19.2 A wireless access point (transmitter) and wireless client device (receiver) with computer data traversing the air using radio frequency

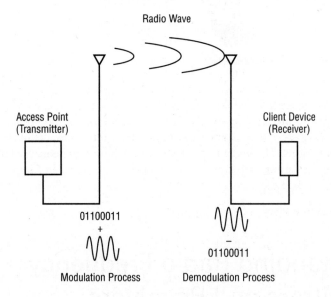

Figure 19.2 sums up the entire process of IEEE 802.11 wireless networking communications discussed throughout this book and is an introduction to understanding the process of troubleshooting.

Steps in the Troubleshooting Process

The following steps are a good rule of thumb to use as part of your troubleshooting process:

1. Establish the cause
2. Test to determine the cause
3. Establish a plan of action
4. Implement the solution

These four steps are typically performed in this order. The components of these steps are described in more detail throughout the remainder of this chapter.

Gathering Information

Gathering information is a key component in troubleshooting any type of problem. This is usually accomplished by asking many questions of those that are experiencing the problems. If you were experiencing a problem with your automobile, you would take it to a repair shop and have it evaluated by a mechanic. The mechanic at the shop would ask you questions about the symptoms you are experiencing. With the answers to these specific questions, the mechanic can evaluate the symptoms to establish the possible cause, perform tests and diagnostics to determine the cause, establish a plan of action, implement the solution, and start the repair process.

Troubleshooting technology issues is no different. The best place to start is with those that are experiencing the problem. Ask them questions to gather enough information and help identify the symptoms. There is no specific method to use; just plain experience will play a role in the entire process.

Chances are that when a user is experiencing a technical problem, they are not the first person to experience it. This is one good reason for accurate and clear documentation of the entire process. If the problems, the findings, and analysis are documented, it will make the next person's job that much easier. Documenting findings, actions and outcomes is discussed later in this chapter.

In addition to gathering information from the user base, there are electronic methods of gathering information, statistics, and details about the components used within the wireless communications area. One such method is the Simple Network Management Protocol (SNMP). You learned about the details and components of SNMP in Chapter 2, "Common Network Protocols and Ports." SNMP, a Layer 7 protocol, will allow a network administrator to use its features to gain valuable knowledge about the health of network devices that can be used for troubleshooting and the associated documentation.

Understanding the Symptoms

Once you have asked the appropriate questions, you should have enough information to start to identify the symptoms of a problem. For example, let's take a look at a wireless LAN problem where a user is complaining that they cannot connect to the wireless network. The network uses WPA2 personal mode to secure a group of client devices. Incorrect security settings can also cause connectivity issues. Although you may get a physical Layer 2 wireless connection to the access point even with incorrect security settings on a wireless client device, you will not get a valid Layer 3 TCP/IP address from the DHCP server. For example, assume an access point is using WPA2 personal mode for security. If a wireless client device has the wrong passphrase or preshared key entered, it will not complete the additional authentication that occurs after the IEEE 802.11 authentication and association process and will not have a valid Layer 3 TCP/IP address. In the Windows operating system, the network adapter would show "Attempting to authenticate" in the network adapter status; it would not be able to complete the Layer 2 connection process and would not receive a valid Layer 3 IP address. Figure 19.3 shows an example of a wireless network adapter with the wrong passphrase set using the Windows 7 client utility.

FIGURE 19.3 The Windows 7 wireless adapter shows "Attempting to authenticate" when a wrong WPA passphrase is entered.

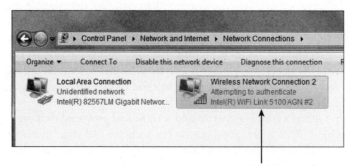

Windows 7 wireless service shows "Attempting to authenticate" because an incorrect passphrase was entered.

Changes to the Environment

Understanding what has changed is a key part to identifying what the problem is. Questioning the user base will sometimes help to discover what changes were made, which may help with a diagnosis. However, it is important to understand that getting this information may not be as easy as it seems. Users may not remember or realize they changed something that had an impact and caused a problem. One of the first questions to ask the user is, "What did you change?" and the response is usually, "I did not do anything," which is no help to anyone. After spending some time troubleshooting, you can sometimes find something the user actually did change that caused a problem. When you point this out the response is usually, "Oh, I forgot about that," which again is not much help to anyone.

Infrastructure changes can also cause problems. These changes are something users have no control over, but they can be affected by them.

Documenting Findings, Actions, and Outcomes

Documentation is a very important component with any part of technology. This holds true with troubleshooting as well. The troubleshooting process and solution should be documented as accurately as possible. When the same problem appears again, the documentation will provide information to whoever is assigned to the task and will hopefully provide assistance to resolve the problem more quickly than if there was no past documentation to go by. There is no reason to reinvent the wheel, as they say.

Help desk software is available from a variety of software manufacturers to help with documenting technical issues, some of which are listed here:

- Freshdesk
- FrontRange Solutions HEAT
- Kayako
- SolarWinds Web Help Desk
- Zendesk

These software packages provide advanced features such as ticket/incident management systems, searchable knowledge bases, and self-service portals, to name a few.

Troubleshooting Specific Problems

There is a wide array of problems that can occur, and there is no way to predict them all. However, there are some problems that commonly come up, and there are some ways to identify them and to address them. The following sections will look at how to troubleshoot some common problems.

Problem: Shortened Battery Life

A user is complaining that the battery in their iPhone does not last as long as they think it should. This may be the result of an aging battery. However, there are options that may help the battery work for longer periods of time between charges:

- Disable unused or battery-hungry applications when not needed.
- Disable services that are seldom used, such as Wi-Fi, Bluetooth, and GPS.

- Disable unnecessary services (such as GPS or Bluetooth) that may have been enabled for individual applications (apps) but are not needed and may affect the battery life.
- Adjust the power save mode to a shorter time.

There are apps available that will tell you what services and features are currently enabled on your phone and are using the battery and how much of it they are using. An app like this would help you identify what you can disable and help conserve battery life. When you replace a battery, I recommend using one that is provided from the original manufacturer. Some third-party replacements have been known to not be as high in quality as the original.

Problem: Synchronization Issues

Keep in mind that many mobile devices do not have the processing power or capability to keep information local (such as email, for example). Therefore, they will need to synchronize with a service or server and rely on a connection using either cellular or Wi-Fi. In some cases users may experience synchronization issues for various reasons. Here are some of the many possible reasons for synchronization issues:

- Cellular and Wi-Fi connections are not available.
- Synchronization software is not available or is misconfigured.
- Services or servers are not available during the attempted synchronization.
- Security or firewall settings are not configured correctly.
- BYOD policy is not enabled or is out-of-date.
- Power adapter or cable is defective.

Using the correct protocols is important. For example, in Chapter 2 you learned that the POP3 protocol will download email messages to the local device. Conversely, with the IMAP protocol, the messages are not downloaded locally to the client device's email software program. Instead the email will remain on the email server for retrieval from any device with an Internet connection and the proper credentials to authenticate to the email server. Many mobile device email or calendaring problems are synchronization-related problems.

Problem: Power Adapter/Supply

You may not expect that the power adapter you use to charge your mobile device may be causing a problem. Keep in mind that adapters plug into a jack on the mobile device. Every time the device is charged (often daily), the connector experiences wear and tear. Here are some of the issues that you may experience with mobile device power adapters:

- Battery is not charging.
- Connector on device or charger is misaligned or defective.

- Cable on power adapter is cracked or broken due to wear.
- Connector on adapter or jack on device is worn from continued use and wear.

> **Real World Scenario**
>
> **USB Connector on Phone Is Defective**
>
> I personally experienced this problem with one of my older smartphones. This particular phone had a cable that plugged into a jack on the phone on one side and on the other side plugged into a standard USB connector, either on a computer or to an adapter that would allow it to be plugged into an AC outlet. When the phone was connected, the battery would charge and the icon on the display would show as such. Apparently the connector plug on the cable for the power adapter had become worn over time and no longer provided a good connection 100 percent of the time. When I plugged the phone into the power adapter, the icon showed as charging. If the phone or connector moved the slightest bit, the connection would be broken and therefore the battery would not charge completely, something I did not immediately notice. I was puzzled as to why my battery did not last nearly as long as it should have even though it showed as charging when it was plugged into a power source. Assuming the battery was getting weak with age, I replaced it. However, the problem persisted. Then I realized the problem was the jack on the power adapter and not the battery in the phone. In the end I replaced the power adapter cable, which solved the problem.

Problem: Password-Related Issues

Passwords are a very important part of technology because they help provide a level of security to ensure that access to resources is available only to the individual with the appropriate credentials. You saw in Chapter 16, "Device Authentication and Data Encryption," that passwords provide a type of access control for mobile and other networking devices. Theoretically, corporate policy will help define requirements of password usage, but we still have our share of issues with passwords. Here are some of the reasons for many troubleshooting issues related to user accounts and passwords.

- Password complexity, such as length, mixed case, and special characters
- Password expiration
- Account restrictions, such as lockout and time of day
- Users forgetting passwords
- Corporate policy

Many password technical support issues are the result of a user typing in an incorrect password or forgetting their password. These may indirectly cause other issues, such as an account lockout. If a user enters a password and they get an error, usually they will enter it again. Depending on the corporate security policy, an account lockout may occur if the retry attempts hit the threshold.

Problem: Device Crash

There are many possible reasons for a device crash:
- Misbehaving application
- Corrupt software or device driver
- Potential viruses or malware
- Battery failure

A device crash can be a big troubleshooting challenge, especially if it is an intermittent problem. This is one place where logging is useful. Many times a log file or crash report will show the last thing that happened prior to the device crash. Analyzing that information may point you in the direction of the issue, easing the troubleshooting process to an extent.

Problem: Power Outage

Because laptop computers, smartphones, tablets, and other mobile devices use batteries to operate, the biggest concern with a power outage is the infrastructure rather than the mobile device. If the device has some battery life and does not need to be recharged, it should operate as it normally does. However, infrastructure services such as the following may not be available:
- Internet access
- Wi-Fi and network access
- Cellular access (if the location is relying on a distributed antenna system, or DAS)

Although a power outage could be more of an infrastructure issue, it can still have a negative effect on productivity if employees cannot get access to network resources or the Internet. An uninterruptible power supply (UPS), which is basically a battery backup, will allow a certain level of infrastructure access even in the event of a power outage. In addition to providing power to the devices, it will protect the system from a crash if an unexpected power outage occurs and provide the opportunity for a graceful shutdown. A UPS will also provide protection for the connected devices against other voltage- or current-related issues, such as voltage spikes or over voltages.

Problem: Missing Applications

You may experience what appears to be missing apps from your smartphone, tablet, or other mobile device. This may occur for several reasons.
- Logged into a different account
- Operating system upgrade

- Disabled app
- Device firmware issue
- Uninstalled, deleted, or erased app

This is a common problem that could be caused by the reasons listed and is usually resolved fairly easily, although it may be time consuming to troubleshoot.

Problem: Email Issues

Trouble with email can be related to several different areas of technology and also depends on if the mobile client device is using a web-based service or an email client program based on Post Office Protocol (POP3). One of the most common problems with email is authentication errors, which can include the email address logon username or password. The following problems can also occur:

- No or poor Wi-Fi signal
- Incorrect email settings, such as POP3 or SMTP server settings
- Incorrect TCP port settings

Email is one area where users will experience many problems. One reason is that it is one of the most commonly used applications. In addition, there are many settings that can either be misconfigured or can change, which will require user interaction. In addition to the potential client-side issues, the server side will play a role in how well email operates. In some cases, email services may be unavailable due to network issues not related to the email programs, ports, or processes.

Problem: Profile Authentication and Authorization Issues

The term *authentication* has different meanings depending on the context in which it is used. IEEE 802.11 open system authentication is performed by the 802.11 protocol and requires no user interaction. User-based authentication, which includes logging into a computer or network, is another form, and software applications such as databases and email clients also require authentication. The authorization process is what a user can do once they have successfully logged into the system or service. Troubleshooting profile authentication and authorization issues will require enough background information to effectively diagnose the following problems:

- Incorrect credentials on the client side
- Insufficient permissions on the network side
- Wrong profile selected

Common Over-the-Air Connectivity Problems

I explained earlier in this chapter that a wireless technical support engineer will encounter all the same problems that occur with wired networking as well as additional problems because wireless technology uses radio frequency for communication and to exchange information. Unlike wired communications, wireless uses an unbounded medium, the air, to send and receive radio frequency signals. With this come many potential problems, which are explained in the following sections.

- Wireless latency
- No cellular signal
- No wireless network connectivity
- Roaming issues
- Cellular device activation issues
- Access point name issues
- Wireless network saturation

LAN Technology vs Cellular Technology

Many connectivity problems pertain to both wireless LAN technology and cellular technology. Even though both are wireless, they use different methods for communication and should not be confused. Depending on the problem, you may need to use a different troubleshooting methodology for each. Here are some considerations and differences between the two technologies:

- Wireless LAN technology is usually part of a local area network, which has a much shorter range than cellular networks.

- Wireless LAN access is limited by the signal strength and coverage area of the infrastructure devices such as access points.

- Cellular users pay a fee to connect and gain access to voice and data services, whereas wireless LAN technology uses unlicensed frequency bands.

- Wireless LAN users may have more control over their network and the device's configuration settings.

- Short-range wireless LAN devices at home allow users to control their own security. Cellular, on the other hand, has a longer range and infrastructure security is provided by the wireless carrier.

- A user who connects to a wireless network at their place of business with a device that is configured to use the corporate security policy will likely use a higher level of security than a home user would use.
- Troubleshooting a wireless LAN is generally performed locally by an administrator or the user. Troubleshooting cellular problems may involve calling the technical support department of the wireless carrier or the service provider for assistance.

Wireless Latency

Network latency is a delay and a measured time value for a packet or frame to get from the source device to a destination device. The application in use on the wireless LAN will determine the effect this will have on communications. In wireless communications, latency will have a big impact on both voice and video communications. Quality of service (QoS) will help lessen the potential impact of latency in wireless LAN communications. Common TCP/IP utilities can be used to measure network latency. The tracert.exe utility and the ping.exe utility are both available in the Microsoft Windows operating system and other operating systems. The proper use and syntax of the commands for these utilities is well documented and available from a simple Internet search and also from help screens within the actual programs. Apps for measuring latency are also available for mobile device operating systems, including Apple or Android devices. Using services across a wired network such as an Internet connection may contribute to the latency that a user experiences. For example, the amount of latency will vary based on the wired communication technology in use, such as, for example, cable modems or DSL. The acceptable levels of latency will vary based on the technology used, such as voice and video applications.

No Cellular Signal

A smartphone or mobile device not showing cellular signal may not necessarily mean there is a problem. The lack of a discernible signal could be because the device is in an area that has a weak signal or no coverage. If the problem occurs indoors, chances are the cellular signal is not strong enough to provide the necessary coverage. In Chapter 3, "Radio Frequency and Antenna Technology Fundamentals," you learned that bringing a cellular signal in from the outside and allowing it to be at an acceptable level in a structure can be accomplished using technology known as a distributed antenna system (DAS). This is becoming more common because the multifunction mobile devices we use need high-quality cellular signals for voice and cellular data communications and Wi-Fi signals for wireless LAN communications in order to operate as designed and to provide the best user experience possible for both technologies. Getting a better-quality cellular signal may also be accomplished by using a cellular repeater. The concept is the same as with DAS, which is bringing a signal into a structure from the outside; however, it is not as complex and is much less expensive because it is similar to installing an IEEE 802.11 access point.

No Wireless Network Connectivity

Several components of a network can cause connectivity issues with wireless networking. Keep in mind that wireless networks are not bound by a physical medium such as Ethernet cable; therefore, issues other than wiring may result in either no connectivity or intermittent connection problems. Wireless LAN technology operates at Layer 1 and Layer 2 of the OSI model and is responsible for getting data to other wireless LAN devices and to the wired distribution system. The "wireless" problem may be only one part of the troubleshooting process. Troubleshooting end-to-end wireless LAN problems that include both wireless and wired networks may be another process altogether. Therefore, knowledge of troubleshooting the wired network will also be beneficial to the network administrator.

In cellular communications, the wireless connection is from the end user device such as a smartphone to the nearest cell tower, and you are limited on what you can do with regard to troubleshooting this type of connection. One example is that you may have a strong enough signal to send a text message but not a strong enough signal to make a voice call. It is important to understand the limitations with cellular communications you may have to deal with as part of the troubleshooting process. One example of such limitations is that chances are you will not have access to the cellular infrastructure devices. With wireless LAN technology, an engineer would have full access to all equipment.

Let's review some of the OSI model information discussed in Chapter 1, "Computer Network Types, Topologies, and the OSI Model." The Physical layer provides the medium for connectivity (the air) in a wireless network and is used to carry radio frequency information that contains the computer data. This layer also includes components such as network adapter cards, which provide an interface to the wireless computer or other wireless device. Layer 1 connectivity provides the capability to transfer information that is sent between devices. Figure 19.4 illustrates the Physical layer and wireless connectivity.

FIGURE 19.4 The lower two layers of the OSI model are responsible for the operation of wireless networks. The Physical layer provides a connection between devices.

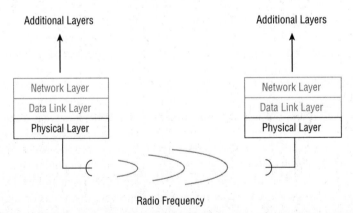

IEEE 802.11 wireless networks require Physical layer connectivity in order to provide successful communications. There are some potential connectivity issues associated with client-side wireless devices. In many cases, no connectivity on the client side is an isolated

issue that will not affect a large number of devices or users. The following problems can cause client-side lack of connectivity:

- Disabled radio or wireless network adapter
- Misconfigured wireless client utility
- Microsoft Windows AutoConfig service not running or not configured
- Protective supplicants (wireless client device side) that can disable the radio in response to specific policy violations

A disabled radio or client adapter in a wireless client device can cause a lack of connectivity. Many devices, such as notebook computers, have physical switches or a combination of keys that can disable a radio. In many cases, a user does not even realize they turned off the physical switch for the wireless LAN radio. A disabled radio cannot provide RF communication between the client device and the wireless infrastructure, such as an access point. With the Microsoft Windows operating system, there are built-in diagnostics and troubleshooting that may help with some wireless LAN connectivity issues. Figure 19.5 shows the wireless connection dialog box as "Not connected." If you click the Troubleshoot link, Windows 7 will go through a series of steps to try to resolve the issue. In this case, the Windows wireless service was not running, and Microsoft was able to determine this and identified the problem.

FIGURE 19.5 Microsoft Windows 7 wireless service not running and resolution using the built-in troubleshooting feature

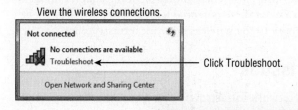

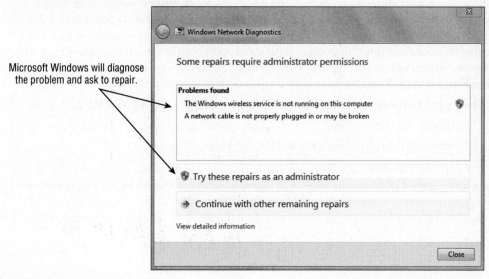

If a client utility is misconfigured—for example, an incorrect SSID is specified—it would cause a lack of connectivity for the wireless client device. Client utilities have the capability to specify the SSID as a parameter. If the client does not see an access point with the specified SSID, it will not be able to connect to the IEEE 802.11 wireless LAN. This feature in a client device is often set by selecting to connect to a preferred SSID only by using a specific profile.

Remember that the service set identifier (SSID) is case sensitive and has a maximum length of 32 characters, or 32 octets, as defined in the IEEE 802.11 standard. Incorrect use of case in the SSID can lead to a lack of connectivity because the SSID will not match. Uppercase and lowercase letters are different ASCII characters. This can also lead to unnecessary help desk calls. Client devices will not connect to the wireless LAN if they have an incorrect SSID specified in the wireless client utility profile. This could be unintentional (accidental) or in some cases intentional, such as for the purpose of getting around a firewall or other SSID restriction.

The Windows built-in wireless client utility is a popular client utility for connecting to IEEE 802.11 wireless networks. The Windows wireless service (Microsoft Windows 7 AutoConfig service) runs as a service or background process in the Microsoft Windows operating systems. If this service is not running or if the client device is not configured correctly to use Windows wireless services, the wireless client device will not be able to connect to the wireless network unless a third-party client utility from the adapter manufacturer is used. Figure 19.6 illustrates some of these potential issues from a wireless client device that may cause a lack of connectivity to the wireless network infrastructure.

Roaming and Transition Issues

You learned in Chapter 6, "Computer Network Infrastructure Devices," that Layer 2 (Data Link layer) transition (roaming) occurs when a computer or other wireless client device moves out of the radio cell of the currently associated access point and connects to a different AP, maintaining Layer 2 connectivity. In a cellular system, this is equivalent to a user moving from one cell tower to another with a disruption of the signal. In a well-designed network, this should be a straightforward process. However, in certain cases it can also be a troublesome area. Roaming issues can be the result of several problems:

- Low signal strength between access points
- Insufficient RF cell overlap
- Incompatible configurations such as SSID or frequency band

It is important to remember that roaming is always decided by the mobile or client device. The criteria used will determine when a device will initiate a roam. This may include roaming aggressiveness or a specific threshold setting on the mobile device. Some wireless LAN manufacturers use proprietary mechanisms that will help with the roaming process. One example is in the single-channel architecture (SCA) model in which all the radio frequency is handled by the controller. In this case, the wireless LAN controller will move the wireless client device transparently as it changes its physical position within the coverage area. In other cases manufacturers may require the client device to roam for load balancing purposes and will direct the device to a different wireless access point.

FIGURE 19.6 Devices and components that make up a wireless LAN showing potential wireless client-side issues

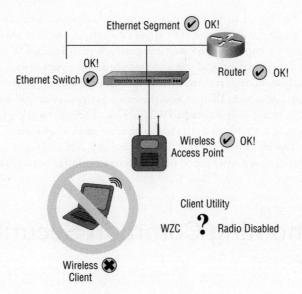

Cellular Device Activation Issues

A cellular device such as a smartphone, tablet, or hotspot will require activation on the wireless carrier's cellular network. The activation of a new or replacement device will allow it to become authorized and usable on the network. Most carriers provide a straightforward activation process, and activation can be accomplished via an activation page by calling into the activation line. If everything is set up with the carrier correctly, the activation should go very smoothly. However, occasionally cellular activation will fail for a variety of reasons. If you have adequate signal, the best option would be to contact the carrier's technical support and have them walk you through the process.

Access Point Name Issues

An access point name (APN) is the name of a gateway between a 3G, 4G, or GPRS cellular mobile network and another computer network, usually the Internet. Do not confuse this with an IEEE 802.11 wireless access point. APNs are fields that must be completed with information to allow a device to have a connection to a wireless carrier's data services. Because cellular communications are carried out via licensed radio frequency bands, access point names are required. Each mobile device that uses the APN, such as a smartphone or a cell phone, needs to be configured with this information. Otherwise, the user may experience connection problems or service problems.

Wireless Network Saturation

With the increasing number of wireless devices (three to five per person), network capacity is an important design consideration. The types of applications that are in use on the network will also have an impact on the wireless network performance. Remember, wireless networks are contention based and use a shared medium. Every device must contend to win the medium to allow them to transmit data. The more devices that are connected to an access point, the more contention there is and the more performance will be degraded.

Network saturation is not only limited to wireless local area networks. Cellular networks also suffer from this problem. Because the number of devices continues to increase, mobile device data traffic is expected to grow at a rate of 66 percent from 2012 to 2017. Capacity planning is just as much of a concern in cellular networks as it is in wireless LANs.

Troubleshooting Common Security Problems

The following are some common security-related issues that pertain to mobile and wireless communications:

- Authentication failures
- Non-expiring passwords
- Expired passwords
- Expired certificates
- Misconfigured firewalls
- False negatives
- False positives
- Content filtering misconfigurations

Authentication Failures

Authentication failures could be the result of a variety of issues. It could be simply that the user mistyped or forgot their password or the result of a potential intrusion. In addition to password issues, certificates can cause authentication failures. In some cases authentication failures could cause an alert to be sent to an administrator from an intrusion detection system (IDS) if the set threshold is reached. Because this type of problem is wide ranging, you should gather as much information as possible from the user to help you effectively narrow down the source of the problem.

Non-expiring Passwords

A password that is set to not expire would be specified in a network security policy. A legitimate reason to allow non-expiring passwords is also important. If a password does not expire, that may pose a security risk because it increases the possibility of an intruder gaining access to a network by discovering a password. This is a practice some network administrators may implement on their own account because they have the ability to do so. If the administrator's account is compromised, then the person that has gained access for the most part has the "keys to the kingdom."

There are some reasons you may want to implement this setting with specific user accounts. One example is an application that needs to log into this system in order to function. In some cases the application requires a network user to run as a service. If the password expired, the user created for the service would not be able to log on; therefore the application would not function correctly.

Expired Passwords

Corporate security policy will determine the company password expiration interval. Requiring passwords to be renewed at specific intervals will enhance network security. Common policy will require users to change passwords every 30 to 90 days. Depending on the organization, this interval could be shorter. An expired password may not allow a user to log in to a system or mobile device.

Expired Certificates

Recall from Chapter 16 that a digital certificate will ensure that only valid authorized users can gain access to the corporate network infrastructure. Client-side digital certificates are used to validate a user's identity and help protect network resources. Keep in mind that in a full public key infrastructure (PKI), servers as well as client devices require certificates. Certificates have an expiration date, and they must be renewed prior to the expiration date. An expired certificate on a web server does not necessarily mean that the website cannot be trusted. There is a good possibility that whoever the certificate was originally issued to simply neglected to renew before the expiration date and that may be an oversight. However, this does reduce the trust of the user who is accessing the site. A user with an expired certificate may not continue to do business with an organization, which can result in lost revenue. From the client side, an expired certificate will cause authentication errors resulting in lost productivity.

Misconfigured Firewalls

Misconfigured firewalls can pose a big security threat and are the cause of most security breaches. Remember that a firewall can consist of either hardware or software that will analyze the contents of a packet and potentially restrict packets from entering or leaving a network. In addition to allowing for a security breach, misconfigured firewalls can cause unnecessary technical support issues, such as a user not able to access the required information because of a blocked port.

False Negatives

When you should be alerted of an event of some type but are not, this is considered a false negative. Here we will look at a false negative in relation to wireless LAN technology. Say you are using a wireless intrusion prevention system (WIPS) solution to monitor and mitigate potential threats to your wireless network. A potential Layer 2 denial of service attack (deauthentication, for example) has occurred and you were not alerted. One reason this could happen is because the threshold setting to alert after a specific number of deauthentication frames were detected may have been set to an incorrect value. Another possibility is that the WIPS software did not have accurate signatures for this type of attack. An experienced intruder may have the knowledge and skill set to be able to elude certain systems and get around thresholds and signatures, which will allow for the false negatives to occur. This can easily become a complex issue. Systems that provide alerts such as WIPS technology must be configured correctly to ensure that they operate as intended.

False Positives

A false positive is the result of something that is identified as a problem even though it is legitimate. One example of this is antispam software. If a user opens an email message that is considered spam and clicks on a link within the message, there is a possibility the user can be driven to an undesirable website and potentially become infected with a virus or malware. To protect against these types of attacks, users and network administrators can use antispam software. The intention of this software is to identify potentially harmful spam messages and protect the user. A false positive is when a valid or legitimate message is inadvertently marked as spam. These messages may simply be marked as such or may be rejected by the email server. In the WIPS example described earlier, there is a possibility that a WIPS may send an alert identifying an event as an issue or problem even though it is normal operation. Accurate configuration, monitoring, and baselines will help lessen the possibility of false positives.

Content Filtering Misconfigurations

The purpose of content filtering is to restrict users from accessing certain web content, including email access from the local network. This is typically accomplished using a software package designed for this specific purpose. Content filtering will provide more granular control over what users are accessing from the corporate network. The ability to access certain types of content could cause a decrease in employee productivity or may cause legal issues for the company. Content filtering is also used on home networks to allow parents to control what is being accessed from their own network. Misconfigurations in this arena could allow users to circumvent corporate policy and cause other issues for an organization.

Summary

In this chapter you learned that troubleshooting and maintenance are important parts of a successful network deployment. A key step in troubleshooting networking problems is to identify whether an issue is a global problem or an isolated (individual) problem. Global problems often include wired and wireless infrastructure components, many client devices, or a complete network failure. An isolated problem will be confined to a single device or user. Understanding this nature of the problem will ease the troubleshooting process.

You also learned that wireless communications uses transceivers to communicate and the unbounded medium must be taken into consideration as a potential source of a problem. Understanding symptoms and changes to the environment are key components to the troubleshooting process, and documentation is a must.

We also explored common problems you come across within the mobile device environment, including battery life, synchronization issues, and problems with power adapters, to name a few. We explored common over-the-air problems including latency, poor or no signals with both cellular and wireless LAN technology, and roaming issues. Finally, you saw some common security-related problems, including authentication failures, password-related issues, and firewall misconfigurations.

Chapter Essentials

Understand the difference between a global problem and an isolated problem. The troubleshooting process can be simplified by first determining if a problem is related to a large number of devices (global) or an isolated (individual) user.

Understand the importance of transceiver communications. Wireless communications, including cellular and wireless networking, use radios to send information over an unbounded medium, which can contribute to mobile device problems. Understand that these devices can transmit or receive data and are known as transceivers.

Understand the four steps of troubleshooting wireless related problems. Know that you must (1) establish the cause of the problem, (2) test to determine the cause, (3) establish a plan of action, and (4) implement the solution. Be familiar with the order in which these steps are performed.

Know that gathering information, understanding symptoms, and recognizing changes to the environment are key components of troubleshooting. Be familiar with these stages of the troubleshooting process and the information obtained will provide the ability to troubleshoot a variety of wireless-related communications problems.

Understand the common problem areas associated with mobile communications. Be familiar with problems related to mobile devices, including problems with battery life, synchronization, power adapters, and others. Common over-the-air problems and common security-related problems include authentication failures and password-related issues.

Appendix

The CompTIA Mobility+ Certification Exam

TOPICS COVERED IN THIS APPENDIX:

- ✓ Preparing for the CompTIA Mobility+ Exam
- ✓ Taking the Exam
- ✓ Reviewing Exam Objectives

This book is intended to provide an introduction to the exciting and emerging world of wireless and mobile technology. Wireless technologies continue to expand at a phenomenal pace, with constant improvements in speed, performance, reliability, and security. As I mentioned in the introduction, this book covers the capabilities of mobile devices and the features of over-the-air technologies, including IEEE 802.11 standards based wireless networking in addition to cellular and other wireless technologies such as Bluetooth and WiMAX. It also discusses the concepts of how to deploy, integrate, support, and manage a mobile environment, ensuring that proper security measures are implemented for devices and platforms while usability is maintained. In addition, you read about topics such as topologies and the OSI model, common network protocols and ports, wireless site surveys, backup and restore, and troubleshooting. One way to demonstrate your ability in these areas is to attain the CompTIA Mobility+ certification.

To attain this certification, you will need to pass the CompTIA Mobility+ exam (MB0-001). The Mobility+ exam is the only mobility technology certification offered by CompTIA. It covers all you need to know to deploy, integrate, support, and manage a mobile environment. The exam certifies that the candidate has the knowledge and skills to understand and research the capabilities of various mobile devices and aspects of over-the-air technologies. Professionals who have passed the Mobility+ exam make sure proper security measures are maintained for devices and platforms to mitigate risks and threats while ensuring usability. The following skills are recommended for the CompTIA Mobility+ candidate:

- CompTIA Network+ or equivalent working knowledge
- At least 18 months of work experience in the administration of mobile devices in the enterprise

This book covers all the exam objectives, plus more, which will enable you to prepare for the certification exam and be successful in your career as a mobile (device management) administrator, mobility engineer, and wireless mobility architect.

Preparing for the CompTIA Mobility+ Exam

Preparation is a key part to being successful. Now that you have read this book and you are ready to take the exam, it is a good idea to consider how the exam will look and feel so that nothing takes you by surprise.

It is highly recommended that you visit the CompTIA website before taking the exam because details about the exam can change. The CompTIA website for this exam is

http://certification.comptia.org/getcertified/certifications/mobilityplus.aspx

The first thing to be aware of is that this certification exam is vendor neutral. The information you have learned in this book is applicable across most, if not all, manufacturers and vendors of wireless infrastructure devices and mobile device technology products.

The exam is 1 hour and 30 minutes long and consists of 100 multiple-choice questions. There are no hands-on exercises or lab simulations contained within the exam questions. The passing score is 720 (on a scale of 100–900) and the exam code is MB0-001.

The exam is broken into the domains of expertise shown in Table A.1.

TABLE A.1 Exam Objective Domain Categories

Domain	% of Examination
1.0 Over-the-Air Technologies	13%
2.0 Network Infrastructure	15%
3.0 Mobile Device Management	28%
4.0 Security	20%
5.0 Troubleshooting	24%

After reviewing the objectives table and the exam objectives, if you feel you are not confident in any particular area, do some more reading and research so that you can go into the exam feeling as prepared as possible.

Taking the Exam

The CompTIA Mobility+ exam can be taken only at Pearson VUE testing centers. Pearson VUE is a leading global training company with more than 5,100 test centers currently in 175 countries.

To schedule your exam, visit www.pearsonvue.com/comptia and follow the online registration procedure. This website enables you to locate your nearest testing center and schedule and pay for the exam. To complete the procedure, you will need a Pearson VUE account (which can be set up on the website), and an electronic payment method or a valid exam voucher. The list price of the exam is $261.00.

After your exam is scheduled, you will need to understand a few important items:

- You will receive instructions regarding appointment and cancellation procedures, ID requirements, and information about the testing center location. In addition, you will receive a registration and payment confirmation email. Exams can be scheduled weeks in advance or, in some cases, even the same day.
- Arrive at least 15 minutes earlier than your scheduled exam time and do not be late for your appointment.
- Remember to bring the required identification that is listed in the email you received when you registered for the exam. You cannot take the exam without providing required identification documentation.

Once you know the location of the testing center and you have the required identification, you are ready for the day of your exam.

On the day of your exam, arrive at the testing center early. Be prepared to surrender all personal electronic devices and other items that are not allowed in the testing room. These must be stored in a secure locker during the exam process.

Reviewing Exam Objectives

Table A.2 provides objective mappings for the MB0-001 exam. You can also find coverage of exam objectives on the book's companion website, www.sybex.com/go/mobilityplus.

TABLE A.2 Exam MB0-001 CompTIA Mobility+ Objectives Map

Objective	Corresponding Book Chapter
1.0 Over-the-Air Technologies	

1.1 Compare and contrast different cellular technologies.
- CDMA
- TDMA
- GSM
  - Edge
  - GPRS
- WiMAX
- UMTS
- CSD
- EVDO
	Chapter 7: Cellular Communication Technology

TABLE A.2 Exam MB0-001 CompTIA Mobility+ Objectives Map *(continued)*

Objective	Corresponding Book Chapter
<ul><li>HSPA</li><li>HSPA+</li><li>LTE</li><li>Roaming & switching between network types</li></ul>	Chapter 7: Cellular Communication Technology
1.2 Given a scenario, configure and implement WiFi client technologies using appropriate options.<ul><li>Bluetooth</li><li>PAN</li><li>802.11a, b, g, n, ac<ul><li>Relevant operating frequencies and channels</li></ul></li><li>SSID</li><li>Broadcast/hidden system</li><li>Authentication methods</li><li>Portable hotspots</li></ul>	Chapter 1: Computer Network Types, Topologies, and the OSI Model Chapter 4: Standards and Certifications for Wireless Technology Chapter 5: IEEE 802.11 Terminology and Technology Chapter 12: Implementation of Mobile Device Technology
1.3 Compare and contrast RF principles and their functionality.<ul><li>RF characteristics<ul><li>Frequencies</li><li>Modulation</li><li>Bandwidth</li><li>Wavelength</li><li>Amplitude</li><li>Phase</li></ul></li><li>Propagation theory</li><li>Absorption</li><li>Refraction</li><li>Reflection</li><li>Attenuation</li><li>Interference</li><li>Antennas</li><li>Omni-directional</li></ul>	Chapter 3: Radio Frequency and Antenna Technology Fundamentals

TABLE A.2 Exam MB0-001 CompTIA Mobility+ Objectives Map *(continued)*

Objective	Corresponding Book Chapter
▪ Semi-directional ▪ Bi-directional ▪ YAGI ▪ Parabolic dish ▪ Faraday cage	Chapter 3: Radio Frequency and Antenna Technology Fundamentals
1.4 Interpret site survey to ensure over the air communication. ▪ Capacity ▪ Coverage ▪ Signal strength ▪ Receive Signal Strength Indicator ▪ Spectrum analysis ▪ Frequency analysis ▪ Site survey documentation / site map ▪ Wireless vs. cellular site survey ▪ Post-site survey	Chapter 5: IEEE 802.11 Terminology and Technology Chapter 8: Site Survey, Capacity Planning, and Wireless Design
2.0 Network Infrastructure	

2.1 Compare and contrast physical and logical infrastructure technologies and protocols.
▪ Topologies
  ▪ Mesh
  ▪ Point-to-point
  ▪ Point-to-multipoint
  ▪ Adhoc
▪ Firewall settings
  ▪ Port configuration
  ▪ Protocols
  ▪ Filtering
  ▪ DMZ
▪ Devices
  ▪ Gateways
	Chapter 1: Computer Network Types, Topologies, and the OSI Model
	Chapter 6: Computer Network Infrastructure Devices
	Chapter 9: Understanding Network Traffic Flow and Control
	Chapter 17: Security Requirements, Monitoring, and Reporting

TABLE A.2 Exam MB0-001 CompTIA Mobility+ Objectives Map *(continued)*

Objective	Corresponding Book Chapter
<ul><li>Proxies</li><li>VPN concentrator</li><li>Autonomous access points</li><li>Wireless LAN</li><li>Controller</li><li>Lightweight AP</li></ul>	Chapter 1: Computer Network Types, Topologies, and the OSI Model Chapter 6: Computer Network Infrastructure Devices Chapter 9: Understanding Network Traffic Flow and Control Chapter 17: Security Requirements, Monitoring, and Reporting

- Services and settings
    - ActiveSync
    - Dynamic VLAN
    - Subnetting

2.2 Explain the technologies used for traversing wireless to wired networks.<ul><li>Bandwidth and user limitations<ul><li>Backhauling traffic</li><li>QoS</li><li>Traffic shaping</li></ul></li><li>Hardware differences</li><li>Traffic routing</li><li>IP addressing<ul><li>TCP</li><li>UDP</li><li>NAT</li><li>DNS</li><li>DHCP</li></ul></li><li>MAC address</li><li>SNMP</li><li>ICMP</li><li>PoE for APs to switches</li></ul>	Chapter 2: Common Network Protocols and Ports Chapter 6: Computer Network Infrastructure Devices Chapter 9: Understanding Network Traffic Flow and Control
2.3 Explain the layers of the OSI model.<ul><li>Layer 1 – Physical</li><li>Layer 2 – Data Link</li><li>Layer 3 – Network</li></ul>	Chapter 1: Computer Network Types, Topologies, and the OSI Model

TABLE A.2 Exam MB0-001 CompTIA Mobility+ Objectives Map *(continued)*

Objective	Corresponding Book Chapter
- Layer 4 – Transport - Layer 5 – Session - Layer 6 – Presentation - Layer 7 – Application	Chapter 1: Computer Network Types, Topologies, and the OSI Model
2.4 Explain disaster recovery principles and how it affects mobile devices. - Server backups - Device backups - Directory services server - Frequency of backups - High availability - DR locations	Chapter 18: Data Backup, Restore, and Disaster Recovery
2.5 Compare and contrast common network ports and protocols for mobile devices. - 20/21 – FTP - 22 – SFTP - 23 – Telnet - 25 – SMTP - 53 – DNS - 80 – HTTP - 110 – POP3 - 135 – MAPI - 143 – IMAP - 389 – LDAP/AD - 443 – SSL - 465 – SSMTP - 587 – Alternate SMTP - 990 – ftps - 993 – IMAP over SSL - 2175 – Airsync - 2195 – APNS - 2196 – Feedback	Chapter 2: Common Network Protocols and Ports

TABLE A.2 Exam MB0-001 CompTIA Mobility+ Objectives Map *(continued)*

Objective	Corresponding Book Chapter
- 3389 – RDP - 4101 – SRP - 5223 – Jabber - 5228-5230 – GCM	Chapter 2: Common Network Protocols and Ports

3.0 Mobile Device Management

Objective	Corresponding Book Chapter

3.1 Explain policy required to certify device capabilities.
- Adherence to IT policies and security policies
    - Balance security with usability
- Differences between vendor default applications
- OS modifications and customization
    - OS vendor
    - OEM
    - Telecommunication vendor
- Backup, Restore and Recovery policies
	Chapter 11: Mobile Device Policy, Profiles, and Configuration

3.2 Compare and contrast mobility solutions to enterprise requirements.
- Mobile Device Management
    - Password strength
    - Remote wipe
    - Remote lock/unlock
    - Application store
- Mobile application management
    - Application store
- Pushing content
- Device platform support
- Infrastructure support
- On-premise vs. SaaS
- Administrative permissions
- Multi-instance
	Chapter 10: Introduction to Mobile Device Management
	Chapter 13: Mobile Device Operation and Management Concepts

TABLE A.2 Exam MB0-001 CompTIA Mobility+ Objectives Map *(continued)*

Objective	Corresponding Book Chapter
■ High availability ■ Device groupings ■ Location-based services 　■ Geo-location 　■ Geo-fencing ■ Monitoring and reporting capabilities and features ■ Interoperability with other products/devices ■ Telecommunication expense management ■ Self-service portal ■ Captive portal	Chapter 10: Introduction to Mobile Device Management Chapter 13: Mobile Device Operation and Management Concepts
3.3 Install and configure mobile solutions based on given requirements. ■ Liaise with appropriate personnel to ensure infrastructure can accept solutions ■ Profile creation ■ Directory services setup ■ Initial certificate issuance ■ EULA ■ Sandboxing ■ Containerization ■ Group profiles based on given requirements 　■ Corporate-owned 　■ BYOD 　■ Executive 　■ Management 　■ Consultant 　■ B2B ■ Initiate pilot, testing and evaluation ■ Create and update documentation	Chapter 11: Mobile Device Policy, Profiles, and Configuration Chapter 12: Implementation of Mobile Device Technology Chapter 17: Security Requirements, Monitoring, and Reporting

TABLE A.2 Exam MB0-001 CompTIA Mobility+ Objectives Map *(continued)*

Objective	Corresponding Book Chapter
• Report feedback post-pilot • SDLC • Approve, train and launch	Chapter 11: Mobile Device Policy, Profiles, and Configuration Chapter 12: Implementation of Mobile Device Technology Chapter 17: Security Requirements, Monitoring, and Reporting

3.4 Implement mobile device on-boarding and off-boarding procedures.
• Device activation on cellular networks
• Mobile hardware that facilitates OTA access
  • Wireless cards, cellular cards, SD cards
• On-boarding and provision process
  • Manual
  • Self-service
  • Batch
  • Remote
  • IMEI or ICCID
  • Device enrollment (SCEP)
  • Profile installations
• Off-boarding and de-provisioning
  • Employee terminations
  • Migrations
  • Applications
  • Content
  • Recycle
  • Proper asset disposal
  • Deactivation
	Chapter 12: Implementation of Mobile Device Technology

3.5 Implement mobile device operations and management procedures.
• Centralized content and application distribution and content management system
• Distribution methods
  • Server-based
  • Content updates/changes
  • Application changes
	Chapter 13: Mobile Device Operation and Management Concepts
	Chapter 14: Mobile Device Technology Advancements, Requirements, and Application Configuration
	Chapter 17: Security Requirements, Monitoring, and Reporting

TABLE A.2 Exam MB0-001 CompTIA Mobility+ Objectives Map *(continued)*

Objective	Corresponding Book Chapter
▪ Permissions	Chapter 13: Mobile Device Operation and Management Concepts
▪ Deployment best practices	Chapter 14: Mobile Device Technology Advancements, Requirements, and Application Configuration
▪ Number of devices	
▪ Number of users	Chapter 17: Security Requirements, Monitoring, and Reporting
▪ Remote capabilities	
▪ Lock/unlock	
▪ Remote wipe	
▪ Remote control	
▪ Location services	
▪ Reporting	
▪ Lifecycle operations	
▪ Certificate expiration/renewal	
▪ Updates	
▪ Upgrades	
▪ Patches	
▪ Change management	
▪ End of life	
▪ OSs	
▪ Devices	
▪ Applications	
3.6 Execute best practice for mobile device backup, data recovery and data segregation.	Chapter 18: Data Backup, Restore, and Disaster Recovery
▪ Device backup for corporate data to corporate server	
▪ Device backup of personal data to vendor/third party server	
▪ Backup to local device: internal storage, SD card, SIM	
▪ Data recovery	
▪ Testing backups	
▪ Restoring corporate data	
▪ Restoring personal data	

TABLE A.2 Exam MB0-001 CompTIA Mobility+ Objectives Map *(continued)*

Objective	Corresponding Book Chapter
3.7 Use best practices to maintain awareness of new technologies including changes that affect mobile devices.<ul><li>OS vendors</li><li>OEMs (hardware)</li><li>Telecommunication vendors</li><li>Third party application vendors</li><li>New risks and threats</li></ul>	Chapter 14: Mobile Device Technology Advancements, Requirements, and Application Configuration
3.8 Configure and deploy mobile applications and associated technologies.<ul><li>Messaging standards<ul><li>MAPI</li><li>IMAP</li><li>POP</li><li>SMTP</li></ul></li><li>Vendor proxy and gateway server settings</li><li>Information traffic topology<ul><li>Third party NOC vs. on-premise vs. hosted</li></ul></li><li>Push notification technologies<ul><li>APNS</li><li>GCM</li><li>ActiveSync</li></ul></li><li>In-house application requirements<ul><li>App publishing</li><li>Platforms</li><li>Vendor requirements</li><li>Certificates</li><li>Data communication</li></ul></li><li>Types of mobile applications<ul><li>Native app</li><li>Web app</li><li>Hybrid app</li></ul></li></ul>	Chapter 14: Mobile Device Technology Advancements, Requirements, and Application Configuration

TABLE A.2 Exam MB0-001 CompTIA Mobility+ Objectives Map *(continued)*

Objective	Corresponding Book Chapter
4.0 Security	

4.1 Identify various encryption methods for securing mobile environments.
- Data in-transit
  - IPSEC
  - VPN
  - SSL
  - HTTPS
  - WPA/TKIP
  - WPA2
  - TLS
  - SRTP
  - RSA
  - WEP
  - SSH
  - RC4
  - CCMP
  - EAP methods
- Data at rest
  - AES
  - DES
  - 3DES
  - Two-Fish
  - ECC
- Full disk encryption
- Block level encryption
- File level encryption
- Folder level encryption
- Removable media encryption
	Chapter 16: Device Authentication and Data Encryption

TABLE A.2 Exam MB0-001 CompTIA Mobility+ Objectives Map *(continued)*

Objective	Corresponding Book Chapter
4.2 Configure access control on the mobile device using best practices.<ul><li>Authentication concepts</li><li>Multifactor<ul><li>Biometric</li><li>Credentials</li><li>Tokens</li><li>Pin</li></ul></li><li>Device access</li><li>Wireless networks<ul><li>Enterprise vs. personal</li></ul></li><li>Application access</li><li>PKI concepts</li><li>Certificate management</li><li>Software-based container access and data segregation</li></ul>	Chapter 16: Device Authentication and Data Encryption
4.3 Explain monitoring and reporting techniques to address security requirements.<ul><li>Device compliance and report audit information</li><li>Third party device monitoring applications (SIEM)</li><li>Monitor appropriate logs pertaining to mobile device activity/traffic</li></ul>	Chapter 17: Security Requirements, Monitoring, and Reporting
4.4 Explain risks, threats and mitigation strategies affecting the mobile ecosystem.<ul><li>Wireless risks<ul><li>Rogue access points</li><li>DoS</li><li>Tower spoofing</li><li>Jamming</li><li>Warpathing</li><li>Man-in-the-middle</li><li>Weak keys</li></ul></li></ul>	Chapter 15: Mobile Device Security Threats and Risks Chapter 17: Security Requirements, Monitoring, and Reporting

TABLE A.2 Exam MB0-001 CompTIA Mobility+ Objectives Map *(continued)*

Objective	Corresponding Book Chapter
<ul><li>Software risks<ul><li>App store usage</li><li>Virus</li><li>Trojans</li><li>Worm</li><li>Malware</li><li>Spyware</li><li>Jailbreak</li><li>Rooting</li><li>Keylogging</li><li>Unsupported OS</li></ul></li><li>Organizational risks<ul><li>BYOD ramifications</li><li>Securing personal devices</li><li>Removable media</li><li>Wiping personal data</li><li>Unknown devices on network/server</li></ul></li><li>Hardware risks<ul><li>Device cloning</li><li>Device theft</li><li>Device loss</li></ul></li><li>Mitigation strategies<ul><li>Antivirus</li><li>Software firewalls</li><li>Access levels</li><li>Permissions</li><li>Host-based and network-based IDS/IPS</li><li>Anti-malware</li><li>Application sandboxing</li><li>Trusted platform modules</li><li>Data containers</li></ul></li></ul>	Chapter 15: Mobile Device Security Threats and Risks Chapter 17: Security Requirements, Monitoring, and Reporting

TABLE A.2 Exam MB0-001 CompTIA Mobility+ Objectives Map *(continued)*

Objective	Corresponding Book Chapter
<ul><li>Content filtering</li><li>DLP</li><li>Device hardening</li><li>Physical port disabling</li></ul>	Chapter 15: Mobile Device Security Threats and Risks Chapter 17: Security Requirements, Monitoring, and Reporting
4.5 Given a scenario, execute appropriate incident response and remediation steps.<ul><li>Incident identification</li><li>Determine and perform policy-based response</li><li>Report incident<ul><li>Escalate</li><li>Document</li><li>Capture logs</li></ul></li></ul>	Chapter 17: Security Requirements, Monitoring, and Reporting
5.0 Troubleshooting	
5.1 Given a scenario, implement the following troubleshooting methodology.<ul><li>Identify the problem<ul><li>Information gathering</li><li>Identify symptoms</li><li>Question users</li><li>Determine if anything has changed</li></ul></li><li>Establish a theory of probable cause<ul><li>Question the obvious</li></ul></li><li>Test the theory to determine cause<ul><li>Once theory is confirmed determine next steps to resolve problem</li><li>If theory is not confirmed re-establish new theory or escalate</li></ul></li><li>Establish a plan of action to resolve the problem and identify potential effects</li></ul>	Chapter 19: Mobile Device Problem Analysis and Troubleshooting

TABLE A.2 Exam MB0-001 CompTIA Mobility+ Objectives Map *(continued)*

Objective	Corresponding Book Chapter
<ul><li>Implement the solution or escalate as necessary</li><li>Verify full system functionality and if applicable implement preventative measures</li><li>Document findings, actions and outcomes</li></ul>	Chapter 19: Mobile Device Problem Analysis and Troubleshooting
5.2 Given a scenario, troubleshoot common device problems.<ul><li>Battery life</li><li>Sync issues</li><li>Power supply problems</li><li>Password reset</li><li>Device crash</li><li>Power outage</li></ul>	Chapter 19: Mobile Device Problem Analysis and Troubleshooting
5.3 Given a scenario, troubleshoot common application problems.<ul><li>Missing applications</li><li>Configuration changes</li><li>App store problems</li><li>Email issues</li><li>Location services problems</li><li>OS and application upgrade issues</li><li>Profile authentication and authorization issues</li></ul>	Chapter 19: Mobile Device Problem Analysis and Troubleshooting
5.4 Given a scenario, troubleshoot common over-the-air connectivity problems.<ul><li>Latency</li><li>No cellular signal</li><li>No network connectivity</li><li>Roaming issues</li><li>Cellular activation</li><li>Email issues</li><li>VPN issues</li><li>Certificate issues</li></ul>	Chapter 19: Mobile Device Problem Analysis and Troubleshooting

TABLE A.2 Exam MB0-001 CompTIA Mobility+ Objectives Map *(continued)*

Objective	Corresponding Book Chapter
- APN issues - Port configuration issues - Network saturation	Chapter 19: Mobile Device Problem Analysis and Troubleshooting
5.5 Given a scenario, troubleshoot common security problems. - Expired certificate - Authentication failure - Firewall misconfiguration - False positives - False negatives - Non-expiring passwords - Expired passwords - Content filtering misconfigured	Chapter 19: Mobile Device Problem Analysis and Troubleshooting

Exam objectives are subject to change at any time without prior notice. Be sure to visit

http://certification.comptia.org/getcertified/certifications/mobilityplus.aspx

for the latest information. Keep in mind that URLs sometimes change. If the URL for the exam objectives has moved, use your favorite search engine to look for it.

Good luck with the Mobility+ certification exam!

Index

Note to the reader Throughout this index boldfaced page numbers indicate primary discussions of a topic. *Italicized* page numbers indicate illustrations.

A

AAA (authentication, authorization, and accounting) services, 586
absolute RF power measurements, 95–96
absorption, 91–92, *92*
acceptable use policies, 466–467
access control, 578–580
access levels in migration strategies, 608
access point names (APNs), 663–664
access points (APs)
   active scanning, 224, *224*
   antenna selection, 399–400
   autonomous, 256
   BBS, 197, *197*
   cloud-based, 268–269, *268*
   consumer-grade, 257–258, *257*
   controller-based, 266–267, *267*
   DAS, 127–128
   direct and distributed connectivity, 278, *278*
   enterprise-grade, 262–266, *263*, *265*
   ESS, 199
   installation limitations, 401
   Layer 2 and Layer 3 connectivity, 278–279, *279*
   name issues, 663–664
   onboarding and provisioning, 498
   overloaded, 213
   overview, 254–256, *255–256*
   remote, 269, *270*
   rogue, 561–562
   site surveys, 368–369, *368*, *378*
   SOHO-grade, 258–261, *259*, *262*
accessibility to disaster recovery plans, 640
Account Options screen for IMAP, 545, *545*
Account Setup screen for IMAP, 543, *543*
Account Type screen for IMAP, 544, *544*
AceBackup product, 632–636, *633–635*
Acronis backup products, 630, 632
actions in troubleshooting, 653
activation
   cellular issues, 663
   devices, 484–485, *485*
Active/Active state configuration, 642
active DAS, 127
active gain of antennas, 105
active mode
   FTP, 49
   power save operations, 245
active scanning in WLANs, 224–225, *224–225*
active site surveys, 371–377, *372–377*
ad hoc connections, 11–12, *12*
ad hoc wireless networks, 191, 561
adapters
   power, 654–655
   WLANs, 486
      cellular data, 493
      Mini PC, 490–493, *490–491*
      SD, 494
      USB, 487–489, *487–489*
adaptive access points, 269
Add An Account screen, 543, *543*
Additional Settings screen, 392, *392*
addresses. *See* device addressing; IP addresses; MAC addresses
adjacent channel interference, 356
adjustable output transmit power, 264
administration in IBSS, 196
administrative permissions, 451–452
administrators, training, 483
Advanced Attributes window, 631
Advanced Encryption Standard (AES) algorithm, 589, 591
Advanced Mobile Phone System (AMPS) technology, 302
Advanced Subnet Calculator program, 431–433, *431–433*
advancements, 532–533
   hardware, 533
   operating systems, 533
   security risks and threats, 535
   third-party vendors, 533–534, *534*
Aerohive Wi-Fi Planner, 387
AES (Advanced Encryption Standard) algorithm, 589, 591
aesthetics in antenna selection, 399
agents in SNMP, 41, *41*
AIDs (association IDs), 245
AirDefense Mobile protocol analyzer, 389
AirMagnet Spectrum XT spectrum analyzer, 348, *349*, 358–361, *358–362*
AirMagnet Survey/Planner, 387
AirSync (Microsoft Desktop AirSync Protocol), 68
alerts in APNS, 64
All Programs window, 26
Amazon Appstore, 534
amplitude of signals, 84, *84*
AMPS (Advanced Mobile Phone System) technology, 302
AND operation for subnet masks, 420, 424–425
Android Device Manager, 514, 516–519, *517–518*
Android Google Play Store, 448
Android operating system
   default device applications, 470
   DLP, 613
   emulator software, 448–451, *449–450*
   firmware upgrades, 523–524, *523–524*
   GCM, 65, *66*
   IMAP with, 542–545, *543–545*
   log files, 605
   mobile devices, 446–447

network gateways, 548–549, *549*
proxy servers, 550, *550*
push notifications, 554–555, *554*
rooting, 570–571
sandboxing, 619
third-party applications, 534, *534*
wiping personal data, 574
antennas
  access points
    enterprise-grade support, **263**
    SOHO-grade support, 259–260
  azimuth and elevation charts, **102**, *102*
  beamwidth, **101**, *101*
  concepts, 99–100
  DAS, **125–129**, *126*, *128*, *130*
  gain, **103–105**, *103–104*
  highly directional, **121–124**, *122–123*
  installation, **134**
    clear communications, 134–137, *136*
    mounting, 140–142, *141–142*
    placement, 137–138
    safety, **140**
    wind and lightning, 138–139, *139*
  lobes, **100**, *100*
  multipath signals, **124–125**, *124*
  omnidirectional, **107–111**, *108–111*
  polarization, **105–107**, *106*
  semidirectional, **111**
    patch/panel, **111–115**, *112–113*, *115*
    placement, **138**
    sector, **115–118**, *116*, *118*
    yagi, **118–120**, *119*, *121*
  site surveys, 342–344, *343–344*, 379, 397–400, *398*
anti-malware, **611**
antivirus (AV) technologies, 606–607
APIPA (Automatic Private IP Addressing) service, 39
APNS (Apple Push Notification Service)
  configuring, 551–554, *552–553*
  overview, 63–65, *64*
app store, 447–451, *449–451*
Apple App Store, 448, 534

Apple Push Notification Service (APNS)
  configuring, 551–554, *552–553*
  overview, 63–65, *64*
Application layer
  Internet protocol suite, 31, *31*
  OSI model, **18**
applications
  configuration, **535**
  development life cycle, 480–481, *481*
  in device deployment, 527
  distribution methods, 510–512
  end of life, 526
  in-house requirements, 535–537
  missing, 656–657
  publishing, 536
  removing, 505–506
  sandboxing, 618–619
  security risks, 570
  site survey factor, 331, 339–340
  third-party vendors, 533–534, *534*
  types, 537–539
  and WLAN capacity, **214**
approving pilot programs, **482**
Apps and Widgets screen
  GCM, 554
  IMAP, 542
APs. *See* access points (APs)
APSD (automatic power save delivery), 246–247
archive attributes, 630–631, *631*
assets
  disposal and recycling, 506
  inventory reviews, 603
association IDs (AIDs), 245
association in WLANs, 230–231, *230*
ATM (automated teller machine) cards, 580
attenuation of signals, **88**
audit information reports, 602–603
authentication
  enterprise-grade access point support, 264
  FTP, 49
  multifactor, 580
  passphrase-based security, 580–581
  PPTP, 286
  RADIUS, 585–586, *585–586*
  troubleshooting, 657, 664–665

  WLANs, **225–226**
    open system, **226–227**, *226–227*
    shared-key, **227–230**, *229–230*
    WPA and WPA2, **582–584**, *583–584*
authentication, authorization, and accounting (AAA) services, 586
authentication servers, 582, *583*
authenticators, 582, *583*
automated teller machine (ATM) cards, 580
automatic power save delivery (APSD), 246–247
Automatic Private IP Addressing (APIPA) service, 39
autonomous access points, **256**
azimuth
  charts, **102**, *102*
  DAS antennas, 129, *130*
  description, **101**, *101*
  highly directional antennas, 123, *123*
  omnidirectional antennas, 107, 111, *111*
  patch/panel antennas, 115, *115*
  sector antennas, 115–117, *118*, *120*, *121*

B

backhauling traffic, **435–436**, *436*
backups
  client devices, **631–636**, *633–635*
  corporate and personal data, 636
  frequency, 637
  local, 636
  overview, **626–627**, *627*
  policies, 467–468
  restoring, 643–644
  server data, **627–631**, *629*, *631*
  testing, 637
badges in APNS, 64
bandwidth
  channel width, 155
  overview, **85–86**, *86*
  proxy servers, 296
  site survey factor, 327, 330
  traffic restrictions, 436–438, *437*

banners in APNS, 64, *64*
bar code scanners, 246
bare metal support, 630
Barracuda backup product, 630
basic service area (BSA), 190, 197
basic service set (BSS)
   802.11r amendment, 158
   overview, **196–198**, *197–198*
basic service set identifiers
   (BSSIDs), **200–201**
battery life, troubleshooting,
   653–654
battery packs for site surveys, 379
beacon frames, 223
beamwidth, 101, *101*
   highly directional antennas,
     121–122
   omnidirectional antennas,
     107
   patch/panel antennas, 112
   sector antennas, 116
   yagi antennas, 118–119
Berkeley Varitronics Systems
   (BV Systems) Swarm
   program, 387
BES (BlackBerry Enterprise
   Server), 69–71
binary number conversions,
   422–424, *422–424*
biometrics, 579
BlackBerry Enterprise Server
   (BES), 69–71
BlackBerry operating system
   mobile devices, **446**
   wiping personal data, **574**
BlackBerry World store, 448, 534
blanketing in single-channel
   architecture, 401
block ciphers, 590–591, *590*
block-level encryption, 596–597
blocked Fresnel zones, 136–137
blueprints for site surveys, 329,
   333–334, 366
Bluetooth technology for PANs,
   160
Braille protocol, 69
bridges
   enterprise-grade access point
     support, 264–265
   overview, 271–272, *272*
   SOHO-grade access point
     support, 260–261
bring your own device (BYOD)
   devices
   employee-owned mobile
     devices profiles,
     474–475
   security risks, 572–573
   site surveys, 324–325

WLANs, 215
broadband wireless, **161**
BSA (basic service area), 190, 197
BSS (basic service set)
   802.11r amendment, 158
   overview, **196–198**, *197–198*
BSSIDs (basic service set
   identifiers), **200–201**
build phase in SDLC, 481, *481*
building age in site surveys,
   325–326, *325*
building materials
   RF effects, 89–94, *90–93*
   WLAN coverage, 212
building-to-building
   connectivity, 178–180, *179*
BumbleBee Handheld spectrum
   analyzer, 348
bus topologies
   description, 8, *8*
   troubleshooting, 9
business requirements as site
   survey factor, 327–328,
   328, 344
BYODs. *See* bring your own
   device (BYOD) devices

C

cables
   connectors, 133–134, *133*
   cost, 132–133
   impedance and VSWR, 131
   length, 132, *133*
   types, 131, *131*
calculator
   decimal-to-binary
     conversions, 422–424,
     422–424
   subnets, 428–433, *429–433*
cameras for site surveys, 379
campus area networks (CANs),
   5–6, *6*
Canada, RF regulations in, 184
capacity
   site survey factor, 340–341
   WLANs, 212–215
captive portals
   hotspots, 170
   MDM, 456–457
   WLANs, 276
carpeted office deployments,
   170–172, *172*
Carrier Sense Multiple Access
   with Collision Detection
   (CSMA/CD) access
   method, 411, *411*

carts for site surveys, 380
CAs (certificate authorities), 595
CCK (complementary code
   keying), 205
CCMP (Counter Mode with
   Cipher Block Chaining
   Message Authentication
   Code Protocol), 589, *589*
CDMA (Code Division Multiple
   Access), 308, *308*
ceiling antenna mounting, 142,
   142
cell tower spoofing, 565
cellular technology
   activation issues, 663
   backhaul diagrams, 436, *436*
   channel access, 305–308,
     306–308
   CSD, 309–310, *309*
   data adapters, **493**
   evolution, 300–305, *301–302*
   packet switching, 310, *310*
   RF spectrum analyses,
     356–361, *357–362*
   roaming, 315–316
   standards and protocols,
     311–314, *314*
   troubleshooting, 659, 663
   WiMAX, 314–315
   vs. WLANs, 658–659
centralized administration,
   274–275
centralized content, 510–512
centralized data forwarding,
   279–280
certificate authorities (CAs), 595
certificate revocation lists, 596
certificates
   authentication, 579
   devices, **537**
   expiration and renewal,
     520–521, *520*
   issuing, 471–472
   onboarding and provisioning,
     496, 498–503, *499–503*
   PKI, 596
   troubleshooting, 665
certification exam
   objectives review
     MDM, 677–681
     network infrastructure,
       674–677
     over-the-air technology,
       672–674
     security, 682–685
     troubleshooting, 685–687
   overview, 670
   preparing for, 670–671
   taking, 671–672

certifications for
    interoperability, **162–166**
Chanalyzer program, 350–355,
    351–354
change management, 525
changes, troubleshooting after,
    652–653
channels
    802.11ac amendment, 155
    architecture, 400–401,
        401–402
    cellular technology, 305–
        308, *306–308*
    DSSS, 203–204, *204*
    IBSS, **194**, *195*
    OFDM, 207–209, *208*
    reuse, 215–216, *216*
    RF, **186–189**, *188*
China
    ISM band, 187
    RF regulations, 184
    U-NII bands, 208–209
chips in CDMA, 308
Choose Components dialog box
    FileZilla client, **55**, *55*
    FileZilla server, **51**, *51*
Choose Destination Location
    dialog box
    AceBackup, 633, *634*
    AirMagnet Spectrum, 359,
        359
    SolarWinds calculator, 430,
        430
Choose Install Location dialog
    box
    Ekahau HeatMapper, 373,
        374
    FileZilla client, 55
    FileZilla server, 51
    YouWave, 449, *449*
Choose Installation Options
    dialog box, 54, *55*
Choose Setup Language screen,
    633, *633*
Choose Start Menu Folder
    dialog box
    FileZilla client, 55
    YouWave, 449
cipher methods, **589–591**, *590*
Circuit Switched Data (CSD)
    technology, 309–310, *309*
classes of IP addresses, **416–417**
classification signatures in PoE,
    282–283, *283*
Clear to Send (CTS) to Self
    option, 248
CLI (command-line interface)
    for enterprise-grade access
    point support, **266**

Client Download page, 54
clients
    backups, **631–636**, *633–635*
    connectivity requirements,
        341–342
    description, 486
    FTP installation, **54–57**,
        54–56
    RDP, 69
    site survey factor, 323,
        378–379
cloning devices, 572
cloud-based access points, 256,
    268–269, *268*
cloud-based distribution models,
    511–512
cloud-managed architectures,
    274–280, *278–279*
clustering, 642
co-channel interference, 356
Code Division Multiple Access
    (CDMA), **308**, *308*
codes
    antenna selection, 400
    installation limitations, **403**
cold sites for disaster recovery,
    641
collisions in LANs, **411–413**,
    411–413
colocation of WLAN devices,
    215–216, *216*
command-line interface (CLI)
    for enterprise-grade access
    point support, **266**
CommView for WiFi protocol
    analyzer, **390–396**, *391–395*
CommView for WiFi Setup
    Wizard, 391, *391*
compatibility in device
    deployment, 527
complementary code keying
    (CCK), 205
computer worms, 569
concentrators, VPN, **284–292**,
    284–285, *287–292*
configuration
    access points, 266
    APNS, **551–554**, *552–553*
    applications, **535**
    devices, **484–485**, *485*
    messaging protocols,
        539–547, *540–541*,
        543–545
    Mini PC cards, **492–493**
    network gateways, **547–549**,
        548–549
    ports, **618**
    proxy servers, **549–551**, *550*
    SDIO cards, **494**

SOHO, 167, *168*, 261
USB devices, **488–489**,
    488–489
WPS, 164
Confirm Attributes Changes
    dialog box, 597
Connect To A Workplace
    window, **289–291**, *290–291*
Connect To Server dialog box,
    52, *52*
Connect VPN Connection
    dialog box, 291, *292*
connection-oriented packet
    switching, 310
connection speeds as site survey
    factor, **338**
connectionless packet switching,
    310
connectionless protocols, 34
connectivity
    site survey factor, **341–342**
    troubleshooting, **658–664**,
        660–661, *663*
connectors, cable, **133–134**, *133*
construction materials as site
    survey factor, **325–326**, *325*
consultant-owned mobile device
    profiles, **475**
consumer-grade access points,
    257–258, *257*
consumer mobile application
    store, **447–448**
contacts in disaster recovery
    plans, **641**
containers
    device deployment, 527
    migration strategies, **619**
content
    distribution, **510–512**
    permissions, **512–513**
content filtering
    migration strategies, **620**
    proxy servers, 296
content management systems,
    510–512
control frames, **221–222**
Control Panel window for
    VPNs, 288
controller-based access points,
    256, **266–267**, *267*
conversions, decimal-to-binary,
    422–424, *422–424*
cordless telephones, 83
corporate data
    backing up, **636**
    removing, **505–506**
    restoring, **644**
corporate-owned mobile device
    profiles, **474**

Counter Mode with Cipher
    Block Chaining Message
    Authentication Code
    Protocol (CCMP), 589, 589
coverage
  site survey factor, 327, 330,
    340–341, 345, 396
  WLANs, 210–212, 211
crashes, device, 656
CSD (Circuit Switched Data)
    technology, 309–310, 309
CSMA/CD (Carrier Sense
    Multiple Access with
    Collision Detection) access
    method, 411, 411
Currently Connected To screen
    for VPNs, 291
Customer Information dialog
    box, 430
customer requirements in
    antenna selection, 399

D

DAS (Distributed Antenna
    System), 125–127, 126
  IEEE 802.11, 127–128
  leaky coax, 127
data
  containers, 516, 619
  encapsulation in OSI model,
    20, 21
  encrypting, 513
  forwarding, 279–280
  restoring, 643–644
data at rest (DAR), 591, 612–613
Data Encryption Algorithm
    (DES), 591
data frames in WLANs, 222
data in motion (DIM), 612
data in transit, 592
data in use (DIU), 612
Data Link layer in OSI model,
    15–16, 16
data loss prevention (DLP),
    612–615, 614
data rates in WLANs, 233–234,
    233–234
dB (decibels), 97, 97
dBd (decibel dipole) unit, 98–99
dBi (decibel isotropic) unit,
    98–99
dBm (decibels relative to
    milliwatts), 96
DDoS (distributed denial of
    service) attacks, 611
deactivation of devices, 505

dead spots in RF spectrum
    analyses, 348
deauthentication, 231–232, 231
decibel dipole (dBd) unit, 98–99
decibel isotropic (dBi) unit,
    98–99
decibels (dB), 97, 97
decibels relative to milliwatts
    (dBm), 96
decimal-to-binary conversions,
    422–424, 422–424
default device applications,
    469–470
delay spread in multipath
    signals, 85
demilitarized zones (DMZs)
  BlackBerry, 70
  overview, 621–622, 621–622
demodulation, 650, 650
denial of service (DoS) attacks
  deauthentication, 232
  overview, 562–563, 563
deployment
  best practices, 526–527
  manufacturer guides, 333
  in SDLC, 481, 481
DES (Data Encryption
    Algorithm), 591
design in SDLC, 481, 481
Destination Folder screen
  Chanalyzer, 351, 351
  RF3D WiFiPlanner2, 383,
    383
Destination Location screen, 392
details section for security
    policies, 466
device addressing, 20–21
  logical, 23–26, 24–26
  physical, 21–23, 22–23
device groups, 453
devices
  cloning, 572
  colocation, 215–216, 216
  compliance, 602–603
  configuration and activation,
    484–485, 485
  crashes, 656
  deactivation, 505
  deployment best practices,
    526–527
  digital certificates, 537
  firmware upgrades,
    522–524, 523–524
  hardening, 615–616
  hardware upgrades,
    522
  limits, 214
  log files, 605–606
  loss and theft, 571

managing. *See* mobile device
    management (MDM)
  migrations, 505
  platforms, 537
  profiles, 470, 471
  pushing content to, 451
  remote wipe, 516
  unknown, 575
DFS (dynamic frequency
    selection), 208
DHCP (Dynamic Host
    Configuration Protocol)
  overview, 35–40, 35–37,
    39–40
  SOHO-grade access point
    support, 261
DHCP acknowledgment
    (DHCPAck) messages, 38
DHCP discover (DHCPDiscover)
    messages, 38
DHCP offer (DHCPOffer)
    messages, 38
DHCP request (DHCPRequest)
    messages, 38
dictionary attacks, 286, 593
differential backups, 628–629
diffraction, 91, 91
diffusion, 92–93, 93
digital cameras for site surveys,
    379
digital certificates. *See*
    certificates
direct AP connectivity, 278, 278
direct sequence spread spectrum
    (DSSS), 202–204, 204
directory services
  device groups, 453
  integration, 471
disabling ports, 616
disassociation in WLANs,
    231–232, 231
disaster recovery
  locations, 641–642
  overview, 637–638
  plans, 638–641
discovery for WLANs,
    222–225, 223–225
disposal of assets, 506
Distributed Antenna System
    (DAS), 125–127, 126
  IEEE 802.11, 127–128
  leaky coax, 127
distributed AP connectivity,
    278, 278
distributed data forwarding,
    279–280
distributed denial of service
    (DDoS) attacks, 611
distribution models, 511–512

distribution systems (DSs)
  BBS, 197
  WLANs, 232–233, 232
DLP (data loss prevention), 612–615, 614
DMZs (demilitarized zones)
  BlackBerry, 70
  overview, 621–622, 621–622
DNS (Domain Name System), 44–45, 45
documentation
  site surveys, 333–337, 334, 336, 379
  technology implementation, 484
  in troubleshooting, 653
DoD model, 31
dollies for site surveys, 380
Domain Name System (DNS), 44–45, 45
DoS (denial of service) attacks
  deauthentication, 232
  overview, 562–563, 563
dotted-decimal addresses, 24, 24, 416, 416
Driver Installation dialog box, 393, 394
DRS (dynamic rate switching), 242
DSs (distribution systems)
  BBS, 197
  WLANs, 232–233, 232
DSSS (direct sequence spread spectrum), 202–204, 204
dwell time in FHSS, 202
dynamic frequency selection (DFS), 208
Dynamic Host Configuration Protocol (DHCP)
  overview, 35–40, 35–37, 39–40
  SOHO-grade access point support, 261
dynamic rate switching (DRS), 242
dynamic routing, 434, 434

E

E-plane fields, 105
EAP (Extensible Authentication Protocol), 582–584, 583–584
earth curvature effects, 137
EAS (Exchange Active Sync) protocol, 68
ECC (Elliptical Curve Cryptography), 591

EDGE (Enhanced Data Rates for GSM Evolution), 312
educational environments
  antenna use considerations, 343–344
  deployments, 173, 174
Ekahau HeatMapper program, 371–377, 372–377
Ekahau HeatMapper Setup wizard, 373, 373
Ekahau Site Survey program, 387
electrical power requirements as site survey factor, 337–339
electrical specifications as site survey factor, 335
electromagnetic interference (EMI), 94
electronic serial numbers (ESNs), 572
elevation with antennas
  charts, 102, 102
  DAS antennas, 129, 130
  description, 101, 101
  highly directional antennas, 123, 123
  omnidirectional antennas, 107, 111, 111
  patch/panel antennas, 115, 115
  sector antennas, 118, 118, 120, 121
Elliptical Curve Cryptography (ECC), 591
email
  filtering, 620
  IMAP, 61–62, 62, 541–545, 543–545
  MAPI, 62–63, 63, 546–547
  POP3, 60–61, 60, 546
  SMTP, 58–59, 59, 540–541, 540–541
  troubleshooting, 657
EMI (electromagnetic interference), 94
employee-owned mobile devices profiles, 474–475
employee termination, 505
emulator software for Android, 448–451, 449–450
encapsulation
  OSI model, 20
  VPNs, 285, 593
encryption
  block-level, 596–597
  CCMP, 589, 589
  cipher methods, 589–591, 590
  data, 513
  folders and file-level, 597

full disk, 596
L2TP, 286, 594
open system authentication, 227
overview, 587
PKI, 595
PPTP, 286
removable media, 598
TKIP, 588, 588
VPNs, 284–285, 593, 593
WEP, 587–588
WPA, 164
end of life (EOL) model, 525–526
end of sale (EOS), 525
End-User License Agreement (EULA), 472, 472
end user training, 483
endpoint injectors in PoE, 282
enforcement section in security policies, 466
Enhanced Data Rates for GSM Evolution (EDGE), 312
enrollment in device deployment, 527
enterprise app store, 447
enterprise-grade access points, 262–266, 263, 265
enterprise networks
  antenna use considerations, 342
  networks deployments, 168–169
  overview, 169
  WPA and WPA2 security, 163, 582–584, 583–584
entities in PKI, 596
environmental factors
  antenna selection, 399
  RF
    absorption, 91–92, 92
    diffraction, 91, 91
    diffusion and free space path loss, 92–93, 93
    Faraday cages, 93–94
    interference sources, 94
    reflection, 89–90, 90
    refraction, 90, 90
    scattering, 91, 92
  site survey factor, 324
  in troubleshooting, 652–653
EOL (end of life) model, 525–526
EOS (end of sale), 525
ERP (extended rate physical)
  overview, 205–206, 206
  protection, 247–248
ESNs (electronic serial numbers), 572
ESSIDs (extended service set identifiers), 200

ESSs (extended service sets), 198–199, *199*
Ethernet installation limitations, 401–402
ETSI (European Telecommunications Standards Institute), 81
EULA (End-User License Agreement), 472, *472*
Europe
 ISM band, 187
 RF regulations, 184
 U-NII bands, 208–209
European Telecommunications Standards Institute (ETSI), 81
evaluating pilot programs, 482
Evaluation Version dialog box, 395
Evolution Data Optimized (EVDO), 312–313
Evolved High Speed Packet Access (HSPA+), 313–314
exam. *See* certification exam
Exchange Active Sync (EAS) protocol, 68
exclusive OR (XOR) process
 CDMA, 308, *308*
 DSSS, 203
 stream ciphers, 589–590
existing networks
 extensions into remote locations, 169
 site surveys, 330, 335–337, *336*, *366*, *367*
expandable antennas, 263
expense management in telecommunications, 456
expiration of certificates, 520–521, *520*
expired certificates, 665
expired passwords, 665
extended rate physical (ERP)
 overview, 205–206, *206*
 protection, 247–248
extended service set identifiers (ESSIDs), 200
extended service sets (ESSs), 198-199, *199*
Extensible Authentication Protocol (EAP), 582–584, *583–584*
Extensible Messaging and Presence Protocol (XMPP), 65
Eye P.A. protocol analyzer, 389

F

facial scans, 579
false negatives and positives, 666
Family Educational Rights and Privacy Act (FERPA), 464
Faraday cages, 93–94
Faraday, Michael, 93
fault tolerance, 452–453, 642–643, *642*
FCC (Federal Communications Commission)
 overview, 80–81
 unlicensed frequency bands, 185–186
FDE (full disk encryption), 596
FDMA (Frequency Division Multiple Access), 306, *306*
Federal Communications Commission (FCC)
 overview, 80–81
 unlicensed frequency bands, 185–186
Federal Information Processing Standards (FIPS), 464
feedback
 in device deployment, 527
 pilot programs, 482
femtocells, 572
FERPA (Family Educational Rights and Privacy Act), 464
FHSS (frequency hopping spread spectrum), 148, 202, *203*
file-level encryption, 597
File Transfer Protocol (FTP), 49–50, *49*
 client installation, 54–57, *54–56*
 server installation, 50–53, *50–53*
File Transfer Protocol Secure (FTPS), 57
file transfer protocols, 48
 FTP, 49–56, *49–56*
 FTPS, 57
 SCP, 58
 SFTP, 57–58
FileZilla FTP
 client installation, 54–57, *54–56*
 server installation, 50–53, *50–53*
filtering
 content, 294–296, *295*, 620
 proxy servers, 296
 troubleshooting, 666–667
fingerprints, 579
FIPS (Federal Information Processing Standards), 464
firewalls

description, 277
DMZs, 621–622, *621–622*
settings, 617
software, 607
troubleshooting, 666
firmware upgrades, 522–524, *523–524*
first generation (1G) cellular technology, 302, *302*
floor plans for site surveys, 329, 333–334, *334*, *336*, *366*
Fluke Networks InterpretAir, 387
folders, encrypting, 597
fourth generation (4G) cellular technology, 304, *304*
frames
 association, 231
 DHCP, 37–38
 shared-key authentication, 229
 WLANs, 221–222
free space path loss (FSPL), 92–93, *93*
frequencies, RF
 overview, 83, *83*
 WLANs, 184–186
Frequency Division Multiple Access (FDMA), 306, *306*
frequency hopping spread spectrum (FHSS), 148, 202, *203*
Freshdesk software, 653
Fresnel zones, 135–137, *136*
FrontRange Solutions HEAT software, 653
FSPL (free space path loss), 92–93, *93*
FTP (File Transfer Protocol), 49–50, *49*
 client installation, 54–57, *54–56*
 server installation, 50–53, *50–53*
FTPS (File Transfer Protocol Secure), 57
full backups, 628
full disk encryption (FDE), 596
full-duplex devices with access points, 255
Full Mini PCIe cards, 491–493, *491*
furnishings as site survey factor, 334–335

G

gain
 antennas, 103–105, *103–104*
 RF, 84

gateways
    configuring, 547–549, *548–549*
    overview, 292–293, *293*
gathering information in in troubleshooting, 651
GCM (Google Cloud Messaging), 65, *66*, 554–555, *554*
general offices
    antenna use considerations for, 342, *343*
    site survey factors, 328, *328*
General Packet Radio Service (GPRS), 312
Genie Backup Manager product, 632
geo-fencing, 454–455, *454*
geo-location, 455–456, *455*
geometry factors in antenna selection, 400
GLBA (Gramm-Leach Bliley Act), 464
Global Positioning System (GPS), 454
Global Standard for Mobile Communications (GSM), 303, 307, **311**
Google Cloud Messaging (GCM), 65, *66*, 554–555, *554*
Google Play Store, 448, 534
Google+ Hangouts, 65
government environments, antenna use considerations for, 343
GPRS (General Packet Radio Service), 312
GPS (Global Positioning System), 454
Gramm-Leach Bliley Act (GLBA), 464
Graphical User Interface (GUI) configuration for enterprise-grade access point support, 266
grid antennas, 121
grounding rods, **139**
group profiles, 473–474, *473*
    consultant and visitor-owned mobile devices, 475
    corporate-owned mobile devices, 474
    employee-owned mobile devices, 474–475
    testing, 475–476
GSM (Global Standard for Mobile Communications), 303, 307, **311**

GUI (Graphical User Interface) configuration for enterprise-grade access point support, 266

H

H-plane fields, 105
hacking tools, 232
half-duplex devices for access points, 255, *256*
Half Mini PCIe cards, **491–493**
hardening devices, **615–616**
hardware
    advancements, 533
    end of life, 526
    security risks, 571–572
    site survey factor, 396–397, *397*
    upgrades, 522
    WLANs, 212, 214
hardware controllers for WLANs, 274–280, *278–279*
HCF (hybrid coordination function), 157
Health Information Technology for Economic and Clinical Health Act (HITECH), 616
Health Insurance Portability and Accountability Act (HIPAA) compliance, 176, **464**
healthcare environments
    antenna use considerations for, 343
    deployments, 175–176, *176*
hiding SSIDs, 158, 193
high availability
    maintaining, 642–643, *642*
    MDM, 452–453
high-density deployments, 177–178
high-rate direct sequence spread spectrum (HR/DSSS), 150, 152, **204–205**
High Speed Downlink Packet Access (HSDPA), 313
High Speed Uplink Packet Access (HSUPA), 313
high-throughput (HT)
    IEEE 802.11 standard, 206–207
    protection mechanisms, 248–249
highly directional antennas
    overview, **121–124**, *122–123*
    placement, **138**
hijack attacks, 565–566, *566*

HIPAA (Health Insurance Portability and Accountability Act) compliance, 176, **464**
HITECH (Health Information Technology for Economic and Clinical Health Act), 616
horizontally polarized antennas, 106, *106*
hospitality environments, antenna use considerations for, 344
host-based IDS/IPS, **609–610**
host part in IP addresses, 24, *24*
hosted network operations center, 556
hostnames in DNS, 44–45, *45*
hot sites for disaster recovery, 641
hotspots
    802.11 deployment scenarios, 169–170, *171*
    antenna use considerations, 344
    cellular connections, 493
    risks, 561
HR/DSSS (high-rate direct sequence spread spectrum), 150, 152, **204–205**
HSDPA (High Speed Downlink Packet Access), 313
HSPA+ (Evolved High Speed Packet Access), 313–314
HSUPA (High Speed Uplink Packet Access), 313
HT (high-throughput)
    IEEE 802.11 standard, 206–207
    protection mechanisms, 248–249
HTTP (Hypertext Transfer Protocol)
    email, 547
    overview, **66–68**, *67*
HTTPS (Hypertext Transfer Protocol Secure), 67–68, 561, 595
hubs in LANs, 413, *413*
human-created disasters, 639
hybrid approach for site surveys, 388
hybrid apps, 538–539
hybrid coordination function (HCF), 157
Hypertext Transfer Protocol (HTTP)
    email, 547

overview, 66–68, *67*
Hypertext Transfer Protocol Secure (HTTPS), 67–68, 561, 595

I

IANA (Internet Assigned Numbers Authority), 417
IBSS. *See* Independent Basic Service Set (IBSS)
ICCID (Integrated Circuit Card Identifier) numbers, 497
ICMP (Internet Control Message Protocol), 45, *46*
identification requirements for certification exam, 672
IDSs (intrusion detection systems), 609–610
IEEE. *See* Institute of Electrical and Electronics Engineers (IEEE)
IETF (Internet Engineering Task Force), 32
ifconfig utility, 25–26
IMAP (Internet Message Access Protocol)
 configuring, 541–545, *543–545*
 overview, 61–62, *62*
IMEI (International Mobile Station Equipment Identity) numbers, 497
impedance of cables, **131**
IMSI (international mobile subscriber identity) numbers, 565
in-house requirements for applications, 535–537
Incoming Server Settings screen, 544, *544*
incremental backups, 628–629
Independent Basic Service Set (IBSS)
 advantages and disadvantages, 195–196
 description, 191, *191*
 RF channels, 194, *195*
 security, 195
 SSIDs, 192–193, *192*, *194*
 terminology, 191–192
industrial deployments, 173–175, *175*
industrial, scientific, and medical (ISM) bands
 channels, 186–187, *188*
 overview, 185–186
industry regulatory compliance, 463–464

information gathering in in troubleshooting, 651
information technology (IT)
 change management, 525
 security policies. *See* security policies
 viruses, 569
information traffic topology, 555–556
infrastructure device redundancy, 277–278
infrastructure factors in site surveys
 connectivity, 337–339
 selection and placement, 396–397, *397*
infrastructure mode for access points, 254
initiation in pilot programs, 482
insertion loss, 134
InSSIDer program
 antenna polarization, 106
 site surveys, 337
installation
 antennas. *See* antennas
 device limitations, 401–402
 Ekahau HeatMapper, 371–377, *372–377*
 FTP client, 54–57, *54–56*
 FTP server, 50–53, *50–53*
 Mini PC adapters, 492–493
 protocol analyzers, 390–396, *391–395*
 RF3D WiFiPlanner2, 382–386, *383–386*
 SDIO cards, 494
 USB adapters, 488–489, *488–489*
instant messaging, XMPP for, 65
Institute of Electrical and Electronics Engineers (IEEE)
 802.3 standard, 280–281
 802.1X standard, 582–584, *583–584*
 802.11 standard, **147–148**. *See also* wireless local area networks (WLANs)
 802.11-2012, 81
 802.11a amendment, 148–149, *149*, 207–208
 802.11ac amendment, 155–156, 207–208
 802.11ad amendment, 156
 802.11b amendment, 150, 204–205, 247–248

802.11e amendment, 157
802.11g amendment, 150–152, *151*, 205–208, *206*, 247–248
802.11i amendment, 157–158
802.11k amendment, 158
802.11n amendment, 153–154, *153*, 206–208, 248–249
802.11r amendment, 158
802.11s amendment, 159
802.11w amendment, 158–159
 access point support, 258–259, *259*, 263
 DAS, 127–128
 licensed bands, 80–81
 summary, 156–157
 802.15 standard, **160**
 802.16 standard, **161**
 working groups, 146–147
Integrated Circuit Card Identifier (ICCID) numbers, 497
intended use factor in site surveys, **322**
interference sources
 RF, 94
 site surveys, 331
 spectrum analyses, 349–350, *350*, 355–356
International Mobile Station Equipment Identity (IMEI) numbers, 497
international mobile subscriber identity (IMSI) numbers, 565
International Organization for Standardization (ISO), 14
International Telecommunication Union – Radiocommunication Sector (ITU-R), 79, *79*
Internet Assigned Numbers Authority (IANA), 417
Internet-based predictive modeling site surveys, 387
Internet Control Message Protocol (ICMP), 45, *46*
Internet Engineering Task Force (IETF), 32
Internet layer, 31, *31*
Internet Message Access Protocol (IMAP)
 configuring, 541–545, *543–545*
 overview, 61–62, *62*
Internet protocol (IP) suite
 addresses. *See* IP addresses

overview, 30–32, *31*
with TCP, 33
Internet Protocol Security
    (IPsec), 286
Internet Society (ISOC), 32
interoperability certifications, **162**
    Wi-Fi Alliance, **162**
    Wi-Fi Voice-Enterprise, **166**
    WMM, **165**
    WMM-PS, **165**–166
    WPA, **162**–163, *163*
    WPA 2.0, **163**–164
    WPS, **164**–165
interviewing stakeholders,
    328–333, *329*, *331*
intrusion detection systems
    (IDSs), **609**–610
intrusion prevention systems
    (IPSs), **610**
inventories, asset, 603
iOS Developer Enterprise
    Program, 536
iOS operating system
    APNS, 64
    application store, 448
    gateways, 549
    mobile devices, **446**
    proxy servers, 550
    push notifications, 551
    sandboxing, 619
IP addresses
    classes, **416**–417
    description, 21
    DHCP, **35**–40, *35–37*, *39–40*
    DNS, **44**–45, *45*
    NAT, **42**–43, *43*
    obfuscation, 296
    physical devices, **23**–25, *24*
    special use, **418**, *418*
    subnets. *See* subnets
IP (Internet protocol) suite
    addresses. *See* IP addresses
    overview, **30**–32, *31*
    with TCP, 33
ipconfig utility
    DHCP information, **35**, *35*,
        39–40, *39–40*
    logical device addresses,
        25, *25*
    physical device addresses,
        23, *23*
    WLAN throughput, **237**–
        238, *237*
iPhone 4, wiping personal data
    from, **574**–575
iPhone Configuration Utility, 605
IPsec (Internet Protocol
    Security), 286
IPSs (intrusion prevention
    systems), **610**

IPv6 for NAT, 42
ISM (industrial, scientific, and
    medical) bands
    channels, **186**–187, *188*
    overview, 185–186
ISO/IEC 27002 standard, 464
ISO (International Organization
    for Standardization), 14
ISOC (Internet Society), 32
isotropic radiators, 98
Israel
    ISM band, 187
    RF regulations, 184
    U-NII bands, 208–209
ITU-R (International
    Telecommunication Union
    – Radiocommunication
    Sector), 79, *79*

J

Jabber protocol, 65
jailbreaking, **570**
jamming
    DoS attacks, **563**, *563*
    RF, **564**–565, *564*
Japan
    ISM band, 187
    RF regulations, 185
    U-NII bands, 208–209
jitter in traffic, **439**
JPerf software program, **236**–
    241, *236–241*

K

Kayako software, 653
key contacts in disaster recovery
    plans, **641**
keylogging, **570**, 611
keys, weak, **567**

L

L2F (Layer 2 Forwarding)
    protocol, 594
L2TP (Layer 2 Tunneling
    Protocol), 286, 594
ladders for site surveys, 380
LANs (local area networks)
    description, 3, *4*
    traffic flow, **411**–414,
        411–414
    wireless. *See* wireless local
        area networks (WLANs)

last-mile data delivery, **177**
latency
    measuring, **659**
    traffic, **439**
law enforcement networks, **178**
Layer 2 Forwarding (L2F)
    protocol, 594
Layer 2 Tunneling Protocol
    (L2TP), 286, 594
LBS (location-based services),
    453–454
    geo-fencing, **454**–455, *454*
    geo-location, **455**–456, *455*
    working with, **514**–515, *514*
LDAP (Lightweight Directory
    Access Protocol), 453
leaky coax DAS, **127**
leases in DHCP, **38**–39, *39*
length of cables, **132**, *133*
licensing
    AceBackup, 633
    AirMagnet Spectrum,
        359
    Chanalyzer, 351, *351*
    CommView for WiFi, 391,
        392
    Ekahau HeatMapper, 373
    EULA, **472**, *472*
    FileZilla, **50**–51, *50*
    RF3D WiFiPlanner2, 384
    SolarWinds calculator,
        429, *429*
    YouWave, 449
life cycle operations, **519**–520
    certificate expiration and
        renewal, **520**–521,
        520
    firmware upgrades, **522**–
        524, *523–524*
    hardware upgrades, **522**
    software patches, **521**–522
    software updates, **521**
lifts for site surveys, 380
lightning arrestors, **139**, *139*
lightning effects with antennas,
    138–139
Lightweight Directory Access
    Protocol (LDAP), 453
line of sight (LoS) in
    communications, **135**
Link layer in Internet protocol
    suite, 31, *31*
LLC (Logical Link Control),
    16, *16*
load balancing, 643
lobes, antenna, **100**, *100*
local area networks (LANs)
    description, 3, *4*
    traffic flow, **411**–414,
        411–414

wireless. *See* wireless local area networks (WLANs)
local backups, **636**
local codes
   antenna selection, 400
   installation limitations, 403
local regulations, 80–81, 184–185
location-based services (LBS), 453–454
   geo-fencing, 454–455, *454*
   geo-location, 455–456, *455*
   working with, 514–515, *514*
locks, remote, 515–516, *515*
logging and log files
   devices, 605–606
   proxy servers, 296
logical addressing, 23–26, *24–26*
logical AND operation for subnet masks, 420
logical exclusive OR (XOR) process
   CDMA, 308, *308*
   DSSS, 203
   stream ciphers, 589–590
Logical Link Control (LLC), 16, *16*
Long Term Evolution (LTE) technology
   CDMA, 308
   overview, 314, *314*
LoS (line of sight) in communications, **135**
loss
   devices, 571
   RF, 84
LTE (Long Term Evolution) technology
   CDMA, 308
   overview, 314, *314*

M

MAC addresses
   description, 16
   device addressing, 21–23, *22–23*
   filtering, 158
   MAC TAL, 504
   spoofing, 563
   switches, 413–414, *414*
   WANs, 415
MAC TAL (Media Access Control Transparent Auto Login), 503–504
maintenance in SDLC, 481, *481*

maintenance section in security policies, 466
malware
   anti-malware, **611**
   overview, 568–570
man-in-the-middle attacks, 566, *567*
management buy-in for disaster recovery plans, **640**
management frames, **221**
Management Information Base (MIB) in SNMP, 41
MANs (metropolitan area networks)
   description, 5, *5*
   wireless, **161**
manual onboarding and provisioning, **496**
manual site surveys
   access points, 368–369, *368*
   active, 371–377, *372–377*
   advantages and disadvantages, 369–370
   blueprints, 366
   existing networks, 366, *367*
   floor plans, 366
   overview, 364–366, *365*
   passive, 370, *371*
   software-assisted, 370
   toolkit, 378–380
manufacturer guidelines
   antenna selection, 399
   site surveys, 333
manufacturing, antenna use considerations for, 342
MAPI (Messaging Application Programming Interface), 62–63, *63*, 546–547
masks, subnet, 419–422, 424–425
mast mounting for antennas, 140–141, *141*
MBSSs (mesh basic service sets), 159, **200**
MC connectors, 134
MCA (multi-channel architecture), 400–401, *401*
MCS (Modulation and Coding Scheme), 155
MCX connectors, 134
MDM. *See* mobile device management (MDM)
measurements
   units, 95–99, *97*
   WLAN throughput, 236–241, *236–241*
measuring devices for site surveys, **379**
Media Access Control (MAC) 802.11n amendment, 154

addresses. *See* MAC addresses
   description, 16, *16*
Media Access Control Transparent Auto Login (MAC TAL), 503–504
medical environments, antenna use considerations for, 343
MEIDs (mobile equipment identifiers), 572
Meraki WiFi Mapper, 387
mesh basic service sets (MBSSs), 159, **200**
mesh mode for enterprise-grade access point support, **265**
mesh topologies
   overview, 10–11, *12*
   wireless, 270, *271*
Messaging Application Programming Interface (MAPI), 62–63, *63*, 546–547
messaging protocols, 58, 539–540
   APNS, 63–65, *64*
   GCM, 65, *66*
   IMAP, 61–62, *62*, 541–545, *543–545*
   MAPI, 62–63, *63*, 546–547
   POP3, 60–61, *60*, 546
   SMTP, 58–59, *59*, 540–541, *540–541*
   XMPP, 65
metropolitan area networks (MANs)
   description, 5, *5*
   wireless, **161**
MIB (Management Information Base) in SNMP, 41
Microsoft Challenge Authentication Protocol (MS-CHAP) version 2, 286
Microsoft Desktop AirSync Protocol (AirSync), 68
Microsoft Point-to-Point Encryption (MPPE-128) protocol, 286, 593
midspan injectors in PoE, **282**
migration strategies
   access levels, **608**
   anti-malware, **611**
   antivirus technologies, 606–607
   content filtering, **620**
   data containers, **619**
   device hardening, 615–616
   devices, 505
   DLP, 612–615, *614*
   DMZs, 621–622, *621–622*
   firewalls, **617**

IDS/IPS, 609–610
overview, 606
permissions, 608–609
physical port disabling, **616**
port configuration, **618**
sandboxing, 618–619
software firewalls, 607
TPMs, **619**
military environments, antenna use considerations for, 343
milliwatts (mW), **96**
MIMO (multiple-input, multiple-output), **154**
Mini PC adapters, **490–493**, *490–491*
Mini PCI Express (Mini PCIe) cards, **490–493**
MINs (mobile identification numbers), 572
missing applications, troubleshooting, 656–657
MMCX connectors, 134
mobile application store, **447–451**, *449–451*
mobile device management (MDM)
  device groups, **453**
  DLP, **614**
  exam review, 677–681
  high availability and redundancy, 452–453
  location-based services, **453–456**, *454–455*
  mobile application store, **447–451**, *449–451*
  on-premise, 444–446
  onboarding and provisioning, **495–504**, *495*, *499–503*
  operating system platforms, 446–447
  overview, 442–443
  permissions, 451–452
  policies. *See* security policies
  portals, 456–458
  profiles. *See* profiles
  SaaS, **444**
  telecommunications expense management, **456**
mobile equipment identifiers (MEIDs), 572
mobile identification numbers (MINs), 572
Mobile Telephone System (MTS) protocol, 301
modulation, 86–87, *87*, 650, *650*
Modulation and Coding Scheme (MCS), **155**
monitoring, SIEM for, 604–605

Motorola AirDefense spectrum analyzer, 348
Motorola LANPlanner, 387
Motorola SiteScanner, 387
mounting antennas, **140–142**, *141–142*
mounting hardware for site surveys, 379
movable carts for site surveys, 380
MPPE-128 (Microsoft Point-to-Point Encryption) protocol, 286, 593
MS-CHAP (Microsoft Challenge Authentication Protocol) version 2, 286
MTS (Mobile Telephone System) protocol, 301
MU-MIMO (multi-user MIMO), 155
multi-channel architecture (MCA), **400–401**, *401*
multi-user MIMO (MU-MIMO), 155
multifactor authentication, 580
multipath signals
  antenna diversity for, **124–125**, *124*
  description, 85
multiple-input, multiple-output (MIMO), **154**
municipal networks, **178**
mW (milliwatts), **96**

N

name issues for access points, 663–664
NAT (Network Address Translation), 42–43, *43*
native apps, **538**
natural disasters, 639
near field communication (NFC) tokens, 164–165
net neutrality, 438
Network Address Translation (NAT), 42–43, *43*
network administrators, training, 483
Network and Sharing Center window for VPNs, 288
network-based IDS/IPS, 609–610
network gateways
  configuring, **547–549**, *548–549*
  overview, **292–293**, *293*
network infrastructure
  exam review, 674–677

site surveys, 326–327
Network Interface layer in Internet protocol suite, 31, *31*
Network layer in OSI model, **16–17**, *17*
network operations centers (NOCs)
  hosted, **556**
  on-premise, **555**
  third-party, **556**
network part in IP addresses, 24, *24*
network protocols
  AirSync, 68
  DHCP, **35–40**, *35–37*, *39–40*
  DNS, **44–45**, *45*
  file transfer, **48–58**, *49–56*
  HTTP, **66–68**, *67*
  ICMP, 45, *46*
  messaging. *See* messaging protocols
  NAT, 42–43, *43*
  overview, 32–33
  PAT, *43*, 44
  RDP, 68–69
  RFCs, 32
  SNMP, **40–42**, *41*
  SRP, **69–71**, *70*
  SSL, 48
  TCP, 33–34, *33*
  UDP, 34, *34*
network topologies, 7–8
  ad hoc connections, **11–12**, *12*
  bus, **8–9**, *8*
  mesh, **10–11**, *12*
  point-to-multipoint connections, 13, *14*
  point-to-point connections, 13, *13*
  ring, 9, *10*
  star, 10, *11*
network types, 2–3
  CANs, **5–6**, *6*
  LANs, 3, *4*
  MANs, 5, *5*
  PANs, **6–7**, *7*
  WANs, **3–4**, *4*
NFC (near field communication) tokens, 164–165
no protection mode, 249
NOCs (network operations centers)
  hosted, **556**
  on-premise, **555**
  third-party, **556**
noise
  RF, **217–218**, *218–219*
  SNR, **219–220**, *220*

Nokia Store, 448
non-expiring passwords, **665**
non-HT mixed mode, **249–250**
Non-Wi-Fi interference sources, **349–350**, *350*
nonmember protection mode, **249**
nonoverlapping channels, 215
note-taking for site surveys, **379**
NovaStor backup product, 630, 632
null function frames, 245
null SSIDs, 224
number system conversions, **422–424**, *422–424*

O

Observer protocol analyzer, 389
obstacles in WLAN coverage, 212
OCR (Office of Civil Rights), 603
octets
   IP addresses, 416, *416*
   subnet masks, 421–422
OEMs (original equipment manufacturers), **469**
OFDM (orthogonal frequency division multiplexing), 148–149, 207–209, *208*
Office of Civil Rights (OCR), 603
OHA (Open Handset Alliance), 446
omnidirectional antennas
   overview, **107–111**, *108–111*
   placement, **138**
OmniPeek protocol analyzer, 389, *390*
on-premise solutions
   MDM, **444–446**
   NOCs, 555
onboarding and provisioning, **495**, *495*
   certificate enrollment methods, **496**, **498–503**, *499–503*
   device deployment, **527**
   manual method, **496**
   miscellaneous methods, **497–498**
   role-based enrollment, **497**
   self-service method, **496**
   voucher methods, **503–504**
one-to-many NAT, 43
one-to-one NAT, 43
Open File – Security Warning dialog box, 50, 54
Open Handset Alliance (OHA), 446

open system authentication, **226–227**, *226–227*
Open Systems Interconnection (OSI) model, **14**
   Application layer, **18**
   data encapsulation, **20**, *21*
   Data Link layer, **15–16**, *16*
   interlayer communication, **19**
   vs. Internet protocol suite, 30, *31*
   memorization tip, **19**
   Network layer, **16–17**, *17*
   peer layer communication, **20**, *20*
   Physical layer, **14–15**, *15*
   Presentation layer, **18**
   Session layer, **17–18**
   Transport layer, **17**, *18*
   VPNs, **592**, *592*
   WLANs, **660**, *660*
operating frequencies for WLANs, **184–186**
operating systems
   advancements, 533
   Android. *See* Android operating system
   default device applications, **469–470**
   in device deployment, 527
   end of life, 526
   iOS. *See* iOS operating system
   mobile devices, **446–447**
   OEM, **469**
   vendors, **468–469**
operation modes in enterprise-grade access point support, **264–265**
organizations, security risks in, **572–575**
original equipment manufacturers (OEMs), **469**
orthogonal frequency division multiplexing (OFDM), 148–149, 207–209, *208*
OSI. *See* Open Systems Interconnection (OSI) model
outcomes in troubleshooting, 653
Outgoing Server Settings screen, 545, *545*
Outlook Web App (OWA), 516
output power for WLANs, 212
over-the-air technology
   connectivity issues, **658–664**, *660–661*, *663*
   exam review, **672–674**
overloaded access points, 213
OWA (Outlook Web App), 516

P

Packet Binary Convolution Code (PBCC), 206
packet switching in cellular technology, **310**, *310*
packets in TCP, 33, *33*
palm prints, 579
PANs (personal area networks)
   description, **6–7**, *7*
   wireless, 160
parabolic dish antennas, 121, *122*
passive DAS, 127
passive gain of antennas, **103–105**, *104*
passive mode FTP, 50
passive scanning of WLANs, **222–223**, *223*
passive site surveys, 370, *371*
passphrase-based security, **580–581**
passwords
   access control, **578–579**
   backups, 636
   FileZilla, 53
   FTP, 49
   security keys, 567
   troubleshooting, **655–656**, 665
PAT (Port Address Translation), 43, *44*
patch/panel antennas, **111–115**, *112–113*, *115*
patches, software, **521–522**
Payment Card Industry (PCI) compliance, 463
PBC (push-button configuration) for WPS, 164
PBCC (Packet Binary Convolution Code), 206
PCF (Point Coordination Function) mode, 157
PCI (Payment Card Industry) compliance, 463
PDs (powered devices) in PoE, **282–283**, *283*
peer layer communication in OSI model, **20**, *20*
peer-to-peer networks
   IBSS, 191
   overview, **11–12**, *12*
performance expectations in site surveys, 324
performance of WLANs, 214
permissions
   content, **512–513**
   MDM, **451–452**
   migration strategies, **608–609**
personal area networks (PANs)
   description, **6–7**, *7*

wireless, 160
personal data
  backing up, 636
  restoring, 644
  wiping, 573–575
personal identification numbers (PINs), 579
personal mode in WPA, 163
personally identifiable information (PII), protecting, 603
phase, signal, 84–85, *85*
physical addressing for devices, 21–23, *22–23*
physical area in WLANs, 211–212
Physical Layer Convergence Protocol (PLCP), 15
Physical layer in in OSI model
  IEEE 802.11 standard. *See* Institute of Electrical and Electronics Engineers (IEEE)
  overview, 14–15, *15*
physical media, encrypting, 596–598
Physical Medium Dependent (PMD), 15
physical ports, disabling, 616
piconets, 484, *485*
pigtail cables, 132, *133*
PII (personally identifiable information), protecting, 603
pilot programs, 482
PIN-based configuration in WPS, 164
ping command
  ICMP, 45, *46*
  for latency, 659
  WLAN throughput, 239
PINs (personal identification numbers), 579
PKI (Public Key Infrastructure), 595–596
plans
  disaster recovery, 638–641
  SDLC, 481, *481*
platforms. *See* operating systems
PLCP (Physical Layer Convergence Protocol), 15
PMD (Physical Medium Dependent), 15
PoE. *See* Power over Ethernet (PoE)
Point Coordination Function (PCF) mode, 157
Point of Sale (PoS) installations, antenna use considerations for, 342–343

point-to-multipoint connections, 13, *14*, 179–180, *179*
point-to-point connections, 13, *13*, 178–180, *179*
Point-to-Point Tunneling Protocol (PPTP), 286, 593
polarization of antennas, 105–107, *106*, 399
pole mounting for antennas, 140–141, *141*
policies. *See* security policies
pools for IP addresses, 35
POP3 (Post Office Protocol 3)
  configuring, 546
  overview, 60–61, *60*
Port Address Translation (PAT), 43, 44
portable battery packs for site surveys, 379
portals
  captive web, 276
  hotspots, 170
  MDM, 456–458
ports
  configuring, 618
  disabling, 616
  FTP, 49, *49*
  overview, 46–48, *47*
PoS (Point of Sale) installations, antenna use considerations for, 342–343
Post Office Protocol 3 (POP3)
  configuring, 546
  overview, 60–61, *60*
post-site surveys, 388
power adapters, troubleshooting, 654–655
power measurements
  absolute, 95–96
  relative, 96–98, *97*
power outages, troubleshooting, 656
Power over Ethernet (PoE)
  benefits, 283–284
  controller support, 275
  installation limitations, 401–402
  overview, 280–281, *280*
  powered devices and classification signatures, 282–283, *283*
  PSE, 281–282
  site survey factor, 332, 338–339
power requirements in site surveys, 337–339
power save operations in WLANs, 244–247, *246*
power-saving devices in WMM-PS, 165–166

power sourcing equipment (PSE) in PoE, 281–282
powered devices (PDs) in PoE, 282–283, *283*
PPTP (Point-to-Point Tunneling Protocol), 286, 593
pre-design audits for site surveys, 344
pre-shared keys, 567
predictive modeling site surveys, 277, 380–387, *381*, *383–386*
Presentation layer in OSI model, 18
private IP addresses, 24, 416
probe request frames, 224, *224*
profiles
  in device deployment, 527
  group, 473–476, *473*
  technology, 470–472, *471–472*
  WLANs, 275–276
project timelines in site surveys, 344
propagation in WLANs, 212
proprietary connectors, 134
protection modes and mechanisms for WLANs, 247–251
protocol analyzers, 388–396, *390–395*
protocols
  cellular technology, 311–314, *314*
  network. *See* network protocols
provisioning. *See* onboarding and provisioning
proxy servers
  configuring, 549–551, *550*
  overview, 293–296, *294–295*
PSE (power sourcing equipment) in PoE, 281–282
pseudorandom streams, 590
Psiber RF3D WiFiPlanner, 387
public access environments, antenna use considerations for, 344
public IP addresses, 24, 416
Public Key Infrastructure (PKI), 595–596
public wireless hotspots, 169–170, *171*, 561
publishing applications, 536
purpose section in security policies, 465
push-button configuration (PBC) for WPS, 164
push technologies, 551
  APNS, 551–554, *552–553*
  in device deployment, 527

GCM, 554–555, *554*
pushing content to mobile devices, 451

Q

Quadrature Amplitude Modulation (QAM), 155
quality of service (QoS)
    description, 277
    site surveys, 332
    traffic, 438–439
Quantum Policy Suite (QPS), 503

R

radiation patterns
    highly directional antennas, 121
    omnidirectional antennas, 107, *108*
    patch/panel antennas, 115, *115*
    sector antennas, 118, *118*
radio chains in MIMO, 154
radio-frequency identification (RFID), 165
radio frequency (RF), 76
    absorption, 91–92, *92*
    antennas. *See* antennas
    attenuation, 88
    bandwidth, 85–86, *86*
    cables, 131–134, *132–133*
    channels, 186–189, *188*, *194*, *195*
    characteristics, 81–85, *82–85*
    coverage requirements, 340–341, 345
    diffraction, 91, *91*
    diffusion and free space path loss, 92–93, *93*
    Faraday cages, 93–94
    fundamentals, 76–79, *77–78*
    interference sources, 94, 331
    jamming, 564–565, *564*
    lobes, 100, *100*
    local regulations, 184–185
    LoS, 135
    measurement units, 95–99, 97
    modulation, 86–87, *87*
    noise, 217–218, *218–219*
    range, 87–88, 189
    reflection, 89–90, *90*
    refraction, 90, *90*
    regulatory authorities, 79–81, *79*

scattering, 91, *92*
signal measurements, 217–220, *217–220*
spectrum management, 277
speed, 87–88
radio frequency (RF) spectrum analyses for site surveys, 341, 344
    cellular, 356–361, *357–362*
    interference sources, 349–350, *350*, 355–356
    overview, 346–347
    WLANs, 347–349, *347–349*
    working with, 350–355, *351–354*
RADIUS (Remote Authentication Dial-In User Service) authentication
    enterprise-grade access point support, 264
    hardware controllers, 276
    overview, 585–586, *585–586*
random password generators, 581
range
    IBSS, 196
    IP addresses, 35
    RF, 87–88, 189
    WLANs, 212
RBAC (role-based access control), 497
RC4 cipher, 590
RDP (Remote Desktop Protocol), 68–69
Ready To Install AirMagnet Spectrum ES screen, 359
Ready To install Chanalyzer screen, 351–352
Ready To Install The Program dialog box
    AceBackup, 634
    SolarWinds calculator, 431
real-time communications (RTC), 34
receive sensitivity, 217, *217*
received signal strength
    in site surveys, 362–364, *363–364*
    WANs, 218–219
received signal strength indicator (RSSI), 218–219, 363
receivers
    description, 78, *78*
    modulation, 87
    troubleshooting, 649–653, *650*, *652*
recovery policies, 467–468
recycling, 506
Redo Backup & Recovery product, 632
redundancy in MDM, 452–453

reflection, 89–90, *90*
refraction, 90, *90*
regulations
    antenna selection, 400
    local, 184–185
    regulatory authorities, 79–81, *79*
relative RF power measurements, 96–98, *97*
remote access points, 269, *270*
Remote Authentication Dial-In User Service (RADIUS) authentication
    enterprise-grade access point support, 264
    hardware controllers, 276
    overview, 585–586, *585–586*
remote control, 519
Remote Desktop Protocol (RDP), 68–69
remote locations, network extensions into, 169
remote management, 513–514
    device wipe, 516
    location services, 514–515, *514*
    lock and unlock, 515–516, *515*
remote wipe, 516
removable antennas
    enterprise-grade access point support, 263
    SOHO-grade access point support, 259–260
removable media
    backups to, 636
    encrypting, 598
    security risks, 573
removing applications and corporate data, 505–506
renewing
    certificates, 520–521, *520*
    leases, 38–39
repeater mode for enterprise-grade access point support, 265
repeaters
    LANs, 412, *412*
    overview, 272–273, *273*
    SOHO-grade access point support, 261
reports
    audit information, 602–603
    remote management, 519
    site surveys, 403, *404*
Request to Send/Clear to Send (RTS/CTS) option, 248
Requests for Comments (RFCs), 32
restoring data

retail installations – security risks

backups, 643–644
policies, 467–468
retail installations, antenna use considerations for, 342–343
retinal scans, 579
revisions section in security policies, 466
RF. *See* radio frequency (RF)
RF3D WiFiPlanner2 tool, 382–386, *383–386*
RFCs (Requests for Comments), 32
RFID (radio-frequency identification), 165
ring topologies, 9, *10*
Rivest, Ron, 590
roaming
  cellular technology, 315–316
  ESS, 199
  site survey factor, 332
  troubleshooting, 662–663, 663
  WLANs, 242–244, *243–244*, 275
robust security network (RSN) compliance, 589
rogue access points, 561–562
role-based access control (RBAC), 497
root access point mode for enterprise-grade access point support, 264
rooting, 570–571
routers
  overview, 433–434, *434*
  WANs, 415, *415*
RP-MMCX connectors, 134
RSN (robust security network) compliance, 589
RSSI (received signal strength indicator), 218–219, 363
RTC (real-time communications), 34
RTS/CTS (Request to Send/ Clear to Send) option, 248
rubber duck antennas, 108–110, *109–110*
Ruckus Wireless access points, 127–128

S

SaaS (Software as a Service), 444
safety in antenna installation, 140
Samsung Store, 448

sandboxing applications, 618–619
SANs (storage area networks)
  description, 3
  high availability, 642, *642*
SANS Technology Institute website, 641
Sarbanes-Oxley Act (SOX), 464
saturation issues, 664
SCA (single-channel architecture) model, 401, 402, 663
scalability in IBSS, 196
scanning WLANs, 222–225, *223–225*
scattering, 91, *92*
SCEP (Simple Certificate Enrollment Protocol), 497–498
scope
  IP addresses, 35
  security policies, 466
SCP (Secure Copy Protocol), 58
SD (Secure Digital) cards
  backups to, 636
  description, 494
SDIO (Secure Digital Input Output) cards, 494
SDLC (System Development Life Cycle), 480–481, *481*
second generation (2G) cellular technology, 303
sector antennas, 115–118, *116*, *118*
Secure Copy Protocol (SCP), 58
Secure Digital (SD) cards
  backups to, 636
  description, 494
Secure Digital Input Output (SDIO) cards, 494
Secure File Transfer Protocol (SFTP), 57–58
Secure Shell (SSH)
  overview, 594
  SFPS, 57–58
Secure Sockets Layer (SSL)
  FTPS, 57
  HTTPS, 67
  overview, 48, 594–595
  SFPS, 57
  SMTP, 59
security
  access control, 578–580
  authentication, 580–586, *581*, *583–586*
  cipher methods, 589–591, *590*
  device compliance and audit information, 602–603
  in device deployment, 527

encryption, 587–589, *588–589*, *596–598*
enterprise-grade access point support, 264, *265*
exam review, 682–685
FTPS, 57
hardware controllers, 276
hotspots, 170
HTTPS, 67–68
IBSS, 195–196
log files, 605–606
migration strategies. *See* migration strategies
PKI, 595–596
SIEM, 604–605
site surveys, 332
SOHO-grade access point support, 260
SSL, 48
troubleshooting, 664–667
tunneling, 591–595, *592–593*
WPS, 165
Security Event Management (SEM), 604
Security Information and Event Management (SIEM) tools, 604–605
Security Information Management (SIM), 604
security keys, weak, 567
security policies
  acceptable use policies, 466–467
  backups, restores, and recovery, 467–468
  framework, 464–465
  implementation and adherence, 465–466, 466
  industry regulatory compliance, 463–464
  overview, 462–463
  testing, 475–476
  usability balancing with, 467
security risks
  from advancements, 535
  cell tower spoofing, 565
  denial of service attacks, 562–563, *563*
  hardware, 571–572
  hotspots, 561
  jamming, 564–565, *564*
  organizations, 572–575
  overview, 560–561, *561*
  rogue access points, 561–562
  security keys, 567
  software, 568–571
  warpathing, 567–568
  wireless hijack attacks, 565–566, *566*

wireless man-in-the-middle attacks, 566, *567*
security tokens, **579**
self-service method for onboarding and provisioning, **496**
self-service portals, **457–458**
SEM (Security Event Management), 604
semidirectional antennas, **111**
   patch/panel, **111–115**, *112–113*, *115*
   placement, **138**
   sector, **115–118**, *116*, *118*
   yagi, **118–120**, *119*, *121*
server-based distribution models, **511**
Server Download page, 50
Server Routing Protocol (SRP), **69–71**, *70*
servers
   authentication, **582**, *583*
   backups, **627–631**, *629*, *631*
   FTP, **50–53**, *50–53*
   proxy, **293–296**, *294–295*, **549–551**, *550*
   RDP, 68
service-level agreements (SLAs), **627**, *627*
service set identifiers (SSIDs)
   access points, 255
   active scanning, 224
   captive portals, 457
   ESS, 199
   vs. ESSIDs and BSSIDs, **200**
   hiding, 158, **193**
   IBSS, **192–193**, *192*, *194*
   onboarding and provisioning, **498–499**
   troubleshooting, **662**
   USB adapters, **489**
Session layer in OSI model, **17–18**
Set Up A Connection Or Network dialog box
   SSIDs, 193, *194*
   VPNs, 289, *289*
Settings screen
   Android Device Manager, 517, *517*
   Android operating system, 523
SFTP (Secure File Transfer Protocol), **57–58**
shaping traffic, **434–439**, *436–437*
shared-key authentication, **227–230**, *229–230*
shared media, 215

shortened battery life, troubleshooting, **653–654**
SIEM (Security Information and Event Management) tools, **604–605**
signal-to-noise ratio (SNR)
   received signal strength, 363
   WLANs, **219–220**, *220*
SIM (Security Information Management), 604
SIM (Subscriber Identification Module) cards
   backups to, 636
   description, 565
Simple Certificate Enrollment Protocol (SCEP), **497–498**
Simple Mail Transfer Protocol (SMTP)
   configuring, **540–541**, *540–541*
   overview, **58–59**, *59*
Simple Network Management Protocol (SNMP)
   overview, **40–42**, *41*
   in troubleshooting, **651**
sine waves, 77, *77*
Singapore, RF regulations in, 185
single-channel architecture (SCA) model, **401**, *402*, 663
site surveys
   antennas, **342–344**, *343–344*, **397–400**, *398*
   application requirements, **339–340**
   building age and construction materials, **325–326**, *325*
   business requirements, **327–328**, *328*
   BYOD acceptance, **324–325**
   capacity requirements, **340–341**
   client connectivity requirements, **341–342**
   client device capabilities, **323**
   coverage requirements, **340–341**, **396**
   device installation limitations, **401–402**
   environmental factors, 324
   hybrid approach, 388
   infrastructure connectivity and power requirements, **337–339**
   infrastructure devices, **326–327**
   infrastructure selection and placement, **396–397**, *397*
   intended use factor, **322–323**

manual process. *See* manual site surveys
   manufacturer and deployment guides, 333
   number of devices, **322–323**
   overview, **320–321**
   performance expectations, 324
   post-site, 388
   predictive modeling, **277**, **380–387**, *381*, *383–386*
   received signal strength, **362–364**, *363–364*
   reports, **403**, *404*
   site-specific documentation, **333–337**, *334*, *336*
   size factors, **322**
   spectrum analyses. *See* radio frequency (RF) spectrum analyses for site surveys
   stakeholder interviews, **328–333**, *329*, *331*
   steps, **344–346**
size factors in site surveys, 322
SLAs (service-level agreements), **627**, *627*
small office, home office (SOHO) networks
   antenna use considerations for, **342**, *343*
   overview, **167**, *168*
   site survey factor, **328**, *328*
smartphone tethering, **493**
SMTP (Simple Mail Transfer Protocol)
   configuring, **540–541**, *540–541*
   overview, **58–59**, *59*
SMTPS, 59
Sniffer protocol analyzer, 389
SNMP Managers, 41
SNMP (Simple Network Management Protocol)
   overview, **40–42**, *41*
   in troubleshooting, **651**
SNR (signal-to-noise ratio)
   received signal strength, 363
   WLANs, **219–220**, *220*
sockets, **47–48**
software
   firewalls, **607**
   patches, **521–522**
   security risks, **568–571**
   updates, **521**
   and WLAN capacity, **214**
Software as a Service (SaaS), **444**
software-assisted site surveys, **370**
SOHO-grade access points, **258–261**, *259*, *262*

SOHO (small office, home
    office) networks
  antenna use considerations
    for, 342, *343*
  overview, **167**, *168*
  site survey factor, 328, *328*
SolarWinds calculator for
    subnets, **428–433**, *429–433*
SolarWinds Web Help Desk
    software, 653
sounds in APNS, 64
SOX (Sarbanes-Oxley Act), 464
spanning single-channel
    architecture, 401
spatial streams, 155
special use IP addresses, **418**, *418*
spectrum analyzers for site
    surveys. *See* radio
    frequency (RF) spectrum
    analyses for site surveys
spectrum management, 277
speed, RF, 87–88
spoofing, cell tower, 565
spread spectrum technology,
    201–202
  DSSS, 202–204
  FHSS, 202, *203*
  HR/DSSS, 204–205
spreading code in DSSS, 203–204
spyware, 569–570
SRP (Server Routing Protocol),
    69–71, *70*
SSH (Secure Shell)
  overview, **594**
  SFPS, 57–58
SSIDs. *See* service set identifiers
    (SSIDs)
SSL. *See* Secure Sockets Layer
    (SSL)
SSMTP, 59
stacking single-channel
    architecture, 401
stakeholders, interviewing,
    328–333, *329*, *331*
standards
  cellular technology, **311–314**,
    314
  IEEE. *See* Institute of
    Electrical and
    Electronics Engineers
    (IEEE)
star topologies, **10**, *11*
Startup Settings dialog box,
    51–52, *51–52*
static IP addresses, 35
static output transmit power in
    SOHO-grade access point
    support, 260
static routing, 434
stations
  ESS, 199

IBSS, 191
storage area networks (SANs)
  description, 3
  high availability, **642**, *642*
storage in disaster recovery
    plans, 640
stream ciphers, **589–590**, *590*
subcarriers in DSSS, 203–204
subnets, **416**, *416*
  creating, **425–433**, *426*,
    428–433
  IP address classes, 416–417
  IP addresses, 418–422
  masks, **419–422**, 424–425
Subscriber Identification
    Module (SIM) cards
  backups to, 636
  description, 565
supplicants in IEEE 802.1x
    standard, **582**, *583*
switches, **413–414**, *414*
Symazntec backup product, 630
symptoms in troubleshooting,
    652, *652*
synchronization issues,
    troubleshooting, 654
System Development Life Cycle
    (SDLC), **480–481**, *481*

T

TACACS (Terminal Access
    Controller Access Control
    System) authentication
    protocol, 586
Taiwan, RF regulations in, 185
TamoGraph Site Survey, 387
target beacon transmission time
    (TBTT), 223
TCP/IP addresses, 23
TCP/IP model, 31
TCP/IP Properties window,
    37, *37*
TCP ports, 46, 47, 49
TCP (Transmission Control
    Protocol), 23, **33–34**, *33*
TDMA (Time Division Multiple
    Access), **307**, *307*
TDoA (Time Difference of
    Arrival) technology, 455
technical support for site
    surveys, 345
technology advancements,
    532–533
  hardware, 533
  operating systems, 533
  security risks and threats, 535
  third-party vendors, **533–
    534**, *534*

technology implementation
  device configuration and
    activation, **484–485**, *485*
  documentation, 484
  offboarding and
    deprovisioning, **504–506**
  onboarding and
    provisioning, **495–504**,
    495, *499–503*
  pilot programs, 482
  SDLC, **480–481**, *481*
  technology launches, **483–484**
  training, 483
  WLAN adapters, **486–494**,
    487–491
technology launches, **483–484**
technology profiles, **470**
  devices, **470**, *471*
  digital certificates, **471–472**
  directory services
    integration, 471
  EULA, **472**, *472*
  testing, **475–476**
telcos, 4
telecommunications, expense
    management for, 456
telephones
  cellular. *See* cellular
    technology
  cordless, 83
Temporal Key Integrity Protocol
    (TKIP), **588**, *588*
10s and 3s rule, **98**
Terminal Access Controller
    Access Control System
    (TACACS) authentication
    protocol, 586
terminating employees, 505
testing
  access point placement,
    368–369, *368*
  antennas, **397–400**, *398*
  backups, 637
  disaster recovery plans, 640
  pilot programs, 482
  policies and profiles, **475–476**
  in SDLC, **481**, *481*
tethering, smartphone, 493
TFTP (Trivial File Transfer
    Protocol), **34**, *34*
TGaf task group, 177
theft, 571
third generation (3G) cellular
    technology, 303
third-party NOCs, 556
third-party vendors
  applications, **533–534**, *534*
  monitoring tools, **604–605**
threats from advancements, 535
3s and 10s rule, **98**
throughput

802.11g networks, **152**
WLANs, **214**, **234–241**, *235–241*
Time Difference of Arrival (TDoA) technology, 455
Time Division Multiple Access (TDMA), **307**, *307*
timelines in site surveys, 344
TKIP (Temporal Key Integrity Protocol), **588**, *588*
TLS (Transport Layer Security), 57, 595
tokens, security, 579
Toolbox dialog box for RF3D WiFiPlanner2, 385
toolkits for site surveys, 378–380
topologies
 information traffic, 555–556
 network. *See* network topologies
TPMs (trusted platform modules), 619
tracert.exe utility, 659
traffic flow, 410
 backhauling, 435–436, *436*
 bandwidth and user restrictions, 436–438, *437*
 LANs, 411–414, *411–414*
 quality of service, 438–439
 routing, 433–434, *434*
 shaping, 434–439, *436–437*
 subnets. *See* subnets
 WANs, 414–415, *415*
training, 483
transition
 cellular technology, 315–316
 ESS, 199
 site survey factor, 332
 troubleshooting, 662–663, *663*
 WLANs, 242–244, *243–244*, 275
translational bridges in distribution systems, 232
Transmission Control Protocol (TCP), 23, 33–34, *33*
transmitters
 description, 78, *78*
 modulation, 86–87
 troubleshooting, 649–653, *650*, 652
Transport layer
 Internet protocol suite, 31, *31*
 OSI model, 17, *18*
Transport Layer Security (TLS), 57, 595
transportation networks, 178
Triple Data Encryption Algorithm (3DES), **591**

Trivial File Transfer Protocol (TFTP), 34, *34*
Trojans, 569
troubleshooting, *663*
 access points, 663–664
 authentication, 657, 664–665
 battery life, 653–654
 bus topologies, 9
 cellular technology, 659, *663*
 certificates, 665
 connectivity, 658–664, 660–661
 device crashes, 656
 email, 657
 exam review, 685–687
 false negatives and positives, 666
 filtering, 666–667
 firewalls, 666
 missing applications, 656–657
 passwords, 655–656, 665
 power adapters, 654–655
 power outages, 656
 process, 648–649, *649*
 roaming, 662–663
 security, 664–667
 steps, 651
 synchronization issues, 654
 transmitters and receivers, 649–653, *650*, *652*
 WLANs, 660–662, *660–661*
trusted platform modules (TPMs), 619
tunnels, 591–592
 L2TP, 286, 594
 PPTP, 286, 593
 SSH, 594
 VPNs, 284, *285*, 592–594, *593*
 web-based, 594–595
20 MHz protection mode, 249
two-factor authentication, 580
256-QAM (Quadrature Amplitude Modulation), 155
Twofish block cipher, **591**

U

U-bolts for antenna mounting, 140, *141*
U-NII (Unlicensed National Information Infrastructure) bands, **148–149**, *149*
 channels, 189
 local regulations, **185–186**
 OFDM, 208, *208*
UAT (user acceptance testing), 607

UDP (User Datagram Protocol), 23, 34, *34*
Ultra High Security Password Generator, 581
UMTS (Universal Mobile Telecommunications System), 312
United States
 ISM band, 187
 RF regulations in, 185
 U-NII bands, 208–209
Universal Mobile Telecommunications System (UMTS), 312
Universal Serial Bus (USB) adapters, **487**, *487*
 features, 488
 installation, 488–489, *488–489*
Universal Serial Bus (USB) drives, backups to, 636
university campus environments, 6, *6*
unknown devices, 575
unlicensed frequency bands, 185–186
Unlicensed National Information Infrastructure (U-NII) bands, **148–149**, *149*
 channels, 189
 local regulations, 185–186
 OFDM, 208, *208*
unlock, remote, 515–516, *515*
updates
 disaster recovery plans, 640
 documentation, 484
 life cycle operations software, 521
upfade, 85
upgrades
 firmware, **522–524**, *523–524*
 hardware, 522
UrBackup product, 632
usability balancing with security policies, 467
USB (Universal Serial Bus) adapters, **487**, *487*
 features, 488
 installation, 488–489, *488–489*
USB (Universal Serial Bus) drives, backups to, 636
user acceptance testing (UAT), 607
User Account Control dialog box
 AceBackup, 633
 AirMagnet Spectrum, 358
 Chanalyzer, 352
 CommView for WiFi, 391, 393–394
 Ekahau HeatMapper, 373

RF3D WiFiPlanner2, 383
SolarWinds calculator, 429
YouWave, 449
User Datagram Protocol (UDP), 23, **34**, *34*
User Experience Improvement Program screen, 351
user limits in WLANs, **214**
user profiles in device deployment, 527
user restrictions for traffic, 436–438, *437*
user training, **483**
usernames
    access control, 578–579
    FTP, 49

V

validation
    device deployment, 527
    disaster recovery plans, 640
VDC (volts of direct current), 281
vendors
    applications, 533–534, *534*
    default device applications, 469–470
    operating systems, 468–469
version control, **513**
vertically polarized antennas, 106, *106*
Very High Throughput (VHT), 155–156
video communications, latency in, 439
virtual local area networks (VLANs), **275**
virtual private networks (VPNs)
    components, 287–288, *287–288*
    concentrators, 284–292, *284–285, 287–292*
    L2TP, 286, **594**
    overview, 592–593, *593*
    PPTP, 286, **593**
    public wireless hotspots, 170
    setting up, 288–292, *289–292*
virtual WLANs, 275–276
virtualization technology, 643
viruses
    antivirus technologies, 606–607
    IT, **569**
visitor-owned mobile device profiles, **475**
VisiWave – AZO Technologies, 387
visual line of sight, **135**

VLANs (virtual local area networks), **275**
voice communications
    latency, 439
    site surveys, 332
voltage standing wave ratio (VSWR), **131**, **134**
volts of direct current (VDC), 281
voucher methods, 503–504
VPNs. *See* virtual private networks (VPNs)
VSWR (voltage standing wave ratio), **131**, **134**

W

wall mounting antennas, **142**
WANs (wide area networks)
    description, 3–4, *4*
    traffic flow, 414–415, *415*
    wireless, **161–162**
warehousing, antenna use considerations for, 342, *343*
warm sites for disaster recovery, 641
warpathing, 567–568
watts, **95**
wavelength, **82**, *82*
WDS (wireless distribution systems), 159
web apps, **538**
web-based security, 594–595
web page caching, 295–296
websites, filtering, 620
WEP. *See* Wired Equivalent Privacy (WEP)
Wi-Fi Alliance Interoperability Certifications
    enterprise-grade access point support, 263
    overview, **162**
    SOHO-grade access point support, 259
Wi-Fi Inspector, 337
Wi-Fi interference sources, 355–356
Wi-Fi Multimedia Power Save (WMM-PS) certification, 165–166
Wi-Fi Multimedia (WMM), 157, **165**
Wi-Fi Protected Access (WPA) certification, 162–163, *163*
Wi-Fi Protected Access 2 (WPA 2.0) certification, 163–164
Wi-Fi Protected Setup (WPS) certification, 164–165
Wi-Fi Voice-Enterprise certification, 166

Wi-Spy spectrum analyses, 348
wide area networks (WANs)
    description, 3–4, *4*
    traffic flow, 414–415, *415*
    wireless, **161–162**
WiFi Analyzer protocol analyzer, 389
wildcard SSIDs, 224
WiMAX (Worldwide Interoperability for Microwave Access)
    overview, 314–315
    WMANs, 161
wind effects on antennas, 138–139
Windows calculator for decimal-to-binary conversions, 422–424, *422–424*
Windows Phone Store, 448
Windows Phone (WP) operating system, **447**
Windows Security dialog box, 374, *374*
wiping devices
    personal data, 573–575
    remote, **516**
WIPSs (wireless intrusion prevention systems)
    enterprise-grade access point support, 264
    hardware controllers, **278**
    RF jamming, 564
    sensor analogy, 610
Wired Equivalent Privacy (WEP)
    802.11i amendment, 157–158
    encryption, 587–588
    open system authentication, 227
    shared-key authentication, 228
    WLAN profiles, 276
wired point-to-multipoint connections, 14
wired point-to-point connections, 13, *13*
wireless access points. *See* access points (APs)
wireless branch routers, remote, 269, *270*
wireless bridges, 271–272, *272*
wireless distribution systems (WDS), 159
wireless hijack attacks, 565–566, *566*
wireless hotspots
    antenna use considerations, 344
    broadband, **493**

cellular connections, 493
deployment scenarios, 169–170, *171*
risks, 561
wireless intrusion prevention systems (WIPSs)
  enterprise-grade access point support, 264
  hardware controllers, 278
  RF jamming, 564
  sensor analogy, 610
wireless local area networks (WLANs)
  adapters, 486
    cellular data, 493
    Mini PC, 490–493, *490–491*
    SD, 494
    USB, 487–489, *487–489*
  association, 230–231, *230*
  authentication, 225–230, *226–227, 229*
  BBS, 196–198, *197–198*
  building-to-building connectivity, 178–180, *179*
  capacity, 212–215
  vs. cellular, 658–659
  channel reuse and device colocation, 215–216, *216*
  connecting to, 220–221
  coverage, 210–212, *211*
  DAS, 127–128
  data rates, 233–234, *233*
  deauthentication and disassociation, 231–232, *231*
  deployment scenarios, 166–167
    carpeted offices, 170–172, *172*
    educational institutions, 173, *174*
    enterprise, 168–169, *169*
    healthcare, 175–176, *176*
    high-density, 177–178
    industrial, 173–175, *175*
    last-mile data delivery, 177
    municipal, law enforcement, and transportation, 178
    public wireless hotspots, 169–170, *171*
    remote location extensions, 169
    SOHO, 167
  discovery, 222–225, *223–225*
  distribution systems, 232–233, *232*
  dynamic rate switching, 242, *242*

ESS, 198–199, *199*
frame types, 221–222
hardware controllers, 274–280, *278–279*
IBSS, 191–196, *191–192, 194–195*
MBSS, 200
operating frequencies and channels, 184–185
  RF channels, 186–189, *188*
  RF range, 189
operating modes, 190
overview, 160–161
physical layer, 201
  extended rate, 205–206, *206*
  high throughput, 206–207
  OFDM, 207–209, *208*
  spread spectrum technology, 201–205, *203–204*
power save operations, 244–247, *246*
profiles, 275–276
protection modes and mechanisms, 247–251
RF signal measurements, 217–220, *217–220*
RF spectrum analyses, 347–349, *347–349*
throughput, 234–241, *235–241*
transition, 242–244, *243–244*
troubleshooting, 660–662, *660–661*
wavelength, 82
wireless man-in-the-middle attacks, 566, *567*
wireless MANs (WMANs), 161
wireless mesh networks, 270, *271*
wireless networking overview
  IEEE 802.16, 161
  LANs, 160–161
  PANs, 160
  types and sizes, 159–160
wireless point-to-multipoint connections, 14
wireless point-to-point connections, 13, *13*
wireless range extenders, 273
wireless repeaters, 272–273, *273*
wireless WANs (WWANs), 161–162
Wireshark packet analyzer, 34, *34*
wiring closets as site survey factor, 338
WLAN Coverage Estimator, 387
WLANs. *See* wireless local area networks (WLANs)

WMANs (wireless MANs), 161
WMM-PS (Wi-Fi Multimedia Power Save) certification, 165–166
WMM (Wi-Fi Multimedia), 157, 165
Worldwide Interoperability for Microwave Access (WiMAX)
  overview, 314–315
  WMANs, 161
worms, 569
WP (Windows Phone) operating system, 447
WPA 2.0 (Wi-Fi Protected Access 2) certification, 163–164
WPA (Wi-Fi Protected Access) certification, 162–163, *163*
WPS (Wi-Fi Protected Setup) certification, 164–165
WWANs (wireless WANs), 161–162

X

XMPP (Extensible Messaging and Presence Protocol), 65
XOR (exclusive OR) process
  CDMA, 308, *308*
  DSSS, 203
  stream ciphers, 589–590
XpressConnect product, 498–503, *499–503, 520, 520*

Y

yagi antennas, 118–120, *119, 121*
Yellowjacket protocol analyzer, 389
YouWave Android Setup dialog box, 449, *449*
YouWave emulator program, 448–451, *449–450*

Z

Zendesk software, 653
Zero Generation (0G) cellular, 301
ZoneFlex 7441 access points, 128

Free Online Learning Environment

Register on Sybex.com to gain access to the free online interactive learning environment and test bank to help you study for your CompTIA Mobility+ certification.

The online test bank includes:

- **Practice Exams** to test your knowledge of the material
- **Digital Flashcards** to reinforce your learning and provide last-minute test prep before the exam
- **Searchable Glossary** gives you instant access to the key terms you'll need to know for the exam
- **Bonus Materials** such as video, software, and whitepapers to assist you in learning concepts and completing exercises

Go to `http://sybextestbanks.wiley.com` to register and gain access to our comprehensive study tools.